Traffic Flow Theory and Characteristics 2018

Transportation Research Record: Journal of the Transportation Research Board

Volume 2672, Issue 20

ISBN: 978-1-5108-7825-9

Printed from e-media with permission by:

Curran Associates, Inc.
57 Morehouse Lane
Red Hook, NY 12571

proceedings.com

Some format issues inherent in the e-media version may also appear in this print version.

Published by Sage Publications USA

ISBN (Print) 978-1-5108-7825-9
ISBN 2018 Printed Set (All Issues) 978-1-5108-7735-1

Printed by Curran Associates, Inc. (2019)

Additional copies of this publication are available from:

Curran Associates, Inc.
57 Morehouse Lane
Red Hook, NY 12571 USA
Phone: 845-758-0400
Fax: 845-758-2633
Email: curran@proceedings.com
Web: www.proceedings.com

Volume 2672 Number 20 December 2018

TRR
JOURNAL OF THE TRANSPORTATION RESEARCH BOARD

Contents

Traffic Flow Theory and Characteristics 2018

Articles

TRB

TRANSPORTATION RESEARCH BOARD

CORE SPONSORS

Transportation Departments of the 50 States and the District of Columbia

Federal Government

U.S. Department of Transportation
- Federal Aviation Administration (FAA)
- Federal Highway Administration
- Federal Motor Carrier Safety Administration (FMCSA)
- Federal Railroad Administration (FRA)
- Federal Transit Administration
- National Highway Traffic Safety Administration (NHTSA)
- Office of the Assistant Secretary of Transportation for Research and Technology (OST-R)

U.S. Air Force Civil Engineer Center (AFCEC)
U.S. Coast Guard (USCG)
U.S. Department of Energy
U.S. Department of the Interior (DOI)
U.S. Environmental Protection Agency (EPA)

Nongovernmental Organizations

American Public Transportation Association (APTA)
Association of American Railroads (AAR)
California Air Resources Board

The National Academies of

SCIENCES · ENGINEERING · MEDICINE

The **National Academy of Sciences** was established in 1863 by an Act of Congress, signed by President Lincoln, as a private, nongovernmental institution to advise the nation on issues related to science and technology. Members are elected by their peers for outstanding contributions to research. Dr. Marcia McNutt is president.

The **National Academy of Engineering** was established in 1964 under the charter of the National Academy of Sciences to bring the practices of engineering to advising the nation. Members are elected by their peers for extraordinary contributions to engineering. Dr. C. D. Mote, Jr., is president.

The **National Academy of Medicine** (formerly the Institute of Medicine) was established in 1970 under the charter of the National Academy of Sciences to advise the nation on medical and health issues. Members are elected by their peers for distinguished contributions to medicine and health. Dr. Victor J. Dzau is president.

The three Academies work together as the National Academies of Sciences, Engineering, and Medicine to provide independent, objective analysis and advice to the nation and conduct other activities to solve complex problems and inform public policy decisions. The Academies also encourage education and research, recognize outstanding contributions to knowledge, and increase public understanding in matters of science, engineering, and medicine.

Learn more about the National Academies of Sciences, Engineering, and Medicine at **www.national-academies.org.**

The **Transportation Research Board** is one of seven major programs of the National Academies of Sciences, Engineering, and Medicine. The mission of the Transportation Research Board is to increase the benefits that transportation contributes to society by providing leadership in transportation innovation and progress through research and information exchange, conducted within a setting that is objective, interdisciplinary, and multimodal. The Board's varied activities annually engage about 7,000 engineers, scientists, and other transportation researchers and practitioners from the public and private sectors and academia, all of whom contribute their expertise in the public interest. The program is supported by state transportation departments, federal agencies including the component administrations of the U.S. Department of Transportation, and other organizations and individuals interested in the development of transportation.

Learn more about the Transportation Research Board at **www.TRB.org**.

Article

TRR
JOURNAL OF THE TRANSPORTATION RESEARCH BOARD

Transportation Research Record
2018, Vol. 2672(20) 1–11

DOI: 10.1177/0361198118773554
journals.sagepub.com/home/trr
SAGE

Model for Planning and Sizing Curbside Parking Lanes in Urban Networks

Chrysanthi Gkini[1], Christina Iliopoulou[1], Konstantinos Kepaptsoglou[1], and Eleni I. Vlahogianni[2]

Abstract

Curbside parking is associated with various adverse impacts on urban traffic networks and is rarely recommended. However, there are cases where parking demand dictates the establishment of on-street parking lanes. Proper planning of the number and type of curbside parking lanes to be located is essential for maximizing roadway capacity and minimizing the resulting impacts of parking operations on the network's performance. This paper develops a bi-level mathematical programming model for planning and sizing curbside parking lanes in an urban network. The model is solved using a genetic algorithm and demonstrated for a medium-sized urban network.

Curbside parking is widespread practice in dense urban areas, where land values, investment costs, and urban space availability do not favor the development of off-street parking facilities. Indeed, parking lanes are a low cost, easy-to-implement option, whose reported advantages include convenience of access, reduction in walking distance, positive effects on retail activities, and provision of a buffer between pedestrians and vehicular traffic (*1*). Nevertheless, negative impacts of on-street parking on traffic conditions and the environment have been widely acknowledged in the literature. For instance, Gantelet and Lefauconnier (*2*) estimated that approximately 70 million hours were spent in search of parking spaces in France each year, while drivers seeking curbside parking accounted for up to 10% of road users in cities and up to 60% in local roads. According to Shoup (*3, 4*), cruising for free parking space leads to an annual fuel consumption of 178,000 liters and the production of 730 tons of carbon dioxide in Los Angeles per year. On-street parking also raises safety considerations, as it is a major source of accidents in local roads (*5*). Most importantly though, on-street parking has an impact on traffic conditions and subsequently on travel times of road users (*6*). The latter can be attributed to traffic lane reduction and parking maneuvering. Evidence show that on-street parking can increase average travel times in urban areas by 15% to 199% and decrease the capacity of adjacent traffic lanes by 6% to 55% (*6*). These figures alone highlight the importance of proper planning for minimizing the impacts of on-street parking on the road network.

This paper presents a model for planning and sizing on-street parking lanes, in such a way that street parking supply is maximized while effects on traffic are minimized. The model is formulated as a bi-level network design problem, in which the upper level determines curbside parking lanes, the type of curbside parking (parallel, angle), and the number of parking spaces per lane, and the lower level estimates traffic impacts derived from a straightforward traffic assignment process. A genetic algorithm is exploited for solving the model, which is demonstrated using a mid-size city network.

The remainder of the paper is organized as follows. The background section presents a brief overview of research related to on-street parking supply and its impacts. Then, the mathematical formulation of the model and the solution are described in detail. Next, an application of the model for a mid-size city network is offered and a sensitivity analysis for key parameters is carried out. Finally, conclusions and future research directions are summarized.

[1]School of Rural and Surveying Engineering, National Technical University of Athens, Athens, Greece
[2]School of Civil Engineering, National Technical University of Athens, Athens, Greece

Corresponding Author:
Address correspondence to Konstantinos Kepaptsoglou: kkepap@central.ntua.gr

Background

Literature on planning of on-street parking lanes is scarce; published work mostly investigates the impacts of on-street parking on network performance. Johnston (*7*) and Yousif and Purnwan (*8*) investigated and quantified maneuver time and space requirements for different parking types (parallel, angle). Sisiopiku (*9*) analyzed on-street parking management techniques and evaluated the advantages and disadvantages of curbside parking. The latter noted that, due to safety concerns arising in the case of angle parking, it should be implemented only when traffic flow is significantly lower than road capacity. Another branch of literature deals with the impacts of on-street parking in the vicinity of signalized intersections. Ye and Xin (*10*) and Cao and Menendez (*11*) evaluated intersection delays caused by parking maneuvers. Similarly, Gu et al. (*12*) estimated the delays caused by the presence of bus stops near signalized intersections, while Gayah et al. (*13*) investigated a wider range of roadway obstructions, including parking maneuvers. Other studies investigated the impact of on-street parking at the network level. Cao and Menendez (*14*) presented a methodology to model macroscopically interactions between urban parking and traffic systems and evaluated associated effects on network performance. Guo et al. (*15*) developed a hazard-based duration model for analyzing the effects of on-street parking on travel time; they concluded that lane width, parking maneuvers, parking turnover, and occupancy rates critically affected traffic conditions. Kladeftiras and Antoniou (*16*) applied microscopic simulation and assessed on-street parking malpractices on average speed and delay in urban networks. Wijayaratna and Wijayaratna (*17*) introduced a model for calculating reductions in road capacity attributed to on-street parking and Vlahogianni et al. (*18*) developed a set of algorithms based on neural networks and survival analysis for predicting on-street parking space occupancy based on real-world parking sensor data.

Planning of curbside parking lanes has scarcely been investigated in the literature. Ceylan et al. (*19*) developed a bi-level optimization model for the identification of links where on-street parking lanes should be allocated. However, as they stated, the proposed model did not consider delays caused by parking maneuvers or capacity reduction caused by on-street parking.

This work contributes by proposing an advanced model for on-street parking planning, which jointly determines parking lanes and type of parking per parking lane, and considers impacts on neighboring traffic. Dynamic modeling of congestion effects and cruising for parking are beyond the scope of this work, which deals with optimal allocation of on-street parking lanes in the context of strategic planning. We perform static traffic assignment as we seek to estimate the total network travel time in the context of cost-benefit analysis for parking provision, rather than a precise representation of spatiotemporal traffic flow patterns.

Methodology

Model Formulation

The proposed model allocates on-street parking lanes (and spaces) to network links and selects the corresponding parking type per lane (parallel, 45° angle, or 90° angle [vertical] to curb), while considering traffic impacts caused by lane reduction and parking maneuvers. A bi-level network design model is developed for that purpose: at the upper level decision variables reflecting network changes (introduction of parking lanes and parking types) are determined so that planning objectives are optimized, while at the lower level traffic impacts are estimated for fixed decision variable values (*20, 21*). In this context, let

- A: Set of road links a
- M: Set of links m where parking is allowed, $M \subseteq A$
- K: Set of routes k between network pairs of nodes
- R: Set of origin nodes r
- S: Set of destination nodes s
- c_a: Capacity of link a
- D_{rs}: Demand between r and s
- $f^k{}_{rs}$: Flow of path k
- L_α: Length of link a
- w_a: Width of link a
- $t^o{}_a$: Free flow travel time of link a
- t_a: Travel time of link a
- v_a: Flow of link a
- y_a: Parking type decision for link a
- $y_{0,a}$: Allowable parking type for link a
- $\delta^{rs}{}_{\alpha,k}$: 1 if path k uses link a and 0 otherwise
- $n_{p,a}$: Number of parking spaces on link a
- β_1, β_2: Upper level objective function weights

The model's decision variable is y_a; it reflects the decision on parking lane and type for link a:

$$y_a = \begin{cases} 0 & \text{No curbside parking} \\ 1 & \text{Parallel curbside parking} \\ 2 & 45^\circ \text{ angle curbside parking} \\ 3 & \text{Vertical curbside parking} \end{cases} \tag{1}$$

The model is formulated as follows:

$$\text{Maximize } Z = \beta_1 \sum_{a \in M} n_{p,a}(y_a) - \beta_2 \sum_{a \in A} v_a(y_a) \cdot t_a(y_a) \tag{2}$$

subject to

$$y_a = \{0,1,2,3\}\ \forall\alpha \in \mathrm{A} \quad (3)$$

$$y_a = 0\forall\alpha \notin \mathrm{M} \quad (4)$$

$$y_a \le y_{0,a}(w_a)\forall\alpha \in \mathrm{M} \quad (5)$$

$$\min W = \sum_{a\in A}\int_0^{v_a} t_a(x, y_a)dx \quad (6)$$

subject to

$$\sum_{k\in K} f_k^{rs} = D_{rs}\,\forall r \in R, s \in S, k \in K_{rs} \quad (7)$$

$$v_a = \sum_{rs}\sum_{k\in K} f_k^{rs}\cdot\delta_{a,k}^{rs}\,\forall r \in R, s \in S, k \in K_{rs} \quad (8)$$

$$f_k^{rs} \ge 0\ \forall r \in R, s \in S, k \in K_{rs} \quad (9)$$

$$t_a = h\left(t_a^0, v_a, c_a(y_a)\right) \quad (10)$$

$$v_a \ge 0\forall a \in A \quad (11)$$

The objective function (Equation 2) jointly considers planning objectives and maximizes the weighted difference (*Z*) between parking supply (parking spaces) and network impacts (total vehicle hours). Constraints (Equations 3–5) set allowable values for decision variable y_a, considering width of links and parking permitted along links. Objective function W given in Equation 6 reflects the minimization of the travel time for each user under the User Equilibrium (UE) assignment. Equations 6 to 11 correspond to the lower level model and represent the user equilibrium formulation for traffic assignment, which yields travel times and flows for network links (*22*).

Parking Type and Lane Sizing

The applicable type of parking (indicated by variable y_0) (parallel, 45° angle, vertical, or not possible) per parking lane is associated to the link's width as follows (*5*):

$$y_{0,a} = \begin{cases} 0 & \text{if } w_a < 6 \text{(curbside parking not possible)} \\ 1 & \text{if } 6 \le w_a < 8.35 \text{(parallel curbside parking)} \\ 2 & \text{if } 8.35 \le w_a < 10 \text{(45° angle curbside parking)} \\ 3 & \text{if } w_a \ge 10 \text{(vertical curbside parking)} \end{cases} \quad (12)$$

For properly sizing parking lanes, vehicles should not park inside an intersection's visibility triangle and therefore no on-street parking should be permitted at a distance *L'* from an intersection. The number of spaces for a parking lane in link α can be calculated as (*5*):

$$n_{p,a} = \begin{cases} \left\lfloor\dfrac{L_a - 2L'}{6}\right\rfloor \text{if } y = 1 \\ \left\lfloor\dfrac{L_a - 2L' - 1.77}{3.54}\right\rfloor \text{if } y = 2 \\ \left\lfloor\dfrac{L_a - 2L'}{2.5}\right\rfloor \text{if } y = 3 \\ 0 \text{ if } y = 0 \end{cases} \quad (13)$$

Equation 13 calculations are based on the typical dimensions of parking spaces according to parking design guidelines (*5*). The link lengths required for parking spaces are 6, 3.54 and 2.5 meters for parallel, angle, and vertical parking, respectively, while an additional 1.77 meters are needed for angle parking. The exact figures in this equation should be modified based on the typical parking space guidelines provided by each country.

Capacity Reduction Factor for On-Street Parking

The Highway Capacity Manual (*23*) introduces a parking capacity reduction factor (HCM factor) for adjacent parking lanes, which is calculated as

$$f_p = \frac{N' - 0.1 - \dfrac{18N_M}{3600}}{N'} \ge 0.5 \quad (14)$$

where *N'* is the number of adjacent lanes to the parking lane and N_M is the number of maneuvers per hour at a parking space in a distance of 75 meters from the closest intersection. For a traffic lane width L_T, the number of remaining adjacent traffic lanes for a link width w and different parking types can be calculated as:

$$N' = \begin{cases} \left\lfloor\dfrac{w - 2.5}{L_T}\right\rfloor & if\ y = 1 \\ \left\lfloor\dfrac{w - 4.85}{L_T}\right\rfloor & if\ y = 2 \\ \left\lfloor\dfrac{w - 5}{L_T}\right\rfloor & if\ y = 3 \\ \left\lfloor\dfrac{w}{L_T}\right\rfloor & if\ y = 0 \end{cases} \quad (15)$$

Also, assuming a parking turnover rate ρ per parking space, the total number of maneuvers is equal to

$$N_M = \rho \cdot n_p \quad (16)$$

Wijayaratna and Wijayaratna (*17*) proposed an alternative, detailed equation for calculating the parking capacity reduction factor (W&W factor), in which they explicitly incorporated the impact of turnover rate and of parking maneuver times on traffic affected in the adjacent lanes, differentiating between entry/exit movements and parking types. The authors considered the reduction in capacity as analogous to the reduction in the effective green time during a signal cycle due to vehicle maneuvering times, instead of estimating the number of remaining adjacent lanes as proposed in the HCM.

$$f_p = \frac{3600 - n_p \cdot \rho \cdot (t_{PI} + kt_{PO})}{3600} \tag{17}$$

where n_p is the number of spaces in a parking lane, ρ is the parking turnover rate, t_{PI} is the time spent on an entry maneuver and t_{PO} is the time spent on an exit maneuver. Coefficient k expresses the percentage of traffic blocked due to the maneuvers required to leave a parking space.

Based on surveys by Yousif and Purnawan (*8*) and Wijayaratna and Wijayaratna (*17*), the following values are available for the parameters of Equation 16:

$$t_{pi} = \begin{cases} 7 \text{ s for vertical parking} \\ 5 \text{ s for } 45^\circ \text{ parking} \\ 24 \text{ s for parallel parking} \end{cases}$$

$$t_{p0} = \begin{cases} 14 \text{ s for vertical parking} \\ 10 \text{ s for } 45^\circ \text{ parking} \\ 7 \text{ s for parallel parking} \end{cases} \tag{18}$$

$$k = \begin{cases} 0 \text{ for parallel parking} \\ 0.5 \text{ for angle parking} \end{cases}$$

The updated link capacity C' after applying the parking capacity reduction factor is equal to:

$$c_a{}' = \frac{c_a}{N} \cdot N_a{}' \cdot f_{p,a} \tag{19}$$

where c_a is the original capacity of link a (without parking lanes), N_i is the original number of lanes before a parking lane is allocated, N_a' is the number of adjacent lanes after a parking lane is allocated and $f_{p,a}$ is the parking capacity reduction factor for link a (calculated by either Equation 14 or 17).

Solution Process

The general network design problem is NP-hard, even for simple linear programming model formulations (*20, 24*). A genetic algorithm is selected for solving the model; genetic algorithms (GA) provide a highly efficient and adaptable approach to solve non-linear combinatorial problems. The components of the proposed GA and the outline of the entire solution procedure are described as follows:

Representation Scheme. Every candidate solution (parking plan) of the problem is represented as a chromosome, where every gene represents the type of parking space allocated to a link. The values of the genes are integer and belong to the range 0 to 3, as specified in Equation 1. An example of the representation scheme is given below:

[3 1 2 3 0 1 0 0 2 0 0........0 1 2 2 3]

In the above solution, vertical parking is assigned to the first link, parallel to the second, and 45°-angle parking to the third link, while no parking is permitted on the fifth link.

Fitness Measure. Each chromosome is evaluated based on a fitness measure, which is the objective function (2) of the mathematical formulation. The number of spaces offered is straightforwardly calculated based on the gene values, while the execution of an external procedure performing traffic assignment is required to estimate the corresponding network cost for each parking plan.

Initialization. The initial population of chromosomes is generated randomly, under link operational constraints, that is, the allocation of parking lanes and types must be feasible or permitted in a link. So, depending on link width, parallel, vertical or 45°-angle parking is initially assigned, while all population members are identical.

Genetic Operators. Two basic genetic operators, uniform crossover and mutation, are applied to generate new chromosomes. The crossover operator creates new parking plans which combine elements from two good quality solutions by swapping genes between them with a fixed probability. The mutation operator randomly modifies the value of a gene within feasibility limits, that is, it changes the type of parking for that link in the specific solution. These operators are applied to selected members of the population, called parents, which are chosen based on a ranking method (*25*). Then, a steady state approach is followed for replacement—worst population members are replaced instead of the whole population. The completion of every iteration yields a pool of candidate solutions, which represent the parking allocation and type for each link.

Frank-Wolfe Algorithm. At the end of each GA iteration, the resulting parking lane allocation and parking type per link alter the capacity of links. A user equilibrium assignment is performed using the Frank-Wolfe algorithm (*26*) to determine optimal link flows and associated travel times per link. Link travel times in this study are given by the formula of the Bureau of Public Records, given below:

$$t(x_a) = t_0{}^a[1 + 0.15\left(\frac{x_a}{c_a}\right)^4] \tag{20}$$

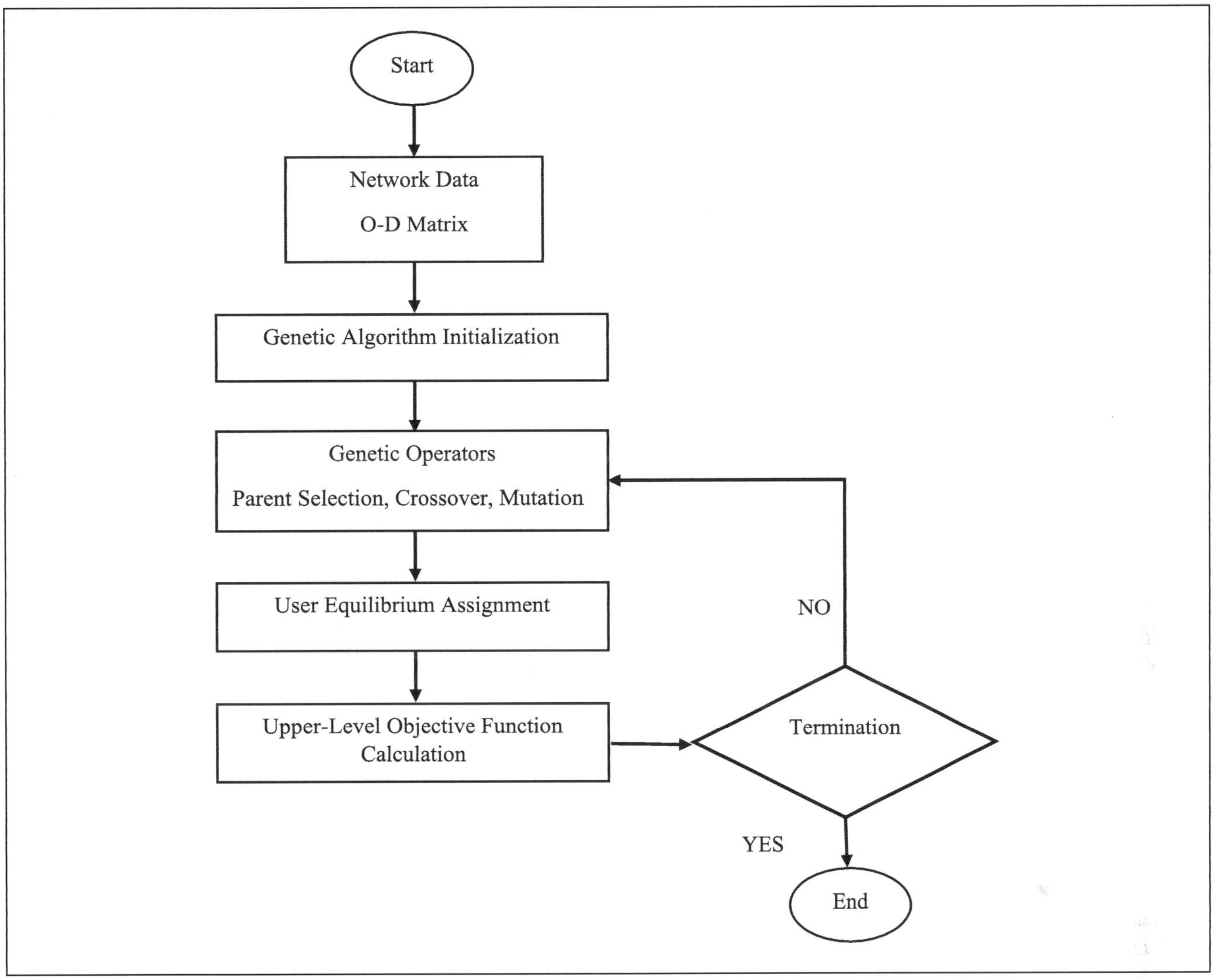

Figure 1. Model structure.

where $t_0{}^a$ is the link free flow travel time, x_a the link flow, and c_a the link capacity.

If the flow-to-capacity ratio exceeds a pre-specified value, curbside parking on that link is not permitted and the solution is corrected accordingly by prohibiting curbside parking for that link, for this candidate solution.

Termination. The termination criterion is set as an improvement of less than 1% in the fitness measure after a specific time duration defined through experimentation and GA calibration.

The overall procedure is outlined in Figure 1.

Application

The model is demonstrated using a mid-size urban network (city of Alexandroupoli, Greece), consisting of 303 links with parking possible on both sides of each link (regardless of whether the link is single or bi-directional). Link capacity values range from 800 to 1,000 vehicles per hour (vph), while free flow travel times were estimated using existing speed limits for each link, ranging from 10 to 60 km/h. The OD matrix was randomly generated and features 10 indicative pairs between the center of the city and its outskirts with values ranging from 500 to 1,100 vph. Palisade Evolver™ for MS-Excel™ is used for implementing the model and both capacity reduction factors (HCM (*23*) and W&W (*17*)) are exploited. Equal importance was attributed to each objective and a total of 10,000 GA iterations were required to satisfy the convergence criterion. Results for both capacity reduction factors are shown in Figures 2 to 5. The results for each capacity factor are presented on two maps, one for traffic in north–south and west–east directions and one for south–north and east–west, to facilitate visualization (Figure 2).

As can be seen in Figure 2, for the HCM case, different parking types appear equally distributed in the

Figure 2. HCM results for north–south and west–east directions.

network, in the north–south and west–east directions. In the opposite directions, vertical parking is more commonly applied (Figure 3). For the W&W case, in certain parts of the network the parking layout appears quite different compared with HCM, with fewer parking spaces allocated (Figures 4 and 5).

Tables 1 and 2 show the comparison of results for the two capacity factors with respect to the parking spaces offered and the total network travel time achieved. Before the allocation of on-street parking lanes the solution of the UE problem yields a network cost of 1,206 vehicle hours. The parking plan generated by the HCM and the W&W equations results in 35% and 41% additional travel cost, respectively. As reported by Wijayaratna and Wijayaratna (*17*), HCM can lead to overestimation of the network's capabilities. Indeed, HCM produces a solution of 9,927 parking spaces to be provided, while W&W generates 9,315 parking spaces, that is, 612 fewer spaces. Further, W&W yields approximately 4% lower total travel time compared with HCM (Table 1).

Table 2 compares the results of the two factors as to the number of links where curbside parking is possible; the HCM factor permits parking on 315 links, while for W&W the total number is 297 links. Again, as Wijayaratna and Wijayaratna (*17*) note, HCM overestimates the capacity of roads in the case of parking lanes and may allow the deployment of parking spaces in links where bottleneck phenomena could result from parking maneuvers. This may be attributed to the fact that the HCM equation disregards parking types (Table 2).

As for the type of parking assigned to each link, it was observed that in some cases, even though the width of the link was sufficient, the model did not assign the parking type which maximized the number of spaces along the link. This phenomenon was observed when the maximum number of parking spaces that a link could provide would reduce road capacity to the extent where either the flow capacity constraint was violated, or a larger increase in network travel time was noted compared with a different type of parking.

Finally, it is noted that the model was executed on a 3.98 GHz desktop with 12 GB of RAM. Computational times ranged from 0.5 to 1.5 minutes, with an average value of 1 minute per experiment. In terms of each of the sub-problems, run time for the upper level was around 0.2 minutes per iteration, while the Frank-Wolfe algorithm (implemented in Python) took about 0.3 to 1 minute to run.

Sensitivity Analysis

Sensitivity analysis was undertaken using the W&W reduction factors, which yielded more conservative results compared with the HCM. Different values of hourly parking turnover rate N_P and reduction factor k

Figure 3. HCM results for south–north and east–west directions.

Figure 4. W&W results for north–south and west–east directions.

Figure 5. W&W results for south–north and east–west directions.

Table 1. Comparison of HCM and W&W Factors' Estimation of Parking Spaces

	Parking spaces	
	HCM	W&W
Parking type		
Parallel	1152	1004
45°angle	2999	2759
Vertical	5776	5552
Total	9927	9315
Difference	612	
Peak time vehicle hours		
Total	1702.72	1630.50
Difference	72.21	

were investigated. Changes in the objective function (Equation 2) for various parking turnover rate values are shown in Figure 6.

According to Figure 6, lower turnover rates yield higher objective values. This is reasonable, since for lower turnover rates duration of parking is longer, thus the impact of maneuvering on adjacent lane capacity is smaller, leading to lower values of total travel cost. Furthermore, the reduction in the objective function is very sharp for turnover rates up to the value of 2, while it is milder beyond that point (Figure 7).

According to Figure 7, lower values for coefficient k lead to higher values for the objective function. This is also expected, since a value of $k = 0$ (formally used for parallel parking [Equation 18]) implies no additional time for exiting a parking spot, whereas greater values imply greater delays for passing vehicles and thus greater reduction in adjacent lane capacity. Further, the decrease in the objective function value observed when increasing coefficient k is rather sharp for all values considered, with a slight change in slope for values higher than 0.5.

Table 2. Number of Links Where On-Street Parking Permitted

	Number of links	
Parking type	HCM	W&W
Parallel	89	76
45° angle	110	109
Vertical	116	112
Total	315	297
Difference	18	

Finally, we present some results for different combinations of weighting factors β1 and β2 for the objective function terms. For the basic scenario, weighting factors were determined as $\beta 1 = 1$ and $\beta 2 = 7$ based on preliminary experiments, in order to create two components of equal magnitude. Figure 8 illustrates some indicative results for different weight ratios.

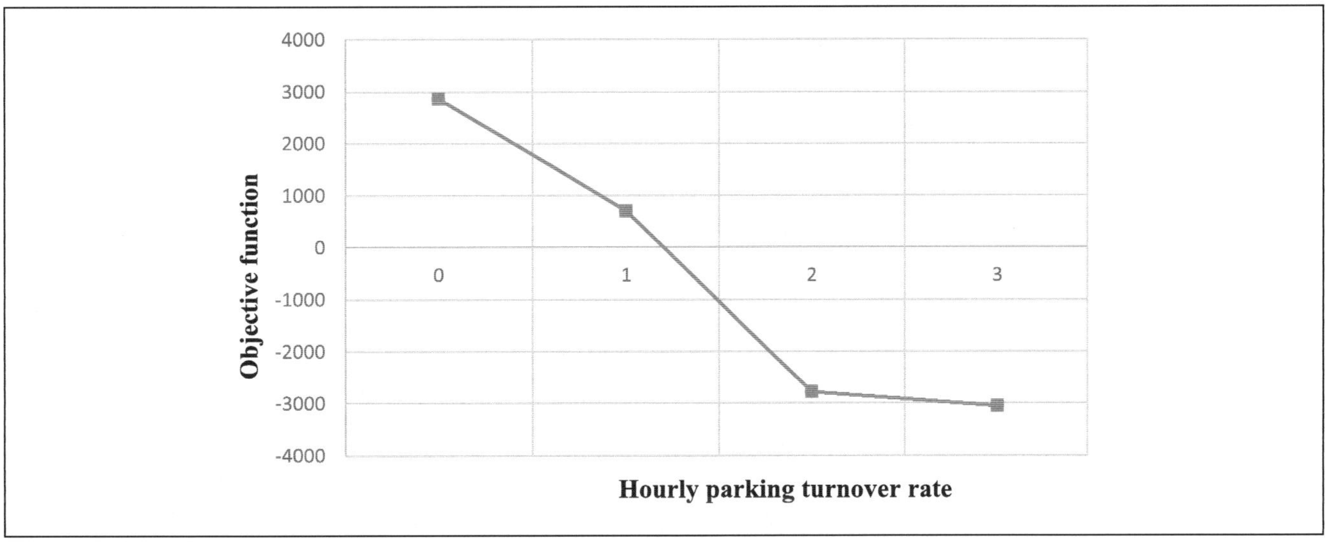

Figure 6. Change in the objective function for different values of parking turnover rates.

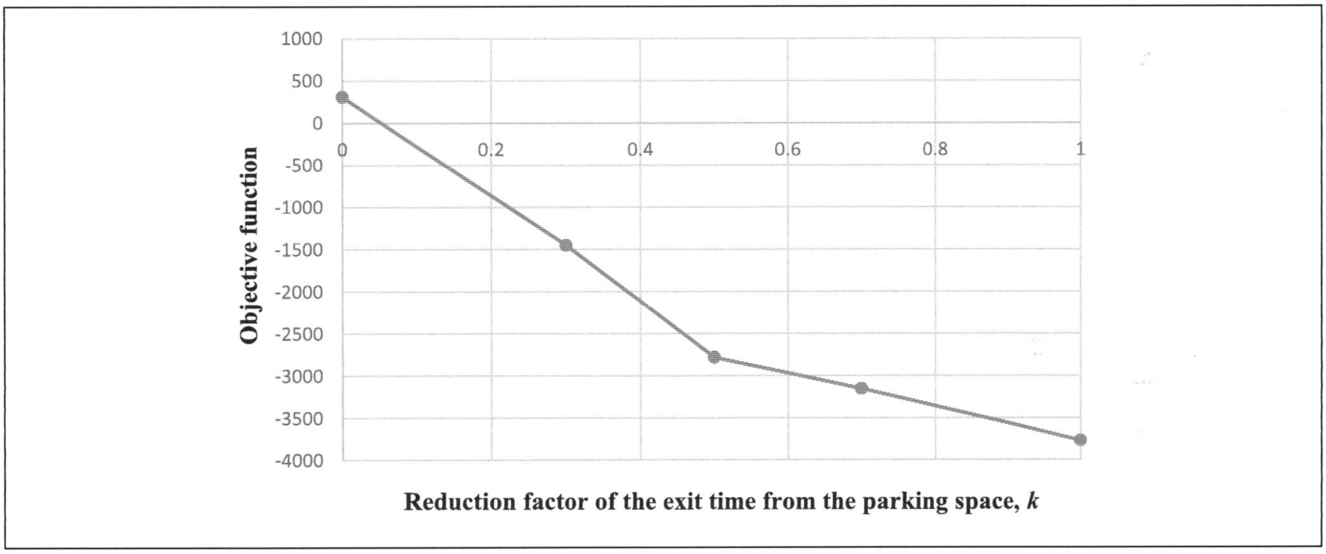

Figure 7. Change of the objective function for different values of *k*.

As can be seen in Figure 8, when the travel time coefficient β2 is seven times greater than the parking supply coefficient β1, total vehicle hours sharply increase, while for ratios over 10 the change is milder. Similarly, for smaller ratios parking supply significantly increases, reaching 11,000 spaces. Generally, decision makers can select the coefficients which best reflect the prioritization of planning preferences and evaluate alternative scenarios to estimate respective trade-offs.

Conclusion

Curbside parking planning and sizing focuses on maximizing space usage without compromising road safety, while at the same time minimizing impacts to traffic due to curbside parking. A bi-level network design model was developed for that purpose, considering availability of space and the effect of vehicle maneuvering on traffic. Two different roadway capacity factors, HCM and W&W, were tested as part of the planning model. The results showed that in some cases, even though the width of the link was sufficient for the allocation of angle parking, which would provide more spaces, the increase in travel costs outweighed the benefit. Also, since vertical and angle parking offer far less visibility than parallel parking, these types of parking should only be permitted in cases where the traffic flow is significantly less than link capacity. Overall, the current results corroborated

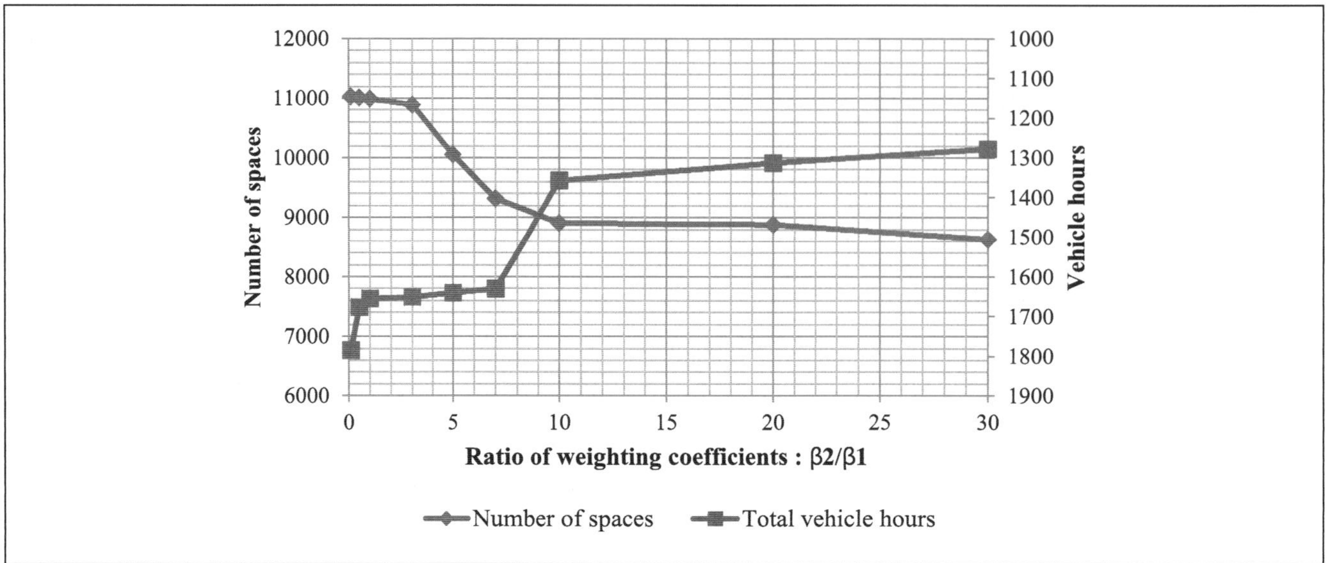

Figure 8. Change of the objective function terms for different weight coefficient combinations.

the observations by Wijayaratna and Wijayaratna (*17*), showing that the equation proposed by HCM (*23*) overestimates operational capacity as it does not consider the type of parking offered. Evidently, there are still ample grounds for research on the quantitative assessment of the impact of on-street parking. For this reason, following application of the model, a full network assessment should be conducted, considering congestion levels and road safety indicators.

Acknowledgments

This work was funded by the General Secretariat for Research and Technology (GSRT) and the Hellenic Foundation for Research and Innovation (HFRI).

Author Contributions

The authors confirm contribution to the paper as follows: study conception and design: Konstantinos Kepaptsoglou, Eleni Vlahogianni; data collection: Chrysanthi Gkini; analysis and interpretation of results: Chrysanthi Gkini, Christina Iliopoulou, Konstantinos Kepaptsoglou, Eleni Vlahogianni; draft manuscript preparation: Chrysanthi Gkini, Christina Iliopoulou. All authors reviewed the results and approved the final version of the manuscript.

References

1. Victoria Transport Policy Institute. Parking Solutions: A Comprehensive Menu of Solutions to Parking Problems. *TDM Encyclopedia.* 2017. http://www.vtpi.org/tdm/tdm72.htm.
2. Fauconnier, A., and E. Gantelet. The Time Looking for a Parking Space: Strategies, Associated Nuisances and Stakes of Parking Management in France. *Presented at European Transport Conference (ETC)*, Strasbourg, France, 2006.
3. Shoup, D. C. *The High Cost of Free Parking.* Planners Press, Chicago, 2005.
4. Shoup, D. Cruising for Parking. *Transport Policy*, Vol. 13, No. 6, 2006, pp. 479–486.
5. Frantzeskakis, I. M., M. H. Pitsiava-Latinopoulou, and D. A. Tsamboulas. *Parking Management*, 2nd ed. Papasotiriou Publications, Athens, 2002.
6. Portilla, A. I., B. A. Orena, J. L. M. Berodia, and F. J. R. Diaz. Using M/M/∞ Queuing Model in On-Street Parking Maneuvers. *Journal of Transportation Engineering-ASCE*, Vol. 135, No. 8, 2009, pp. 527–535.
7. Johnston, B. K. *Angle vs. Parallel Curb Parking: Time and Street Width Required for Maneuvering.* University of California, 1960.
8. Yousif, S., and Purnawan. A Study into On-Street Parking: Effects on Traffic Congestion. *Traffic Engineering and Control*, Vol. 40, No. 10, 1999, pp. 424–427.
9. Sisiopiku, V. P. On-street Parking on State Roads. *Presented at ITE Annual Meeting and Exhibit, Chicago, Ill.*, 2001.
10. Ye, X., and J. Chen. Traffic Delay Caused by Curb Parking set in the Influenced Area of Signalized Intersection. *Proc., 11th International Conference of Chinese Transportation Professionals: Towards Sustainable Transportation Systems, ICCTP 2011*, Nanjing, China, American Society of Civil Engineers, 2011, pp. 566–578.
11. Cao, J., and M. Menendez. Generalized Effects of On-Street Parking Maneuvers on the Performance of Nearby Signalized Intersections. *Transportation Research Record: Journal of the Transportation Research Board*, 2015. 2483: 30–38.
12. Gu, W., V. V. Gayah, M. J. Cassidy, and N. Saade. On the Impacts of Bus Stops near Signalized Intersections: Models

of Car and Bus Delays. *Transportation Research Part B: Methodological*, Vol. 68, 2014, pp. 123–140.
13. Gayah, V. V., S. Ilgin Guler, and W. Gu. On the Impact of Obstructions on the Capacity of Nearby Signalised Intersections. *Transportmetrica B: Transport Dynamics*, Vol. 4, No. 1, 2016, pp. 48–67.
14. Cao, J., and M. Menendez. System Dynamics of Urban Traffic Based on its Parking-Related-States. *Transportation Research Part B: Methodological*, Vol. 81, 2015, pp. 718–736.
15. Guo, H., Z. Gao, X. Yang, X. Zhao, and W. Wang. Modeling Travel Time under the Influence of On-Street Parking. *Journal of Transportation Engineering-ASCE*, Vol. 138, No. 2, 2012, pp. 229–235.
16. Kladeftiras, M., and C. Antoniou. Simulation-Based Assessment of Double-Parking Impacts on Traffic and Environmental Conditions. *Transportation Research Record: Journal of the Transportation Research Board*, 2013. 2390: 121–130.
17. Wijayaratna, S., and K. P. Wijayaratna. Quantifying the Impact of On-Street Parking on Road Capacity: A Case Study of Sydney Arterial Roads. Presented at 95th Annual Meeting of the Transportation Research Board, Washington, D.C., 2015.
18. Vlahogianni, E. I., K. Kepaptsoglou, V. Tsetsos, and M. G. Karlaftis. A Real-Time Parking Prediction System for Smart Cities. *Journal of Intelligent Transportation Systems*, Vol. 20, No. 2, 2013, pp. 192–204.
19. Ceylan, H., O. Baskan, C. Ozan, and G. Gulhan. Determining On-Street Parking Places in Urban Road Networks Using Meta-Heuristic Harmony Search Algorithm. In *Computer-Based Modelling and Optimization in Transportation*, Springer International Publishing, Cham, 2014, pp. 163–174.
20. LeBlanc, L. J., and D. E. Boyce. A Bilevel Programming Algorithm for Exact Solution of the Network Design Problem with User-Optimal Flows. *Transportation Research Part B: Methodological*, Vol. 20, No. 3, 1986, pp. 259–265.
21. Ben-Ayed, O., D. E. Boyce, and C. E. Blair. A General Bilevel Linear Programming Formulation of the Network Design Problem. *Transportation Research Part B: Methodological*, Vol. 22, No. 4, 1988, pp. 311–318.
22. Wardrop, J. G., and J. I. Whitehead. Correspondence. Some Theoretical Aspects of Road Traffic Research. *ICE Proceedings: Engineering Divisions*. Vol. 1, No. 5, 1952, pp. 767–768.
23. *Highway Capacity Manual*. TRB, National Research Council, Washington, D.C., 2000.
24. Gao, Z., J. Wu, and H. Sun. Solution Algorithm for the Bi-Level Discrete Network Design Problem. *Transportation Research Part B: Methodological*, Vol. 39, No. 6, 2005, pp. 479–495.
25. Pternea, M., K. Kepaptsoglou, and M. G. Karlaftis. Sustainable Urban Transit Network Design. *Transportation Research Part A: Policy and Practice*. Vol. 77, 2015, pp. 276–291. https://doi.org/10.1016/j.tra.2015.04.024.
26. Frank, M., and P. Wolfe. An Algorithm for Quadratic Programming. *Naval Research Logistics*, Vol. 3, No. 1–2, 1956, pp. 95–110.

The Standing Committee on Traffic Flow Theory and Characteristics (AHB45) peer-reviewed this paper (18-00818).

TRR
JOURNAL OF THE TRANSPORTATION RESEARCH BOARD

Article

Transportation Research Record
2018, Vol. 2672(20) 12–24

DOI: 10.1177/0361198118792997
journals.sagepub.com/home/trr
SAGE

Off-Ramp Coupling Conditions Devoid of Spurious Blocking and Re-Routing

Najmeh Salehi[1], Julia Somers[1], and Benjamin Seibold[1]

Abstract

When modeling vehicular traffic flow on highway networks via macroscopic models, suitable coupling conditions at the network nodes are crucial. Frequently, the evolution of traffic flow on each network edge is described in a lane-averaged fashion using a single-class Lighthill-Whitham-Richards model. At off-ramps, split ratios (i.e., what percentage of traffic exits the highway) are prescribed that can be drawn from historical data. In this situation, classic FIFO coupling conditions yield unrealistic results, in that a clogged off-ramp yields zero flux through the node. As a remedy, non-FIFO conditions have been proposed. However, as we demonstrate here, those lead to spurious re-routing of vehicles. A new coupling model, FIFO with queue (FIFOQ), is presented which preserves the desirable properties of non-FIFO models while not leading to any spurious re-routing.

One way to model vehicular traffic flow on a highway network is to describe interchanges and ramps as nodes (vertices) of the network, and the roads between them as edges of the network. The flow on an edge in a given direction is then described via a lane-aggregated Lighthill-Whitham-Richards (LWR) model (*1, 2*), and suitable coupling conditions are formulated on each node that describe the flow balance of vehicles between in-going and out-going edges. The classical FIFO coupling conditions were introduced by Daganzo (*3*). Mathematical proofs of well-posedness of the resulting hyperbolic conservation law network flow have been provided in several works (*4–7*) and references therein.

A key shortcoming of modeling FIFO conditions is that a clogged off-ramp will result in zero flow through the node. Clearly, on multi-lane highways this is unrealistic, as a queue forming from the off-ramp will generally be restricted to the right-most lane, and vehicles that do not wish to exit can pass the queue (within certain limitations). To remedy this shortcoming, non-FIFO coupling conditions were proposed (*8*). These allow for a nonzero flow past a clogged off-ramp, however, at the expense of violating route choices of drivers; some vehicles that in reality form a queue waiting to exit will instead be accounted for as flow continuing along the highway. Other more recent off-ramp models possess similar re-routing effects (*9*).

What is needed are coupling conditions that remedy the shortcomings of FIFO, but without producing spurious re-routing of vehicles. One possible way to achieve this is by formulating multi-commodity models, such as by Daganzo (*10*) or Bressan and Nguyen (*11*). These models replace LWR by a system of conservation laws, that is, they explicitly track different types of vehicles. This simplifies some challenges in formulating diverge models, for example, no split ratio need be provided. The practical challenge is that (in the present day) it is usually not known in advance which vehicles intend to exit the highway and which do not. Instead, typical split ratios tend to be known from historical data. We therefore consider here the situation in which a single-class lane-aggregated LWR model is to be used on the edges, as for instance was done in the Mobile Millennium project (*12*).

This paper presents how this goal can be achieved by the introduction of a vertical queue that keeps track of the excess vehicles of a certain type (exiting versus non-exiting) that may queue up by more than the other vehicle type does. While vertical queues have been used in macroscopic traffic models for the "storage" of vehicles that wish to enter the network (*11, 13*), their use for the purpose of balancing splitting flows is novel. The vertical queue could be present on either of the two out-roads (the highway or the ramp), however, it is always designed to be minimal, that is, two queues cannot be active simultaneously.

This paper is organized as follows. First, the mathematical definitions and notations are introduced, and a

[1]Department of Mathematics, Temple University, Philadelphia, PA

Corresponding Author:
Address correspondence to Benjamin Seibold: seibold@temple.edu

discussion of split ratios versus turn ratios is provided. Then, the existing coupling models, FIFO and non-FIFO, are presented, and their modeling shortcomings discussed. After that, the new model is presented, including how it can be implemented in a cell transmission model (CTM) framework. The three models are then systematically compared in a representative example describing a first forming and then clearing off-ramp queue. The example illustrates how the new model remedies the weaknesses of both existing models. The paper closes with conclusions and future research suggestions.

Modeling Foundations

We consider models for road networks that are represented by a directed graph, whose edges represent the roads (all lanes going in one direction aggregated), and nodes (vertices) represent the interchanges or ramps (sometimes called "junctions"). An edge i of the network is an interval $I_i = [a_i, b_i]$. This paper specifically focuses on 1-in-2-out nodes modeling highway off-ramps. Therefore, the discussion is restricted to one node with three edges: one in-road (I_1) and two out-roads (I_2 and I_3).

On each edge i, the evolution of the traffic density, $\rho_i(x,t)$, is described by the LWR model

$$\partial_t \rho_i + \partial_x f(\rho_i) = 0\ , \quad \text{where}\,(x,t) \in I_i \times \mathbb{R}^+\ . \tag{1}$$

The flux function $f = f(\rho) = \rho v(\rho)$ encodes the fundamental diagram (FD) of traffic flow, where $v = v(\rho)$ is the bulk velocity of traffic. In this paper, we use the Greenshields flux

$$f(\rho) = v^{\max}\rho\left(1 - \frac{\rho}{\rho^{\max}}\right),$$

corresponding to an affine linear density–velocity relationship, with $v^{\max}$ the speed limit and $\rho^{\max}$ the jamming density. However, it is important to stress that the model, its CTM implementation, and its analysis, apply to any concave down flux function, including triangular FDs. The critical density at which the flow is maximized is denoted with σ.

Riemann Problem and Cell Transmission Model

The key building block for finite volume discretizations of the LWR model (Equation 1) is the Riemann problem (RP), which is a Cauchy problem with initial data

$$\rho(x,0) = \rho_0(x) = \begin{cases} \rho_L & x < 0 \\ \rho_R & x \geq 0\ . \end{cases}$$

By the standard theory of hyperbolic conservation laws (*14*), the RP (with concave flux) has the following unique entropy solution:

- If $\rho_L < \rho_R$, then $f'(\rho_L) > f'(\rho_R)$, and then the solution

$$\rho(x,t) = \begin{cases} \rho_L & x < st \\ \rho_R & x \geq st \end{cases}$$

- consists of a shock, that is, a traveling discontinuity in which vehicles brake (e.g., the upstream end of a traffic jam). The shock speed $s = \frac{f(\rho_R) - f(\rho_L)}{\rho_R - \rho_L}$ is given by the secant slope in the FD.
- If $\rho_L \geq \rho_R$, then $f'(\rho_L) \leq f'(\rho_R)$, and the solution

$$\rho(x,t) = \begin{cases} \rho_L & x < f'(\rho_L)t \\ (f')^{-1}\left(\frac{x}{t}\right) & f'(\rho_L)t \leq x < f'(\rho_R)t \\ \rho_R & x \geq f'(\rho_R)t \end{cases}$$

- is a rarefaction wave, in which vehicles gradually accelerate.

The RP is the key building block of the Godunov method (*15*), which divides each edge into finite volume cells, and updates the average density in each cell by the numerical fluxes across cell boundaries. Those fluxes are the RP solutions, evaluated at the cell interface. The CTM (*16*) is equivalent to the Godunov method, applied to the LWR model.

An important conceptual interpretation of the Godunov fluxes is in terms of supply and demand functions. Given a concave down flux function $f(\rho)$ with critical density σ, the supply and demand functions

$$\gamma^s(\rho) = \begin{cases} f(\sigma) & 0 \leq \rho < \sigma \\ f(\rho) & \sigma \leq \rho \leq 1 \end{cases} \quad \text{and} \quad \gamma^d(\rho) = \begin{cases} f(\rho) & 0 \leq \rho < \sigma \\ f(\sigma) & \sigma \leq \rho \leq 1 \end{cases} \tag{2}$$

are the non-increasing and non-decreasing components of the flux function, respectively.

The flux of vehicles through an interface between two cells is then the maximum possible value that does not exceed the demand (on road capacity) imposed by the upstream cell (L), and the supply (of road capacity) that the downstream cell (R) provides:

$$F(\rho_L, \rho_R) = \min\left(\gamma^d(\rho_L), \gamma^s(\rho_R)\right). \tag{3}$$

For the implementation of a Godunov scheme, and in turn a CTM, the interface flux (Equation 3) is all that is needed. However, the full solution of the RP can also be constructed, as follows. On each cell (upstream and downstream), there are two possibilities: if the flux (Equation 3) matches the flux $f(\rho)$ in that cell, then the constant state remains as it is; otherwise, a new state $\hat{\rho}$ emerges at the cell interface that reproduces the interface flux, that is, $f(\hat{\rho}) = F$. Of the two solutions that this equation generally possesses, the one is chosen that results in a wave that travels away from the cell interface, that is, the congested state $\hat{\rho}>\sigma$ on the upstream cell and the free-flow state $\hat{\rho}<\sigma$ on the downstream cell.

Generalized Riemann Problem

In a Godunov/CTM discretization of a road network, a network node can be treated in the same fashion. A generalized Riemann Problem (GRP) is given by a constant density state on each edge. For a general node, the demands of all in-roads and the supplies of all out-roads are computed, and by a route choice matrix that determines how the in-fluxes wish to distribute into the out-fluxes, the resulting vehicle flows are constructed so that they never exceed their respective supply/demand values.

In this paper, we focus on the 1-in-2-out case. The GRP considers a constant density ρ_1 on the in-road I_1 (highway), a constant density ρ_2 on the out-road I_2 (highway), and a constant density ρ_3 on the out-road I_3 (off-ramp). A node coupling model (or coupling condition) Φ is then a mapping from those three densities to three new fluxes $\Phi(\rho_1, \rho_2, \rho_3) = (\Gamma_1, \Gamma_2, \Gamma_3)$, where the flux on the in-road matches the sum of the two out-road fluxes, $\Gamma_1 = \Gamma_2 + \Gamma_3$.

Note that, by the same construction as in the simple RP, one also obtains three new states $(\hat{\rho}_1, \hat{\rho}_2, \hat{\rho}_3)$ emanating at the node's position on each edge, but as we are interested in Godunov/CTM discretizations, the construction of the fluxes Γ_i suffices.

Split Ratio

In a 1-in-2-out node, drivers make route choice decisions. Therefore, realistic coupling models must require an additional piece of information. This is commonly assumed to be a "split ratio" which prescribes what ratios of the incoming flow proceed onto which of the two out-roads. For a general node with multiple in- and out-roads, a split ratio matrix is needed (*7*). In the 1-in-2-out case, the split ratios are given by two numbers, α_2 and α_3, with $\alpha_2 + \alpha_3 = 1$, corresponding to the two out-roads I_2 and I_3, respectively.

While the notion of a split ratio (matrix) is easy to accept mathematically, its practical/modeling rationale in fact calls for a careful discussion. If both out-roads are in free flow, then what truly determines how many vehicles exit at a given time is the composition of the incoming traffic flow into "type 2" (intending to continue on the highway) and "type 3" (intending to exit) vehicles. Unfortunately, this "type ratio" is generally not known. (The type may be known for a few vehicles; and future vehicle-to-everything [V2X] connectivity may substantially increase that knowledge; but, for now, the type ratio is not known.) Therefore, a historical "exit ratio" is used as a proxy for the unknown "type ratio." Assuming that traffic behaves relatively similarly from week to week, and assuming that exit ratios evolve slowly (relative to the flow dynamics) in time, historic data on how many vehicles have exited at a certain time of day can be used to define the (quasi-constant-in-time) split ratio $\alpha_2 + \alpha_3 = 1$.

The problem with this approach is that the type ratio and the exit ratio are not necessarily the same. As detailed below, they are identical if both out-roads are in free flow, or if the split happens in a way that passing of other vehicles is impossible. However, at a highway off-ramp, neither of these assumptions needs to be satisfied. As an extreme example, consider an off-ramp that is completely clogged ($\rho_3 = \rho_3^{\max}$) due to an incident. Hence, no vehicle flow occurs on the ramp, and type 3 vehicles will start to queue up on the highway. However, a multi-lane highway generally allows type 2 vehicles to pass this queue [to some extent, compare the empirical study by Munoz and Daganzo (*17*)]. Consequently, the type ratio of vehicles that are upstream of the ramp will change and gradually shift toward more and more type 3 vehicles. However, this is not due to an actual change in upstream traffic, but rather due to the clogged (downstream) off-ramp.

In the following, we present two classical coupling models, both of which fail to capture this situation correctly (for different reasons). We then present a new model that remedies the problems.

Existing Models

As above, we consider a GRP at a 1-in-2-out node, that is, states ρ_1, ρ_2, and ρ_3 are given. Using Equation 2 we obtain the in-road demand γ_1^d, and the out-road supplies γ_2^s and γ_3^s. Then, using the split ratio $\alpha_2 + \alpha_3 = 1$, the "partial demands" (*18*) are given as

$$\gamma_{21}^d = \alpha_2 \gamma_1^d \quad \text{and} \quad \gamma_{31}^d = \alpha_3 \gamma_1^d .$$

Below we always assume that traffic actually "splits", that is, $\alpha_2>0$ and $\alpha_3>0$.

FIFO Model

The FIFO ("first-in-first-out") coupling model (Equation 3) is based on the assumption that the actual exit ratio equals the split ratio under all circumstances. Hence, the new fluxes satisfy

$$\Gamma_2 = \alpha_2\Gamma_1 \quad \text{and} \quad \Gamma_3 = \alpha_3\Gamma_1 \,. \tag{4}$$

As described above, the GRP requires that the new fluxes do not exceed the respective supplies/demands on the edges, that is, $0 \leq \Gamma_1 \leq \gamma_1^{\mathrm{d}}$, $0 \leq \Gamma_2 \leq \gamma_2^{\mathrm{s}}$, and $0 \leq \Gamma_3 \leq \gamma_3^{\mathrm{s}}$. Using Equation 4, the latter two conditions can be rewritten as $0 \leq \Gamma_1 \leq \frac{1}{\alpha_2}\gamma_2^{\mathrm{s}}$ and $0 \leq \Gamma_1 \leq \frac{1}{\alpha_3}\gamma_3^{\mathrm{s}}$. Maximizing the flux through the node under those constraints determines the fluxes as

$$\Gamma_1 = \min\left(\gamma_1^{\mathrm{d}}, \frac{\gamma_2^{\mathrm{s}}}{\alpha_2}, \frac{\gamma_3^{\mathrm{s}}}{\alpha_3}\right), \quad \Gamma_2 = \alpha_2\Gamma_1 \,, \quad \Gamma_3 = \alpha_3\Gamma_1 \,.$$

The FIFO model ensures that the resulting fluxes are always distributed according to the prescribed split ratio. Therefore, in the case of a clogged off-ramp I_3 (but free flow I_2), FIFO would result in zero flow through the node, $\Gamma_1 = \Gamma_2 = \Gamma_3 = 0$. Clearly, this is unrealistic for highway off-ramps, whose multiple lanes allow vehicles to by-pass queues (to some extent), and vehicles waiting to pass through the node do not necessarily hold up all traffic. In other words, highways are clearly not FIFO.

Non-FIFO Model

Using partial demands, the out-fluxes in FIFO can equivalently be written as

$$\Gamma_j = \min\left(\gamma_{j1}^{\mathrm{d}}, \frac{\alpha_j}{\alpha_2}\gamma_2^{\mathrm{s}}, \frac{\alpha_j}{\alpha_3}\gamma_3^{\mathrm{s}}\right), \quad j = 2, 3 \,.$$

The idea of the non-FIFO model, proposed by Lebacque and Khoshyaran (*8*), is to associate the supply constraints of each out-road only with the flux on that respective road, leading to the model

$$\Gamma_2 = \min(\gamma_{21}^{\mathrm{d}}, \gamma_2^{\mathrm{s}}) \,, \quad \Gamma_3 = \min(\gamma_{31}^{\mathrm{d}}, \gamma_3^{\mathrm{s}}) \,, \quad \Gamma_1 = \Gamma_2 + \Gamma_3 \,.$$

A physical interpretation of this model is that, before reaching the node, drivers are already presorted according to the respective out-road that they plan to take. Then, respective portions of the road width are allocated to the drivers according to the split ratio. Thus, each type of driver can pursue their destination without impediment from the other type.

By construction, the non-FIFO model does not incur the blockage problem that FIFO incurs. In the case of a clogged off-ramp (i.e., $\gamma_3^{\mathrm{s}} = 0$), one has $\Gamma_3 = 0$ (as it has to be), but $\Gamma_2 > 0$ in general. Type 2 vehicles can pass the off-ramp queue, and the flux Γ_2 is determined solely by the demand and supply of the highway segments.

Unfortunately, the non-FIFO model suffers from a different modeling problem. It assumes that the split ratios are given and independent of the actual density states and fluxes. However, because the out-fluxes are generally *not* distributed according to the split ratio (i.e., $\frac{\Gamma_2}{\Gamma_3} \neq \frac{\alpha_2}{\alpha_3}$), vehicles of one type will actually become more prevalent upstream of the node than vehicles of the other type. The non-FIFO model ignores this fact: once the clogged off-ramp becomes free flow again, vehicles of type 3 will start flowing again; however, the fact that a queue of type 3 vehicles should be present is ignored. This results in a spurious re-routing of vehicles; vehicles that should have taken the off-ramp will be allocated to outflow on the highway instead.

To recap, the non-FIFO model remedies the unrealistic blockage imposed by the FIFO model. However, it does so at the expense of unphysical re-routing effects. It should be stressed that under certain circumstances, drivers in reality may in fact change their route choices based on the actual traffic state; however, this is not always possible, and a model should not produce re-routing as an unwanted side effect. To describe intentional route changing, models that allow the split ratios to depend on the traffic state have been proposed (*19*). However, these models do not remove the fundamental flaws of FIFO and non-FIFO. A methodology that would resolve the re-routing problem is to explicitly track and evolve the split ratios (or more accurately, the type ratios) as they move along the incoming edge, as proposed, for instance, in (*10, 11*). In addition to being computationally substantially more demanding, such explicit multi-commodity models are held back by the aforementioned fundamental challenge that the route choices of vehicles are generally not accessible in advance.

The question is therefore, what can be done to remedy the problems of both models (FIFO and non-FIFO) within the framework of single-class, lane-aggregated, macroscopic models? The new model should still be based on historic split ratios; it should allow for passing of queued vehicles; it should not lose any vehicles (that are waiting in queues); and it should account for the type of queued vehicles. As the formation of a queue, caused by reduced supply of one out-road, introduces non-local-in-time effects (vehicles queued up initially may still wait later in time), imbalances in the vehicle type composition among queued vehicles must, in some form, be tracked. Next, we construct a model that does so, with the minimal amount of additional information stored, namely the excess of vehicles of a certain type, relative to the actual split ratio.

New FIFOQ Model

We now derive a new model that remedies both shortcomings of FIFO and non-FIFO, called FIFOQ ("FIFO with queue"). As argued above, the proper evolution of waiting and passing vehicles requires the model to be augmented by some additional variable that accounts for the composition of backed-up vehicle types. We choose to introduce a local vertical queue as that additional variable. Because vertical queues improperly capture the non-local impact of true vehicle jamming, we derive a model that minimizes the impact of the vertical queue, so that its sole purpose is to account for excess vehicles of a certain type. It should be stressed that the use of vertical queues in traffic models is not novel at all. For instance, vertical queues are a common means to implement vehicles that enter the network (*11, 13*). In contrast, the usage of queues for the purpose of properly tracking vehicle types upstream of an off-ramp is a novel concept.

The fundamental dynamics of a vertical queue are that its rate of change equals inflow into its upstream end, Γ_{in}, minus outflow out of its downstream front, Γ_{out}, that is, $\dot{m} = \Gamma_{\text{in}} - \Gamma_{\text{out}}$, where m denotes the number of vehicles stored in the queue (*20*).

To derive the new model, we start with a preliminary setup, a non-FIFO model with queues. Recall that the non-FIFO model can be interpreted as traffic being presorted and vehicle types given a share of the in-road width according to the split ratio. We therefore think of separating the in-road into two parts and consider two independent 1-in-1-out nodes with queues m_2 and m_3, see Figure 1*a*. These queues could now be used, for example, to prevent backward going shocks on the in-road. However, this is not an adequate model because it does not guarantee a preservation of the split ratio. Moreover, two queues may form simultaneously.

We therefore extend the model by adding additional components. As shown in Figure 1*b*, a 1-in-2-out FIFO node serves to sort the vehicles into two groups based on their destinations. Next downstream, a free-flow section follows, which consists of two independent pipes with flux functions $f_{1i} = v_1^{\max}\rho(1 - \frac{\rho}{\rho_1^{\max} c_i})$ for $i = 2, 3$, where c_2 and c_3 are the road sharing ratios of different vehicle types. In many situations, one highway lane will be associated with the off-ramp, while the remaining lanes are associated with the flow past the off-ramp traffic, and these geometric considerations could be incorporated into the model. In this paper, however, we use a simplifying assumption, namely that the road is divided precisely according to the split ratio, that is, $c_2 = \alpha_2$ and $c_3 = \alpha_3$. The reason why we can assume this two-pipe region to be in free flow is that the two queues, m_2 and m_3, can be used to absorb any congestion that emanates from the two out-roads. Of course, these individual components are for model derivation purposes only; in the end, the whole model is "collapsed to zero length" to yield a single coupling condition.

Thanks to the queues, this model allows for traffic to pass through the junction while still respecting the split ratio, even when one out-road is clogged. However, the model can be improved further. In the case when both out-roads are clogged (or provide sufficiently low supply), this current model would develop two queues, one in each pipe. In turn, it would never create a congested state on the in-road. That is unsatisfactory, as the impact that a real traffic jam would have on the road conditions further upstream would not be seen. We therefore modify the model to allow for backward propagating shocks on the in-road, as long as that congested state is composed according to the split ratio. We therefore introduce a quantity μ which is removed from the flux into the node, so that at most one queue is active at any instant in time. That one active queue then tracks the excess of vehicles of a certain type that are more prevalent near the off-ramp than the split ratio would dictate.

Let $\tilde{\Gamma}_1$, $\tilde{\Gamma}_{12}$, $\tilde{\Gamma}_{13}$, $\tilde{m}_2$, and $\tilde{m}_3$ denote the fluxes and queue values before the removal of μ. We assume that the two-pipe region is in free flow initially. Thus, the supplies of the pipes, γ_{12}^{s} and γ_{13}^{s}, are always maximal.

Because we here assume that traffic divides into two pipes whose width ratio equals the split ratio, we simply have $\tilde{\Gamma}_1 = \gamma_1^{d}$ (if the pipe widths were different, $\tilde{\Gamma}_1 < \gamma_1^{d}$ could occur). Moreover, $\tilde{\Gamma}_{12} = \alpha_2\tilde{\Gamma}_1 = \alpha_2\gamma_1^{d}$ and $\tilde{\Gamma}_{13} = \alpha_3\tilde{\Gamma}_1 = \alpha_3\gamma_1^{d}$.

Collapsing the pipes to have length zero implies that the flux out of them equals the flux into them. Therefore,

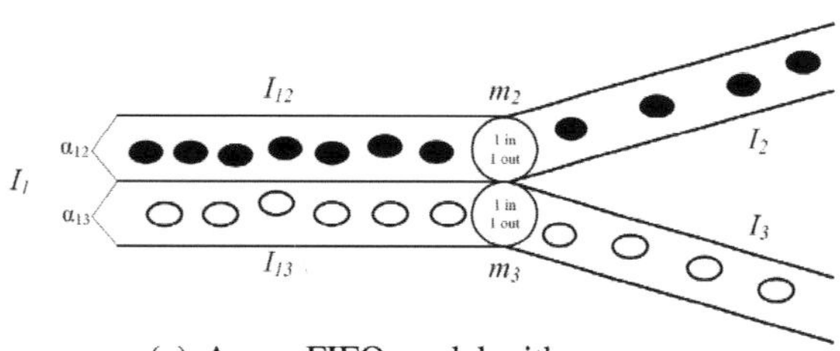

(a) A non-FIFO model with queues.

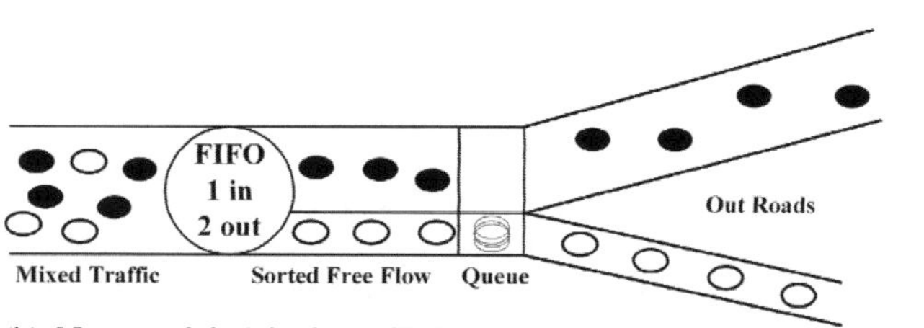

(b) New model: 1-in-2-out FIFO node with queue ("FIFOQ").

Figure 1. Interpretation/derivation of two models with queues.

the fluxes into the 1-in-1-out nodes (or queues) are $\tilde{\Gamma}_{12}$ and $\tilde{\Gamma}_{13}$, respectively. If there are no queues, the out-road fluxes are

$$\Gamma_2 = \min(\alpha_2\gamma_1^d, \gamma_2^s) \quad \text{and} \quad \Gamma_3 = \min(\alpha_3\gamma_1^d, \gamma_3^s). \tag{5}$$

In turn, if a queue is active, that is, $\tilde{m}_i>0$, then the corresponding out-flux is maximal, that is, $\Gamma_i = \gamma_i^s$. The evolution of the queues is described by the differences in the fluxes: $\dot{\tilde{m}}_2 = \tilde{\Gamma}_{12} - \Gamma_2 = \alpha_2\tilde{\Gamma}_1 - \Gamma_2$ and $\dot{\tilde{m}}_3 = \tilde{\Gamma}_{13} - \Gamma_3 = \alpha_3\tilde{\Gamma}_1 - \Gamma_3$.

As in the prior discussion, we now determine μ such that at most one queue is active at any time. If one queue is active, for example, $\tilde{m}_2>0$, and the other queue is filling, that is, $\dot{\tilde{m}}_3>0$, then $\mu = \frac{\dot{\tilde{m}}_3}{\alpha_3}$. With $\Gamma_1 = \gamma_1^d - \mu$, we obtain for the in-flux

$$\Gamma_1 = \min\left(\gamma_1^d, \frac{\gamma_3^s}{\alpha_3}\right).$$

Conversely, if no queue is active, that is, $\tilde{m}_2 = \tilde{m}_3 = 0$, we have

$$\mu = \min\left(\frac{\dot{\tilde{m}}_2}{\alpha_2}, \frac{\dot{\tilde{m}}_3}{\alpha_3}\right). \tag{6}$$

By combining Equations 5 and 6, we have

$$\begin{aligned}\mu &= \min\left(\gamma_1^d - \min\left(\gamma_1^d, \frac{\gamma_2^s}{\alpha_2}\right), \gamma_1^d - \min\left(\gamma_1^d, \frac{\gamma_3^s}{\alpha_3}\right)\right)\\ &= \gamma_1^d - \min\left(\gamma_1^d, \max\left(\frac{\gamma_2^s}{\alpha_2}, \frac{\gamma_3^s}{\alpha_3}\right)\right),\end{aligned}$$

and with $\Gamma_1 = \gamma_1^d - \mu$, we obtain for the in-flux

$$\Gamma_1 = \min\left(\gamma_1^d, \max\left(\frac{\gamma_2^s}{\alpha_2}, \frac{\gamma_3^s}{\alpha_3}\right)\right).$$

Therefore the complete model reads as:

$$\text{If } m_2 = m_3 = 0: \begin{cases}\Gamma_1 = \min\left(\gamma_1^d, \max\left(\frac{\gamma_2^s}{\alpha_2}, \frac{\gamma_3^s}{\alpha_3}\right)\right)\\ \Gamma_2 = \min(\alpha_2\gamma_1^d, \gamma_2^s)\\ \Gamma_3 = \min(\alpha_3\gamma_1^d, \gamma_3^s)\end{cases}$$

$$\text{If } m_2>0: \begin{cases}\Gamma_1 = \min\left(\gamma_1^d, \frac{\gamma_3^s}{\alpha_3}\right)\\ \Gamma_2 = \gamma_2^s\\ \Gamma_3 = \min(\alpha_3\gamma_1^d, \gamma_3^s)\end{cases}$$

$$\text{If } m_3>0: \begin{cases}\Gamma_1 = \min\left(\gamma_1^d, \frac{\gamma_2^s}{\alpha_2}\right)\\ \Gamma_2 = \min(\alpha_2\gamma_1^d, \gamma_2^s)\\ \Gamma_3 = \gamma_3^s\end{cases}$$

$$\dot{m}_2 = \alpha_2\Gamma_1 - \Gamma_2$$

$$\dot{m}_3 = \alpha_3\Gamma_1 - \Gamma_3$$

By construction, this model never generates more than one queue to be active (given that at most one queue is active initially): if $m_3>0$, then the model dynamics automatically imply that $\dot{m}_2 = 0$, and vice versa. Therefore, as soon as one queue becomes active, the other one remains constant at zero. Moreover, in the case of a singular split ratio, such as $\alpha_3 = 0$, we let $\mu = \frac{\dot{m}_2}{\alpha_2}$, in which case the model reduces to the standard FIFO model for a 1-in-1-out node.

Cell Transmission Discretization

We now describe how the new model is discretized into a CTM, by suitably augmenting the Godunov scheme (*15*) with a treatment of the queue evolution.

Approximation along Edges

Let space and time be discretized via a regular grid, where

(i) Δx is the cell size;
(ii) Δt is the time step, adhering to the Courant–Friedrichs–Lewy (CFL) conditions $\Delta t \leq \frac{\Delta x}{\max|f'(\rho)|}$, where the maximum is taken over all flux functions in the network; and
(iii) $(x_j, t^n) = (j\Delta x, n\Delta t)$ are the space-time grid points, and ρ_j^n denotes the (average) density on cell j at time t_n.

Using $\lambda = \frac{\Delta t}{\Delta x}$, the Godunov scheme along an edge reads as

$$\rho_j^{n+1} = \rho_j^n - \lambda(F(\rho_{j+1}^n, \rho_j^n) - F(\rho_{j-1}^n, \rho_j^n)), \tag{7}$$

where, for internal cell boundaries, the numerical flux

$$F(\rho_L, \rho_R) = \begin{cases}\min\limits_{\rho_L \leq \rho \leq \rho_R} f(\rho) & \text{if } \rho_L \leq \rho_R\\ \max\limits_{\rho_R \leq \rho \leq \rho_L} f(\rho) & \text{if } \rho_R \leq \rho_L\end{cases}$$

equals the standard CTM flux. At the terminal cells of an edge, that is, adjacent to a network node, the in-flux or out-flux in Equation 7 are replaced by the respective flux Γ_i defined by the coupling model. Moreover, at boundaries that are not connected to a node, ghost cells are used (cf. [6]).

Treatment of Queues

The proper time-stepping of the model with queues must take into account that a queue may deplete during a time step (*13*). For simplicity of notation, we describe the situation of a 1-in-1-out node with a vertical queue m, that is, $\dot{m} = \Gamma_1 - \Gamma_2$. At each time, we must determine the

Table 1. Total Vehicle Flow after $t = 25$ Min

Model	Total in-flux	Total out-flux on I_2	Total off-ramp flux on I_3	Ratio of out-fluxes
FIFO	133 veh	111 veh	22 veh	5:1
Non-FIFO	196 veh	174 veh	22 veh	7.81:1
FIFOQ	200 veh	167 veh	33 veh	5:1

new length of the queue. If the queue is filling, the increment is simply added to m^n. However, if the queue is emptying, we must calculate the time of queue depletion, $\bar{t} = t^n + \frac{m^n}{\Gamma_2 - \Gamma_1}$, and compare it with the time $t^{n+1} = t^n + \Delta t$. We have the following cases:

- If $\Gamma_1 \geq \Gamma_2$, then: $m^{n+1} = m^n + \Delta t(\Gamma_1 - \Gamma_2)$.
- If $\Gamma_1 < \Gamma_2$, then: $m^{n+1} = \begin{cases} m^n + \Delta t(\Gamma_1 - \Gamma_2) & \text{if } \Delta t \leq \bar{t} - t^n \\ 0 & \text{if } \Delta t > \bar{t} - t^n \end{cases}$.

We now show the important property that if a queue empties in a given time step, it remains empty until the end of the step. If $\dot{m} < 0$, the GRP at the junction has a switching point when $m = 0$. Thus, we consider not only the initial states ρ_1, ρ_2 and the final states $\hat{\rho}_1, \hat{\rho}_2$, but also intermediate states $\bar{\rho}_1, \bar{\rho}_2$, that is, the states at $\bar{t}$. While the queue is still emptying, we have $\Gamma_1 = f(\rho_1)$ and $\Gamma_2 = \gamma_2^s(\rho_2)$. Then, at time $\bar{t}$, when $m = 0$, we consider the GRP with initial states $\bar{\rho}_1$ and $\bar{\rho}_2$. Since by the definition of the demand function we have: $f(\bar{\rho}_1) \leq \gamma_1^d(\bar{\rho}_1)$, and because the queue was depleting, we will have $f(\bar{\rho}_1) < \gamma_2^s(\bar{\rho}_2)$. Therefore, we have that $f(\bar{\rho}_1) \leq \min(\gamma_1^d(\bar{\rho}_1), \gamma_2^s(\bar{\rho}_2))$. Thus, the queue does not begin to fill again.

We also need to modify the Godunov scheme in the case of an emptying queue. We divide the time step into two sub-intervals, $(t^n, \bar{t})$ and $(\bar{t}, t^{n+1})$, where $\Delta t_a = \bar{t} - t^n$ and $\Delta t_b = t^{n+1} - \bar{t}$. Then, we solve two different RPs. For Δt_a, we solve the classic Godunov scheme, and for Δt_b, we solve another RP with fluxes as given by the case when $m = 0$. Thus, the total fluxes over the full time step add up to

$$\Gamma_1 = \frac{\Delta t_a}{\Delta t} f_1(\rho_1) + \frac{\Delta t_b}{\Delta t} \min(\gamma_1^d(\bar{\rho}_1), \gamma_2^s(\bar{\rho}_2)) ,$$

$$\Gamma_2 = \frac{\Delta t_a}{\Delta t} \gamma_2^s(\rho_2) + \frac{\Delta t_b}{\Delta t} \min(\gamma_1^d(\bar{\rho}_1), \gamma_2^s(\bar{\rho}_2)) .$$

Model Comparison and Examples

We construct a specific example that highlights the key differences of the three models: FIFO, non-FIFO, and FIFOQ. In this example, the off-ramp is set to be completely clogged at the initial time ($t = 0$), and after some fixed time it is (artificially) set to free flow. Then, at the final time, the total tally of vehicles exiting and passing the off-ramp is taken.

We use the Greenshields flux function with a critical density of 40 veh/km/lane, and a maximum velocity of 100 km/h, which yields a capacity of 2000 veh/h/lane. For the three edges, we use

$$\rho_1^{\max} = \rho_2^{\max} = 320 \,\text{veh/km} \quad \text{and} \quad \rho_3^{\max} = 80 \,\text{veh/km} ,$$
$$v_1^{\max} = v_2^{\max} = v_3^{\max} = 100 \,\text{km/h} ,$$

representing a four-lane highway and a single-lane off-ramp (for simplicity, with the same speed limit as the highway). Moreover, a split ratio of $\alpha_2 = \frac{5}{6}$ and $\alpha_3 = \frac{1}{6}$ is assumed, and the initial densities are $\rho_1 = 0.4\rho_1^{\max}$, $\rho_2 = 0$, and $\rho_3 = \rho_3^{\max}$, representing a free-flow highway with a completely clogged off-ramp. This situation is let to evolve until $t = 9$ min. Then, we perform an idealized maneuver (for the sake of simplicity) in which we set $\rho_3 = 0$, that is, we instantaneously remove all vehicles from the off-ramp. We then let this new situation evolve until $t = 25$ min. At that final time, we tally the total (time-integrated) fluxes I_2 and I_3 onto the highway and the off-ramp, respectively.

Table 1 shows the total vehicle fluxes, obtained by the different models. At the end of this experiment, the queue that forms in the FIFOQ model has emptied. One can clearly see the key shortcomings of the existing models. The FIFO model respects split ratios, but it predicts zero total flux up until $t = 9$ min, resulting in severely reduced highway flux. In turn, the non-FIFO model produces reasonable highway flows, but it fails to respect the split ratio: due to ignoring the off-ramp queue that arises in reality, it falsely re-allocates vehicles that intended to exit into the highway flow.

The new FIFOQ model remedies both of those shortcomings. Unlike FIFO, it produces a nonzero highway flux before $t = 9$ min. However, unlike non-FIFO, it tracks the accumulation of type 3 vehicles via the growing vertical queue m_3. Then, after $t = 9$ min, this queue depletes, releasing those queued type 3 vehicles onto the off-ramp. Therefore, the total off-ramp flux produced by the FIFOQ model is noticeably larger than with the other two models; which is more realistic, because it remedies the spurious re-routing of the non-FIFO model.

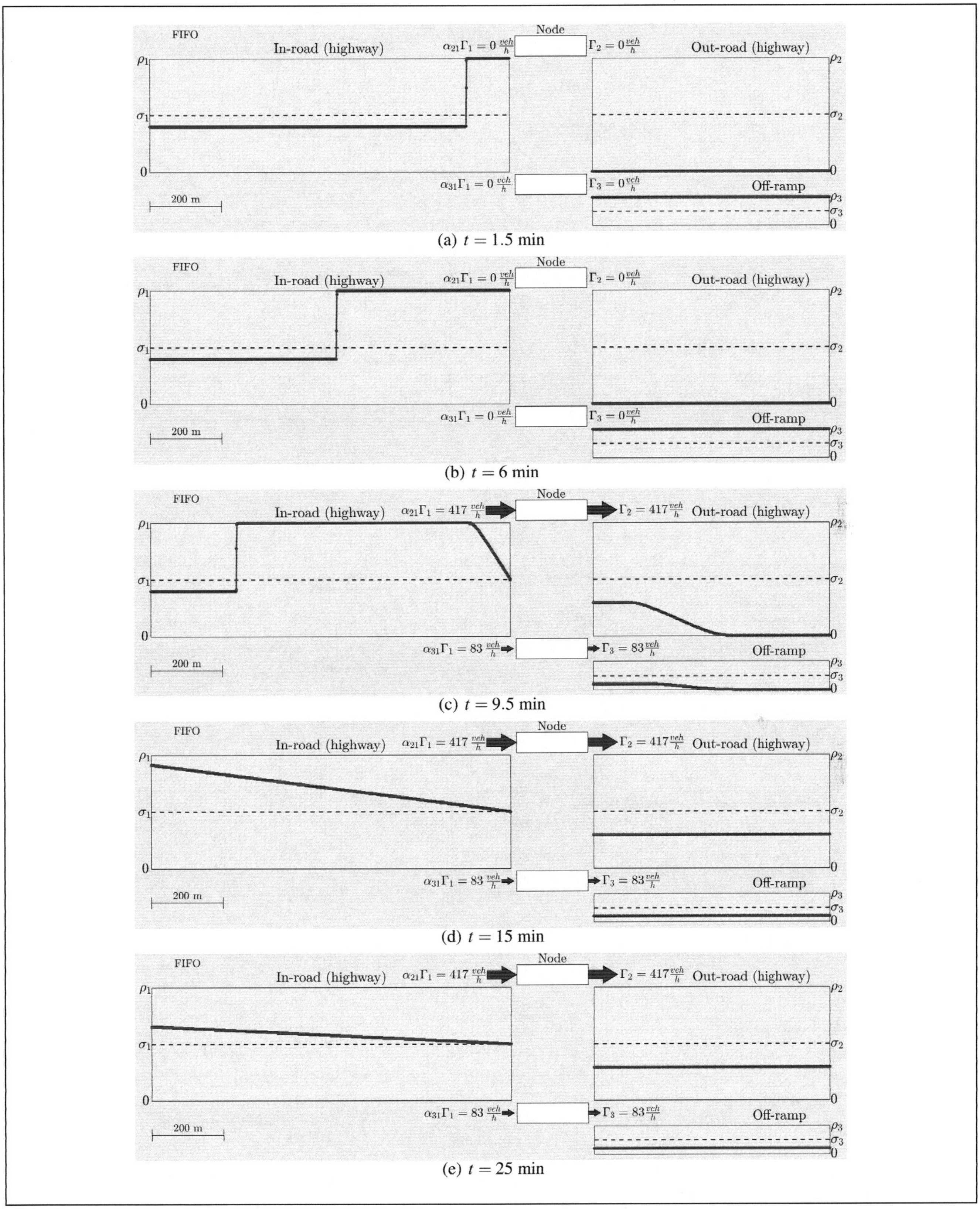

Figure 2. Model comparison: Time evolution of the solution of the FIFO model.

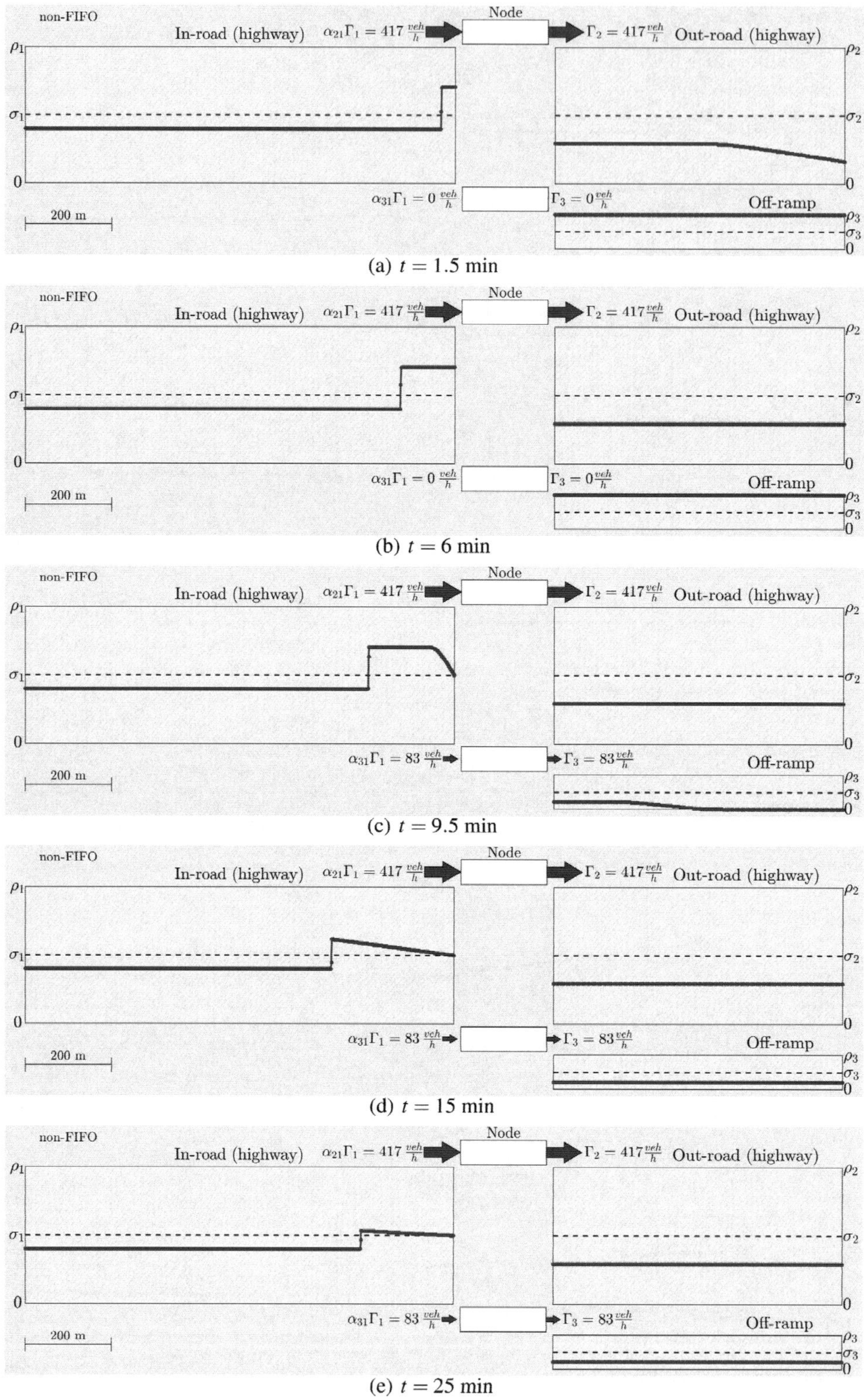

(a) $t = 1.5$ min

(b) $t = 6$ min

(c) $t = 9.5$ min

(d) $t = 15$ min

(e) $t = 25$ min

Figure 3. Model comparison: Time evolution of the solution of the non-FIFO model.

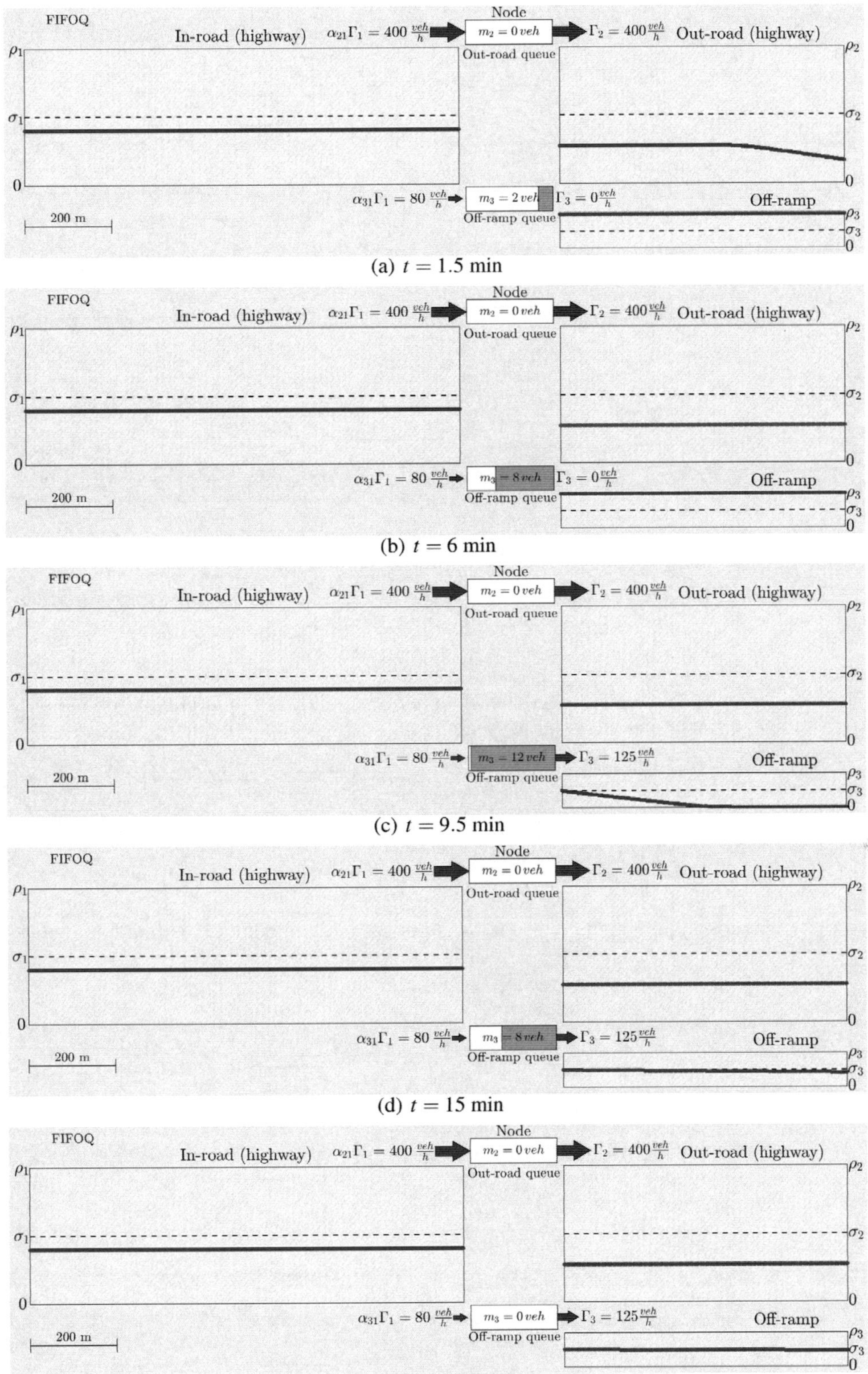

(a) $t = 1.5$ min

(b) $t = 6$ min

(c) $t = 9.5$ min

(d) $t = 15$ min

(e) $t = 25$ min

Figure 4. Model comparison: Time evolution of the solution of the FIFOQ model.

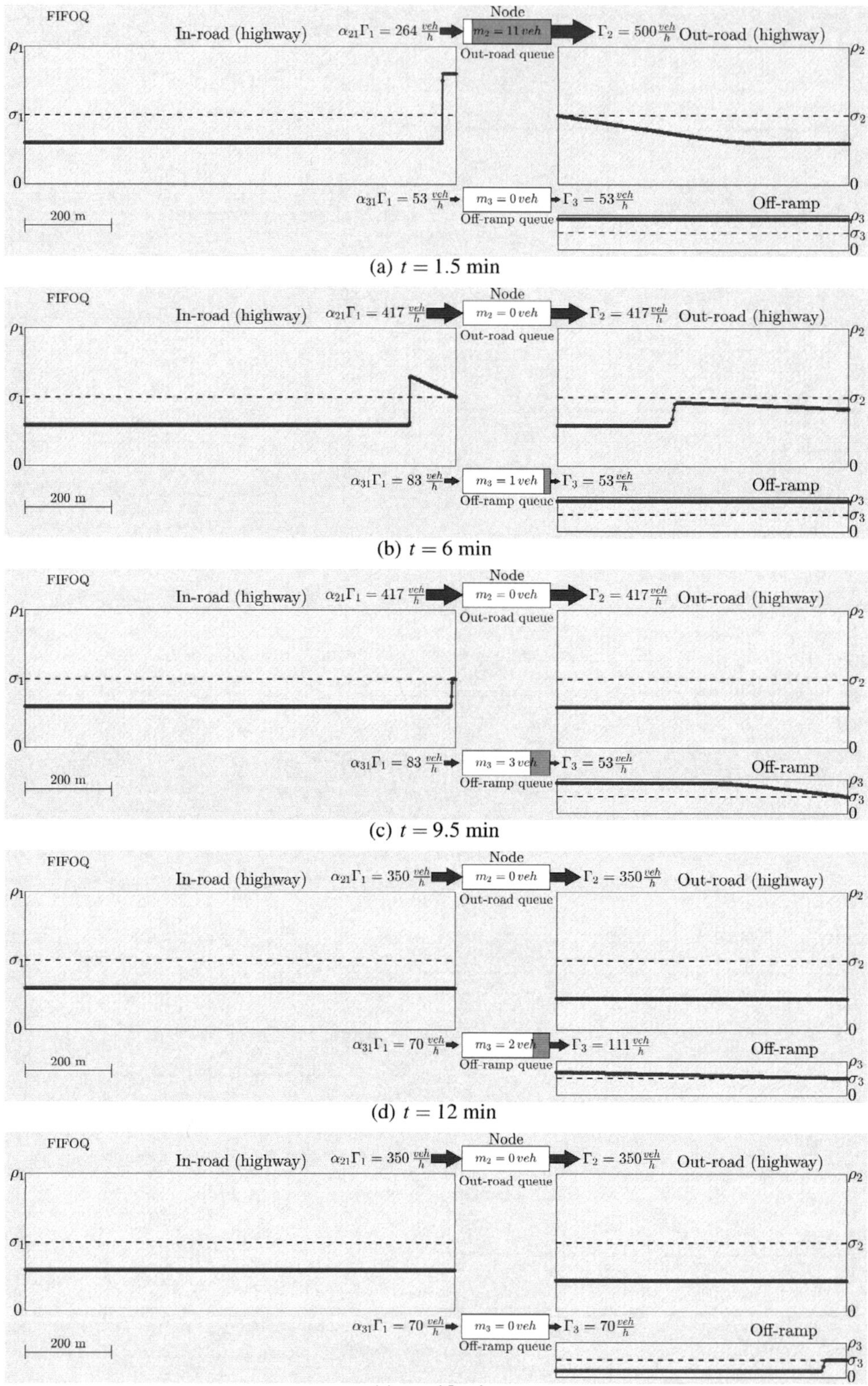

Figure 5. More complex example: Time evolution of the FIFOQ solution.

Figures 2 to 4 show the time evolution (at five representative snapshots in time) of the solutions produced by the three models, computed by a highly resolved Godunov/CTM discretization (with proper queue treatment). For the FIFOQ model, Figure 4 shows the magnitude of the queue m_3 in the lower queue box. The blue arrows visualize the flows into and out of the respective queues. For consistency, empty queue boxes are also shown for the models without queues.

For both classical models, Figure 2*a* and 2*b*, and Figure 3*a* and 3*b* show the initial backward propagating shock that arises due to the clogged off-ramp. Clearly, with FIFO, the level of congestion on the in-road is much greater (in fact, completely jammed) than with non-FIFO. For the FIFOQ model (Figures 4*a* and 4*b*) the accumulation of type 3 vehicles is instead tracked in the queue.

Then, for all models, Figures 2*c*, 3*c*, and 4*c* show the system state right after the clearing of the off-ramp (the idealized maneuver). For both FIFO and non-FIFO, the flow turns maximal, resulting in a rarefaction fan on the in-road. In turn, in the FIFOQ model the queue m_3 has started to decrease.

Finally, for both FIFO and non-FIFO, Figure 2*d* and 2*e*, and Figure 3*d* and 3*e* show the gradual approach of the system toward a uniform state on the in-road. In contrast, for FIFOQ the in-road has remained in free flow the whole time. Figure 4*d* and 4*e* show the shrinking and eventual depletion of the queue. Right after $t = 25$ min, the outflow states will change to what the other two models would also yield as $t \longrightarrow \infty$.

In order to demonstrate the capabilities of the FIFOQ model, we study it in a more complicated second example. We use the same road and split parameters as the previous example, but modify the initial and boundary data as follows: $\rho_1 = 0.3\rho_1^{\max}$, $\rho_2 = 0.3\rho_2^{\max}$, and $\rho_3 = 0.88\rho_3^{\max}$, representing a free-flow highway with a congested but not clogged off-ramp. Boundary conditions of the same values are prescribed on the in-road and off-ramp. Moreover, the highway queue is initialized with $m_2 = 17$ veh. This situation is let to evolve until $t = 9$ min. Then, the outflow conditions at the ramp are changed to maximum capacity ($\rho_3 = \sigma_3$), and the solution is tracked until $t = 25$ min. Figure 5 shows time evolution of the solution. Figure 5*a* shows how m_2 is decreasing and a shock travels backward on the in-road. In this example, the queue m_2 depletes, and queue m_3 starts forming. Figure 5*b* shows the situation soon after that instance. Figure 5*c* shows the state of the system after the outflow conditions of the ramp have been changed. A rarefaction arises on the ramp, and once the density at the beginning of the ramp has decreased sufficiently the queue m_3 shrinks again. Figure 5*d* and 5*e* show the shrinking and eventual depletion of the off-ramp queue.

Conclusions and Outlook

In this paper, we have introduced a new coupling model for macroscopic lane-aggregated traffic flow at 1-in-2-out nodes modeling off-ramps, FIFOQ. The new model demonstratively remedies fundamental modeling shortcomings of the existing FIFO and non-FIFO models (blockage and spurious re-routing, respectively). It achieves this goal by means of a vertical queue that tracks the excess of vehicles of a certain type. A cell transmission discretization of the new model has been presented, and applied in two representative examples.

The new model is clearly devoid of the fundamental shortcomings of existing models. Of course that does not necessarily imply that the new model is descriptive of reality; all we know is that it is "less bad" than existing models. A model validation with real traffic data will be an important step for future research. FIFOQ model calibration, as well as state estimation, can be conducted in the same fashion as with existing LWR-based models [cf. (*12*)]. A simple comparative model validation (e.g., FIFOQ versus non-FIFO) can be conducted in terms of total fluxes (alike Table 1), if flow-rate sensor data for a ramp, as well as upstream and downstream on the highway, are available. One key validation task is to investigate the accuracy of the models for varying queue lengths.

On the modeling side, one key simplification assumed in this paper is that upstream of the off-ramp, traffic divides the width of the highway according to the split ratio. This is clearly unrealistic in most situations, where off-ramp queues commonly are restricted to the right-most lane. Future work will be devoted to extending the model to the more general situation. In fact, on real highways the road width available for passing an off-ramp queue does depend on the queue length itself, as long queues tend to affect other lanes and not just the right-most lane. Again, such modeling extensions will be explored in the future.

Author Contributions

The authors confirm contribution to the paper as follows: study conception and design: Salehi, Somers, Seibold; data collection: N/A; analysis and interpretation of results: Salehi, Somers, Seibold; draft manuscript preparation: Salehi, Seibold. All authors reviewed the results and approved the final version of the manuscript.

References

1. Lighthill, M. J., and G. B. Whitham. On Kinematic Waves. II. A Theory of Traffic Flow on Long Crowded Roads. *Proceedings of the Royal Society of London. Series A*, Vol. 229, No. 1178, 1955, pp. 317–345.

2. Richards, P. I. Shock Waves on the Highway. *Operations Research*, Vol. 4, 1956, pp. 42–51.
3. Daganzo, C. F. The Cell Transmission Model, Part II: Network Traffic. *Transportation Research Part B: Methodological*, Vol. 29, 1995, pp. 79–93.
4. Holden, H., and N. H. Risebro. A Mathematical Model of Traffic Flow on a Network of Unidirectional Roads. *SIAM Journal on Mathematical Analysis*, Vol. 26, No. 4, 1995, pp. 999–1017.
5. Herty, M., and A. Klar. Modelling, Simulation and Optimization of Traffic Flow Networks. *SIAM Journal on Computing*, Vol. 25, No. 3, 2003, pp. 1066–1087.
6. Coclite, G. M., M. Garavello, and B. Piccoli. Traffic Flow on a Road Network. *SIAM Journal on Mathematical Analysis*, Vol. 36, No. 6, 2005, pp. 1862–1886.
7. Garavello, M., and B. Piccoli. *Traffic Flow on Networks*. American Institute of Mathematical Sciences, 2006.
8. Lebacque, J. P., and M. M. Khoshyaran. First Order Macroscopic Traffic Flow Models for Networks in the Context of Dynamic Assignment. *Transportation Planning*, 2002, Vol. 64, pp. 119–140.
9. Lebacque, J. P., X. Louis, and B. Schnetzler. A Multilane Junction Model. *Transportmetrica*, Vol. 8, No. 4, 2012, pp. 243–260.
10. Daganzo, C. F. A Continuum Theory of Traffic Dynamics for Freeways with Special Lanes. *Transportation Research Part B: Methodological*, Vol. 31, No. 2, 1997, pp. 83–102.
11. Bressan, A., and K. T. Nguyen. Conservation Law Models for Traffic Flow on a Network of Roads. *Networks and Heterogeneous Media*, Vol. 10, No. 2, 2015, pp. 255–293.
12. Mobile Millennium Project. *Website*, http://traffic.berkeley.edu.
13. Delle Monache, M. L., J. Reilly, S. Samaranayake, W. Krichene, P. Goatin, and A. M. Bayen. A PDE-ODE Model for a Junction with Ramp Buffer. *SIAM Journal on Applied Mathematics*, Vol. 74, No. 1, 2014, pp. 22–39.
14. LeVeque, R. J. *Numerical Methods for Conservation Laws*. 2nd ed. Birkhäuser, 1992.
15. Godunov, S. K. A Difference Scheme for the Numerical Computation of a Discontinuous Solution of the Hydrodynamic Equations. *Matematicheskii Sbornik*, Vol. 47, 1959, pp. 271–306.
16. Daganzo, C. F. The Cell Transmission Model: A Dynamic Representation of Highway Traffic Consistent with the Hydrodynamic Theory. *Transportation Research: Part B*, Vol.28, 1994, pp. 269–287.
17. Munoz, J. C., and C. F. Daganzo. The Bottleneck Mechanism of a Freeway Diverge. *Transportation Research Part A: Policy and Practice*, Vol. 36, No. 6, 2002, pp. 483–505.
18. Jin, W. L., and H. M. Zhang. On the Distribution Schemes for Determining Flows Through a Merge. *Transportation Research: Part B*, Vol. 37, No. 6, 2003, pp. 521–540.
19. Herty, M., and A. Klar. Modeling, Simulation, and Optimization of Traffic Flow Networks. *SIAM Journal on Computing*, Vol. 25, No. 3, 2003, pp. 1066–1087.
20. Vickrey, W. S. Congestion Theory and Transport Investment. *American Economic Review*, Vol. 59, No. 2, 1969, pp. 251–260.

The Standing Committee on Traffic Flow Theory and Characteristics (AHB45) peer-reviewed this paper (18-05222).

Article

TRR
JOURNAL OF THE TRANSPORTATION RESEARCH BOARD

Transportation Research Record
2018, Vol. 2672(20) 25–38

DOI: 10.1177/0361198118786473
journals.sagepub.com/home/trr
SAGE

Resurrecting the Lost Vehicle Trajectories of Treiterer and Myers with New Insights into a Controversial Hysteresis

Benjamin Coifman[1], Lizhe Li[2], and Wen Xiao[3]

Abstract

The 1974 paper by Treiterer and Myers is a seminal work in traffic flow theory. This longevity is in part because of the impressive collection of manually extracted vehicle trajectories. To date, only a few studies have rivaled the scale of the empirical vehicle trajectory data used in Treiterer and Myers. Their data collection used high-speed aerial photography and manual data reduction to follow hundreds of vehicles. In spite of the Herculean collection effort, the trajectory data set was never released and has since been lost. Fortunately, the plots survive and the present work re-extracts the vehicle trajectory data from the time–space diagrams. The discussion places the value of the data in context and then uses the data to put an end to decades of misinterpretation that started with Treiterer himself. The central thesis of Treiterer and Myers generated considerable interest: a hysteresis whereby drivers exhibit different fundamental behavior depending on whether they are entering or exiting a disturbance. There has been extensive debate about the authors' findings in the literature, but without the original data set any interpretation has required considerable speculation. With the resurrected trajectories, this work reexamines the vehicles underlying the hysteresis and finally quells the speculation. Rather than arising from car following behavior, it turns out that the enigmatic progression arose from a combination of lane change maneuvers and unremarkable transitions into or out of the congested regime. On publication, the re-extracted data from this paper will be released to the research community.

Very few transportation papers achieve a longevity that lasts decades, and Treiterer and Meyers is among them, frequently cited more than 40 years after publication (receiving over 20 citations in 2016 alone) (*1*). This longevity is due in part to the microscopic vehicle trajectory data used in the study. To date, only a few studies have rivaled the scale of the empirical vehicle trajectory data used in Treiterer and Myers. Joseph Treiterer led a 9-year-long study with many collaborators that used high-speed aerial photography to follow hundreds of vehicles, and then they used manual extraction to track the individual vehicles over time and space at 1-s intervals. The details of this process were presented over a series of lengthy reports (see [*2*] for a good overview). Treiterer's reports reference tracking at least 10 different platoons on freeways, but they ultimately show time–space diagrams for only a few of these platoons. Treiterer and Myers used the most interesting of the platoons presented in the reports (reproduced in Figure 1*a*). This platoon passed through a stop wave propagating upstream in the median lane of southbound I-71 in Columbus, Ohio. The raw data were collected just over 50 years ago, on July 25, 1967, starting at 7:45 a.m. Treiterer and Myers focused strictly on the median lane, but the research reports show the corresponding vehicle trajectory plots for all three lanes involved (*3*).

In spite of the Herculean collection effort, the trajectory data set was never released and has since been lost. Fortunately, the time–space diagrams survive. Our research seeks to recover the vehicle trajectory data from the time–space diagrams used in Treiterer and Myers and the adjacent lanes. The input for our work are Figures 1.1–1.5 in Treiterer et al. (*3*), as reproduced in Figure 1. The recovery steps consist of (i) rectifying the images; (ii) finding all curves in a given figure; (iii) eliminating the grid lines; (iv) connecting successive points on a given trajectory; (v) cleaning up distortions caused by noise, exclusion of trajectory points and/or inclusion non-trajectory points; then (vi) calculating speed and acceleration for the trajectories (on publication, the re-extracted trajectory data from this paper will be available at Coifman [*4*]).

[1]Joint Appointment with the Department of Civil, Environmental, and Geodetic Engineering, and the Department of Electrical and Computer Engineering, The Ohio State University, Columbus, OH
[2]Department of Electrical and Computer Engineering, The Ohio State University, Columbus, OH
[3]School of Engineering, Newcastle University, UK

Corresponding Author:
Address correspondence to Benjamin Coifman: Coifman.1@OSU.edu

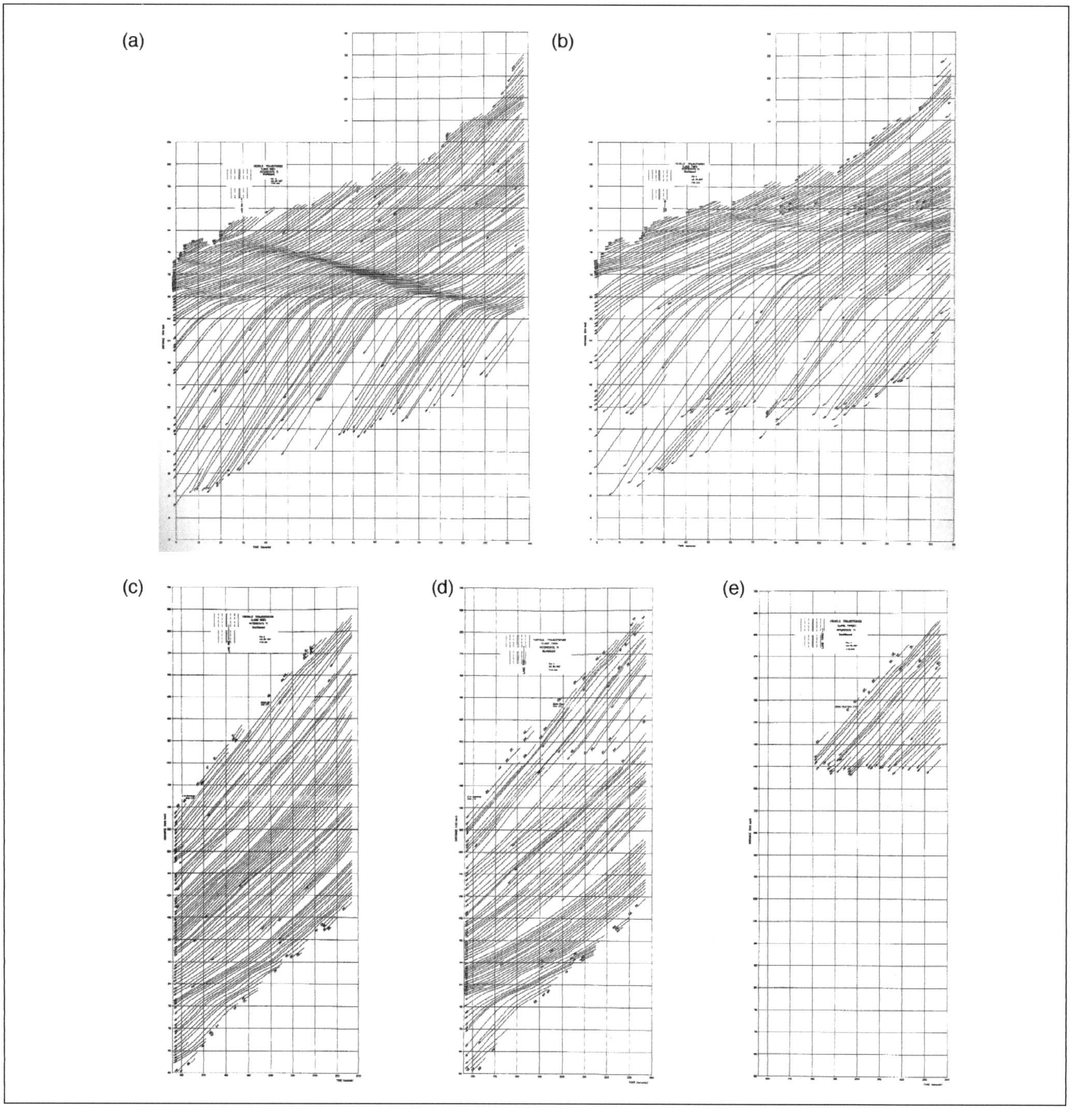

Figure 1. Vehicle trajectories for time 0 to 160 s in (*a*) lane 1 and (*b*) lane 2; vehicle trajectories for time 160 to 240 s in (*c*) lane 1, (*d*) lane 2, and (e) lane 3, all reproduced from Treiterer et al. (*3*).

After presenting the recovery process, this paper places the value of the data in a contemporary context, and then uses the data to put an end to decades of misinterpretation that started with Treiterer himself. The central thesis of Treiterer and Myers has generated considerable interest: a hysteresis phenomenon in traffic flow, whereby drivers exhibit different fundamental behavior depending on whether they are entering or exiting a disturbance (*1*). There has been extensive debate about their findings in the literature, but without the original data set any interpretation has required considerable speculation. With the resurrected trajectories in hand, this paper can finally quell the speculation.

The remainder of this section reviews the literature to place the vehicle trajectory work of Treiterer in context. The next section describes our process of re-extracting the trajectory data from the time–space diagrams. The third section analyzes the data to gain deeper insight into the traffic conditions observed. This paper closes with a discussion and conclusions.

Vehicle Trajectory Literature Review

Although Treiterer's work remains remarkable to this day, it was not the first effort to extract vehicle trajectories from orthorectified imagery. Treiterer's effort was preceded by a few prior works: in 1962 Leutzbach and Egert (as presented in Newell [*6*]) induced a stop wave on a one-lane segment of an Autobahn and traced the vehicle trajectories in the following platoon as filmed from a helicopter at 24 frames per second (fps) (*5*). In 1968, Forbes and Simpson studied freeway deceleration waves after tracing vehicle trajectories in platoons filmed from an airplane at 1 fps and noted that the study was a continuation of their earlier work presented in 1964 (*7*). Regardless, there have only been a few studies to extract empirical vehicle trajectory data on a scale that rivals Treiterer's work, for example, Turner-Fairbank Highway Research Center empirical trajectory data (*8*), and the Next Generation Simulation (NGSIM) project (*9*).

Of all of these vehicle trajectory data sets, only the NGSIM data are widely used. Since the release of the NGSIM data sets, there has been a growing minority of researchers who have found unrealistic relationships in the NGSIM data and now question the accuracy of the NGSIM trajectories, (e.g., Thiemann et al.; Hamdar and Mahmassani; Duret et al.; Montanino and Punzo [*10–13*]). Coifman and Li returned to the original video to manually re-extract a portion of the NGSIM data and conclusively showed that the NGSIM errors are beyond anything that could be corrected strictly through cleaning and interpolation of the reported NGSIM data (*14*).

Regardless, even if NGSIM were perfect, the traffic flow community is in great need of many more microscopic data sets. Treiterer's trajectory data set resurrected in our paper is the only one to track so many vehicles over such a long distance. Treiterer tracked over 200 vehicles along 3 mi of roadway. There are data sets with more vehicles but shorter distance, for example, NGSIM only covers a tenth of that distance; and there are probe-vehicle-based data sets that cover more distance but only monitor the ambient traffic, yielding trajectories from a handful of vehicles at any given instant (e.g., Coifman et al. [*15*]). There is a valuable place for all of these coverage strategies.

Methodology

The first step in the research was to digitize the original figures. To this end, Figures 1.1–1.5 from Treiterer and Research Staff were scanned at 600 dpi and imported into Matlab as grayscale bitmaps (*3*). Figure 2*a* shows a detail of the bitmap for lane 1 as input to the processing (the full diagram is in Figure 1*a*). These original plots include grid lines intended to help manual interpretation of the curves, but the grid will disrupt our automated processing. To identify the vertical grid lines, we take a moving average of 401 rows with the output placed in the center row. This range of 401 rows is just under the spacing of the horizontal grid lines. In this fashion, any feature that exists in many rows will have a strong return in the moving average. Any pixel that is at least 60% black in the moving average is identified as a vertical feature, for example, as shown in Figure 2*b* for the small example region. This process is repeated, taking the moving average of 401 columns to find the horizontal grid lines. Unfortunately, during the stop wave the trajectories of interest are also horizontal and are returned as false positives by this test, as shown in Figure 2*c*. All of the short line segments caused by the stopped traffic are manually eliminated in this stage (not shown), with only the long horizontal grid lines retained.

Although the grid lines may disrupt the automated extraction, they proved invaluable for removing distortions, from the original printing and subsequent scanning. The intersection between the horizontal and vertical grid lines are found, as shown with circles in Figure 2*d*. The grid size is computed and the coordinates of the intersection points are unified. The grid size is determined by averaging the spacing between all of the intersection points in the given plot. Next, the average x value of all the grid points in the first column and the average y value of all the grid points in the first row are taken as references. The grid points in the following rows and columns are then unified with respect to the grid size. The whole image is then undistorted by transforming the original coordinates to the unified control coordinates by piecewise linear transformation (PLT) that breaks up the image into local piecewise linear regions in which an affine transformation maps the original points to the control points (*16*). Other transformation methods exist, such as local weighted mean transformation; however, based on our experiments PLT was proven effective in the case when the distortion varies locally at different parts of the image (*17*). Then, the images coming from the same lane (e.g., Figure 1, *a* and *c*, from lane 1) are stitched together to extract the complete trajectories.

The grid line detection was then repeated on the rectified images, except now the locations of the vertical and horizontal grid lines are set to white, Figure 2*e*. The next step is to find all of the discrete black areas in the bitmap. This process is done twice. First, to the bitmap in Figure 2*a* (after rectification) with the grid lines taken column by column. Within a given column, each distinct, contiguous "dark area" (80% gray or darker) is found and the vertical center is recorded for the dark area. The process is then repeated on the bitmap in Figure 2*e* without any gridlines. Any center point that is found to be identical in both bitmaps is retained, resulting in the points in Figure 2*f*. Whereas points that shift position or disappear altogether between the two bitmaps are discarded because they have some interaction with the grid lines that were eliminated in Figure 2*e*.

The points in Figure 2*f* show the extracted trajectory elements. At this stage, only the instantaneous vehicle positions have been identified in each vertical column of pixels (i.e., a

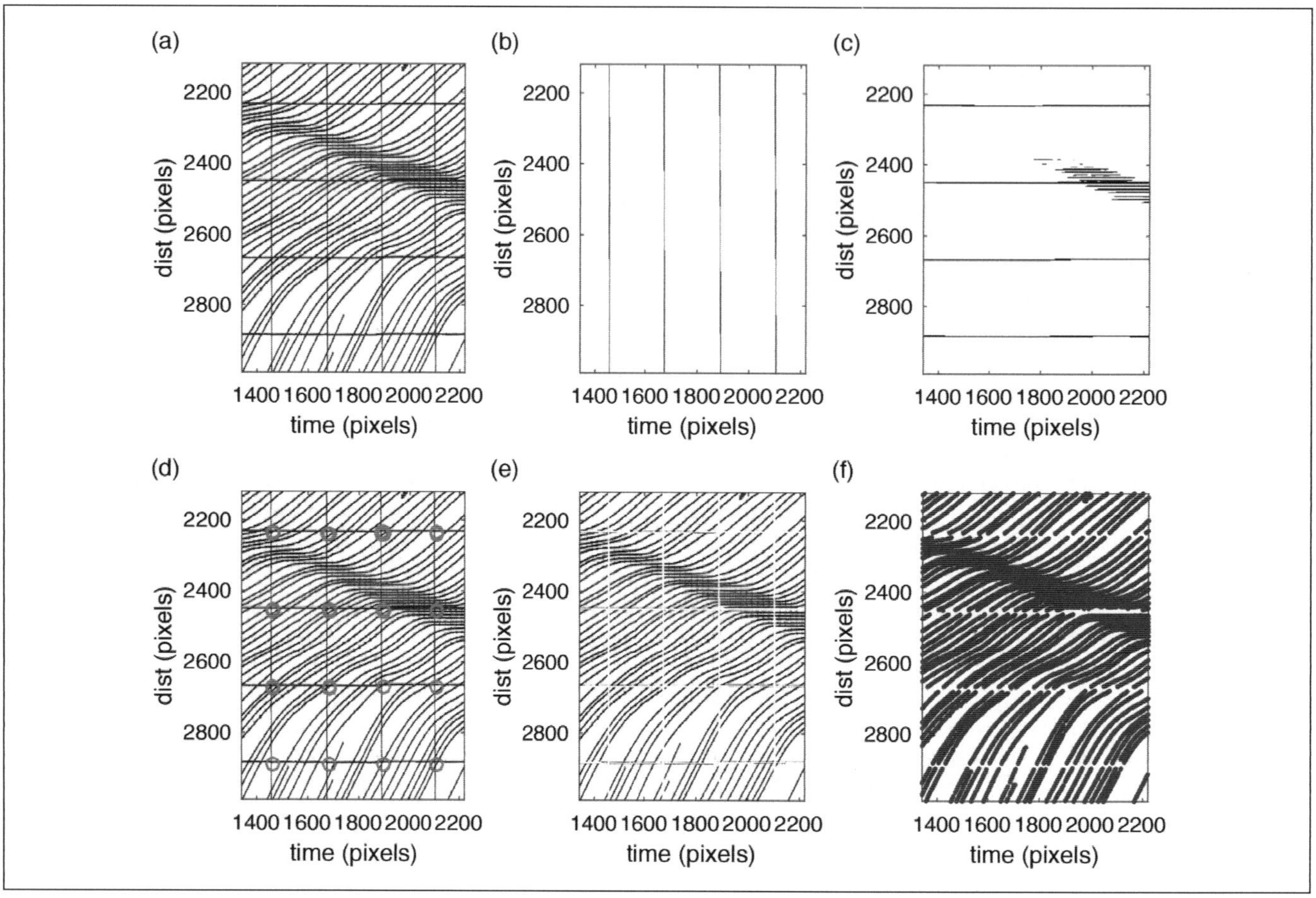

Figure 2. (*a*) Detail of the scanned bitmap shown in Figure 1*a*; (*b*) the identified vertical edge features; (*c*) the identified horizontal edge features (predominantly gridlines, but a few stopped trajectories were also caught and subsequently manually removed); (*d*) the intersections of the detected grid lines; (e) removal of most of the grid lines from the bitmap in part (a); (*f*) the resulting center-points of dark regions taken column by column that are consistent between parts (a) and (e). At the scale of the plot, the discrete points visually blur together into curves, but the actual points are discrete at this stage.

collection of points in each column); the association from one column to the next has not yet been made. Noise is still present in the vehicle positions, including jitter, blank areas in which the grid lines were eliminated, an occasional non-trajectory element, and so forth.

The next step is to associate the vehicle positions across columns into discrete trajectories while reducing the remaining noise. Figure 3*a* shows a detail from Figure 2*f*, superimposed on the corresponding region of rectified Figure 2*a*, and at this resolution the discrete points are visible. Ripples in the original artwork are evident, as well as a few stray noise points among the vehicle positions. Since the regions with the grid lines were removed, there are also noticeable gaps in the extracted vehicle positions, (i.e., the points). A human reviews the points to eliminate non-vehicle trajectory points, for example, the two stray points around time 1,690. Note that at several stages of the processing a human is in the loop. Since the present extraction tools are envisioned only to be used on a few special cases, rather than developing a robust automated system that can handle any situation, like Coifman et al, (*15*) the present work relies on a human to do the assessments that are difficult for a computer; however, even when the human is in the loop, we use graphical interfaces to improve their efficiency.

A simple automated process is then run to match each vehicle position in one column with its respective observation in the next column when the subsequent point is present, as shown in Figure 3*b* where the points have now been joined into partial trajectories. A related automated process is then run to join trajectories across gaps, as shown with the two bold curves in Figure 3*c*. A large gap remains in the trajectory that starts around distance 2,700. At this stage, the human again verifies the automated tracking and corrects any errors, thereby ensuring that the correct trajectories are associated across the gaps or when needed, manually associating the portions of the same trajectory, for example, the bold curve in Figure 3*d*. This comparison is done on top of the original image (not shown in Figure 3*d* simply for clarity

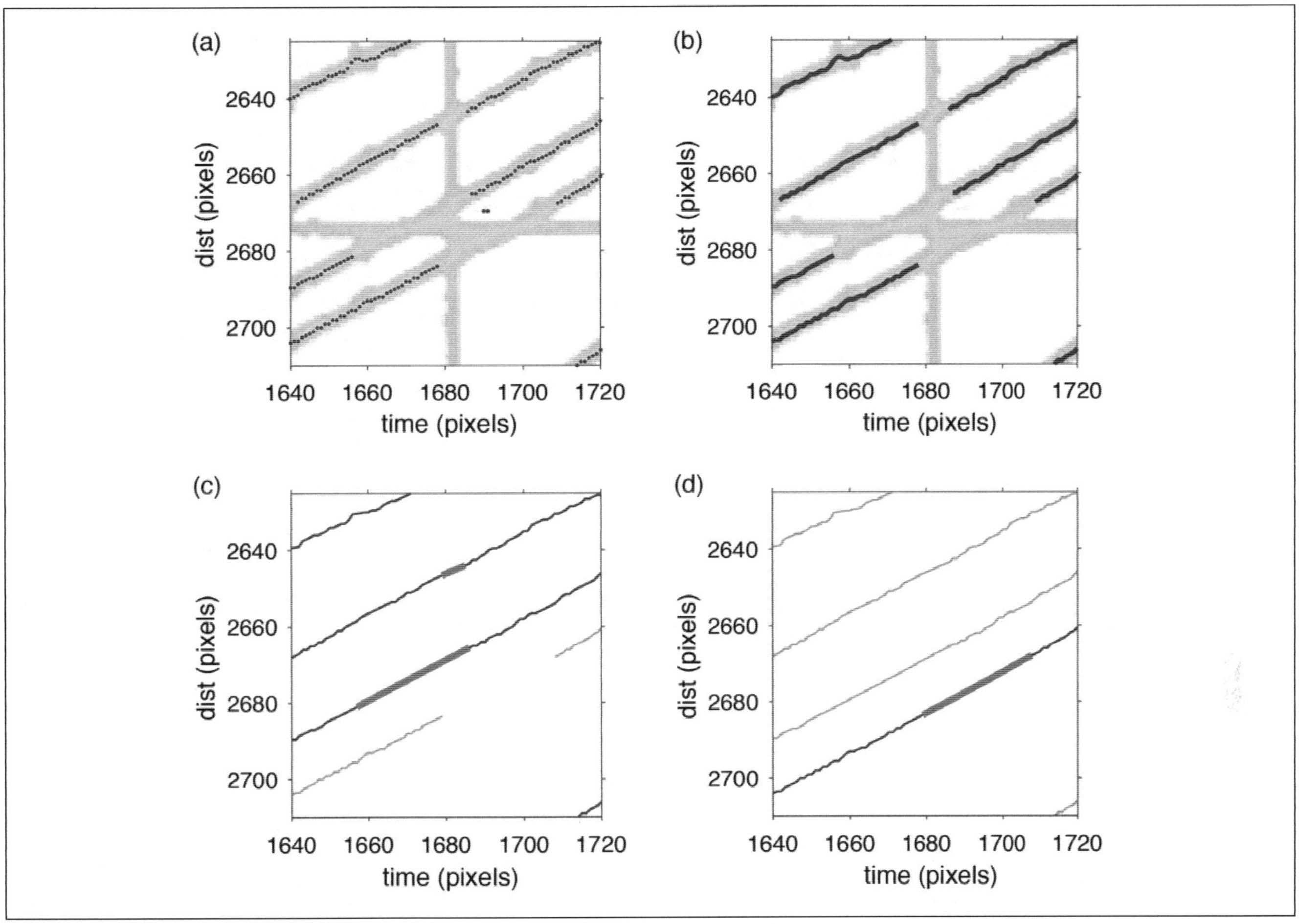

Figure 3. (*a*) Detail of Figure 2*f* showing the detected discrete points with gaps caused by the omitted grid lines, superimposed on top of the original bitmap; (*b*) the points joined automatically into partial trajectories; (*c*) most of the partial trajectories automatically joined across the gaps (bold curves); (*d*) a few needing manual intervention (bold curves).

of presentation). In some cases, it is not enough to simply join the start and end points across a gap. If a trajectory undergoes an acceleration while crossing over a grid line, the human may insert additional points so that the infill portion of the trajectory follows the original image. On rare occasions, it is also necessary for the human to move the extracted points from Figure 2*f* closer to the trajectory in the bitmap.

Calculating Speed and Acceleration

After the manual cleanup, the trajectories are fully associated and the coordinate system is changed from "pixels" to "feet" and "seconds". However, the jitter from column to column in a given trajectory remains and is amplified in any derivatives, for instance, speed and acceleration. The process of calculating speed and acceleration from noisy trajectory data is borrowed from Coifman and Li (*14*) and is summarized in this section.

Figure 4*a* shows the extracted raw trajectories for several vehicles as they pass through the stop wave, and the jitters are readily apparent. This position noise is not surprising given the resolution of the original scan, in which trajectories over 3 mi of roadway are presented in under 10 in. of printed media (Figure 4*c* shows the same region from the scan; note that the line width for each trajectory is several feet along the roadway and so we kept our extracted trajectories in the middle of the printed curves). Following the trajectory smoothing from Coifman and Li, the first step is finding the time series speed for each raw trajectory, which is calculated using the median difference over multiple time steps, as follows (*14*). First, for each trajectory the time series speed is separately calculated using a range of time steps via Equation 1, where i denotes the current frame, n is the time step, and $x(i + n)$ is the vehicle's position in frame $i + n$. Thus, $\hat{v}_n(i)$ is the estimated average speed of the vehicle over $2*n$ frames ($n/10$ s) bounding the current frame. This work simultaneously evaluates Equation 1 with integer n from 1 to 7, yielding seven different measures of speed at a given instant. The vehicle's median speed at frame i is then calculated by taking the median of all seven time steps. Figure 4*b* shows the

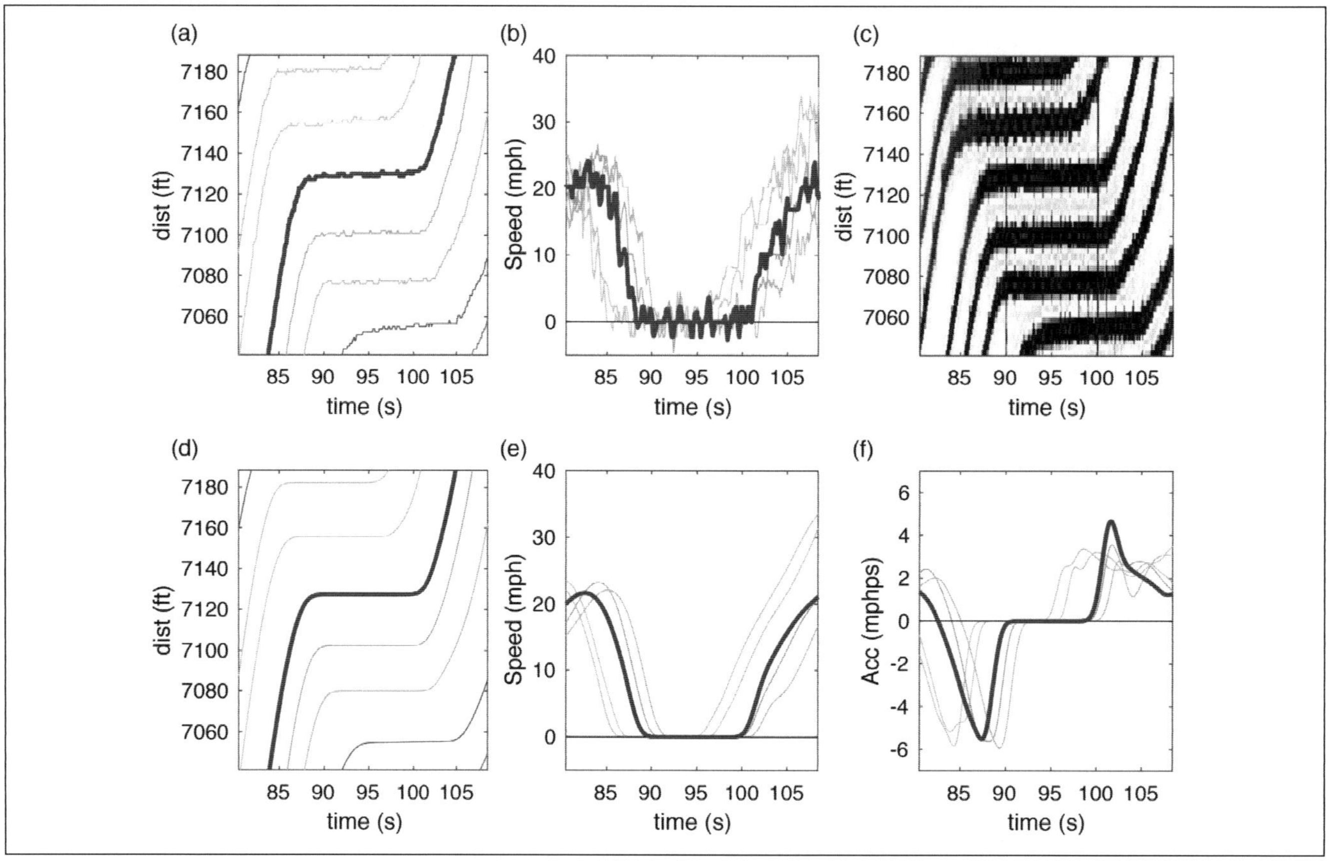

Figure 4. (*a*) Extracted raw trajectories for several vehicles as they pass through the stop wave (note the apparent jitters); (*b*) the resulting time series speed for the bold trajectory from part (a) as well as the two vehicles ahead and the two vehicles behind shown with fainter curves; (*c*) region of the original scan that corresponds to part (a); (*d*) the resulting cleaned trajectories, as calculated from part (e); (e) the smoothed time series speeds from part (b); (*f*) the resulting time series acceleration calculated from the speeds in part (e).

resulting time series median speed for the bold trajectory from Figure 4*a*, as well as the two vehicles ahead and the two vehicles behind (shown with fainter curves):

$$\hat{v}_n(i) = \frac{x(i+n) - x(i-n)}{2 \cdot nT} \tag{1}$$

This processing could be viewed as a non-causal, non-linear low pass filter. As a vehicle comes to a stop, the relative impact of the filter can exceed the magnitude of the actual speed. To avoid inducing large relative errors at extremely low speeds, we specifically split the trajectory into the moving periods (median speed above 5.9 ft/s, or 1 distance axis pixel per 1 time axis pixel in the scans) and stopped periods (median speed is below 0.2 ft/s, or 0.14 mph). The moving periods are processed independently of the stopped periods, as follows.

All vehicles have at least one period in which speeds are classified as moving. Some vehicles will come to a stop, in which case there will be a period of moving, followed by a period of stopped, followed by another period of moving. Between each moving and stopped period there is a brief transition through a period classified as almost-stopped that serves as a buffer for the filtering. For each contiguous period classified as stopped, the final speed is set to zero. For each contiguous period classified as moving, a Savitzky-Golay filter is applied to the median speed to smooth out any remaining noise. In this fashion, the filter samples from the almost-stopped period but it does not include any points where the vehicle was actually stopped, which in turn would distort the filtered time series in the vicinity of the transition between moving and stopped. In the almost-stopped periods, to ensure a smooth transition, the speed is interpolated using a cubic spline between the moving speed on one end and the zero stopped speed on the other end of the almost-stopped period (the cubic spline fitting may disrupt the dynamics of how exactly a vehicle comes to a stop, but those dynamics are already beyond the resolution of the original data). Finally, a Gaussian kernel smoothing is applied to each time series speed to reduce any residual ripples. The resulting smoothed speeds from Figure 4*b* are shown in Figure 4*e*.

With the smoothed speeds, acceleration is calculated using a simpler approach. First, for each trajectory the time series acceleration is calculated using a single time step via Equation 2, where n is set to 1 and $v_{revised}(i)$ is the final speed

Table 1. Origin Destination Patterns for the Vehicle Trajectories in Figures 5 and 6

		Destination						
		Lane 1	Lane 2	Lane 3	Cook Rd.	N. Bdwy.	Webber	Sum
Origin	Lane 1	61	14	1	1	0	0	77
	Lane 2	9	26	9	5	1	0	50
	Morse Rd.	14	18	0	1	0	0	33
	Cook Rd.	12	11	4	n/a	0	0	27
	N. Bdwy.	0	6	11	n/a	n/a	0	17
	Webber	0	0	5	n/a	n/a	n/a	5
	Sum	96	75	30	7	1	0	209

Note: Rd. = Road; N. = North; Bdwy. = Broadway.

for frame i. Then the acceleration time series is smoothed according to the median speed classification for the given period using the same filtering strategy: applying a Savitzky-Golay filter during each contiguous moving period, and a cubic spline interpolation across each contiguous almost-stopped period. Figure 4*f* shows the resulting time series acceleration for the speed curves in Figure 4*e*. Finally, the smoothed speeds are integrated to find the cleaned trajectory for each vehicle. The resulting cleaned trajectories for the ongoing example are shown in Figure 4*d*.

$$\hat{a}_1(i) = \frac{v_{revised}(i+1) - v_{revised}(i-1)}{2 \cdot T} \qquad (2)$$

Obviously, this processing disrupts internal consistency between the three time series of position, speed, and acceleration. Given the unavoidable measurement errors from the relatively low resolution plots in Treiterer et al. (*3*), it is impossible to achieve internal consistency without inducing errors on at least two of the three time series. We chose to minimize the errors individually on position, speed, and acceleration; but this approach comes at the expense of no longer maintaining internal consistency. Fortunately, if internal consistency is critical for a given application, the reader can use the speed from our re-extracted data and then recalculate position and acceleration by summing and differencing the speed, respectively.

Analysis

Figures 5 and 6 show the resulting cleaned trajectories extracted from Figure 1. The schematic to the left of the plot shows the location of overpasses, underpasses, and ramps based on Treiterer et al. (*3*). Lane change maneuvers are indicated with triangles. Ramp locations were refined based on the locations where trajectories started or ended in the outside lane.

The most prominent feature in lane 1 (Figure 5) is the stop wave. Throughout the time period, the vehicles in lane 1 appear to slow down as they approach approximately 6,200 Ft, suggesting that free-flowing vehicles are reaching the tail of a downstream queue, and the location of the tail does not change much over the period of observation. Past the stop wave the lane 1 vehicles gain speed, but they never achieve the higher speeds (steeper slopes) seen upstream of 6,200 Ft. Several smaller disturbances are evident between 6,200 Ft and 12,000 Ft. The stop wave originates downstream of the tracked vehicles and when it is first seen the disturbance starts out as simply a nondescript slow wave. After about 10 s the disturbance grows and traffic comes to a complete stop as the wave passes. Later, the stop wave returns to a slow wave just after 130 s. There are many vehicles seen entering the lane between 8,000 and 11,000 Ft; if this rate is typical of conditions right before the data collection then the unseen entering vehicles are quite possibly the source of the stop wave.

No stop waves are seen in lane 2 (Figure 6), but there are many small disturbances between 7,000 Ft and 10,000 Ft. Speeds generally drop downstream of 7,000 Ft in the lane 2 data (lower slopes) and recover past the on-ramp at 9,000 Ft where 27 vehicles entered (for reference, Table 1 shows that only 50 vehicles originated in lane 2, but another 33 vehicles entered from the on-ramp at 3,500 Ft). Although there are many gaps upstream of 7,000 Ft, the speeds seen in lane 2 in this region appear to be generally slower than those in lane 1 for the same location.

Figure 7*a* shows the median speed across all of the trajectories in the given lane as a function of time. The median speeds fluctuate simply due to where the platoon is at the given instant. Figure 7*b* repeats the comparison by lane as a function of location. Consistent with the observations of Figures 5 and 6, the tail of a queue appears to be around 6,000 Ft in lanes 1–2, and then the median speed slowly increases as vehicles travel further downstream, but the median speed remains below that seen upstream of 6,000 Ft until past the start of lane 3. Upstream of 6,000 Ft, lane 1 is about 10 mph faster than lane 2. The median acceleration (not shown) stays close to zero throughout, as should be expected, and thus, does not indicate any problems with the extracted data.

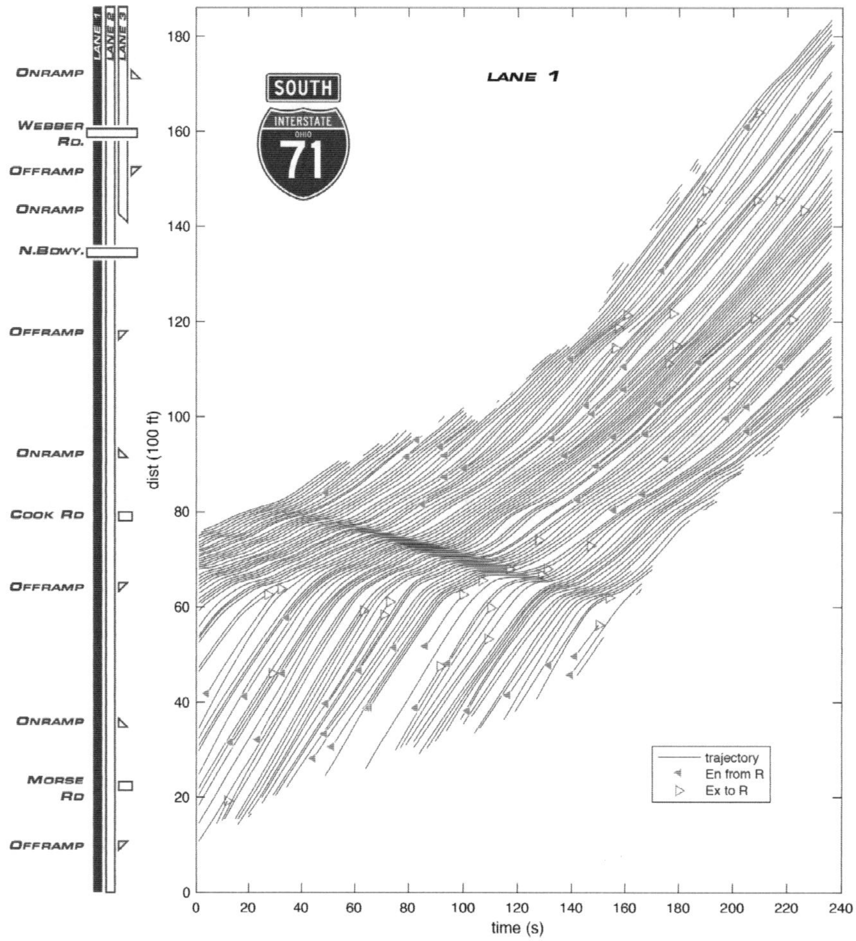

Figure 5. The cleaned extracted trajectories from Figure 1, *a* and *c*, in lane 1 on I-71 southbound, with a schematic to the left of the plot showing the location of cross streets and several ramps. The triangles in the plot indicate the direction of lane change maneuvers, with solid triangles for entrances and open triangles for exits.

Discussion

Old versus New

One might intuitively argue that there should be little value in 50-year-old trajectory data, but intuition is a fickle thing. Treiterer's data are like an arctic ice core; this work yields valuable measurements from a past time. The fact is that accurate, empirical microscopic data remain in scarce supply, so even old data have value to help understand how people drive, and the emergent traffic dynamics. This scarcity is more acute given the extreme deficiencies recently demonstrated in the NGSIM data (*14*). Although it is certainly plausible that people now drive with different response times or preferred spacing (a point we will return to in the next paragraph), the underlying equations explaining how we drive should not have changed, for example, as implicitly assumed in the continued use of hydrodynamic and car following models that predate Treiterer's data collection, including Lighthill and Whitham, Richards, Chandler et al., and Gazis et al. (*18–21*).

Briefly exploring the relevance of such old data, the light cloud of points in Figure 7*c* shows the individual speed spacing for all points in lane 1 of Treiterer's data, whereas the darker curve shows the median spacing across the points as a function of speed. Comparing these results to several contemporary empirical microscopic data sets, this process is repeated in Figure 7*d* for the raw NGSIM data from I-80, lane 3, 4:00–4:15 p.m period. Coifman and Li (*14*) manually re-extracted the vehicle trajectories for a portion of the same data set. Figure 7*e* repeats the analysis for the newly extracted NGSIM data, once more in lane 3 of I-80 during the same period, but over a shorter span of the roadway. Next, Figure 7*f* repeats the analysis for the instrumented probe vehicle data from Coifman et al. (*15*). Interestingly, these probe-based data include the portion of I-71 used by Treiterer. Figure 7*g* compares the median curves from the four data

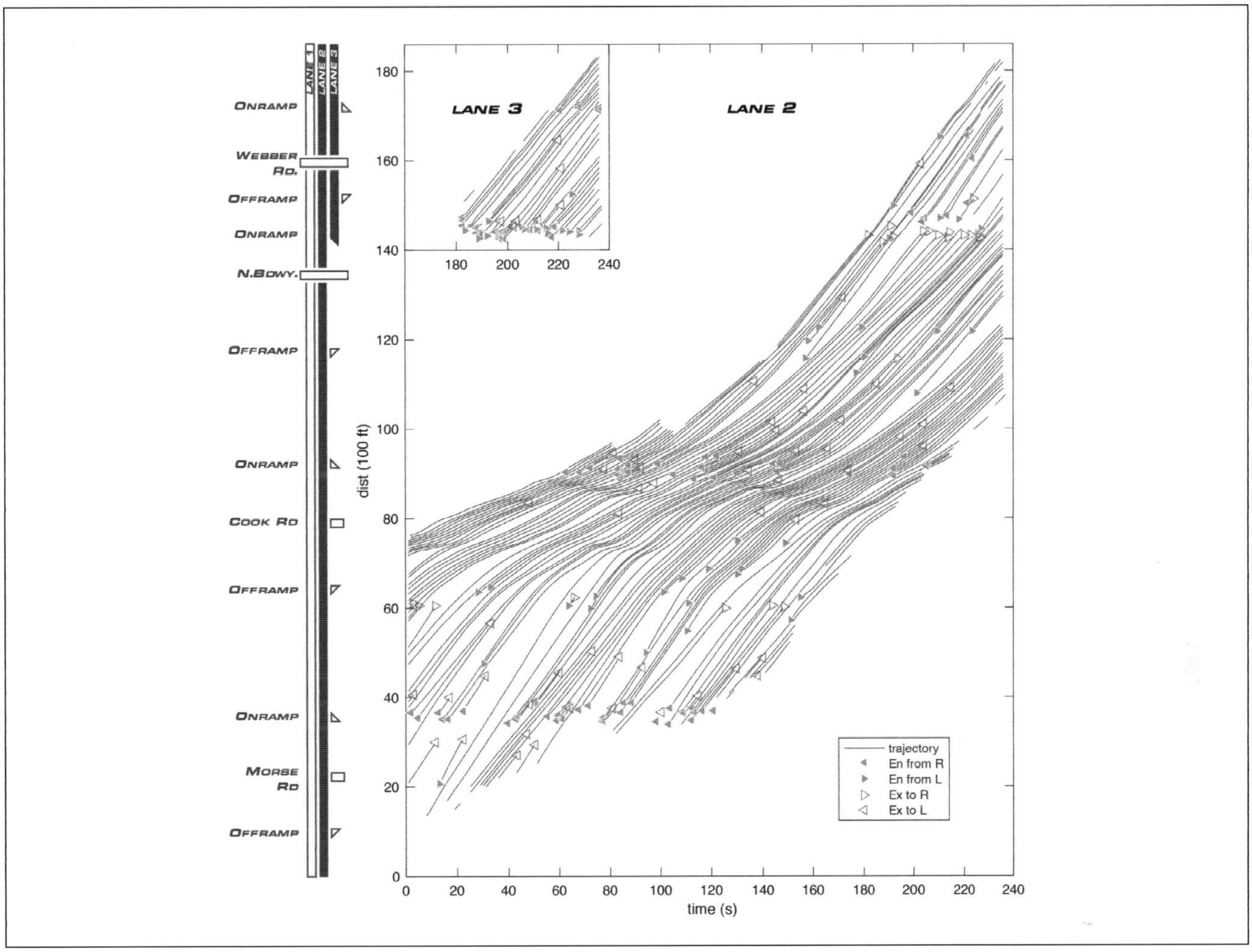

Figure 6. The cleaned extracted trajectories from Figure 1, *b* and *d*, in lane 2, and inset Figure 1e in lane 3 on I-71 southbound, with a schematic to the left of the plot showing the location of cross streets and several ramps. The triangles in the plot indicate the direction of lane change maneuvers, with solid triangles for entrances and open triangles for exits.

sets. The trends are consistent across the four curves. The raw NGSIM and probe vehicle data consistently exhibit spacings about 8 Ft larger than the other two data sets, whereas the spacing from the cleaned NGSIM data is very similar to Treiterer's data. A comparison is shown in Figure 7*h*, only this time it is against the single vehicle passage loop detector data for (effective length) 18–22 Ft vehicles from Coifman (*22*) denoted "Loop 1997", and Coifman (*23*) denoted "Loop 2000". Once more Treiterer's data are similar to contemporary empirical data sets. In spite of all of the advances in vehicular technology, the general car following behavior does not seem to have changed much over 50 years.

The Hysteresis Goes Round and Round

The darker trajectories in Figure 8*a* capture "platoon 123", which was used to demonstrate a hysteresis in Treiterer and Meyers (*1*), as recreated in Figure 8*b*. This processing verifies that Treiterer measured density and space-mean speed at every instant for the platoon and then calculated flow via the fundamental relationship. This particular hysteresis has generated considerable attention over the years, in part for the simple fact that the drop and recovery do not follow the same path, but equally quizzical is that the progression in the flow-density plane starts out with a clockwise loop and finishes with a counterclockwise loop. The lack of source data has led many to speculate about the mechanisms at work. The clockwise loop should be expected as drivers will decelerate when their spacing is lower than preferred, pushing traffic states above the equilibrium flow-density curve as speeds drop, and they will accelerate when the spacing is larger than preferred, pushing traffic states below the equilibrium flow-density curve as speeds rise. The existence of such clockwise loops in the flow-density plane was anticipated by Newell and earlier works (*24*). However, the counterclockwise loop is unexpected. As noted by Daganzo, this counterclockwise loop has been "a

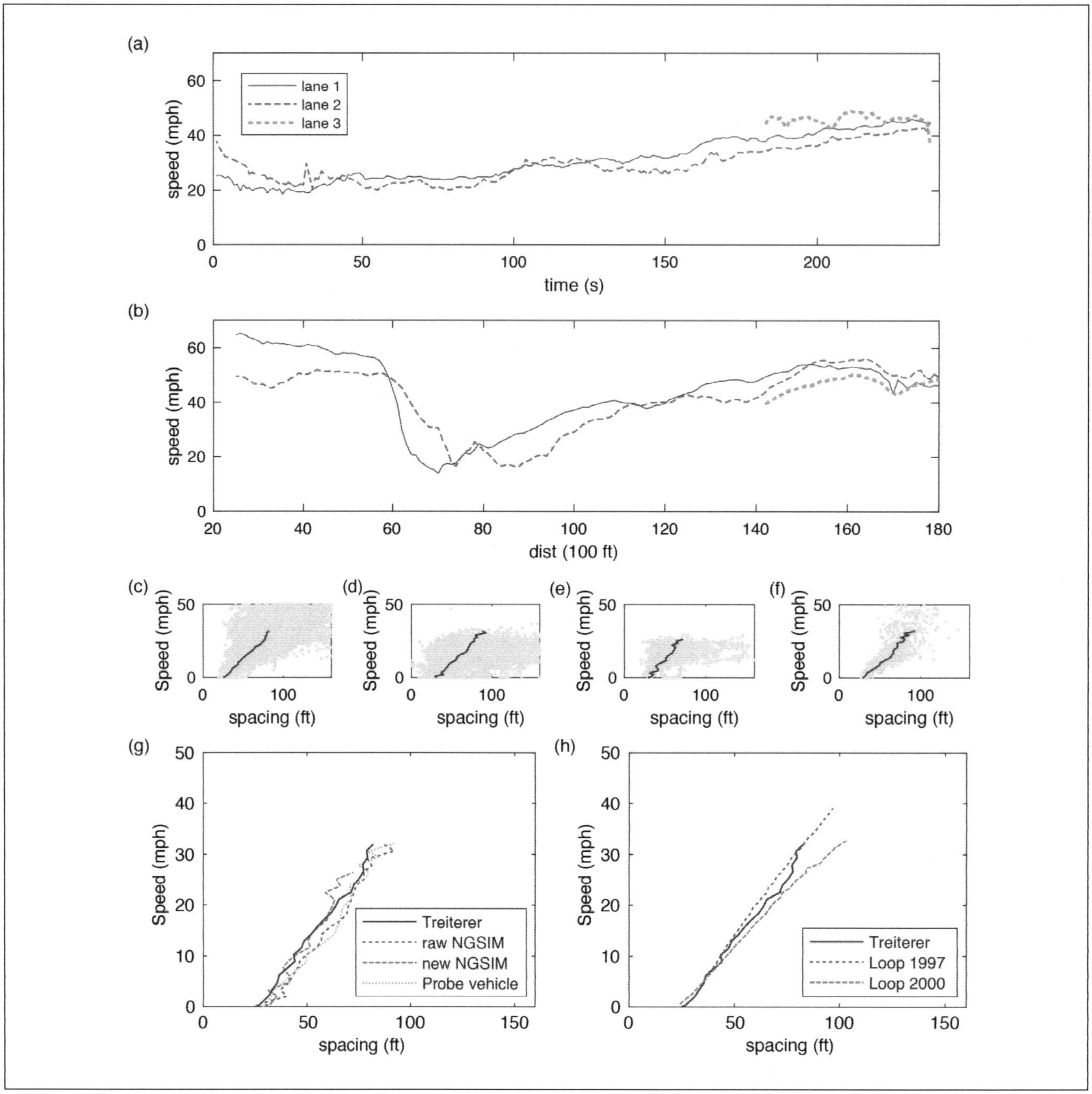

Figure 7. Median speed across all trajectories in each lane as a function of (*a*) time and (*b*) distance; (*c*) the light cloud of points shows the individual speed spacing and the darker curve the median speed spacing across all of the points for lane 1 of Treiterer's data; (*d*) lane 3 of I-80 NGSIM; (*e*) lane 3 of re-extracted I-80 NGSIM; (*f*) instrumented probe vehicle data; (*g*) comparison of the median curves from (c)–(e); (*h*) median speed spacing from Treiterer's data against two loop detector data sets.

longstanding source of speculation in the transportation literature, often used to justify questionable models" (*25*).

Using the resurrected data, it is finally possible to investigate the underlying dynamics. First, we add "x" markers every 10 s to Figure 8*b* to convey the progression of time. The lower loop took just over 30 s, whereas the upper loop took over 2 min. Figure 8*c* shows the time series speeds for all of the vehicles in the platoon (thin, solid curves) and the space-mean speed of the entire platoon (bold dashed curve). It should be clear that before $t = 67$ s, the vehicles were joining a queue. The time series traffic states corresponding to this period are marked with "o" in Figure 8*d* and, in this

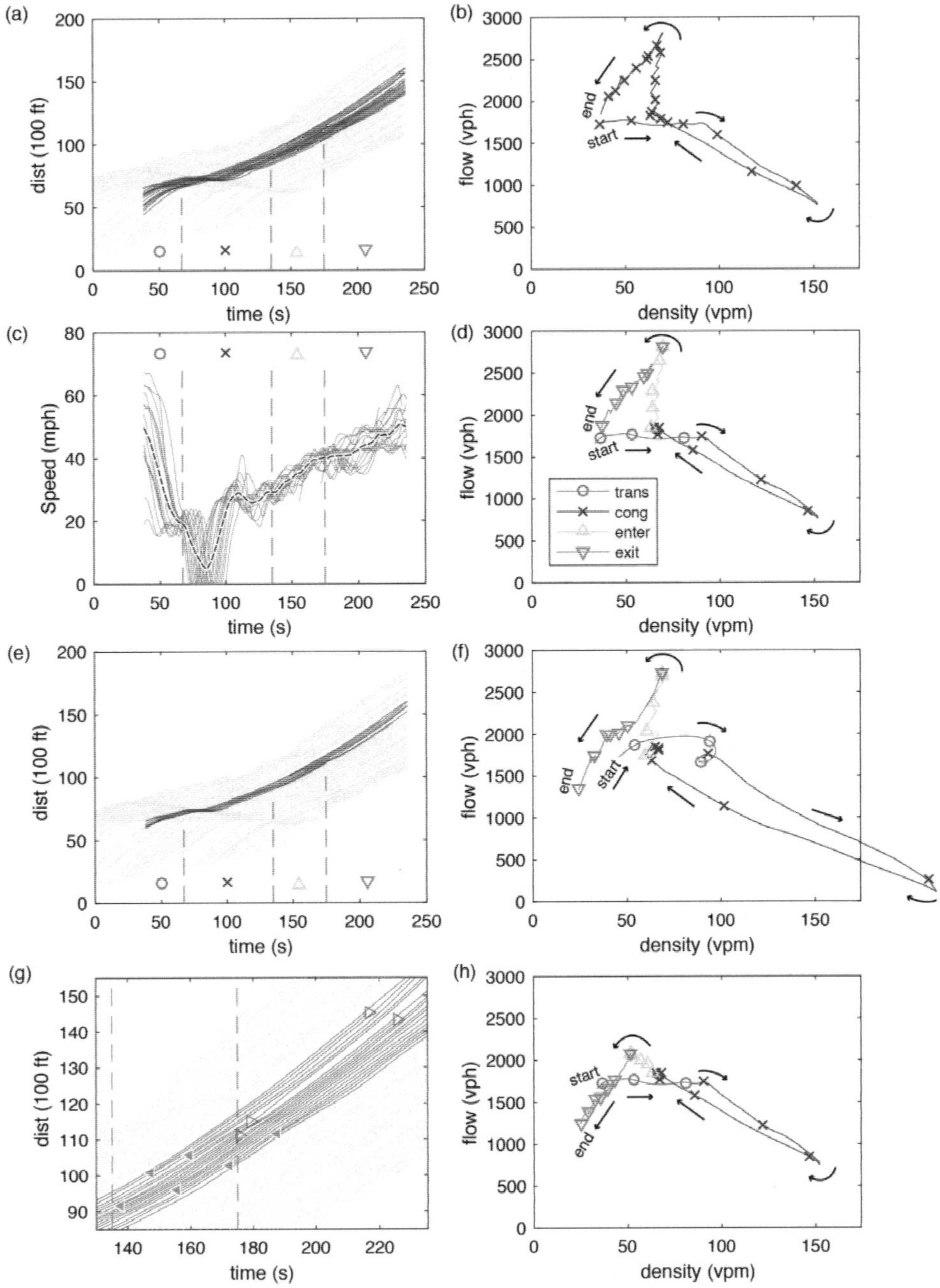

Figure 8. (*a*) "platoon 123" from the resurrected data of the 21 vehicles used by Treiterer and Meyers to demonstrate the hysteresis; (*b*) the hysteresis reproduced from (a); (*c*) the corresponding speeds for the individual vehicles (solid curves); and space-mean speed (bold dashed curve); (*d*) repeating (b) only now splitting into the four time periods as shown in (c); (e) repeating (a) but only for the first 10 vehicles; (*f*) repeating (b) but only for the first 10 vehicles; (*g*) detail of (a) showing the entering and exiting vehicles; (*h*) repeating (b) only now excluding all six of the entering vehicles.

context, it clearly shows a transition from the uncongested regime to the congested regime. Then, between $t = 67$ s and $t = 135$ s, we see speeds drop and then recover along the expected clockwise path, as indicated with "x" in the same subplot (again, the markers are plotted every 10 s). The fact that this period is not stationary because of the range of speeds evident in Figure 8*c* is not in itself problematic. Assuming a triangular fundamental curve in the flow-density plane, the average traffic state across many different stationary traffic states within the congested regime should still fall on the equilibrium flow-density curve (see, e.g., Wang and Coifman [*26*]), and in fact any sample of traffic containing more than one vehicle is likely to be non-stationary in the strictest definition of stationarity (see, e.g., Coifman [*27*]). Treiterer and Myers note that six vehicles joined the platoon after the stop wave, and inspection of Figure 8*a* shows that all but one of these entrances (denoted as dark points with white rings) occurred between t = 135 s and t = 175 s (*1*). During this period, density remains roughly constant whereas flow increases by about 1,000 vehicles per hour (vph), as shown with "Δ" in Figure 8*d*. The progression around the counterclockwise loop is much slower. Then Treiterer and Myers go on to note that five vehicles left the platoon (denoted as white points with dark rings in Figure 8*a*) (*1*). All of the exiting maneuvers occurred after t = 175 s, and the progression of traffic states for this period are denoted with "∇" in Figure 8*d*. Figure 8*c*, shows the speeds in this period remain below those seen before the vehicles joining the queue, so in spite of the positive sloped transition, the traffic state might not be purely free flowing; rather, both density and flow drop in response to the newly formed holes left by the exiting vehicles. The boundaries between the four time periods are superimposed on the four plots in the left column of Figure 8, and except for part g, the symbol denoting the respective time periods are also shown.

The preceding analysis does not support the assertion by Treiterer and Myers that the counterclockwise loop represents the traffic recovering to the state before the disturbance. The specific shape of both loops is sensitive to the region of averaging in the time–space plane, for example, Figure 8, *e* and *f*, repeat the analysis using just the first half of the platoon. Now the number of vehicles is small enough that almost all of the vehicles are stopped for a moment and space-mean speed drops below 1 mph. The clockwise loop expands in length and width. The counterclockwise loop is much smaller simply because of when and how the entering vehicles affect the traffic state.

Daganzo offers one of the few assessments that correctly surmised that the counterclockwise loop arose from lane change maneuvers; however, there is little evidence to support his detailed speculation that, "the aggressive followers in the platoon sensed that the (timid) leader was about to move out of the way and they followed more tightly so as to prevent drivers in the neighboring lane from 'cutting in' the queue" (*25*). Given that Treiterer and Myers showed the trajectories from only one lane (with the other lane only presented in obscure reports, e.g., Treiterer et al. [*3*]) and did so in print at a scale of about 1:20,000, considerable speculation was necessary to interpret their findings (*1*). Only now, with the resurrected data, can we finally look closer. Figure 8*g* shows a detail of the platoon during the period when the entrances occurred. The entering drivers chose large pre-existing gaps that had persisted for over half a minute and there was no apparent evidence of aggressive would-be-followers trying to prevent these vehicles from entering ahead. As evidenced by a cluster of "x" at about 66 vehicles per mile (vpm) in Figure 8*d*, the traffic state was in fact stable for more than 30 s before the first entrance. Although some of the entering drivers in Figure 8*g* show evidence of relaxation after taking a short spacing (*26*, *28*), the chosen gaps were large enough that followers behind the entrances generally did not exhibit much (if any) relaxation in response to their new leaders. Daganzo surmised that after the entrances, the "pattern changed sharply and suddenly. The speed continued to increase but flows decreased and the platoon expanded. It is hard to believe that the sharp reversal could have been caused by anything other than the clearance of the lane by the slow platoon leader" (*25*). Figure 8*g* shows that the platoon leader never changed and the slow speeds resulted from conditions further downstream of the leader. Although the entering vehicles exhibited some relaxation, as noted above, the followers did not appear to relax, and the platoon did not expand much. Instead, the flow and density simply dropped because of the vehicles exiting the lane. If one repeats this exercise, except now excluding all of the vehicles that entered the platoon via lane change maneuvers (Figure 8*h*), the resulting progression of the traffic state looks very different. Between t = 135 s and t = 175 s, the state simply continues to increase with the same slope seen in the earlier recovery, reaffirming the absence of aggressive drivers trying to impede the entering vehicles. After t = 175 s, flow and density simply drop in response to the exiting vehicles.

Conclusions

Treiterer and Meyers have demonstrated a remarkable longevity, due in part to the microscopic vehicle trajectory data used in the study (*1*). In spite of the Herculean collection effort, the data set was never released and has since been lost. Fortuitously, the plots survive and the present work resurrects the vehicle trajectory data from the time–space diagrams used in Treiterer and Myers and the adjacent lanes (*1*).

Although the traffic used in Treiterer and Myers was filmed over 50 years ago, the vehicle trajectories remain of great value (*1*). To date, only a few studies have rivaled the scale of the empirical vehicle trajectory data used in Treiterer and Myers and the distance spanned by Treiterer and Myers is 10 times greater than any of the NGSIM data sets. Although roadway design standards and vehicle performance have improved over the intervening years, the basic physics of driving remains, and Figure 7, *g* and *h*, show that the speed-spacing relationships are similar to five contemporary empirical data sets. Given the relative scarcity of empirical vehicle trajectory data for advancing traffic flow theory, simulation, and control and poor quality of the NGSIM data (*14*), one of our goals is to share this data set with the research community (on publication, the resurrected trajectory data from this paper will be available at Coifman [*4*]).

As noted by Daganzo (*25*), the hysteresis presented in Treiterer and Myers (*1*) has been "a longstanding source of speculation in the transportation literature, often used to justify questionable models." With the resurrected trajectories, this work reexamined the vehicles underlying the hysteresis and it finally quells the speculation. Rather than arising from car following behavior, it turns out that the enigmatic progression arose from a combination of lane change maneuvers and unremarkable transitions into or out of the congested regime.

Acknowledgments

This material is partly based on work supported in part by the National Science Foundation under Grant No. 1537423. The authors would like to thank Zona Kahkonen Keppler, a specialist in Research Services at the Ohio Department of Transportation, for her help in tracking down original copies of Treiterer's reports, which were the source for the scans used in this research.

Author Contributions

The authors confirm contribution to the paper as follows: study conception and design: Coifman; data collection: Coifman and Li; analysis and interpretation of results: Coifman, Li, and Xiao; draft manuscript preparation: Coifman, Li, and Xiao. All authors reviewed the results and approved the final version of the manuscript.

References

1. Treiterer, J., and J. Myers. The Hysteresis Phenomenon in Traffic Flow. *Proc., Sixth International Symposium on Transportation and Traffic Theory* (D. J. Buckley, ed.), 1974, pp. 13–38.
2. *Transportation Research Circular E-C149: Treiterer's Legacy: Traffic Flow Measurements and Characteristics.* Transportation Research Board of the National Academies, Washington, D.C. 2011. http://onlinepubs.trb.org/Onlinepubs/circulars/ec149.pdf
3. Treiterer, J., and Research Staff. *Investigation of Traffic Dynamics by Arial Photogrammetry Techniques*. Interim Report 2, EEB 278-2. Experiment Station, Ohio State University, Columbus, 1969.
4. Coifman, B. Data Sets, 2018. http://www.ece.osu.edu/~coifman/documents/
5. Leutzbach, W., and P. Egert. Untersuchungen über die Auswirkung von teilweisen Sperrungen auf der Autobahn auf den Verkehrsablauf, mimeographed report 1961, Abstract in Highway Research Abstracts. Vol. 32, No. 9, 1962, pp. 18–19.
6. Newell, G. F. Instability in Dense Highway Traffic, A Review. *Proc., of the Second International Symposium on the Theory of Traffic Flow*. London, 1963, pp. 73–83.
7. Forbes, T., and M. Simpson. Driver-And-Vehicle Response in Freeway Deceleration Waves. *Transportation Science, Operations Research Society of America*, Vol. 2, No. 1, 1968, pp. 77–104.
8. Smith, S., and E. Mark. Creation of Data Sets to Study Microscopic Traffic Flow in Freeway Bottleneck Sections. *Transportation Research Record: Journal of the Transportation Research Board*, 1985. 1005: 121–128.
9. Kovvali, V., V. Alexiadis, and L. Zhang. Video-Based Vehicle Trajectory Data Collection. Presented at 86th Annual Meeting of the Transportation Research Board, Washington, D.C., 2007.
10. Thiemann, C., M. Treiber, and A. Kesting. Estimating acceleration and Lane-Changing Dynamics from Next Generation Simulation Trajectory Data. *Transportation Research Record: Journal of the Transportation Research Board*, 2008. 2088: 90–101.
11. Hamdar, S., and H. Mahmassani. Driver Car-Following Behavior: From Discrete Event Process to Continuous Set of Episodes. Presented at 87th Annual Meeting of the Transportation Research Board, Washington, D.C., 2008.
12. Duret, A., C. Buisson, and N. Chiabaut. Estimating Individual Speed-Spacing Relationship and Assessing Ability of Newell's Car-Following Model to Reproduce Trajectories. *Transportation Research Record: Journal of the Transportation Research Board*, 2008. 2088: 188–197.
13. Montanino, M., and V. Punzo. Trajectory Data Reconstruction and Simulation-based Validation Against Macroscopic Traffic Patterns. *Transportation Research Part B*, Vol. 80, 2015, pp. 82–106.
14. Coifman, B., and L. Li. A Critical Evaluation of the Next Generation Simulation (NGSIM) Vehicle Trajectory Dataset. *Transportation Research Part B*, Vol. 105, 2017, pp. 362–377. [see paper for dataset information]
15. Coifman, B., M. Wu, K. Redmill, and D. Thornton. Collecting Ambient Vehicle Trajectories from an Instrumented Probe Vehicle- High Quality Data for Microscopic Traffic Flow Studies. *Transportation Research Part C*, Vol. 72, 2016, pp. 254–271. [see paper for dataset information]
16. Goshtasby, A. Piecewise Linear Mapping Functions for Image Registration. *Pattern Recognition*, Vol. 19, 1986, pp. 459–466.
17. Goshtasby, A. Image Registration by Local Approximation Methods. *Image and Vision Computing*, Vol. 6, 1988, pp. 255–261.
18. Lighthill, M., and G. Whitham. On Kinematic Waves II. a Theory of Traffic Flow on Long Crowded Roads. *Proceedings of the Royal Society of London, Part A*, Vol. 229, No. 1178, 1955, pp. 317–345.
19. Richards, P. Shock Waves on the Highway. *Operations Research*, Vol. 4, No. 1, 1956, pp. 42–51.
20. Chandler, R. E., R. Herman, and E. W. Montroll. Traffic Dynamics: Studies in Car Following. *Operation Research*, Vol. 6, No. 2, 1958, pp. 165–184.
21. Gazis, D. C., R. Herman, and R. W. Rothery. Nonlinear Follow the Leader Models of Traffic Flow. *Operations Research*, Vol. 9, No. 4, 1961, pp. 545–567.
22. Coifman, B. Revisiting the Empirical Fundamental Relationship. *Transportation Research- Part B*, Vol. 68, 2014, pp. 173–184. [see paper for dataset information]
23. Coifman, B. Empirical Flow-Density and Speed-Spacing Relationships: Evidence of Vehicle Length Dependency. *Transportation Research- Part B*, Vol. 78, 2015, pp. 54–65. [see paper for dataset information]
24. Newell, G. F. Theories of Instability in Dense Highway Traffic. *Journal of Operation Research Society of Japan*, Vol. 5, No. 1, 1962, pp. 9–54.

25. Daganzo, C. F. A Behavioral Theory of Multi-lane Traffic Flow. Part I: Long Homogeneous Freeway Sections. *Transportation Research Part B*, Vol. 36, No. 2, 2002, pp. 131–158.
26. Wang, C., and B. Coifman. (2008). The Effect of Lane-Change Maneuvers on a Simplified Car-following Theory. *IEEE Transactions on Intelligent Transportation Systems*, Vol. 9, No. 3, 2008, pp. 523–535.
27. Coifman, B. Jam Occupancy and Other Lingering Problems with Empirical Fundamental Relationships. *Transportation Research Record: Journal of the Transportation Research Board*, 2014. 2422: 104–112.
28. Xuan, Y., and B. Coifman. Identifying Lane Change Maneuvers with Probe Vehicle Data and an Observed Asymmetry in Driver Accommodation. *Journal of Transportation Engineering, ASCE*, Vol. 138, No. 8, 2012, pp. 1051–1061.

The Standing Committee on Traffic Flow Theory and Characteristics (AFB45) peer-reviewed this paper (18-00020), which was selected to receive the 2018 Greenshields Prize awarded by the committee honoring the contributions of Bruce Greenshields to the field of traffic flow theory.

TRR
JOURNAL OF THE TRANSPORTATION RESEARCH BOARD

Article

Transportation Research Record
2018, Vol. 2672(20) 39–50

DOI: 10.1177/0361198118790346
journals.sagepub.com/home/trr
SAGE

Real-Time Estimation of Lane Change Risks Based on the Analysis of Individual Vehicle Interactions

Hyunjin Park[1], Cheol Oh[1], and Jaepil Moon[2]

Abstract

Lane-changing vehicles continuously interact with adjacent vehicles by adjusting longitudinal and lateral maneuvering. Safe lane changing can be regarded as an outcome of safe vehicle interactions during lane changes. An effective safety assessment of a vehicle changing lanes is essential to developing various countermeasures to prevent crashes. This study developed a methodology for estimating the risks of changing lanes in real time using vehicle trajectory data. Both a real-time risk exposure level (RREL) and a real-time risk severity level (RRSL) were devised to quantify risk levels associated with lane changing based on an analysis of vehicle interactions. Then, fault tree analysis was adopted to integrate the two indicators, RREL and RRSL. This study demonstrated the feasibility of the proposed methodology by conducting VISSIM simulation experiments. The outcome of this study can be used to not only evaluate the effectiveness of traffic management strategies in terms of lane change safety but also support the development of various advanced driving assistance systems.

According to the U.S. NHTSA report (*1*), approximately 450,000 accidents involving lane changes occur every year. Vehicles changing lanes continuously interact with adjacent vehicles by adjusting their speed and maneuvering laterally. Therefore, understanding the characteristics of vehicle interactions during lane changes is essential to developing various countermeasures for crash prevention. Although a variety of studies dealing with lane change events have been conducted in the field of transportation, less effort has been made to identify real-time crash risks based on individual vehicle interactions due to the limited availability of high resolution vehicle maneuvering data. However, the recent advancement of vehicle sensor technologies for automated driving allows us to obtain detailed vehicle maneuvering data for surrounding vehicles in addition to the subject vehicle. Therefore, a valuable research opportunity to analyze individual vehicle interactions encourages traffic researchers and engineers to develop novel methods to estimate crash risks. This has also motivated our study to develop a methodology for estimating risks while performing lane changes in real time.

This study developed a methodology for estimating the risk of changing lanes in real time using vehicle trajectory data. A novel feature of the proposed methodology was to incorporate crash occurrence potential and crash severity. Both a real-time risk exposure level (RREL) and a real-time risk severity level (RRSL) were devised to quantify lane-changing risk levels based on an analysis of vehicle interactions. Then, a fault tree analysis (FTA) was performed to integrate the two indicators RREL and RRSL. The RREL was developed to reflect a situation where the likelihood of a crash occurring increases when a subject vehicle is exposed to a dangerous situation. The RRSL was developed to reflect those situations where a higher collision speed leads to an increase in the severity of a crash. This study also demonstrated the feasibility of the proposed methodology by conducting VISSIM simulation experiments. Vehicle trajectory data obtained from a simulated freeway work zone in VISSIM were used to estimate the risks posed by lane-changing vehicles in real time.

Literature Review

The literature review was focused on safety issues regarding lane change events because the main purpose of this

[1]Department of Transportation and Logistics Engineering, Hanyang University Erica Campus, Ansan, Republic of Korea
[2]Korea Institute of Construction Technology, SOC Research Institute, Goyang, Republic of Korea

Corresponding Author:
Address correspondence to Cheol Oh: cheolo@hanyang.ac.kr

study was to develop a new method to estimate lane change risks. Existing studies can be categorized into two groups. The first group includes modeling efforts attempting to relate traffic flow parameters obtained from traffic surveillance systems, such as inductive loop detectors, with crashes. Pande and Abdel-Aty (*2*) and Xu et al. (*3*) identified contributing factors affecting lane change crashes using driver characteristics, road geometric conditions, and traffic conditions. Lee et al. (*4*) investigated the impact of traffic flow parameters on drivers' lane change behavior in terms of traffic safety based on the analysis of binary logistic regression. It was identified that flow-related variables were related with the sideswipe crash. Ling et al. (*5*) used a multilevel Bayesian logistic regression to model safety at freeway weaving segments where lane changes occur frequently. Although these methods are useful, there exists a limitation that individual vehicle interactions are not considered in the safety evaluation because aggregated traffic parameters are used.

The other group of existing studies uses vehicle trajectory data. Because vehicle trajectory data for both the subject and surrounding vehicles are obtainable by vehicle sensor technologies, individual vehicle interactions can be analyzed while the lane change event occurs. Several studies have been conducted to identify the relationship between safety and lane-changing events. Bascunana (*6*) analyzed the characteristics of safe and hazardous lane changes using vehicle trajectory data. The minimum lateral spacing for safe lane changes was derived by Julia et al. (*7*). Additionally, a Bayesian network method was adopted by Schubert et al. (*8*) to evaluate the safety of lane changes using vehicle sensor data. Recently, a study was conducted to determine vehicle trajectories for safe lane changes to support driver decision making (*2*). It should also be noted that surrogate safety measures (SSM) play an important role in fully exploiting the benefits of vehicle trajectory data for safety evaluations (*9*). In addition to time-to-collision (TTC), i.e., the time remaining to avoid an accident from the time the driver takes action to the point where the accident occurs (*10*), various SSMs including post encroachment time, maximum deceleration, deceleration rate to avoid crash, and stopping distance index (SDI) have been widely used to evaluate safety indirectly (*11–15*). These SSMs, which were originally developed for the purpose of identifying rear-end crash potential, are not appropriate for lane change risks. Meanwhile, the surrogate safety assessment model presented by the FHWA allows for analyzing both traffic conflict frequency and severity (*16*). However, this methodology only deals with the interaction of two vehicles and does not consider both the conflict frequency and severity at the same time. In addition, an important issue is whether SSMs can explain the occurrence of the crash. As a validation method, the extreme value theory, which is a method for modeling an event that occurs at a very low frequency, has recently been applied to identify the relationship between actual crashes and SSMs (*17–20*).

We have identified based on the literature review that the capability of analyzing vehicle interactions would provide a valuable research opportunity for the development of a novel method to quantify crash risks. In addition, we are not aware of any study to estimate lane change risks in real time by incorporating the amount of exposure time to potential crashes and the expected crash severity level, which motivates our study.

Methodology

Overall Framework

This study attempts to develop a method for the systematic and scientific estimation of the risk of changing lanes in real time, where the risk is an outcome of improper interactions. Consider that the trajectory of a vehicle is the change in vehicle position over time. The proposed methodology continuously evaluates, in real time, the risk of changing lanes using FTA. The overall procedure, consisting of four steps, for estimating lane change risk in real time is presented in Figure 1.

The first step is to collect individual vehicle trajectory data, including vehicle positions and speeds. A total of four adjacent vehicles affect the risk of an accident when a subject vehicle (k_S) changes lanes, as shown in Figure 2. The four adjacent vehicles include the front (k_F) and rear (k_R) vehicles in the starting lane and the leading (k_{Le}) and lagging (k_{La}) vehicles in the target lane.

The second step is to derive the RREL. The SDI is an index for determining the rear-end collision risk based on the stopping sight distance (SSD), which is, for example, derived based on k_S and k_F where the front spacing is $S_t(1)$ at the current time step t. Then, the time remaining until the SDI becomes zero (TSO), which is newly proposed in this study, is calculated. TSO is the remaining time until SDI becomes 0, where it is assumed that the traveling vehicle maintains a constant speed at the current time t. An SDI less than 0 indicates a hazardous situation in which k_S is not able to make the proper evasive maneuver to avoid a collision with k_F due to insufficient spacing between the two vehicles. In other words, the TSO is the time remaining until the vehicle enters a state where it is impossible to avoid a dangerous situation ahead. The calculated TSO was applied to the exponential decay function (EDF) to calculate the probability of the occurrence of an accident from 0 to 1; this value was defined as the real-time risk exposure in this study.

The third step is to determine the RRSL. To derive RRSL, time to collision with deceleration condition

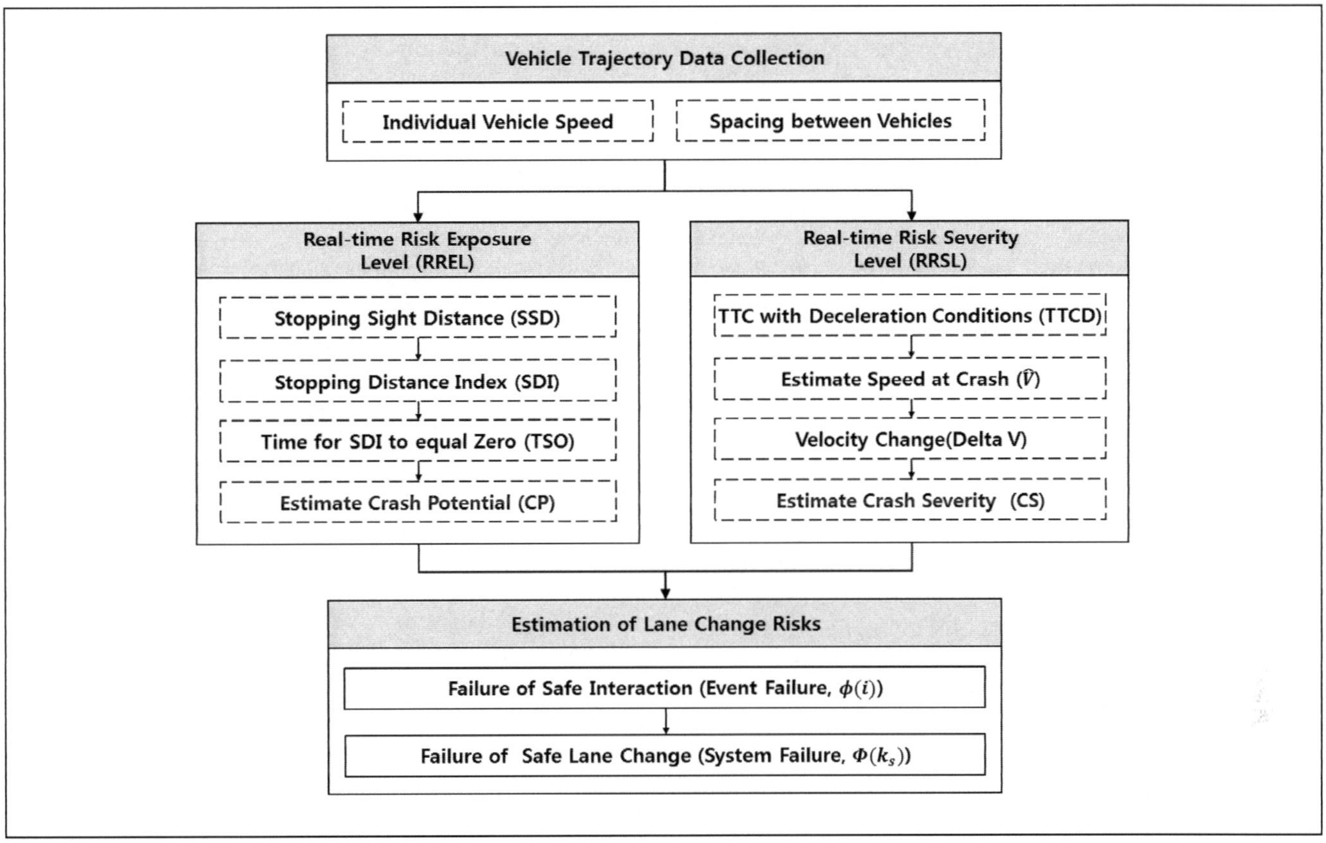

Figure 1. Overall evaluation framework.

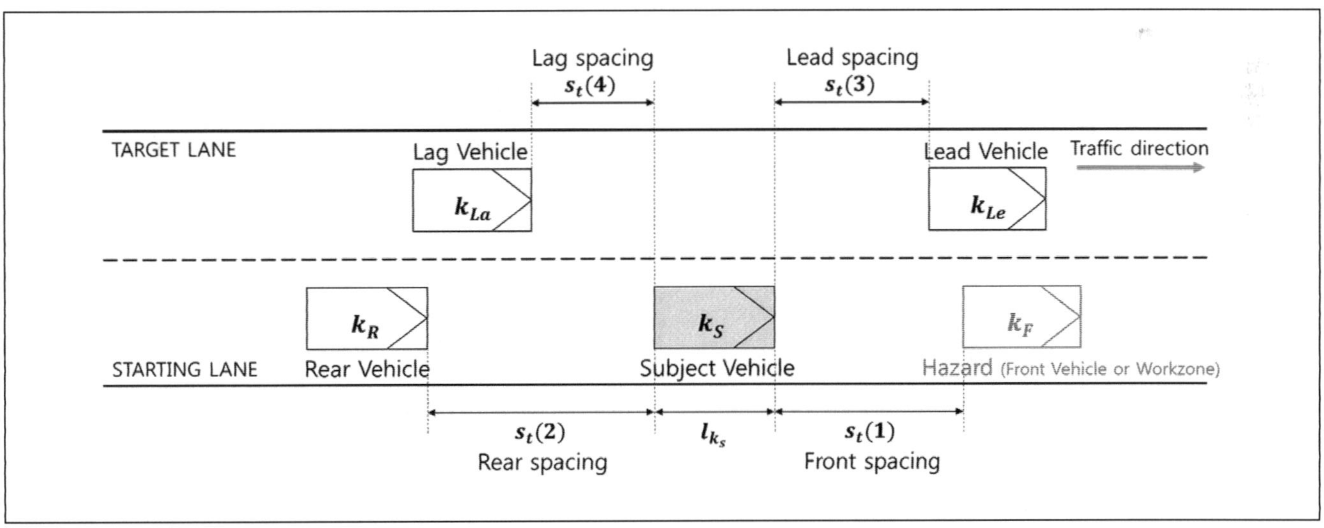

Figure 2. Definition of subject vehicle and surrounding vehicles.

(TTCD), which is a newly proposed element in this study, defined as the time remaining from when the two vehicles start braking at the current time t and the distance between the two vehicles becomes zero, is calculated. Then TTCD is further used for obtaining Delta V, which is the difference in speed at the moment ($\hat{V}$) when the subject vehicle collides with the interacting vehicle. Utilizing the actual observed Delta V and the theoretically observable maximum velocity difference (Delta V_{MAX}^{obs}), the RRSL is expressed as a probabilistic measure between 0 and 1.

The last step is to perform an FTA. Four interaction events with adjacent vehicles (k_F, k_R, k_{Le}, k_{La}) are evaluated in terms of lane change safety based on the RREL and RRSL. As a result, the failure of a safe vehicle interaction ($\varphi(i)$) for k_S and each adjacent vehicle $i(i = 1, 2, 3, 4)$ is quantified. Then, the integration of $\varphi(i)$ is conducted to estimate the risk level represented by a probabilistic measure of k_S failing to perform a safe lane change ($\Phi(k_s)$). More details of the derivation of the RREL and RRSL, along with how to apply FTA, are presented in the following sections.

Real-Time Risk Exposure Level (RREL)

The RREL was developed to reflect a situation where the likelihood of a crash occurring increases when a subject vehicle is exposed to a dangerous situation. SSD and SDI can be obtained using Equations 1 and 2. SSD is used to compute minimum stopping distance (*21*), and SDI is used for determining whether a given car-following event is safe or not (*15*). An SDI greater than 0 represents a situation in which k_S is able to stop safely when k_F makes a sudden stop. On the other hand, an SDI less than 0 indicates a hazardous situation in which k_S is not able to make the proper evasive maneuver to avoid a collision with k_F due to insufficient spacing between the two vehicles. Therefore, in this study, we define dangerous driving behavior to occur when the SDI is less than or equal to 0, and safe driving behavior to occur when it is greater than 0.

When the TSO is small, a collision cannot be avoided. As shown by Equation 3, the TSO can be calculated by dividing the SDI at the current time t by the difference in speeds between the traveling vehicles. It is assumed that the speeds of the traveling vehicles are constant. If the TSO is 0, it can be inferred that an accident has occurred:

$$SSD = \frac{V^2}{254 \times (f \pm g)} + t_r \times V \times 0.278 \tag{1}$$

where

SSD = stopping sight distance;

V = vehicle speed (km/h);

f = coefficient of friction, typically on poor, wet pavement;

g = grade, decimal; and

t_r = perception-reaction time, typically 2.5 s.

$$SDI_t(i) = S_t + SSD_t^{LV} - SSD_t^{FV} \tag{2}$$

where

i = interaction event with k_S and each adjacent vehicle (k_F, k_R, k_{Le}, k_{La}), for example, $i = 1$ represents "k_S and k_F";

$SDI_t(i)$ = stopping distance index for leading vehicle and following vehicle at time step t;

S_t = spacing between leading vehicle and following vehicle at time step t;

SSD_t^{LV} = leading vehicle's stopping sight distance;

SSD_t^{FV} = following vehicle's stopping sight distance; and

t = time step.

$$TSO_t(i) = \frac{SDI_t(i)}{V^{k_s}} \times 3.6 \tag{3}$$

where

$TSO_t(i)$ = remaining time until SDI becomes 0 between subject (k_S) and adjacent vehicles (k_F, k_R, k_{Le}, k_{La}) at time step t; and

V^{k_s} = speed of subject vehicle (k_S).

Finally, the calculated TSO was input to the EDF. A generalized EDF can be described by $y = a + b \times \exp(-z/c)$. To use this function to estimate the potential of a crash based on TTC, the parameters a and b should be set to $a = 0$ and $b = 1$ since the crash potential ranges from 0 to 1. The characteristics of the monitoring location will affect the determination of c. For example, it is desirable for a road segment with a higher frequency of crashes to have larger value for c. The form of the exponential decay model in this study can be described by Equation 4. If the TSO is less than 0, the SDI will be less than 0. In this case, P (Crash | TSO) = 1 because the event cannot be avoided when an event occurs. In this study, the crash potential (CP) derived from the above calculation is defined as the RREL (Equation 5):

$$CP_t(i) = \begin{cases} \text{if } TSO_t(i) \geq 0, & P(Crash_t(i)|TSO_t(i)) = \exp\left(-\frac{TSO_t(i)}{c}\right) \\ \text{if } TSO_t(i) < 0, & P(Crash_t(i)|TSO_t(i)) = 1 \end{cases} \tag{4}$$

where

$CP_t(i)$ = crash potential for subject vehicle (k_S) adjacent vehicles (k_F, k_R, k_{Le}, k_{La}) at time step t; and

c = shape parameter for EDF.

$$RREL(i) = CP_t(i) \tag{5}$$

Real-Time Risk Severity Level (RRSL)

The RRSL was developed to reflect a situation where a higher collision speed leads to an increase in the severity of a crash. The RRSL utilized Delta V, which represents the difference in speed between the preceding vehicle and the following vehicle (*16*). The equation for calculating Delta V is given by Equation 6. In this study, only SDI is considered to be less than 0. This is because Delta V can be estimated only when the SDI is less than 0, since there is no possibility of an accident if the SDI is greater than 0 even if a dangerous condition exists:

$$\text{Delta } V_t(i) = \hat{V}^{FV}_{t_{TTCD}} - \hat{V}^{LV}_{t_{TTCD}} \tag{6}$$

where

Delta $V_t(i)$ = the estimated speed difference at time step t_{TTCD} between following vehicle and leading vehicle at time step t;

$\hat{V}^{FV}_{t_{TTCD}}$ = the estimated speed at time t_{TTCD} of following vehicle; and

$\hat{V}^{LV}_{t_{TTCD}}$ = the estimated speed at time t_{TTCD} of leading vehicle.

If two vehicles start braking at time t, the estimate of their relative speed at the moment of the collision, time t_{TTCD}, is given by Equation 7. The collision speed can be estimated using the rate of deceleration for the remaining time until the collision occurs. The time TTCD is the time remaining from when the two vehicles start braking at the current time t and the distance between the two vehicles becomes zero, which is newly proposed in this study. When the distance between two vehicles is zero, it means that a collision occurred:

$$\begin{cases} \hat{V}^{FV}_{t_{TTCD}} = V^{FV}_t + a^{FV} \times \text{TTCD}(i) \\ \hat{V}^{LV}_{t_{TTCD}} = V^{LV}_t + a^{LV} \times \text{TTCD}(i) \end{cases} \tag{7}$$

where

V^{FV}_t = the speed of following vehicle at time step t;

V^{LV}_t = the speed of leading vehicle at time step t;

a^{FV} = the decelerations of following vehicle;

a^{LV} = the decelerations of leading vehicle; and

TTCD(i) = the time remaining until the two vehicles start braking at the current time t and the distance between the two vehicles becomes zero.

The time remaining until collision, TTCD, is summarized by Equation 8. The two vehicles will collide within the braking distance so the TTCD can be derived using an equation that calculates the vehicle's distance and each stopping distance:

$$\text{TTCD} = \frac{-(V^{FV}_t - V^{LV}_t) - \sqrt{(V^{FV}_t - V^{LV}_t)^2 - (a^{FV} - a^{LV}) \times 2\{S_t(i) + (V^{FV}_t - V^{LV}_t) \times tr\}}}{a^{FV} - a^{LV}} \tag{8}$$

When TTC is greater than t_r, the two vehicles will collide while stopping after t_r has elapsed. Therefore, TTCD can be calculated to estimate the speed at which the two vehicles collide. Meanwhile, if TTC is less than or equal to t_r, the two vehicles collide within the distance traversed during t_r because the collision occurs before t_r has elapsed. Therefore, the collision speed is assumed to be the speed at the present time. Equation 9 shows how to determine Delta V:

$$\text{Delta } V(i) = \begin{cases} \text{if } \text{TTC}_t(i) > t_r, & \hat{V}^{FV}_{t_{TTCD}} - \hat{V}^{LV}_{t_{TTCD}} \\ \text{if } \text{TTC}_t(i) \le t_r, & V^{FV}_t - V^{LV}_t \end{cases} \tag{9}$$

where

$\text{TTC}_t(i)$ = time to collision for subject vehicle (k_S) adjacent vehicles (k_F, k_R, k_{Le}, k_{La}) at time step t.

Using the estimated Delta V and Delta V^{obs}_{MAX} ratios in the above procedure, the real-time incident severity is expressed as a probabilistic measure between 0 and 1 provided by Equation 10. This is defined as RREL in this study. Delta V^{obs}_{MAX} is set to180 km/h by assuming that the spacing between two interacting vehicles is 0 m and the speed of the following vehicle is 180 km/h. A threshold value, 180 km/h, is used to detect an outlier during the data processing for Korean freeways (*22*):

$$\text{RRSL}(i) = \frac{\text{Delta } V(i)}{\text{Delta } V^{obs}_{MAX}} \tag{10}$$

Lane Change Risks Index

In this study, the safety evaluation of lane changes was carried out using FTA to determine the RREL and RRSL. FTA is a method for conducting risk analysis that is widely used in the domain of reliability engineering. Reliability evaluations can be performed to identify the possibility of system failure in a quantitative manner. In the transportation field, several studies have used FTA to understand the contributing factors affecting the occurrence of a crash. In the transportation field, research was carried out applying FTA to grasp the causes and problems resulting in traffic accidents. Kronprasert and Thipnee (*23*) constructed a fault tree using various causes of traffic accidents. They proposed a monitoring system and operation management strategy for preventing traffic accidents and improving road safety. However, they were limited to using FTA to identify the causes of traffic accidents and suggesting solutions. Meanwhile, Joo and Oh (*24*) proposed an integrated evaluation index for evaluating bicycling environments via FTA using instrumented probe bicycle data. More detailed information on FTA can be found in Gardoni (*25*).

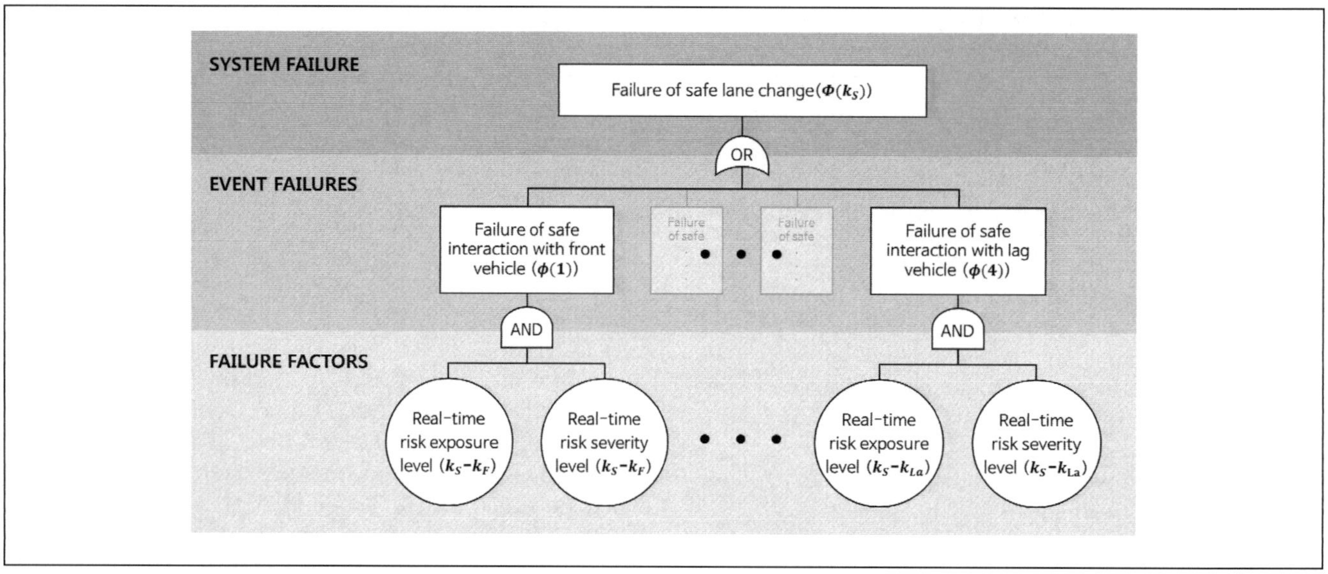

Figure 3. Fault tree structure of failure to perform safe lane change.

We set two failure factors. The first failure factor, RREL, uses the SDI between the subject vehicle and the surrounding interacting vehicles and applies the residual time to the EDF until collision with a value between 0 and 1. When the subject vehicle is likely to have an accident while the lane change is being made, RREL = 1 is defined as the real-time risk exposure factor. The second failure factor, RRSL, was expressed as the ratio of Delta V to Delta V_{MAX}^{obs} using the TTC and TC of the subject vehicle and the surrounding interacting vehicles. The larger the speed difference, the more dangerous the situation. In the case where RRSL = 1, it is defined as the real-time risk severity factor.

This study considers a lane change as a process to be analyzed in terms of crash risks. Event failure is defined as the failure of safe vehicle interaction between a subject vehicle and surrounding vehicles, which is denoted by $\varphi(i)$. The failure of safe vehicle interaction between adjacent vehicles (k_F, k_R, k_{Le}, and k_{La}) is denoted by $\varphi(1)$, $\varphi(2)$, $\varphi(3)$, and $\varphi(4)$, respectively. As shown in Equation 11, the fault of event $\varphi(i)$ is calculated based on a combination of failure factors, including the RREL and RRSL.

System failure is defined as the failure of a subject vehicle to perform a safe lane change and is denoted by $\Phi(k_s)$. The probability that a subject vehicle fails to conduct a safe lane change, which is referred to as the lane change risk index (LCRI) $\Phi(k_s)$, can be calculated via the integration of the failure of safe vehicle interaction, as shown by Equation 12. A smaller value of the probability means that a subject vehicle is changing lanes more safely. Figure 3 illustrates the FTA diagram designed for this study to estimate the probability of failure when performing a lane change.

$$\varphi(i) = \mathrm{RREL}(i) \times \mathrm{RRSL}(i) \tag{11}$$

where φ = event failure (failure of safe vehicle interaction).

$$\Phi(k_s) = 1 - \prod_{i=1}^{4}[1 - \varphi(i)] \tag{12}$$

Where

Φ = system failure (probability of failure to perform safe lane change: LCRI);
φ = event failure; and
k_s = subject vehicle.

The objective of this study is to evaluate whether a lane change event is safe or not, rather than simply estimating the probability of a lane change crash. When evaluating the safety of lane change events, we can think of them from two different perspectives. The first is a situation where the severity of a crash is high when it occurs although the possibility of a crash occurring is low. The other is a situation where the possibility of a crash occurring is high although its severity is low. A reasonable question is then "Which situation is more dangerous?" The answer to this question is not simple because we need to take both situations into consideration when evaluating safety. Additionally, unlike the existing SSMs, the LCRI proposed in this study not only considers the longitudinal safety of the subject vehicle but also takes the forward vehicle and the rear vehicle motion in the driving lane into account, as well as the front vehicle and the rear vehicle as the lane is changed. To meet this requirement, we adopted FTA as a method to integrate both RREL and RRSL.

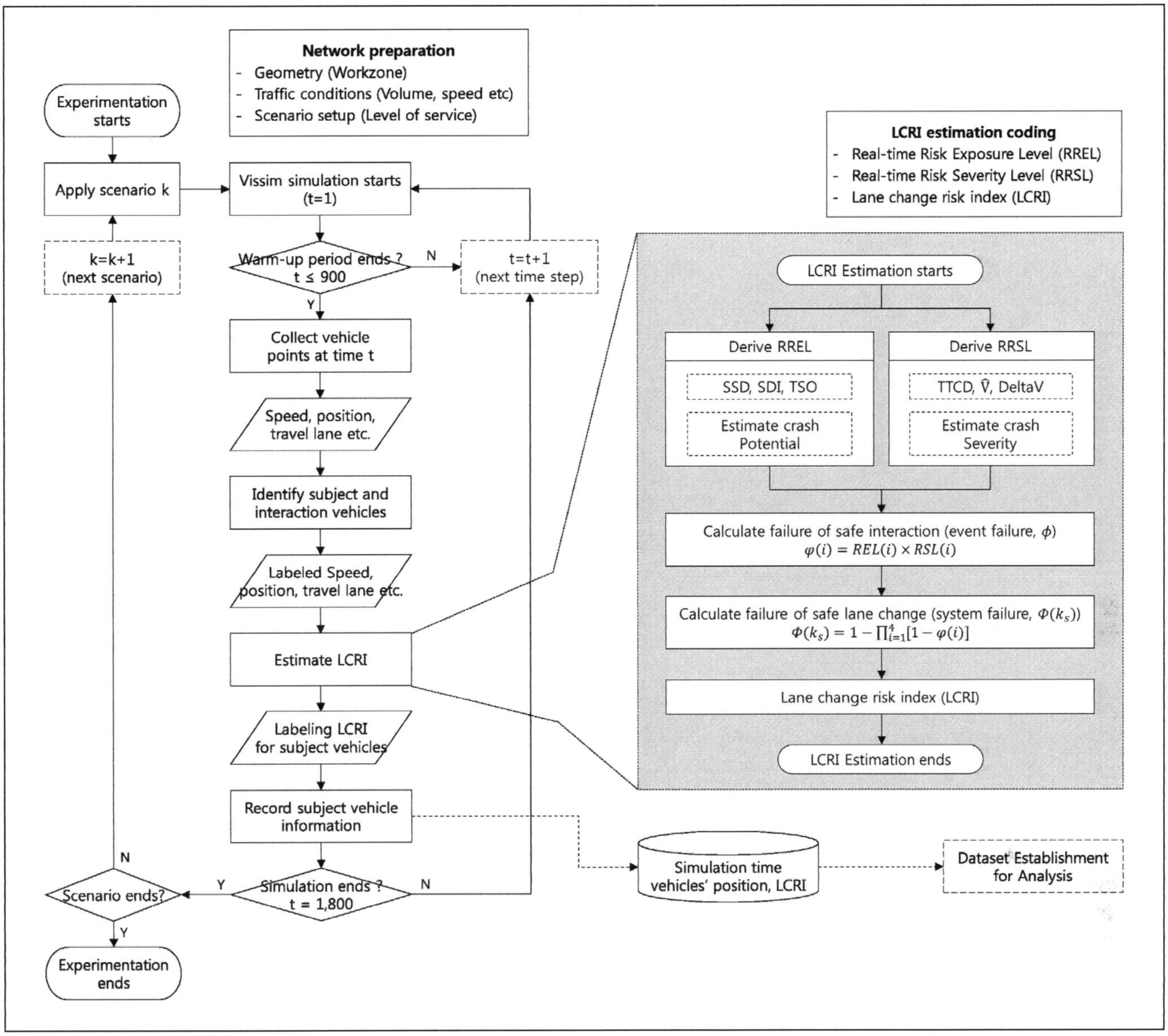

Figure 4. Implementation of proposed methodology with VISSM.

Results of Simulation Analysis

We evaluated the methodology developed in this study using the traffic flow simulation program VISSIM. To do this, the one-way second-lane section near the Gonjiam IC on the Jungbu Expressway was simulated. The speed limit is 110 km/h, and the capacity is approximately 2100 pcphpl (passenger cars per hour per lane). We analyzed five scenarios of highway level of service (LOS) A ~ E by applying the methodology presented in this research. LOS F scenario was not analyzed because dangerous situations representing high crash severity were not readily observable due to low vehicle speeds under congested traffic conditions. The analysis time for a simulation was set to 15 min and the warm-up period was set to 15 min. To increase the reliability of the simulation results, simulations of five scenarios were repeated 10 times, and the average LCRI was computed by scenarios. To determine which LOS scenario is more dangerous in the work zone section, we defined the representative value as the average LCRI calculated by repeating various simulation environments.

Figure 4 shows the overall simulation process for deriving the LCRI. The simulation network environment was implemented, and the LCRI estimation process was coded using Com-Interface. After the warm-up period, we selected the scenarios to analyze and collected all the vehicle speed and location data at every time-step. After

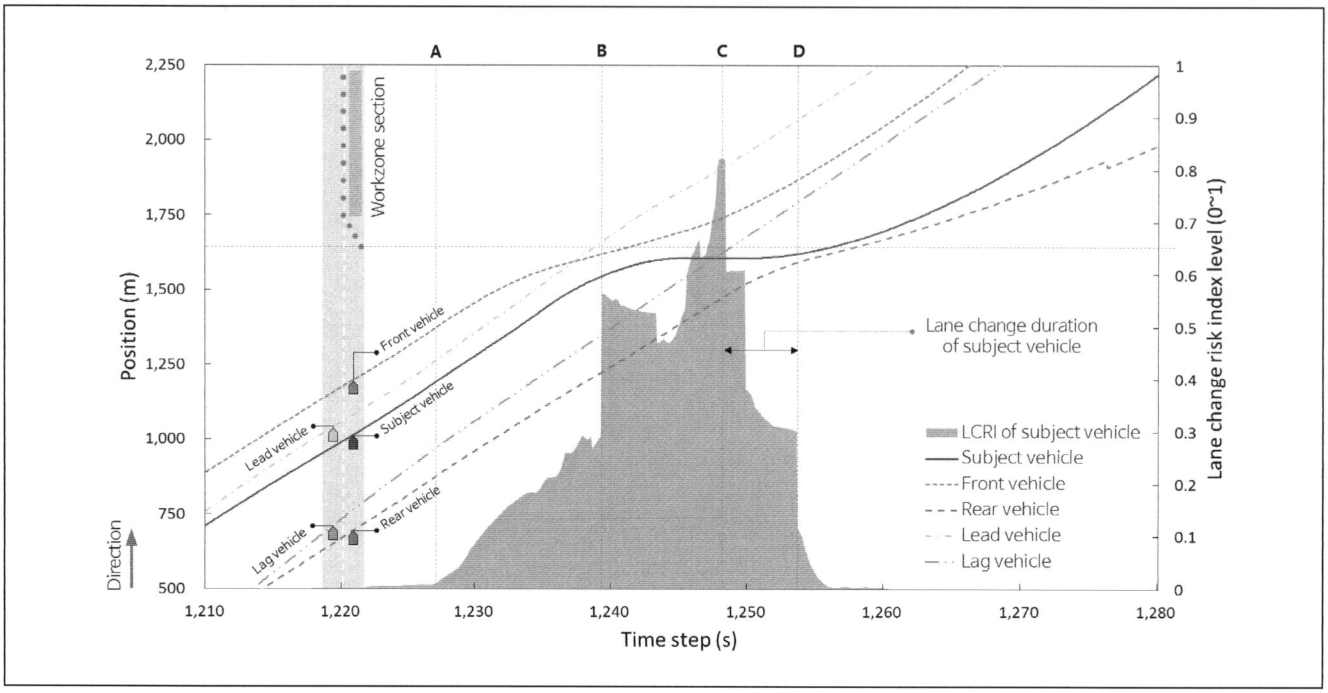

Figure 5. Example of LCRI profile during lane changing.

defining the subject vehicle and the interacting vehicles, the RRSL and RREL were calculated using the speed and position data collected based on the interrelationship of the defined vehicles. Then, the $\varphi(i)$ of the subject vehicle and the interacting vehicles were derived, and the resulting LCRI ($\Phi(k_s)$) was estimated. When a series of processes was completed, a new scenario was set up, and the simulation analysis was repeated.

Figure 5 shows an illustrative example of LCRI profiles while a subject vehicle changes lane. The simulated work zone starts at 1640 m. The subject vehicle indicated by the solid line travels on the right lane and attempts to change its path to the left lane due to the work zone. From the position A, the front vehicle on the starting lane changed the lane while decelerating due to the leading vehicle on the target lane, which led to the increase in LCRI. Once the front vehicle completed the lane change, the subject vehicle was required to change lane to avoid lane blocking by the work zone. This led to a sharp increase in LCRI of the subject vehicle from the position B. The subject vehicle was not able to change lane between B and C because the lag vehicle on the target lane came quickly. The highest LCRI, which resulted from interactions with the lane block, the lag vehicle, and the rear vehicle, was observed near the position C. LCRI showed a decreasing pattern from the position C, and then gradually disappeared after completing the lane change.

Figure 6 shows the average LCRIs by different LOS. The highest LCRI was observed at LOS B, while the lowest LCRI was observed at LOS E. This can be interpreted that traffic conditions corresponding to LOS B represent relatively high driving speeds and the existence of traffic volume leading to frequent conflicts. The vehicles driving on the right lane are more frequently exposed to the dangerous situation because they are required to conduct compulsory lane changes due to the lane-block by the work zone.

Figure 7 shows the spatio–temporal patterns of LCRIs, where the x-axis represents time and the y-axis represents position. When LCRI becomes large, the LCRI is indicated by the red. For example, in the left lane of LOS A, an LCRI of approximately 0.5 was generated at 1150 s and a location of 1400 to 1600 m, where it was found that approximately 50 s was maintained.

The LCRI was 0.8 at about 1150 s and 1400 $\sim$ 1600 m on the left lane under LOS A conditions; it was maintained for about 50 s. In the case of the LOS A, a higher level of LCRI occurred in the right lane than in the left lane. This was due to the influence of vehicles that made lane changes from the right lane, which was blocked. In the case of LOS B, an LCRI of 0.7 or more was observed in the left lane in the upstream part of the work zone section. LCRI then disappeared as vehicles approach the work zone section and disappeared by the approach of the work zone section. However, in the right lane, LCRI,

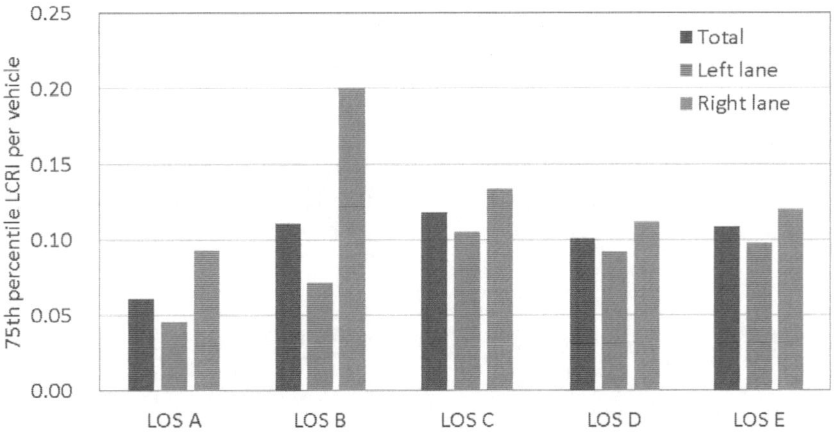

Figure 6. 75th percentile LCRI per vehicle.

which was higher than 0.7, occurred at 900 and 1500 m, and was maintained for 250 s.

In the left lane, the vehicles continued to pass the work zone, but in the right lane, the higher LCRI was generated by vehicles that failed to change lanes. In LOS B, the LCRI began approximately 1200 m upstream of the work zone section. Then, the congestion generated after approximately 1100 s in LOS B transitioned from the left and right lanes of the work zone section to the upstream section, and the LCRI also transferred to the upstream section. After 1600 s of LOS B, the LCRI reoccurred as the congestion resolved from the beginning of the work zone section. Below LOS C, a similar pattern of LCRI was observed due to persistent congestion. A high level LCRI rarely occurred, and the LCRI was generated and extinguished 200 m ahead of the work zone. In particular, unlike LOS A and LOS B, when LCRI occurs under LOS C, the hazardous situation represented by the estimated LCRI which occurred at the beginning of the work zone was moved to the upstream section of the work zone. This is because the degree of freedom of vehicle maneuvering is severely restricted.

Discussion of Results and Conclusion

The methodology proposed by this study evaluates real-time lane change risk. Unlike existing SSMs that are used for longitudinal safety such as the rear-end collisions risk, the proposed LCRI is able to evaluate lateral safety such as potential collisions during a lane change event based on the interaction with nearby vehicles. An FTA method was adopted to incorporate the crash occurrence potential and crash severity. An unsafe lane change state is defined as a system failure by the FTA, and the driving safety of the subject vehicle is quantified. The outcome of this study can be effectively used for evaluating the effectiveness of traffic safety for various traffic operations and management strategies. In addition, it can support the safe lane-changing capabilities of advanced safety vehicles through real-time monitoring, and in particular, it is expected to contribute to the development of Advanced Driver Assistance System (ADAS) related technologies. Finally, from the real-time monitoring perspective, it will contribute to establishing a traffic management strategy to provide active and effective warning information.

Further research should be conducted to fully implement this methodology in practice. First, the RREL and RRSL proposed in this study assume that the characteristics of the driver and the performance of the driving vehicle are the same. However, in an actual traffic situation, it is necessary to overcome these limitations because the driver's cognitive response and the performance of the driving vehicle vary. Second, LCRI was estimated by simulation, and the simulated unsafe lane changes were analyzed. The results of evaluating different scenarios varied depending on the risk threshold. Therefore, a validated process for setting a threshold is required. For example, based on c determining the form of the EDF, the trigger point for providing warning information depends on the criteria for determining the riskiness of the situation. Therefore, it is necessary to determine the appropriate level of c, and realistic verification results based on field data are needed. In the long term, we need to compare and analyze the results of the indicators presented in this study with the extreme value theory analysis of traffic accident data. Third, it is worthwhile to incorporate the collision angle into the proposed methodology because the collision angle would give a significant impact on the crash severity. Fourth, in order for the methodology presented in this study to be applied in

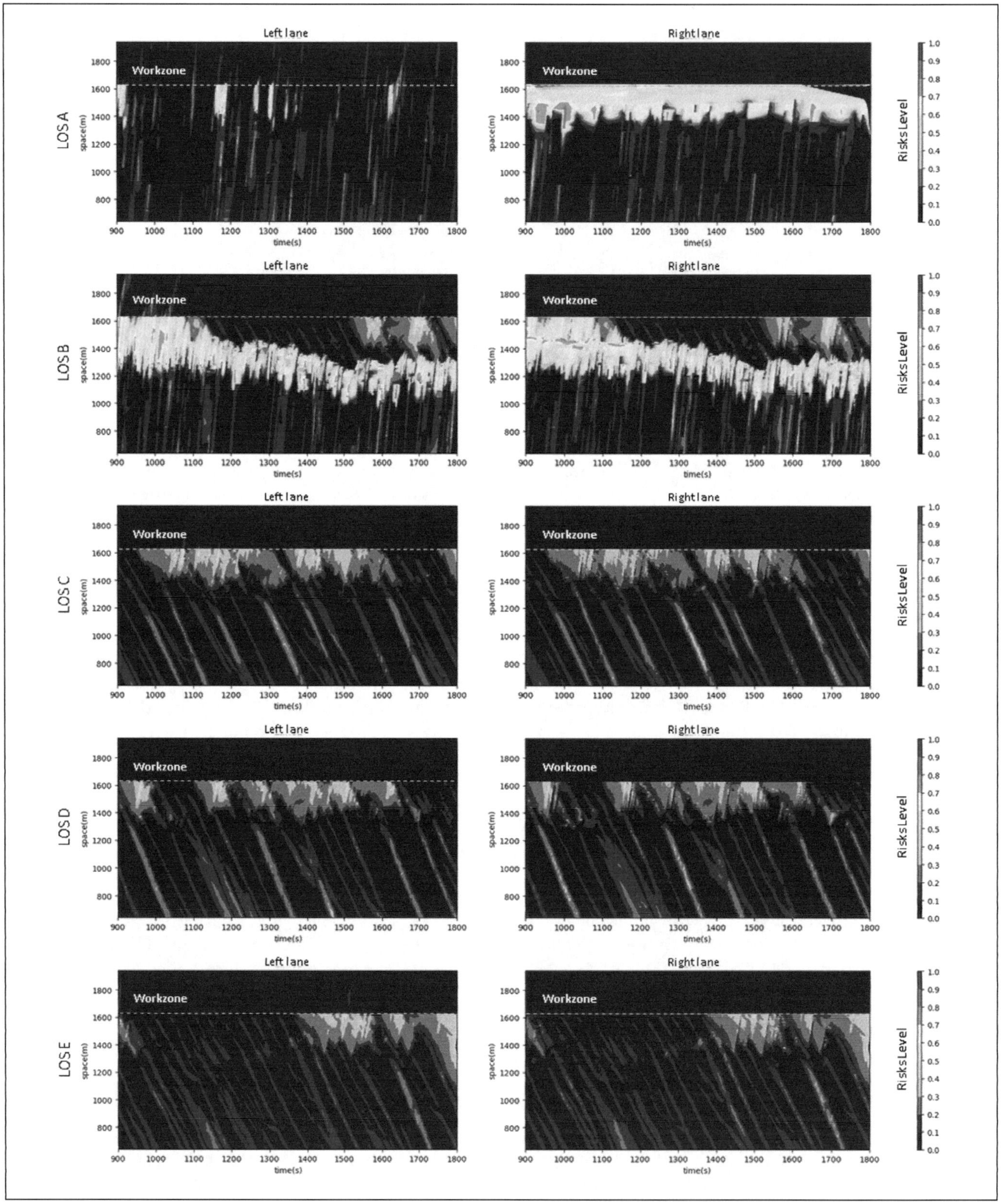

Figure 7. LCRI contour map.

the field, the performance of the sensor and the technical issues of traffic data processing techniques should be studied. Despite the development of traffic data collection technology, it is not enough to process the trajectory data of both the subject and interacting vehicles. However, the advancement of vehicle sensors to monitor vehicle maneuvering and driver status for automated driving would provide an opportunity to overcome current limitations associated with data collection. Finally, it is necessary to develop the warning information strategy for the effective operation of the traffic management strategy proposed in this study.

Acknowledgments

This research was supported by a grant from Development of Accident Risk Mitigating Technology for Vertical Structure Collision on Road and Road Workers (Grant Number 18TLRP-C096228-04) funded by the Ministry of Land, Infrastructure and Transport, Republic of Korea.

Author Contributions

The authors confirm contribution to the paper as follows: study conception and design: HP, CO, and JM; data collection: HP; analysis and interpretation of results: HP and CO; draft manuscript preparation: HP and CO. All authors reviewed the results and approved the final version of the manuscript.

References

1. National Highway Traffic Safety Administration. 2015 Motor Vehicle Crashes: Overview. *Traffic Safety Facts Research Note*, Vol. 2016, 2016, pp. 1–9.
2. Pande, A., and M. Abdel-Aty. Assessment of Freeway Traffic Parameters Leading to Lane-Change Related Collisions. *Accident Analysis and Prevention*, Vol. 38, No. 5, 2006, pp. 936–994.
3. Xu, G., L. Liu, Y. Ou, and Z. Song. Dynamic Modeling of Driver Control Strategy of Lane-Change Behavior and Trajectory Planning for Collision Prediction. *IEEE Transactions on Intelligent Transportation Systems*, Vol. 13, No. 3, 2012, pp. 1138–1155.
4. Lee, C., P. Y. Park, and M. Abdel-Aty. Lane-by-Lane Analysis of Crash Occurrence Based on Driver's Lane-Changing and Car-Following Behavior. *Journal of Transportation Safety and Security*, Vol. 3, No. 2, 2011, pp. 108–122.
5. Wang, L., M. Abdel-Aty, Q. Shi, and J. Park. Real-Time Crash Prediction for Expressway Weaving Segments. *Transportation Research Part C: Emerging Technologies*, Vol. 61, 2015, pp. 1–10.
6. Bascunana, J. L. *Analysis of Lane Change Crash Avoidance*. Proceedings of the Future Transportation Technology Conference and Exposition, Costa Mesa, CA. Society of Automotive Engineers, SAE Technical Paper Vol. 951895, 1995, pp. 34–41.
7. Jula, H., E. B. Kosmatopoulos, and P. A. Ioannou. Collision Avoidance Analysis for Lane Changing and Merging. *IEEE Transactions on Vehicular Technology*, Vol. 49, No. 6, 2000, pp. 2295–2308.
8. Schubert, R., K. Schulze, and G. Wanielik. Situation Assessment for Automatic Lane-Change Maneuvers. *IEEE Transactions on Intelligent Transportation Systems*, Vol. 11, No. 3, 2010, pp. 607–616.
9. Hydén, C. The Development of a Method for Traffic Safety Evaluation: The Swedish Traffic Conflicts Technique, No. 70. Lund Institute of Technology, Lund, Sweden, 1987.
10. Hayward, J. C. *Near Misses as a Measure of Safety at Urban Intersections*. Pennsylvania State University, Department of Civil Engineering, Old Main, State College, PA, 1971.
11. Cooper, D. F., and N. Ferguson. Traffic Studies at T-Junctions. 2. A Conflict Simulation Record. *Traffic Engineering and Control*, Vol. 17, No. 7, 1976, pp. 306–309.
12. Allen, B. L., B. T. Shin, and P. J. Cooper. Analysis of Traffic Conflicts and Collisions. *Transportation Research Record: Journal of the Transportation Research Board*, 1978. 677: 67–74.
13. Hupfer, C. Deceleration to Safety Time (DST)—A Useful Figure to Evaluate Traffic Safety? *10th ICTCT Conference Proceedings of Seminar 3*. Department of Traffic Planning and Engineering, Lund University, Lund, Sweden.
14. Gettman, D., and L. Head. Surrogate Safety Measures from Traffic Simulation Models. *Transportation Research Record: Journal of the Transportation Research Board*, 2003. 1840: 104–115.
15. Oh, C., S. Park, and S. G. Ritchie. A Method for Identifying Rear-End Collision Risks using Inductive Loop Detectors. *Accident Analysis and Prevention*, Vol. 38, No. 2, 2006, pp. 295–301.
16. Gettman, D., L. Pu, T. Sayed, and S. G. Shelby. *Surrogate Safety Assessment Model and Validation*. Publication FHWA-HRT-08-051. FHWA, U.S. Department of Transportation, Washington, D.C., 2008.
17. Farah, H., and C. L. Azevedo. Safety Analysis of Passing Maneuvers using Extreme Value Theory. *International Association of Traffic and Safety Sciences Research*, Vol. 41, No. 1, 2017, pp. 12–21.
18. Azevedo, C. L., I. C. Cardoso, and M. E. Ben-Akiva. Probabilistic Safety Analysis using Traffic Microscopic Simulation. Presented at 94th Annual Meeting of the Transportation Research Board, Washington, D.C., 2015.
19. Zheng, L., K. Ismail, and X. Meng. Freeway Safety Estimation using Extreme Value Theory Approaches: A Comparative Study. *Accident Analysis and Prevention*, Vol. 62, 2014, pp. 32–41.
20. Songchitruksa, P., and A. P. Tarko. The Extreme Value Theory Approach to Safety Estimation. *Accident Analysis and Prevention*, Vol. 38, No. 4, 2006, pp. 811–822.
21. American Association of State Highway and Transportation Officials. *A Policy on Geometric Design of Highways and Streets*, 6th ed. AASHTO, Washington, D.C., 2011.
22. Kim, K., H. Lee, S. Kim, and S. Namkoong. Introduction to an Expressway Traffic Data Monitoring System using Traffic Quality Information and Visualization. *Transportation Technology and Policy*, Vol. 10, No. 3, 2013, pp. 69–76.

23. Kronprasert, N., and N. Thipnee. Use of Evidence Theory in Fault Tree Analysis for Road Safety Inspection. *Proc., International Conference on Belief Functions*, Springer, Cham, Switzerland, 2016, pp. 84–93.
24. Joo, S., and C. Oh. A Novel Method to Monitor Bicycling Environments. *Transportation Research Part A: Policy and Practice*, Vol. 54, 2013, pp. 1–13.
25. Gardoni, P. *Risk and Reliability Analysis: Theory and Applications*. Springer, International Publishing, New York, NY, 2017.

The Standing Committee on Traffic Flow Theory and Characteristics (AHB45) peer-reviewed this paper (18-02577).

Article

TRR
JOURNAL OF THE TRANSPORTATION RESEARCH BOARD

Transportation Research Record
2018, Vol. 2672(20) 51–62

DOI: 10.1177/0361198118799166
journals.sagepub.com/home/trr
SAGE

Application of a Discontinuous Form of Macroscopic Gazis–Herman–Rothery Model to Steady-State Freeway Traffic Stream Observations

Ishtiak Ahmed[1], Billy M. Williams[2], and M. Shoaib Samandar[1]

Abstract

In this study, a two-regime, steady-state, traffic stream model is developed by applying the macroscopic Gazis–Herman–Rothery model to fixed sensor data on freeways. The uncongested and congested regimes are modeled discontinuously with an overlap range defined in terms of density. The overlap is important as various phenomena related to the change in traffic state can be modeled by introducing this overlap. Two empirical tools for removing non-stationary, mixed-state, and erroneous observations are applied at different stages of the model development process. Three constraints justified by the *Highway Capacity Manual* (HCM) were applied to fit the model so that the fitted parameters have reasonable and physically interpretable values. The proposed model is applied to one year of data (2013) obtained from fixed sensors located at five freeway basic segments near Raleigh, North Carolina. The resulting fundamental diagrams show that the fitted models reasonably represent the steady-state observations. Two forms of the freeway flow model described in the HCM were applied to the same observations to provide a continuous model comparison. Two statistical performance measures, mean squared error of flow rate and Bayesian Information Criterion, verify that the proposed model is preferable to the HCM models both in terms of fit alone and when considering the tradeoff between fit and model complexity. It is expected that the proposed discontinuous steady-state model will be useful to researchers and practitioners to study various site-specific freeway traffic stream characteristics.

Models representing the so-called fundamental diagram of traffic flow are widely used by researchers and practitioners for applications such as operational analysis, capacity estimation, and bottleneck identification. Within the field of traffic flow theory, several macroscopic steady-state models have been developed (*1–3*). The mathematical bridge between macroscopic steady-state models and the microscopic car-following models was first established for the widely known Gazis–Herman–Rothery (GHR) car-following model (*4*, *5*). The full version of the GHR car-following model yields a general feasible regime steady-state model for specific ranges of two key model parameters. Various versions of the GHR model have been developed and proposed based on investigation and analysis of empirical traffic stream data (*6–8*). In addition, the continuity of the fundamental diagram between uncongested and congested flow has been under debate. Although the literature includes studies that acknowledge a ubiquitously observed discontinuity between uncongested and congested flow (*6*, *9–11*), others have argued against discontinuity on empirical and theoretical grounds as mentioned in the literature review that follows. Nonetheless, there are important phenomena related to the change in traffic state, such as the widely accepted queue-discharge capacity drop, that can be modeled by introducing discontinuity. The *Highway Capacity Manual* (HCM) (*12*) does not directly speak to traffic state discontinuity in its basic discussion of freeway segment analysis methodology, but does acknowledge and allow for a capacity drop at queue discharge in its freeway facilities methodology.

One of the practical challenges that lies behind empirically fitting a steady-state macroscopic model is identifying non-stationary and mixed-state observations in addition to identifying outliers (*13*). If not removed

[1]Institute for Transportation Research and Education (ITRE), North Carolina State University, Raleigh, NC
[2]North Carolina State University, Raleigh, NC

Corresponding Author:
Address correspondence to Ishtiak Ahmed: iahmed2@ncsu.edu

before or during model fitting, such observations will introduce bias in the model parameters and resulting estimates. Although many outlier-detection algorithms are available in the literature (*13*, *14*), the selection of algorithm should be specific to the modeling objective and data characteristics. In this study, a two-regime, discontinuous, steady-state macroscopic model is developed for basic freeway segments (*12*) with explicit identification and removal of non-stationary and outlying observations.

A review of past studies on the relevant subject is presented in the next section. Following that, the methodology for fitting the two regimes independently is described. The model is then applied to the data collected from five different sensors operated by Traffic.com (*15*). A continuous and a discontinuous speed–flow model, both currently in practice, are selected and applied to the same data. Comparison of outcomes from the three models is made based on the statistical performance measure. The key findings from fitting the models are summarized in the last section.

Literature Review

Edie (*6*) developed a complementary theory for steady-state conditions by combining the models proposed by Greenberg (*2*) and Underwood (*3*). It was the first study to observe a sharp speed drop in a small density range in some observed speed–density phase plots and proposed a two-regime phase diagram to model it. The frequent appearance of this discontinuity in several cases was mentioned in this study, which evidenced that the discontinuity did not appear because of random factors or by circumstances upstream from a bottleneck. The Lincoln Tunnel data used by Greenberg and Gazis et al. (*4*) was fitted in this study discontinuously to the proposed model and showed that the regression lines have better fits than the continuous model.

Similarly, Drake et al. revealed that the two-regime model produced the best fit with the lowest standard error compared with the single-regime model (*16*). The term "inverted lambda-shaped flow–density curve" was first used by Koshi et al. (*11*) who analyzed traffic data from the Tokyo Expressway and acknowledged the presence of the discontinuity.

More recently, Zhang and Kim proposed a modified car-following theory to model capacity drop and traffic hysteresis phenomena (*10*). The Pipe's theory of car-following model (*17*) was modified and four different models were proposed. One of the models produced a fundamental diagram that resembles the mirrored image of the reversed lambda hypothesized by Koshi et al. (*11*). The model was developed by incorporating the concept varying "gap-time" which is defined as the time required for the follower to travel a gap-distance.

However, several past studies modeled the relationship between traffic stream parameters continuously. The speed–flow relationship for base conditions described in the HCM (*12*) can be interpreted as a three-regime continuous model (although as mentioned above, the HCM freeway facilities methodology incorporates a capacity-drop discontinuity). According to Chapter 10 and Chapter 25 of HCM, the flow–density curve for the oversaturated regime assumes a simplified linear relationship. Others have argued against the existence of a discontinuity in the fundamental diagrams as a permanent characteristic and advocated the continuous flow–density relationship (*18*, *19*). Cassidy's (*18*) empirical observation from the re-scaled curves of cumulative arrival number versus time and cumulative occupancy versus time led to the assumption that the flow–density relationship is continuous under steady-state traffic flow conditions. The existence of a gap that appeared in the observed loop detector data is explained in this study as an absence of density values in some ranges. Li and Zhang (*19*) theoretically showed that a discontinuous fundamental diagram is non-differentiable at the discontinuous point and results in infinite characteristic wave speeds.

Past outlier-detection algorithms based on various traffic data are also instructive. To obtain only valid data points from PeMS (*20*) archived data, a study adopted an empirical rule of excluding observation days for which PeMS reported a functionality of 80% or less (*14*). In addition, this study detected outliers even from the valid observation days by plotting box plots of observed flow rate over multiple days and imposing a threshold of 1.5 times the interquartile range.

Xu et al. (*13*) developed a method for fitting the HCM oversaturated model to fixed sensor data by filtering steady-state observations only. Mixed-state and inconsistent data points were removed by imposing three thresholds. Detailed about these thresholds are described in the Methodology section.

The use of the robust regression technique in dealing with erroneous data has gained popularity in recent times (*21*, *22*). This technique iteratively dampens the influence of outlying observations to provide a better fit to the majority of the data. However, the application of robust regression or a similar technique in handling traffic-related data is limited in the literature.

From the above review, it can be said that the discontinuity in the steady-state fundamental diagrams has been recognized from as early as the 1960s. Nonetheless, modern computational tools can improve the development of such models by using data collected over an extended time period. In contrast, a few studies have challenged the existence of fundamental diagram discontinuity. Finally, there are opportunities for including state-of-art techniques of outlier detection in empirical traffic data modeling.

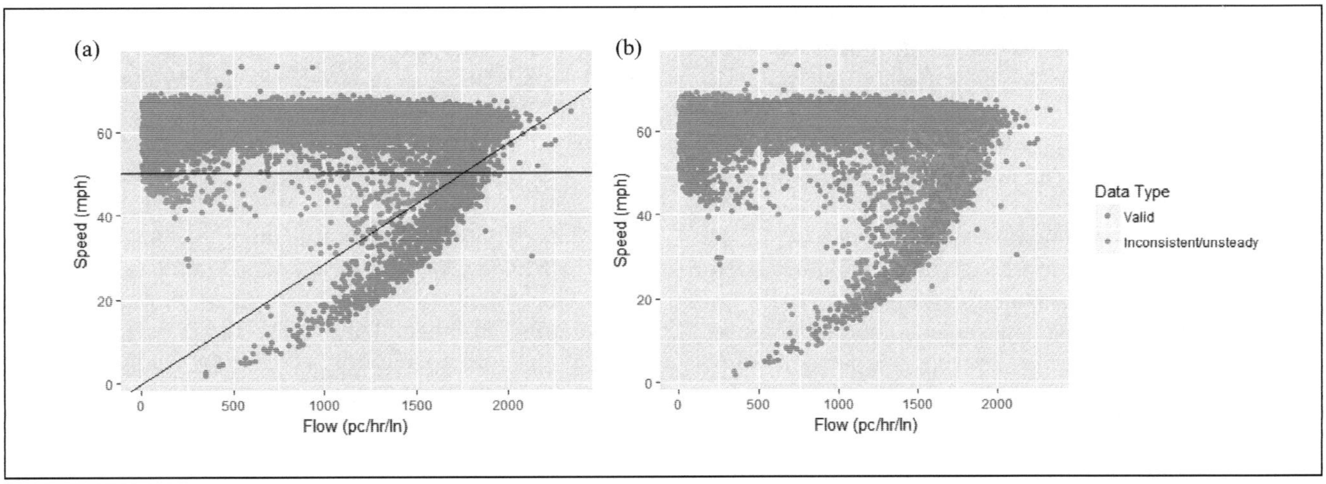

Figure 1. (*a*) Inconsistent data filtered by CST and CDT and (*b*) non-stationary data filtered by SFDT.

Methodology

Initial Cleaning of Data

For developing the proposed discontinuous steady-state model, traffic stream data from fixed-location sensors are needed for a long span of time (preferably one year). As the model development is data driven, inconsistent and mixed-state data were removed in two stages of the process. This section presents the first stage, which involves applying three thresholds that are introduced in an earlier study (*13*). In this study, the thresholds are modified and applied in a slightly different manner to conform to the study objectives.

The first two thresholds applied to detect inconsistent observations are known as the critical speed threshold (CST) and critical density threshold (CDT). The combined application of these two thresholds removes the low-speed observations associated with low volume. Such observations exist because of inclement weather, work zones, incidents, or any other sort of capacity-drop phenomena, as well as from observations that include both congested and uncongested flow. CST is used to identify low-speed (congested) observations. Analysis of traffic data reveals that 10 mph below the speed limit is a reasonable threshold to determine congested observations. CDT is used to identify low-volume observations. A density of 35 pc/mi/ln, the threshold that separates level of service D and E for urban freeways in the HCM is defined as the CDT threshold. Observations that fall below both CST and CST thresholds are tagged as inconsistent (low-speed and low-volume) data points. Figure 1 shows speed–flow data obtained from Traffic.com for calendar year 2013 and for sensor ID 23774 WB. The inconsistent data points are shown in Figure 1*a* bounded by the CST and CDT threshold lines (green points).

To detect time interval (in this case, 5 min) data points that represent non-stationary traffic conditions, the third threshold, termed speed first difference threshold (SFDT), is applied. SFDT excludes observations whose speeds differ from the previous 5-minute observation by more than 10 mph. These non-stationary observations for sensor 23774 WB are shown in Figure 1*b* by the green points. It should be noted that outliers generating from system measurement error may still exist even after applying these thresholds.

Fitting a Two-Regime GHR Steady-State Model

The process of fitting the proposed steady-state model is explained in this section. The starting point is the flow–density relationship for the single-regime full GHR model (*4*) as shown in Equation 1:

$$q_{M,i} = k_i * u_f * \left(1 - \left(\frac{k_i}{k_j}\right)^{l-1}\right)^{\frac{1}{1-m}} \quad (1)$$

Where,

$i = 1,2,3 \ldots$ Observation index
$q_{M,i}$ = Model flow for observation i (pc/hr/ln)
u_f = Free-flow speed (mph)
k_i = Density for observation i (pc/mi/ln)
k_j = Jam density (pc/mi/ln)
l = Distance headway exponent
m = Speed difference exponent

In a two-regime (uncongested and congested regime) GHR model, the same form of Equation 1 is used for each regime, and the regime boundary is determined by a density break point (k_b). In the proposed model, a transition range is introduced by allowing the two regimes to overlap each other, which produces the recognized inverted lambda shape when plotted in the flow–density domain. The major challenge is to model this transition range. Empirical observations of traffic stream data depict that the data points in the transition range follow either the uncongested or the congested regime's characteristics. In other words, the transition range (defined in terms of density) includes a mixture of observations from both regimes. In this study, it is proposed to model each data point within the overlap by the regime model, either uncongested and congested, that results in the smallest absolute error.

The algorithm for modeling the uncongested and congested regimes along with the overlap range is defined by Equations 2, 3, and 4:

Where,

r = 1, 2 for uncongested and congested regime, respectively.

In Equation 2 through 4, the two density breakpoints (kb_1 and kb_2) define the upper and lower limit of the transition range, respectively.

Requirement for Additional Constraints

Before discussing the constraints added to fit the proposed model, it is important to explain the limitations of distinguishing the two regimes and defining the overlap based on empirical data. All data in this study are from HERE roadside sensors (formerly Traffic.com) in the Raleigh, North Carolina urban area. Figure 2 shows an example application of the algorithm described above to clarify the limitations. In this figure, the observed data are for HERE sensor ID 23771 WB for the calendar year of 2013 aggregated at a 5-minute interval. The model was fitted in MATLab software using a nonlinear optimization tool with an objective function of minimizing the sum of squared error of flow. The red points show the fitted model with no additional constraints, and the green ones show the same but with the inclusion of several constraints described later.

Several important aspects of unconstrained fitted model shown in Figure 2 are noted below:

- The uncongested curve of the unconstrained model (red line) does not capture a group of data points with high flow (>2,000 pc/hr/ln). Instead, the slope of the uncongested regime curve flattens out at a flow value near 2,000 pc/hr/ln. It is apparent that this model is poorly fitting several important and valid observations. Assuming a free-flow

Observed density	Formula for model speed	
$k_i \leq k_{b,r=2}$	$q_{M,i,r=1} = k_i{*}u_{f,r=1}{*}\left(1 - \left(\frac{k_i}{k_{j,r=1}}\right)^{l_{r=1}-1}\right)^{\frac{1}{1-m_{r=1}}}$	(2)
$k_i \geq k_{b,r=1}$	$q_{M,i,r=2} = k_i{*}u_{f,r=2}{*}\left(1 - \left(\frac{k_i}{k_{j,r=2}}\right)^{l_{r=2}-1}\right)^{\frac{1}{1-m_{r=2}}}$	(3)
$k_{b,r=2} < k_i < k_{b,r=1}$	$q_{M,i,r} = \begin{cases} q_{M,i,r=1}\ if\ \lvert q_{M,i,r=1} - q_i\rvert < \lvert q_{M,i,r=2} - q_i\rvert \\ q_{M,i,r=2}\ Otherwise \end{cases}$	(4)

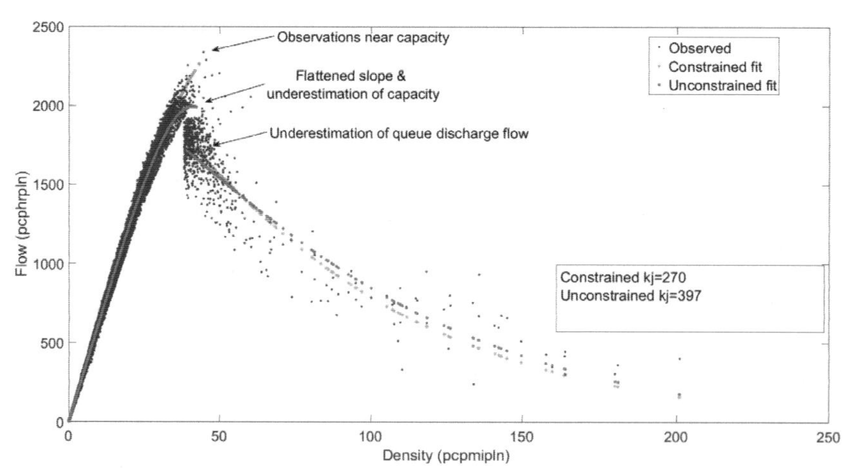

Figure 2. An example of the proposed flow–density model with and without any constraint.

speed of 65 mph (posted speed limit at this location was 60 mph), the capacity of the road section should be 2,350 pc/hr/ln according to HCM. Therefore, the fit of the uncongested curve can be improved by constraining the slope at capacity using this information.

- The queue-discharge flow rate (or post-breakdown flow rate) appears to be 1,700 pc/hr/ln from Figure 2. Approximating the pre-breakdown flow rate from the observed data as 2,350 pc/hr/ln, it is found that the drop in post-breakdown flow rate is about 28% of the pre-breakdown flow rate. This percentage is significantly higher than the usual range mentioned in HCM (2–20%).
- The jam density for the congested regime estimated by the model is 397 pc/mi/ln. The physical interpretation of this value is that the average length of the vehicles in the traffic stream must be unrealistically small (about 13 feet).

$$q_{\text{pre}} = k_{b,r=1} * u_{f,r=1} * \left(1 - \left(\frac{k_{b,r=1}}{k_{j,r=1}}\right)^{l_{r=1}-1}\right)^{\frac{1}{1-m_{r=1}}} \quad (5)$$

$$q_{\text{post}} = k_{b,r=2} * u_{f,r=2} * \left(1 - \left(\frac{k_{b,r=2}}{k_{j,r=2}}\right)^{l_{r=2}-1}\right)^{\frac{1}{1-m_{r=2}}} \quad (6)$$

Constraint: $q_{\text{pre}}*0.98 \geqslant q_{\text{post}} \geqslant q_{\text{pre}}*0.8$

Constraint on Slope at Capacity. It is shown earlier in Figure 2 that the capacity is underestimated by the unconstrained model as the slope of the uncongested regime flattens out, and the regime continues to a high-density value. The slope of the flow–density curve at any point of the uncongested regime can be obtained by differentiating the model flow equation with respect to density:

$$\frac{dq_{M,i,r=1}}{dk_i} = u_{f,r=1}\left(\left(1 - \left(\frac{k_i}{k_{j,r=1}}\right)^{l_{r=1}-1}\right)^{\frac{1}{1-m_{r=1}}} - k_i\left(\frac{1}{1-m_{r=1}} * \left(1 - \left(\frac{k_i}{k_{j,r=1}}\right)^{l_{r=1}-1}\right)^{\frac{m_{r=1}}{1-m_{r=1}}} * \frac{l_{r=1}-1}{k_{j,r=1}} * \left(\frac{k_i}{k_{j,r=1}}\right)^{l_{r=1}-2}\right)\right) \quad (7)$$

The observations highlighted above are also present in the datasets for the other traffic monitoring stations used in this study. The essence of the above discussion is that fitting the proposed empirical model without additional constraints may result in unusual and infeasible values that deviate from the findings of past research cited in the HCM. To resolve these issues, three constraints were added to properly fit the model. Although these constraints are selected empirically, they are justified by the guidelines provided in HCM. The resulting model after adding these constraints is shown in Figure 2 by the green points.

Constraint on Queue-Discharge Flow Rate. The post-breakdown flow rate is usually lower than the pre-breakdown flow rate, resulting in a significant loss of freeway throughput during congestion. According to a past study cited in Chapter 10 and Chapter 26 of HCM (*23*), the average difference between the post-breakdown and the pre-breakdown flow rates varies widely from as little as 2% to as much as 20%. In light of this information, the limit of post-breakdown flow rate is set as 80–98% of the pre-breakdown flow rate. The mathematical expressions for pre-breakdown and post-breakdown flow rates are shown in Equations 5 and 6:

Substituting $k_i = k_{b,r=1}$ in Equation 7 yields the slope at capacity. To resolve the issue of decreasing slope at capacity, the slope at capacity value in HCM is used as a guideline. According to Chapter 10 and Chapter 25 of HCM, the equations for flow in a basic freeway segment at different density regimes are shown in Equation 8:

$$q_{\text{HCM},i} = \begin{cases} k_i * u_f \ \forall k_i \in [0, k_b] \\ q_b + \frac{\sqrt{1-4*\beta*q_b*k_i + 4*\beta*u_f*k_i^2-1)}}{2*\beta*k_i} \forall k_i \in (k_b, 45] \\ q_c*\left(1 - \frac{k_i-45}{k_j-45}\right) \ \forall k \in (45, k_j] \end{cases} \quad (8)$$

Where,

$q_{\text{HCM},i}$ = HCM model flow for observation i

β = A coefficient defined as a function of free-flow speed

q_b = Flow breakdown (pc/hr/ln) as a function of free-flow speed

q_c = Capacity of the segment (pc/hr/ln) as a function of free-flow speed

k_b = Density at breakpoint (pc/mi/ln)

To obtain the slope at capacity equation, Equation 8 for density range k_b to 45 pc/mi/ln is differentiated with respect to the observed density and the following expression is obtained:

$$\frac{dq_{\mathrm{HCM},i}}{dk_i} = \frac{1 + \frac{k_i(-4q_b\beta + 8\beta u_f k_i)}{2\sqrt{1-4q_b\beta k_i + 4\beta u_f k_i^2}} - \sqrt{1 - 4q_b\beta k_i + 4\beta u_f k_i^2}}{2\beta k_i^2} \tag{9}$$

In Equation 9, plugging $k_i = 45$ gives the slope at capacity value according to the HCM model. The proposed constraint here is that the slope at capacity according to the proposed model must be greater than or equal to the HCM slope at capacity value.

Constraint on Jam Density. Jam density is defined by the HCM as "the maximum density that can be achieved on a segment." Jam density occurs under congested flow conditions as speed approaches zero. According to the HCM, the default value for jam density is 190 pc/mi/ln. However, the HCM recommends that jam density be based on site-specific data if such data are available. The recommended range of site-specific jam density is given in the HCM as 150–270 pc/mi/ln. Therefore, the fitted jam density for the congested regime in the proposed model is capped at 270 pc/mi/ln.

Robust Regression

Although the initial stage of the model development process involves removing clearly inconsistent and non-stationary data from the raw datasets, outlying observations are likely to remain in the data after the stage one filters. As the method described above is data driven, it is imperative to fit the model through valid observations only. A customized application of robust regression method is proposed in this study. According to this approach, the standard error for each data point is estimated by fitting the model. Data points with a standard error higher than a certain threshold are removed from the original dataset. Then, the model is fitted again with the updated dataset. The process is continued until the maximum standard error becomes lower than the threshold.

In this regard, the selection of threshold is critical. First, there is a significant difference in the number of data points in the uncongested and congested regime. Therefore, it is important to distinguish the two regimes and estimate the standard error for each of them separately. The estimation of the standard error for each regime is shown in Equation 10:

$$SE_{i,r} = (q_{M,i,r} - q_i)/Std_r \tag{10}$$

Where,

Std_r = Standard deviation of flow error for regime r

The second issue with robust regression is that there are two types of observations that need to be removed: mixed-state observations and observations with extreme measurement error. The mixed-state data points are valid observations that represent time intervals in which there was a switching between congested and uncongested regimes. Following the initial cleaning described above, such observations that remain are expected to be more prevalent in the congested regime than in the uncongested regime, as illustrated in Figure 1*b*. The distribution of the mixed-state observations will be asymmetric around the flow–density curve of the congested regime (lying on the left side of the curve). Because these observations do not result from measurement error, they are likely to be more prevalent than measurement errors. For example, for every flow breakdown, there are likely to be two pronounced mixed-state observations, one for the time interval when the queue formation shock wave passed by and one for the time interval when the queue clearance shock wave passed by. Considering these facts, a symmetric threshold for the standard error of ± 3.5 is applied for the uncongested regime. This symmetric threshold is expected to remove the observations with high measurement error in the uncongested regime. For the congested regime, an asymmetric threshold of $+2$ is applied for all but the final step of the robust regression. This threshold is expected to remove the remaining mixed-state data points and the measurement error outliers on the left side of the flow–density curve. In the final step of robust regression, a threshold of -3.5 is applied to exclude any remaining outliers from the congested regime on the right side of the flow–density curve.

The symmetric threshold of ± 3.5 standard error applied on the uncongested regime observations represents a confidence interval of approximately 99.95% (assuming that the errors have Gaussian distribution). On the other hand, a less conservative threshold is applied to the congested regime because of the potential presence of mixed-state observations. With these sets of thresholds, the robust regression process is expected to converge without removing excessive data points.

The HCM Model: As a Continuous Fundamental Diagram

As shown earlier in Equation 6, if queue-discharge capacity reduction is ignored, the HCM models the freeway fundamental diagrams as a three-regime continuous model.

According to this model, the breakpoint, capacity, and the coefficient β is expressed as a function of free-flow speed. Jam density has a default value of 190 pc/mi/ln, but the HCM recommends calibration when field data are available.

As the comparison between the proposed and HCM model is an integral part of this study, the parameters need to be calibrated in a manner that provides a best fit model. Here, it is proposed to calibrate the capacity and jam density using the observed flow by minimizing the sum of squared error. Free-flow speed is estimated as the average speed of observations with $Flow \leq 1,000 pc/hr/ln$ as per guidance provided in HCM.

In this study, the proposed discontinuous steady-state model described above is to be compared with the HCM model based on two statistical parameters: mean standard error (MSE) of fitted flow rates and Bayesian Information Criterion (BIC) (*24*). The MSE value represents the average of squared error for each observed flow. However, the MSE value can be misleading in judging two models if the parameterization of the two models varies greatly. BIC is an effective tool to resolve this issue because it penalizes models for having additional parameters.

Description of Data and Study Area

To illustrate the application of the proposed model, traffic stream data were used from five HERE side-fire radar stations located near the Triangle Area of Raleigh, North Carolina for the calendar year of 2013. The downloadable data are aggregated in 5-minute intervals. Each sensor is located adjacent to a basic freeway segment according to the HCM segmentation rules.

Results

To fit the traffic stream model by imposing the constraints described above, a nonlinear optimization tool available in MATLab was used. The run time for each model fitting varied from 1 to 3 minutes. Results from data filtering, fitted models, interpretation, and sensitivity of the parameters obtained are described in the following subsections.

Application of CST, CDT, and SFDT

This section summarizes the two stages of cleaning inconsistent, mixed-state, and outlying observations. Figure 3 shows the percentage of data reduced by the initial thresholds (CST, CDT, and SFDT) and by robust regression (Figure 3).

The most striking observation from Figure 3 is the higher percentage of data reduced by the initial thresholds for sensor 23786 EB (7%) compared with that for other sensors. Data exclusion by robust regression is lower than that by the initial filters across all the sensors. The highest proportion of data reduced by it is only about 3%. The proportion of outlier detection may also depend on the health of the detectors. Furthermore, in no case was the number of outliers more than 10% in total.

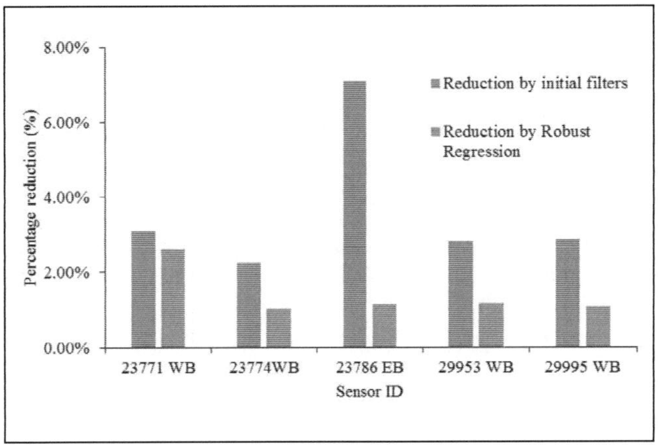

Figure 3. Percentage of data reduction in different sensors.

Fitted Models

The final parameter values used to plot the flow–density and speed–flow fundamental diagrams as shown in Figure 4 are the parameters resulting from convergence of the robust regression algorithm. The estimated values along with their standard deviations are listed in Table 1. Standard deviation for such nonlinear optimization models is calculated using a method described in a past study (*25*).

In general, the plots in Figure 5 illustrate that the fitted models reasonably follow through the steady-state observations. Here, the value of the parameters that have physical interpretation needs to be discussed. The distance head exponent (l) and speed exponent (m) for both regimes do not have any straightforward physical meaning. The same statement applies to the jam density and free-flow speed for the uncongested and congested regime, respectively. From the fitted speed–flow plots in Figure 4, the free-flow speed for the uncongested regime ($ffs_{r=1}$) is evident. The density breakpoint for the uncongested regime ($kb_{r=1}$) is less than HCM default of 45 pc/mi/ln for all but sensor 29953 WB. This parameter and $ffs_{r=1}$ are of importance for estimating the capacity of the segment at the sensor locations. According to the HCM default method, the capacity of a basic roadway segment varies between 2,320 to 2,350 pc/hr/ln for the given range of $ffs_{r=1}$ (62–65 mph). However, the site-specific capacities from the plots in Figure 4 are apparently less for sensor 23771 WB and 23774 WB and more for sensor 23786 EB and 29953 WB than the HCM defaults. The close agreement of HCM default capacity to the data of only one out of five sensors (29995 WB) underscores that the national average capacity values provided by HCM need to be calibrated with field data if high fidelity analysis is desired.

The difference between the two density break points represents the overlap range. This range appears to be a

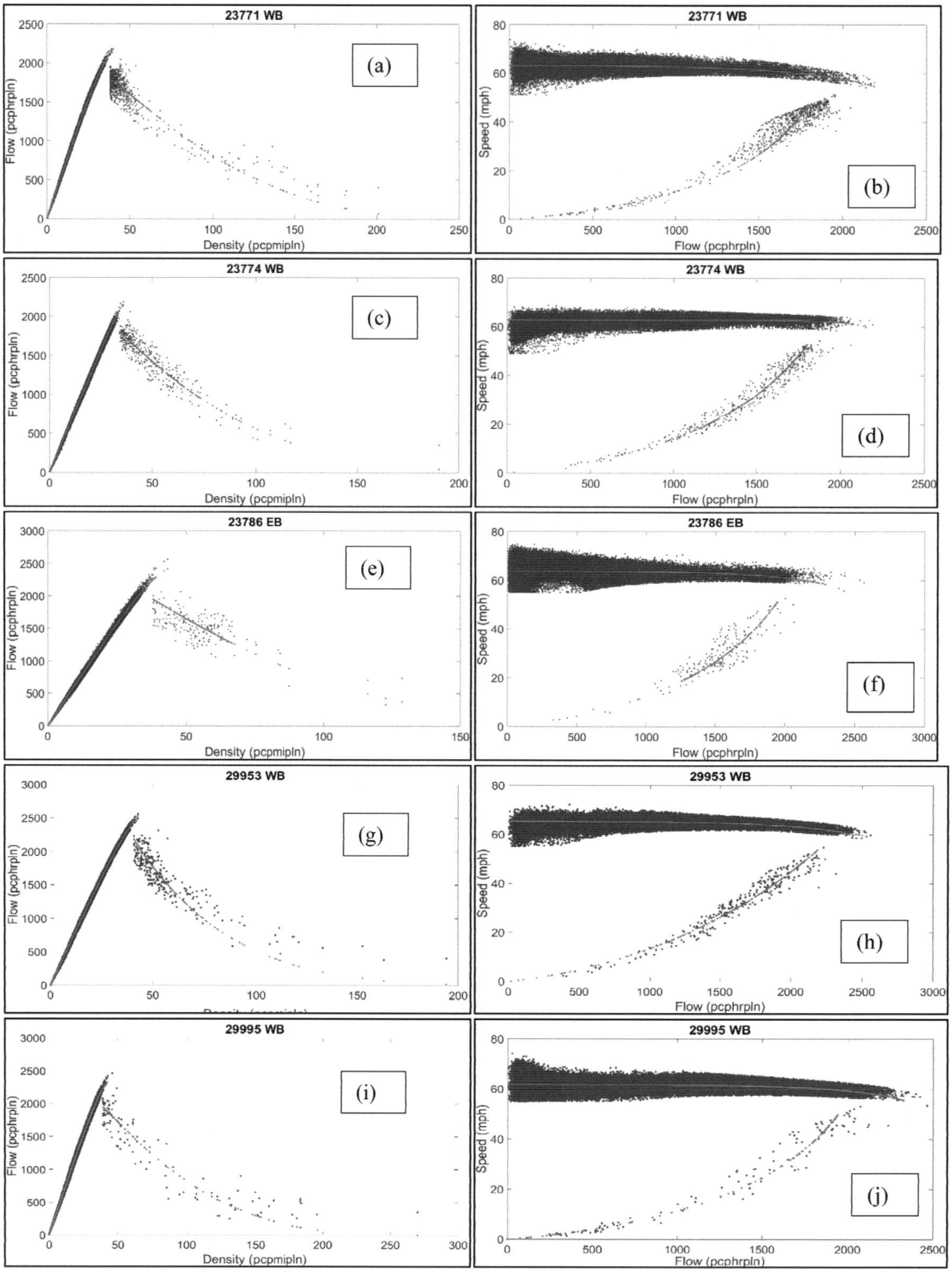

Figure 4. Flow–density (left column) and speed–flow (right column) diagrams for five sensors. The corresponding sensor IDs are shown in each plot.

Table 1. Parameter Estimates and Standard Deviations (in Parentheses) for Different Sensors Obtained from the Fitted Model

ID	$ffs_{r=1}$ (mph)	$k_{b,r=1}$ (pc/m/l)	$k_{j,r=1}$ (pc/m/l)	$l_{r=1}$	$m_{r=1}$	$ffs_{r=2}$ (mph)	$k_{b,r=2}$ (pc/m/l)	$k_{j,r=2}$ (pc/m/l)	$l_{r=2}$	$m_{r=2}$
23771 WB	63 (5×10^{-3})	39.9	231 (0.29)	4.2642 (1×10^{-3})	0.9776 (9×10^{-5})	10^5 (298)	37.4	270 (0.56)	1.0288 (4×10^{-6})	0.6212 (10^{-4})
23774 WB	63 (4×10^{-3})	36.4	79 (0.29)	11.8203 (4×10^{-3})	0.9957 (1×10^{-4})	10^5 (403)	33.4	270 (0.65)	1.0465 (10^{-5})	0.6834 (2×10^{-4})
23786 EB	64 (7×10^{-3})	44	412 (0.69)	5.1581 (2×10^{-3})	0.9993 (4×10^{-6})	5×10^5 (4500)	38	270 (1.43)	1.0246 (9×10^{-6})	0.6658 (3×10^{-4})
29953 WB	65 (5×10^{-3})	48	522 (0.55)	4.7793 (1×10^{-3})	0.9992 (3×10^{-6})	2×10^6 (19177)	40	270 (0.98)	1.0300 (6×10^{-6})	0.7300 (2×10^{-4})
29995 WB	62 (5×10^{-3})	42	208 (0.21)	5.4970 (2×10^{-3})	0.9932 (3×10^{-5})	5×10^5 (7354)	38.5	269.4 (2.56)	1.0185 (1×10^{-5})	0.6338 (6×10^{-4})

unique characteristic for the five locations based on the proposed approach modeled. The fitted overlap in density varies from 2.5 to 8 pc/mi/ln across the five sensors. This site-specific modeling of overlap along with the drop in capacity have many potential applications such as estimating pre-breakdown and queue-discharge flow rate, identifying recurring freeway bottlenecks, and so forth.

As described earlier, the jam density for the congested regime ($kj_{r=2}$) is capped at the maximum feasible value mentioned in HCM. As it was observed in Figure 1 that this value tends to go beyond the limit of feasibility, all the fitted values of $kj_{r=2}$ are suppressed to this maximum limit by the constraint. However, it was noted that the overall fit and the key parameter values were not significantly affected after introducing this constraint.

The standard deviations of the estimates, shown within parenthesis in Table 1, are of small magnitude. This reveals that most of these estimates are significant and the fitted models are sensitive to these parameters. It should be noted that the standard deviations for the two breakpoints $k_{b,r=1}$ and $k_{b,r=2}$ cannot be estimated because these are only classifiers of the regimes. Standard deviation for $ffs_{r=2}$ is very high because the fitted models are insensitive to this parameter.

Comparison with HCM Models

Two forms of steady-state HCM traffic stream model, with and without capacity drop, were applied to the data from the five sensors discussed above. A capacity drop between 2% to 20% is acknowledged by HCM when traffic state transfers from uncongested to congested regime. As an example, the fundamental diagrams for sensor 29995 WB obtained by fitting both forms of HCM model are shown in Figure 5.

The free-flow speed, density breakpoint, capacity, jam density, and capacity drop obtained from the HCM models for five sensors are shown in Table 2. One important observation from Figure 5, which is also found for other sensors, is that the linear flow–density relationship for the congested regime is unable to capture most of the data points. In the continuous model, this lack of fidelity is caused both by the linear shape and by the constraint imposed by the continuity with the uncongested regime. The poor fit results in an underestimation of capacity for three sensors compared with the HCM defaults (top three sensors in Table 2). The discontinuous model balances this underestimation of capacity to some extent. The jam density is bounded to its minimum reasonable value (150 pc/mi/ln) for three sensors for both forms of model. The capacity drop is constrained at its maximum limit for four sensors. As this discontinuous model fitting process is applied to more freeway data sets, factors related to capacity drop can be studied in more depth. The free-flow speed values estimated by HCM do not differ much from what was obtained from the proposed model.

To compare the goodness of fit and information criteria of the HCM models with the proposed one, Figure 6*a* illustrates the MSE of flow for the three models in question. Figure 6*b* shows the comparison of these models based on the BIC value obtained by them for the sensors under consideration.

It is apparent from Figure 6 that the MSE value is highest for the continuous HCM model and lowest for the proposed model across all five sensors. Similar trend exists in terms of the BIC values of the models. Although the proposed model is penalized because of the additional fitted parameters, which is reflected by the *BIC* values, the *BIC* values across all the sensors are lowest for the proposed model. The HCM discontinuous model is better than the continuous one. Nonetheless, the proposed model would be considered preferable to the HCM models in terms of both information criterion and squared error.

Summary and Conclusion

This paper has proposed an empirical process for fitting a discontinuous two-regime steady-state model to data

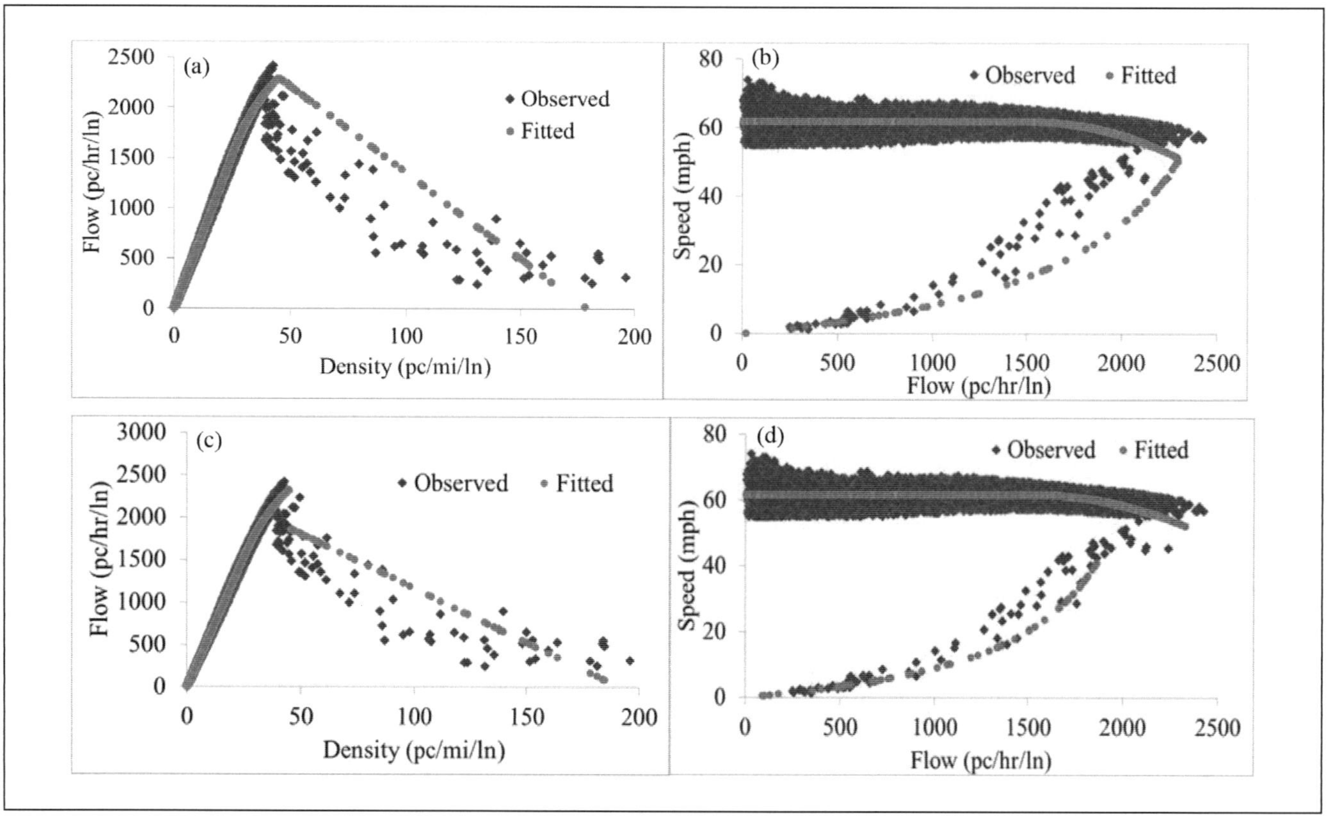

Figure 5. Fundamental diagrams for sensor 29995 WB fitted by HCM models: (*a*) continuous flow–density; (*b*) continuous speed–flow; (*c*) discontinuous flow–density; (*d*) discontinuous speed–flow.

Table 2. Parameter Values for Five Sensors Obtained by the HCM Models

	Continuous				Discontinuous				
ID	*ffs mph*	*kb pc/m/l*	*Cap. pc/h/l*	*kj pc/m/l*	*ffs mph*	*kb pc/m/l*	*Cap. pc/hr/l*	*kj pc/m/l*	*Cap. drop %*
23771 WB	63.3	23.2	1857	165.4	63.3	23.2	1920	183.2	17.0
23774 WB	61.8	24.7	1840	150.0	61.8	24.7	1949	150.0	20.0
23786 EB	64.4	22.1	2123	150.0	64.4	22.1	2223	150.0	20.0
29953 WB	64.6	22.0	2389	150.0	64.6	22.0	2509	150.0	20.0
29995 WB	61.7	24.8	2300	179.3	61.7	24.8	2336	191.4	20.0

from basic freeway segment sensors. The transition range, also known as the overlap range, is modeled by comparing the deviation of the two regimes from the observed data. Three HCM-justified constraints were added to reasonably fit the model. Non-stationary, mixed-state, and outlying data points were addressed explicitly in two stages. Three initial thresholds and a tailored application of robust regression were applied to filter the invalid observations.

The proposed model was applied to five basic freeway segments near Raleigh, North Carolina. Fundamental diagrams for each of the sensors show acceptable fits for the proposed model. Discussions on the interpretation of the parameters revealed that all the parameters obtained from the fitted models represent reasonable values. The site-specific capacity values apparent from the observed data and echoed in the fitted models show deviation from the national average default values provided by the HCM.

Two forms of HCM model, continuous (without capacity drop) and discontinuous (with capacity drop) are fitted to the same five sensors data in the best possible way and the fitted fundamental diagrams are shown. It appears that the linear flow–density relationship in the

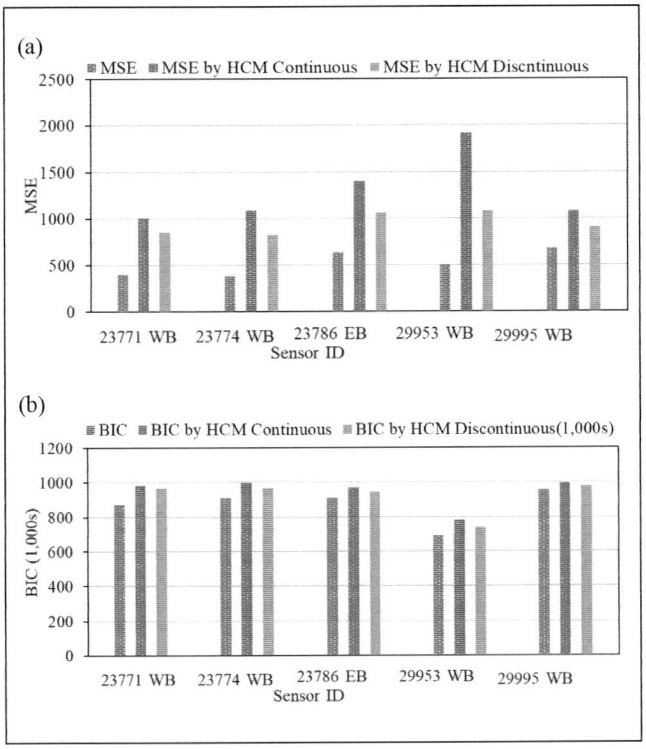

Figure 6. Comparison of (*a*) MSE and (*b*) BIC values obtained from the proposed and HCM models.

congested regime by the continuous HCM model could not capture the observed data points properly, with this poor fit compounded by the constraint of continuity between the two regimes. Moreover, the capacity resulting from most of the continuous models is substantially lower compared with both the national defaults and with what is obtained from the proposed model. The insertion of discontinuity in the HCM model minimizes these deviations to some extent. Statistical comparison of the results showed that the MSE of flow is highest for HCM continuous and lowest for the proposed model across all five sensors. Although the BIC penalizes the proposed models for additional parameters, it nonetheless indicates a preference over all of the HCM models.

It is expected that the proposed discontinuous steady-state model will be useful to researchers and practitioners for estimating capacity, identifying bottlenecks, and other freeway operational analyses. Moreover, observed phenomena related to traffic state transition such as queue-discharge capacity drop can be modeled with the proposed technique as well.

However, the study does not come without limitations. The nonlinear optimization technique used in this study may require an extensive period of time for computation if traditional license-based (e.g., Excel) or open source (e.g., R) software is used. Also, it is recognized by the authors that the first-stage removal of unwanted data points recommended in this study may remove some valid data points. In future, attempts will be taken to combine the two-stage data-reduction process into a single robust process that is more efficient and consistent.

Author Contributions

The authors confirm contribution to the paper as follows: study conception and design: IA, BMW, MSS; data collection: IA, MSS; analysis and interpretation of results: IA, BMW, MSS; draft manuscript preparation: IA, BMW, MSS. All authors reviewed the results and approved the final version of the manuscript.

References

1. Greenshields, B. D., W. Channing, and H. Miller. A Study of Traffic Capacity. In *Highway Research Board Proceedings*, Vol. 1935. National Research Council (USA), Highway Research Board, 1935.
2. Greenberg, H. An Analysis of Traffic Flow. *Operations Research*, Vol. 7, No. 1, 1959, pp.79–85.
3. Underwood, R. T. *Speed, Volume, and Density Relationships*. Pennsylvania State University, University Park, PA, 1960, pp. 141–188.
4. Gazis, D.C., R. Herman, and R. W. Rothery. Nonlinear Follow-the-Leader Models of Traffic Flow. *Operations Research*, Vol. 9, No. 4, 1961, pp. 545–567.
5. May, A. D. Traffic Flow Fundamentals. Prentice Hall, Upper Saddle River, NJ, 1990, pp. 290–303.
6. Edie, L. C. Car-Following and Steady-State Theory for Noncongested Traffic. *Operations Research*, Vol. 9, No. 1, 1961, pp. 66–76.
7. May, A. D., and E. M. Harmut. Non-Integer Car-Following Models. *Highway Research Record*, Vol. 199, No. 1, 1967, pp. 19–32.
8. Ceder, A., and A. D. May. Further Evaluation of Single- and Two-Regime Traffic Flow Models. *Transportation Research Record: Journal of the Transportation Research Board*, 1976. 567: 1–30.
9. Kerner, B. S. Experimental Features of Self-Organization in Traffic Flow. *Physical Review Letters*, Vol. 81, No. 17, 1998, p. 3797.
10. Zhang, H. M., and T. Kim. A Car-Following Theory for Multiphase Vehicular Traffic Flow. *Transportation Research Part B: Methodological*, Vol. 39, No. 5, 2005, pp. 385–399.
11. Koshi, M., M. Iwasaki, and I. Ohkura. Some Findings and an Overview on Vehicular Flow Characteristics. *Proc., 8th International Symposium on Transportation and Traffic Flow Theory*, Vol. 198, No. 1, University of Toronto, ON, Canada, 1983, pp. 403–426.
12. *Highway Capacity Manual: A Guide for Multimodal Mobility Analysis*, 6th ed. Transportation Research Board, Washington, D.C., 2016.
13. Xu, Y., B. Williams, N. Rouphail, and R. Chase. Development of an Oversaturated Speed-Flow Model Based on the Highway Capacity Manual. *Transportation Research Record: Journal of the Transportation Research Board*, 2013. 2395: 41–48.

14. Li, X., Z. Li, J. Han, and J. G. Lee. Temporal Outlier Detection in Vehicle Traffic Data. *Proc., 2009 IEEE 25th International Conference on Data Engineering*, Shanghai, China, IEEE, New York, NY, 2009, pp. 1319–1322.
15. Traffic.com. *ITIP Stakeholder Applications*. www.stakeholder.traffic.com. Accessed July 30, 2017.
16. Drake, J. S., J. L. Schofer, and A. D. May. A Statistical Analysis of Speed Density Hypothesis. *Highway Research Record*, Vol. 154, 1967, pp. 53–87.
17. Pipes, L. A. Car Following Models and the Fundamental Diagram of Road Traffic. *Transportation Research*, Vol. 1, No. 1, 1967, pp. 21–29.
18. Cassidy, M. J. Bivariate Relations in Nearly Stationary Highway Traffic. *Transportation Research Part B: Methodological*, Vol. 32, No. 1, 1998, pp. 49–59.
19. Li, J., and H. M. Zhang. Modeling Space–Time Inhomogeneities with the Kinematic Wave Theory. *Transportation Research Part B: Methodological*, Vol. 54, 2013, pp. 113–125.
20. Caltrans. *PEMS*, http://pems.dot.ca.gov/. Accessed July 31, 2017.
21. Leroy, A. M., and P. J. Rousseeuw. *Robust Regression and Outlier Detection*. Wiley Series in Probability and Mathematical Statistics. Wiley, New York, NY, 1987.
22. Bary, M. N. A. Robust Regression Diagnostic for Detecting and Solving Multicollinearity and Outlier Problems: Applied Study by Using Financial Data. *Applied Mathematical Sciences*, Vol. 11, No. 13, 2017, pp. 601–622.
23. Hu, J., B. Schroeder, and N. Rouphail. Rationale for Incorporating Queue Discharge Flow into Highway Capacity Manual Procedure for Analysis of Freeway Facilities. *Transportation Research Record: Journal of the Transportation Research Board*, 2012. 2286: 76–83.
24. Schwarz, G. Estimating the Dimension of a Model. *The Annals of Statistics*, Vol. 6, No. 2, 1978, pp. 461–464.
25. Smith, R. C. *Uncertainty Quantification: Theory, Implementation, and Applications*, Vol. 12. SIAM, Philadelphia, PA, 2013, pp. 141–150.

The Standing Committee on Traffic Flow Theory and Characteristics (AHB45) peer-reviewed this paper (18-06577).

TRR
JOURNAL OF THE TRANSPORTATION RESEARCH BOARD

Article

Transportation Research Record
2018, Vol. 2672(20) 63–74

DOI: 10.1177/0361198118787987
journals.sagepub.com/home/trr
SAGE

Evaluating Recurring Traffic Congestion using Change Point Regression and Random Variation Markov Structured Model

Emmanuel Kidando[1], Ren Moses[1], Thobias Sando[2], and Eren E. Ozguven[1]

Abstract
This study develops a probabilistic framework that evaluates the dynamic evolution of recurring traffic congestion (RTC) using the random variation Markov structured regression (MSR). This approach integrates the Markov chains assumption and probit regression. The analysis was performed using traffic data from a section of Interstate 295 located in Jacksonville, Florida. These data were aggregated on a 5-minute basis for 1 year (2015). Estimating discrete traffic states to apply the MSR model, this study established a definition of traffic congestion using Bayesian change point regression (BCR), in which the speed–occupancy relationship was explored. The MSR model with flow rate as a covariate was then used to estimate the probability of RTC occurrence. Findings from the BCR model suggest that the morning peak congested state occurs once speed is below 58 miles per hour (mph), whereas the evening peak period occurs at a speed below 55 mph. Evaluating the dynamics of traffic states over time, the Bayesian information criterion confirmed the hypothesis that a first-order Markov chain assumption is sufficient to characterize RTC. Moreover, the flow rate in the MSR model was found to be statistically significant in influencing the transition probability between the traffic regimes at 95% posterior credible interval. The knowledge of RTC transition explained by the approaches presented here will facilitate developing effective intervention strategies for mitigating RTC.

The growing need for mobility coupled with traffic incidents on roadways in urban areas has resulted in the saturation of transportation facilities with vehicles. As a result, congestion, excessive delays, high risks of crash occurrence, pollution, stress, and wastage of fossil energy have become the primarily associated problems (*1, 2*). These problems underline inefficiencies related to the transportation systems, and therefore the necessity of developing effective countermeasures to address them has been recognized. Establishment and implementation of practical solutions to relieve traffic congestion require a better understanding of traffic conditions' stochastic evolution process and the associated relevant factors. In this context, mathematical models and algorithms play a significant role in understanding, evaluating, and predicting the traffic congestion characteristics as input to the congestion mitigation strategies.

In modeling freeway traffic conditions, two major regimes—free-flow and congested regimes—are usually identified and examined (*2, 3*). In order to analyze these regimes, the breakdown phenomenon becomes the most relevant process to describe traffic congestion occurrence. The breakdown event is characterized by an abrupt and significant drop in the operating speed and the traffic throughput (*4–6*). Several researchers who have evaluated the characteristics of this event have declared it to be highly stochastic in nature (*7–10*). Basically, traffic breakdown can occur at any traffic volume level ranging from uncongested to congested conditions (*11–13*). The occurrence of the breakdown at various flow volumes is described by the probability theory (*6, 9, 10*). Under similar prevailing conditions, several reasons have been hypothesized to cause breakdown to occur at various volumes, one of them being driving behavior (*14*).

[1]Department of Civil and Environmental Engineering, FAMU-FSU College of Engineering, Tallahassee, FL
[2]School of Engineering, The University of North Florida, College of Computing, Engineering, and Construction, Jacksonville, FL

Corresponding Author:
Address correspondence to Emmanuel Kidando: ek15f@my.fsu.edu

Many empirical studies use the traffic speed–flow relationship with a concept of a speed threshold to detect breakdown (*7, 8, 15–18*). For instance, Laflamme and Ossenbruggen and Yeon et al. explored this relationship and found that the speed threshold is 50 miles per hour (mph) (*17, 18*). The studies by Kim et al. and Dong and Mahmassani applied a speed drop of 10 mph from free-flowing speed as the threshold in the analysis (*7, 8*). In general, the estimated threshold is then used to record the immediate traffic flow before the breakdown event, which is incorporated in the analysis of the roadway capacity (*6, 14*). In light of the stochastic definition of the breakdown event, probabilistic models are normally used to characterize the breakdown process and estimate the distribution of the stochastic roadway capacity. Most of the studies apply the analogy of survival concept to model the breakdown event (*7, 8, 15, 19*). Among the various survival models that have been applied in some previous studies, the Weibull distribution was found to offer a superior fit of the empirical data than other survival models, such as log-logistic, Gumbel, lognormal, and exponential distributions (*4, 19, 20*). Also, Geistefeldt and Brilon revealed that censored data provide more accurate data than uncensored data (*21*).

Most of the previous investigations attempting to identify and analyze the breakdown phenomenon use the idea of fixed speed threshold and speed drop to define breakdown. Usually, these thresholds are estimated by examining the speed–flow diagram. However, as the actual traffic data are highly random and scattered (*15*), establishing the speed threshold using this relationship may be challenging, especially when there is no statistical approach to estimate the threshold value. Therefore, it is necessary to apply the analytical procedures that account for the data variations to determine the speed threshold.

The change point model is well developed in time series problems to estimate a threshold value where patterns before and after are significantly different (*22*). The change point problem has also been analyzed in the regression context, whereby the position of a significant trend shift before and following the change point in the stochastic data is required (*22*). Some of the areas where this approach has been successfully applied include medical research, financial modeling, manufacturing areas, and climatology (*22–24*). In medical diagnosis, for example, this approach has been used to compare drug reaction times between patient groups with age as a covariate in the model (*23*). Conversely, the application of the change point approach in traffic studies is not very common, particularly in analyzing highway safety and traffic patterns (*25–27*).

The objective of this research is to develop a probabilistic framework that evaluates the dynamic evolution process of recurring traffic congestion (RTC) over time on a basic freeway segment. The study establishes the definition of traffic congestion using the Bayesian change point regression (BCR) model. The model particularly explores the relationship between traffic speed and occupancy to estimate the speed threshold. The estimated value is assumed to divide traffic data into two homogeneous groups: congested and uncongested regimes. The random variation Markov structured regression (MSR) model, a model with the transition probability estimated using probit regression, is then used to explain the dynamic evolution of RTC. The dynamic evolution process here involves any of the following conditions: (i) the transition from the free-flow to the congested regime (traffic breakdown), (ii) the process of staying in the free-flow regime, (iii) the recovery process, and (iv) the process of staying in the congested regime. Before establishing the MSR model, the study estimated the appropriate memory length (order of the Markov chains) given the dataset. The Markov chains' memory length provides the number of the previous time steps that influence the probability of the traffic conditions evolution process (*28*).

For this study, traffic data collected for a year from a basic freeway segment on Interstate 295 located in Florida were used in the modeling process. Parameters of the MSR model were estimated using the Markov Chain Monte Carlo (MCMC) simulation approach. The contribution of this work to the existing literature is the integration of the BCR and the MSR models to classify and then describe the dynamic transition characteristics of RTC. Unlike previous studies, the present work explains the dynamic evolution process beyond the transition from the free-flow to the congested state (i.e., breakdown event). The next section of this paper describes the data preparation process. The third section explains the methodological framework involving the BCR and the MSR regression. The fourth section shows the model results beginning with validation of the BCR model using the Monte Carlo simulated data followed by the results of empirical data analysis. Next, the MSR results are presented. The last section delivers study conclusions and recommendations.

Data Description

Traffic data for the analysis were obtained from a busy 4.8-miles freeway basic segment. This corridor is located on Interstate 295 in Jacksonville, Florida with a 65 mph posted speed limit (Figure 1*a*). The field traffic data collected from January 1, 2015 through December 31, 2015 were obtained from the Regional Integrated Transportation Information System. These data were collected using five microwave vehicle detectors on a 5-minute basis with three major attributes: spot speed,

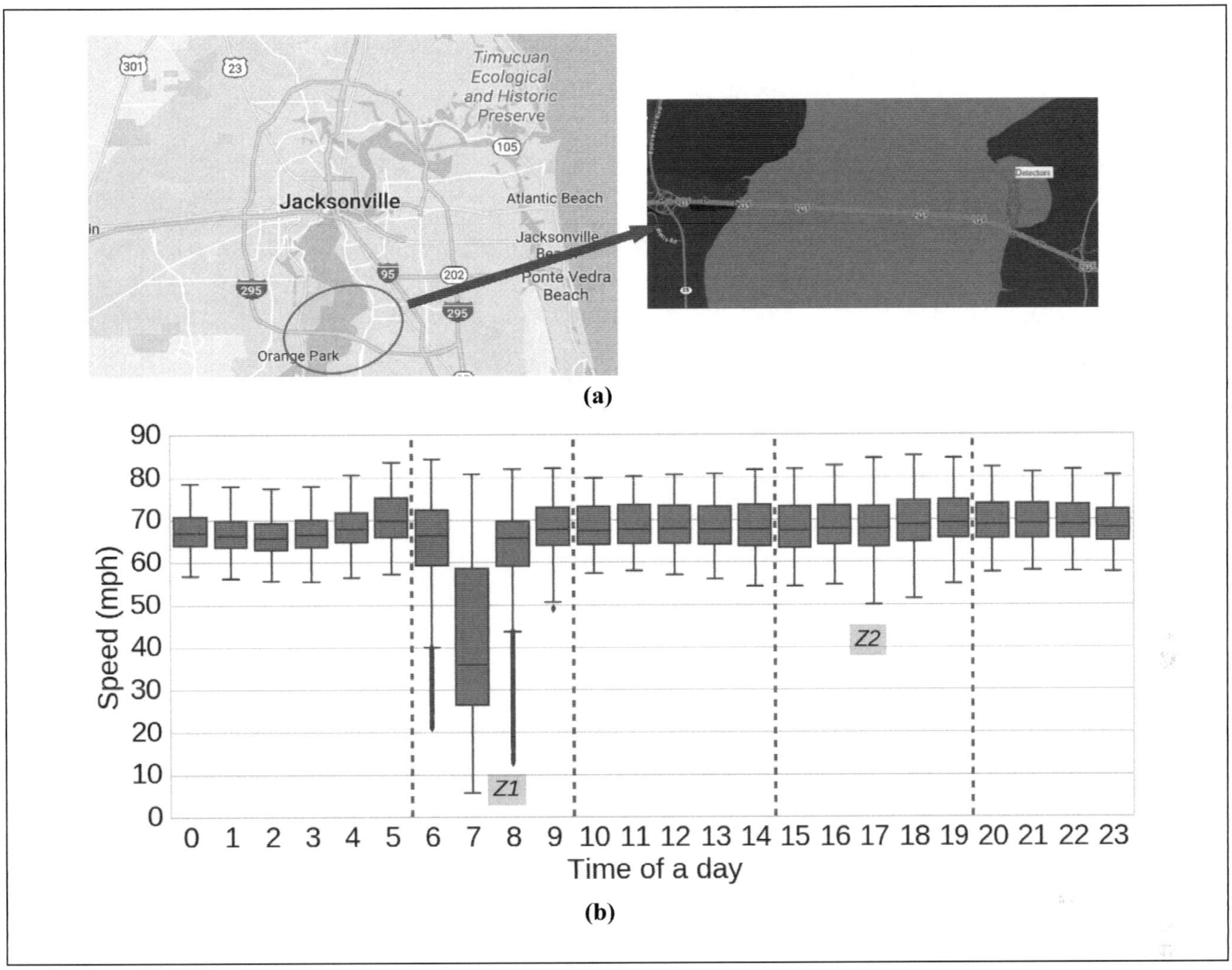

Figure 1. Study corridor and its hourly speed summary: (*a*) Study corridor and detector locations on I-295 southbound (Source: Google map) and (*b*) Hourly speed profile for the study segment.

traffic volume, and occupancy. In order to analyze only RTC, weekend, holidays, and periods in which incidents (crashes, work zones, etc.) happened were omitted from the analysis.

The boxplot of the 5-min traffic data in Figure 1*b* indicates that 6 a.m. to 9 a.m. is the morning peak period of the segment, and the evening peak period mainly occurs from 3 p.m. to 7 p.m. Figure 1*b* also depicts that morning peak period is usually more congested than evening peak period. Traffic data are non-stationary when a long time interval is analyzed (*29, 30*). This characteristic is associated with the distinct difference in traffic parameters (speed, volume, and density) between peak and off-peak periods. The traffic characteristics are approximately stationary during either peak or non-peak period. As a result, this study divided traffic data into blocks representing peak and off-peak periods to approximate the stationary section of the dataset. This approach has been applied in previous studies in developing predictive models (*29–31*). The two interval analyses that were identified are the morning peak (Z1) and the evening peak period (Z2) shown in Figure 1*b*.

Modeling Framework

Based on the assumption that the model is discrete in both state (free-flow and congested states) and time (5-minute interval), the first task was to transform traffic data to discrete states. The occupancy–speed relationship was used for this purpose, after which the resulting discrete states were used in the MSR model to analyze the dynamic evolution process of RTC.

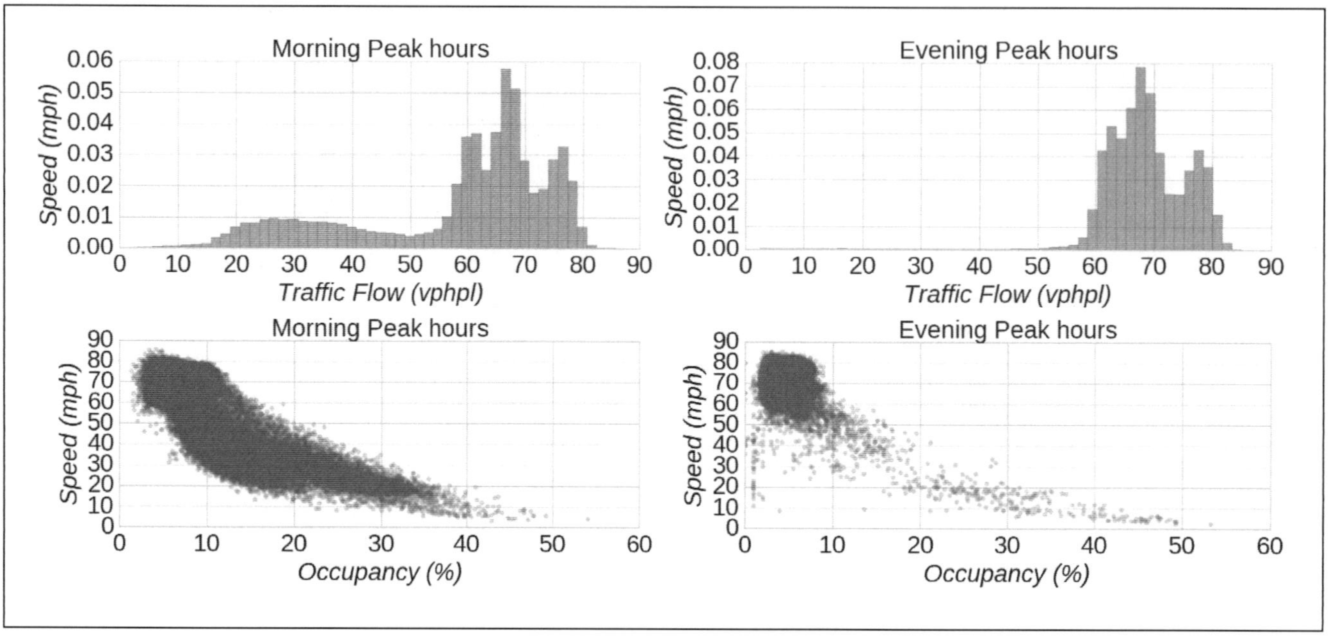

Figure 2. Speed distribution and scatter plot of occupancy and speed.

Traffic Congestion Definition

To conform to the *Highway Capacity Manual* methodology and to avoid recording short-term speed fluctuations as congestion, traffic breakdown was assumed to occur when speed below the estimated speed threshold is sustained for at least 15 minutes (*7, 8, 15*). As there is no standard method for estimating the speed threshold in the literature, which separates free-flow and congested conditions using the traffic stream parameters, this study attempts to establish the quantitative measure of congestion using the proposed mathematical model.

The speed distributions during the evening and the morning peak periods indicate two predominant subpopulations (bimodal). It can be seen from Figure 2 that the morning peak period evidently shows bimodal characteristics, whereas in the evening peak period, because of few data points in lower speeds, the number of components are not noticeably visible. A close examination of the occupancy–speed relationships in Figure 2 identifies a non-linear trend with an inverse relationship. Based on this trend, this study hypothesizes that two linear models can fit the non-linear relationship between speed and occupancy; one to represent the free-flow regime and the second component to be a congested regime. The two regression models are separated by the critical point called speed threshold.

Bayesian Regression to Detect Speed Threshold

In computing the unknown change point using the BCR approach, the assumption about parameter estimation is needed. The problem in this study was assumed to have a linear regression with errors following Gaussian distribution (*32*). The model outputs of the BCR are the unknown change point, slopes, and intercepts. The following equation indicates the BCR model with the change point V_c :

$$\begin{aligned} Y &\sim \text{Normal}(\alpha_1 + X\beta_1, \sigma) \; if \; X \leq V_c \\ Y &\sim \text{Normal}(\alpha_2 + X\beta_2, \sigma) \; if \; X > V_c \end{aligned} \quad (1a)$$

where,

$X$is the speed

β_1 *and* β_2 are the occupancy coefficients,

α_1 *and* α_2 are the intercept coefficients,

Y is the occupancy, and

σ is the standard deviation (std.) of the data.

The following are the priors for model parameters, which were taken as non-informative:

$$\begin{aligned} &\beta_1 \sim Normal(\text{mean} = 0, \text{std} = 100), \\ &\beta_2 \sim Normal(\text{mean} = 0, \text{std} = 100), \\ &\alpha_1 \sim Normal(\text{mean} = 0, \text{std} = 100), \\ &\alpha_2 \sim Normal(\text{mean} = 0, \text{std} = 100), \\ &\sigma. \sim Normal(\text{mean} = 0, \text{std} = 100), \\ &\text{and } V_c \sim DiscreteUniform(\text{min\_speed}, \text{max\_speed}). \end{aligned} \quad (1b)$$

The prior distribution of the change point was assumed to follow a discrete uniform distribution with equal probability of being at any recorded speed between the minimum and maximum value in the dataset. In order to accommodate for the discrete distribution of the change point during the modeling process, all posterior distributions of the model parameters were estimated using the Metropolis–Hastings sampling step. Before sampling these distributions, the optimization technique through the maximum a posteriori was used to determine the initial parameters.

Markov Chains Model

The Markov chains model is a probabilistic computational model that has been applied in modeling the dependence process over time. Transition probabilities describe the evolution process of the Markovian system. The simplest Markov chains model, and perhaps the most widely used in modeling of the stochastic systems, is the first-order Markov chain. This model assumes that the current state of the system only depends on the immediately preceding state. Assume a system is described by a sequence of the finite states Y_t evolving with a discrete time t, its first-order Markov process can be expressed as (*33*):

$$P(Y_t) = P(Y_t|Y_{t-1}) \quad (2a)$$

The higher order (ℓ^{th}-order) Markov chains model incorporates the historical memory of the observed states and is referred to as the memory-based Markov model. The transition process of this model depends on the sequence of the previously observed states (*33*) (see Equation 2b):

$$P(Y_t) = P(Y_t|Y_{t-\ell}, Y_{t-\ell+1}, \ldots, Y_{t-1}) \quad (2b)$$

where,
P represents probability
Y_t is the present observed state, and
$Y_{t-1}$is the immediate previous observed state,
$Y_{t-\ell}$ is the ℓth previous observed state.

Investigation of the Markov Chains' Memory Length

Before estimating the transition probability of the Markov chains model, the order of the model given the two estimated traffic states was identified. Notably, the number of time steps in the previous states that influence the dynamic transition of RTC was estimated using the Bayesian information criterion (BIC) approach. The BIC approach penalizes higher order models in a way that it balances the model complexity and the predictive accuracy to avoid the overfitting problem. The penalty term of the BIC comprises a number of observations n and the degrees of freedom (*33*). In the given ℓ^{th} order Markov chain model, the BIC score is estimated as:

$$BIC(\ell) = -2*ln(L) + \left(|S|^m - |S|^\ell\right)(|S|-1)*ln(n) \quad (3)$$

where,
L is the maximized likelihood of the model,
S is the number of states,
m is the highest order tested against lower order models in the analysis, and
n is the number of observations.

The model with the lowest BIC score is preferred and taken as the best over other competing models. From Figure 3, it is evident that the BIC approach favors lower order models compared with higher order Markov chains models. The morning and the evening peak periods are best explained by the first-order Markov chain process. That is to say, the current traffic condition is only influenced by the immediate previous traffic state and conditional independence of the other historical traffic states. For this reason, the first-order Markov chains assumption was used in implementing the MSR in the next section.

Random Variation Markov Structured Model

Using the assumption presented in Equation 1a that there is one change point in the regression model, the

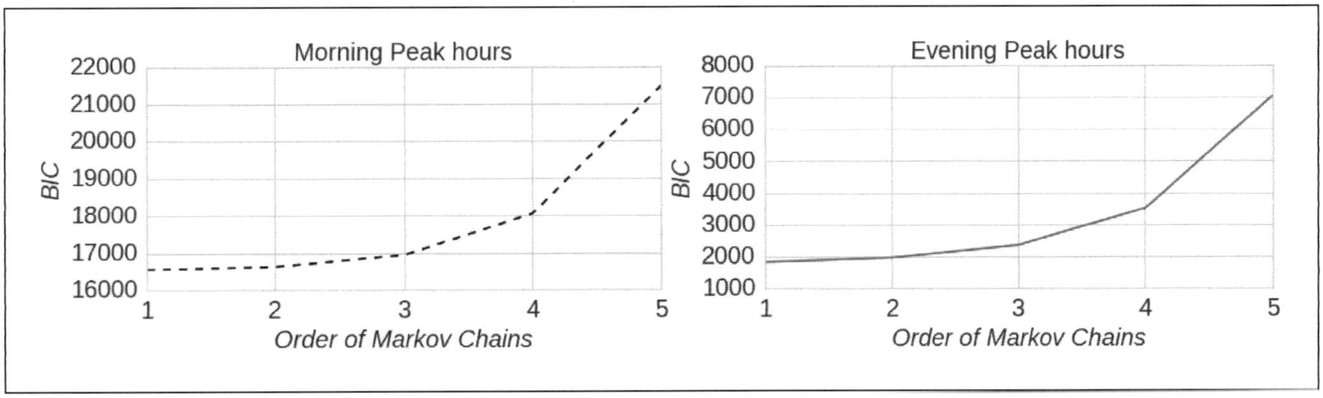

Figure 3. Results of the BIC and Markov chains length of memory.

resulting states are free-flow (0) and congested condition (1). The dynamic evolution of these states over time can be described using the transition probability matrix presented in Equation 4.

$$P_{ij} = \begin{pmatrix} P_{00} & P_{01} \\ P_{10} & P_{11} \end{pmatrix} \quad (4)$$

where,

$$\sum_{j=1}^{2} P_{ij} = 1(sum\ of\ rows = 1)$$

P_{ij} is the probability of transmitting from state *i* to *j*,

P_{00} is the probability of staying in state 0,

P_{01} is the probability of changing from state 0 to 1 (breakdown probability),

P_{10} is the probability of changing from state 1 to 0 (recovery probability),

P_{11} is the probability of staying in state 1.

The first-order Markov chains model in Equation 2*a* implicitly assumes that the evolution process of recurring traffic regimes is constant at any observed time. This type of model is also known as the stationary or time-homogeneous Markov model. In modeling traffic conditions, however, such an assumption may be unrealistic because of several time-varying factors associated with the transition process of the traffic conditions over time. Examples of such factors are driving behavior, vehicle mix, and demand variation, to name a few. It is essential to develop statistical approaches that can capture these time-heterogeneity characteristics. Thus, the transition probability matrix shown in Equation 4 for the first-order Markov model is fitted using a probit regression. The application of the probit regression addresses the time-variant random variation because of the estimated probability changes over time, which is described by the traffic flow variation over time (Equation 5a). Accordingly, the traffic flow in the transition matrix greatly enhances the flexibility of the developed model.

$$P(x) = P(Y_t = 1 | Y_{t-1} = y, X_{t-1} = x) \quad (5a)$$

where,

y is the immediate previous traffic state, either free-flow (0) or congested condition (1), and *x* is the immediate previous explanatory variable, traffic flow.

The transition probability matrix specified in Equation 4 fitted by probit regression is expressed as follows:

$$P_{01} = P(Y_t = 1) | Y_{t-1} = 0, X_{t-1} = x) = \Phi(\beta_0 + \beta X_{i,t-1}) \quad (5b)$$

$$P_{11} = P(Y_t = 1) | Y_{t-1} = 1, X_{t-1} = x) = \Phi(\beta_0 + \beta X_{i,t-1}) \quad (5c)$$

$$P_{00} = 1 - P_{01}\ and\ P_{11} = 1 - P_{10} \quad (5d)$$

where,

$X_{i,t-1}$ represents traffic flow record *i* at the previous time $t - 1$,

β_0 and β are model parameters, and

Φ is the standard normal cumulative function.

The posterior distribution densities of the unknown parameters of the MSR model presented in Equation 5b and 5c were estimated using the MCMC simulations. In particular, No-U-Turn Sampler was used to estimate the regression parameters β_0 and β. The prior distributions for these parameters were assumed to be non-informative with *Normal* (mean = 0, std. = 100).

Results and Discussions

The parameters' distributions estimation for both the BCR and the MSR models was performed using PyMC3 package, an open source probabilistic programming framework implemented in Python (*34*). In the analysis, two chains each with 20,000 simulations (iterations) were used and it was found that 10,000 simulations out of the total were stable and adequate to draw the inference of the posterior distributions in both the BCR and MSR models. The convergence of the simulations in the models was assessed using the autocorrelation, trace, and density plots. Also, the Gelman–Rubin statistic and Markov chain errors were applied in the analysis.

Simulation Analysis and BCR Model Validation

As the actual speed threshold is unknown, to verify the BCR with actual value is somehow difficult. This study used Monte Carlo simulated data as an alternative the actual data to validate the model. The simulations were conducted to examine the robustness of the BCR model estimates against the actual parameters. The parameters which were examined include slopes, intercepts, and change points. Table 1 shows parameters that were used in the simulation study and the results of the BCR model.

According to Table 1, three experiments were conducted with randomly selected parameters. Fitting the simulated data with the BCR resulted in the parameters which are estimated with a minimal error. All model parameters—slope (β), intercept (α), and deviation of the data—are statistically significant at the 95% credible interval. Examining the change point estimates of the regression models, the change point values in all experiments were estimated close to 100% accuracy.

Empirical Results of the BCR Model

The results of the BCR regression model are shown in Figure 4. It can be observed from Figure 4*a* that the

Table 1. Change-Point Regression Results of the Simulated Data

Parameters name	Actual parameters	BCR estimated parameters: Posterior mean	Posterior std.	95% credible interval: Lower	Upper
		Experiment 1 (1, 000 data points)			
Change point	50	50	0	50	50
$\alpha 1$	9.98	9.98	0.07	9.86	10.12
$\alpha 2$	5.08	5.08	0.24	4.63	5.56
$\beta 1$	−0.08	−0.08	0	−0.09	−0.08
$\beta 2$	−0.02	−0.02	0	−0.03	−0.01
$\sigma 1$	0.8	0.8	0.02	0.75	0.85
$\sigma 2$	0.79	0.79	0.03	0.74	0.85
		Experiment 2 (1, 000 data points)			
Change point	20	20.67	0.71	19	22
$\alpha 1$	30	29.95	0.14	29.69	30.2
$\alpha 2$	40	40.09	0.11	39.87	40.29
$\beta 1$	0.6	0.6	0.01	0.58	0.62
$\beta 2$	0.1	0.1	0	0.1	0.1
$\sigma 1$	1	1.03	0.05	0.94	1.12
$\sigma 2$	1	0.99	0.03	0.94	1.04
		Experiment 3 (10, 000 data points)			
Change point	20	20	0	20	20
$\alpha 1$	30	29.91	0.09	29.73	30.07
$\alpha 2$	40	45.04	0.06	44.91	45.16
$\beta 1$	0.6	0.6	0.01	0.59	0.61
$\beta 2$	−0.1	−0.1	0	−0.1	−0.1
$\sigma 1$	2	2.03	0.03	1.98	2.1
$\sigma 2$	2	2.01	0.02	1.98	2.05

speed thresholds during the morning and evening peak periods are 58 mph and 55 mph, respectively. Using the approach applied by Kim et al. (*7*), Dong and Mahmassani (*8*), and Kondyli et al. (*16*) of 10 mph speed drop from the free-flow speed, the estimated threshold for southbound is approximately 65 mph (75 mph–10 mph = 65 mph) and 67 mph for northbound traffic. The free-flow speeds were estimated by computing the 85th percentile of the speeds when the flows are less than or equal to 1,000 vehicles per hour per lane (vphpl). These estimates deviate from the speed thresholds estimated by the BCR model. The advantage of using the BCR method is the fact that the approach accounts for uncertainty associated with traffic data variation.

Moreover, higher uncertainty is explicitly demonstrated by the broadly spread of the posterior predictive lines in Figure 4*a*. Such an observation can be attributed to the higher variability of the data in the congested regime for the evening peak than the morning peak period. On the other hand, the estimated speed thresholds show no significant difference between the morning and evening peak periods.

Apart from the BCR model being validated using the Monte Carlo simulated data, this study also used the hypothesis test to verify whether the trends before and after the speed threshold are credibly different. The estimated model parameters (i.e., slope, intercept, and data variance) were investigated for this purpose. The null hypothesis was assumed to be: parameters of the regression prior to the speed threshold are the same with those of the regression after the threshold. The alternative hypothesis was: parameters in these regression models are credibly different. The hypothesis test was implemented using the MCMC inference samples that created the posterior distributions of the model parameters in the two regimes (congested and free-flow regimes) to assess if they are credibly different at 95% highest density interval (HDI). Previous studies have used this approach in the hypothesis test (*26, 35*). The HDI summarizes the belief uncertainty, such that every value inside the density interval has higher credibility than any other point outside the interval (*35*).

The results of the hypothesis test suggest that there is a statistically significant difference between the regression before and after the speed threshold (Figure 4, *b* and *c*). The 95% HDI is presented by the horizontal line near the *x*-axis. As shown in Figure 4, *b* and *c*, the 95% HDI parameter differences do not cross zero (0); that is, they are either negative or positive. Thus, it is concluded that the variable parameter difference equal to zero (0) is not one of the most credible values in the 95% density interval. Therefore, the parameters between the two regimes

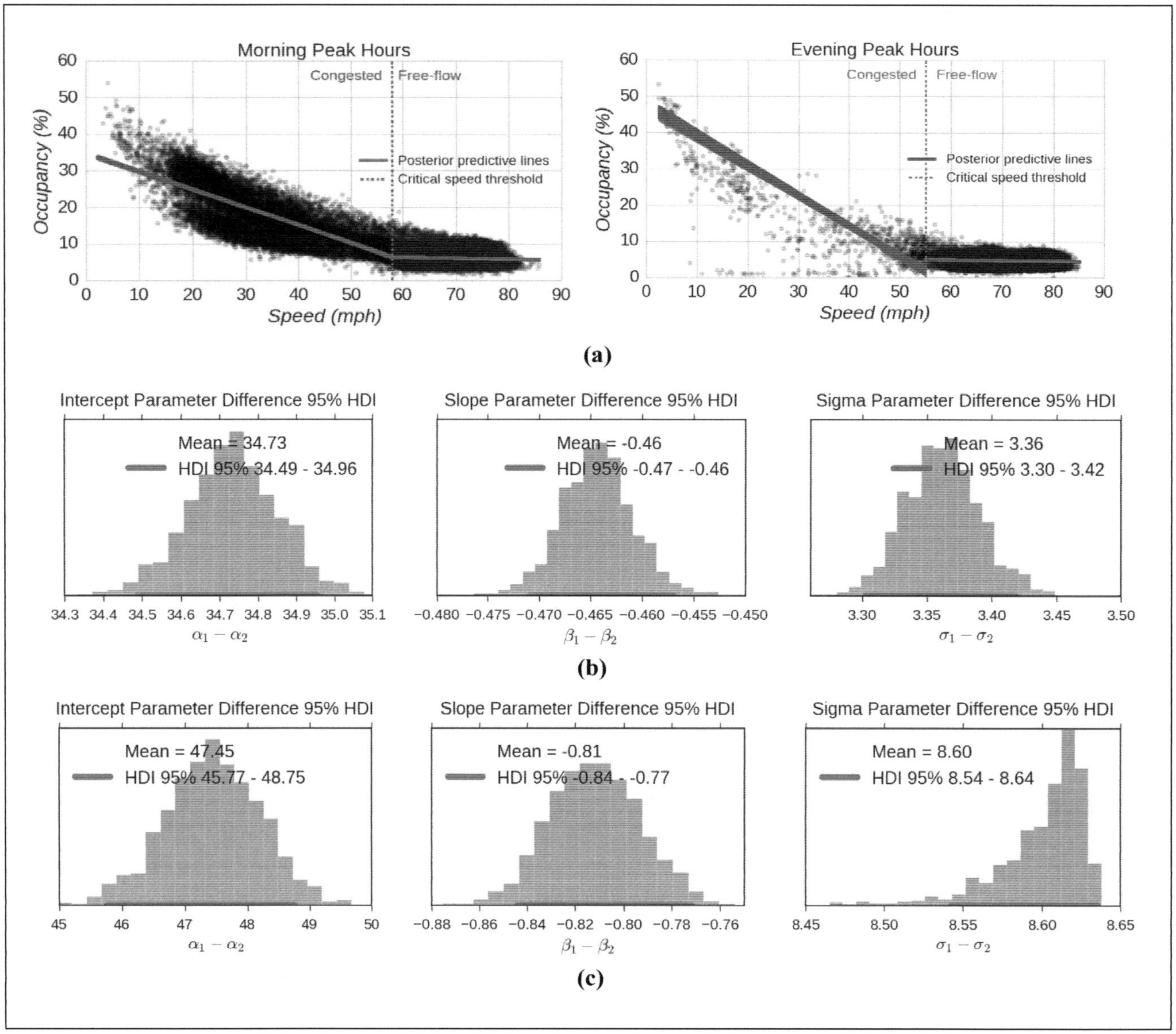

Figure 4. Results of the traffic congestion identification: (*a*) Speed thresholds in the occupancy and speed diagram, (*b*) Parameter posterior distributions differences for the morning peak period, and (*c*) Parameter posterior distributions differences for the evening peak period.

are significantly different and so too the patterns between the two regressions. Based on the results of the hypothesis test, the estimated speed thresholds during both the morning and evening peak periods were considered to be the speed thresholds, which distinguish between the free-flow and congested regimes.

If traffic speed lower than the speed threshold is sustained for at least 15 minutes in the sequential readings, the breakdown was assumed to occur in that interval. That is, if speed in time interval t is higher or at a speed threshold, the next time interval $t + 1$ is below the speed threshold and speed stays below for at least 15 minutes (i.e., until $t + 3$ or more), then the speed at $t + 1$ was considered to be at congested state (*5, 7, 8, 36*). Using these conditions, the traffic speed data were transformed to discrete data and the corresponding flow rate was recorded and utilized in the MSR model for dynamic evolution process analysis.

Results of Random Variation Markov Structured Model

For the morning peak period, the model results indicate that traffic flow is a significant variable in influencing the evolution process of RTC at 95% posterior credible interval. Figure 5, *a* and *b*, shows the pattern how traffic flow influences the stochastic evolution process of traffic

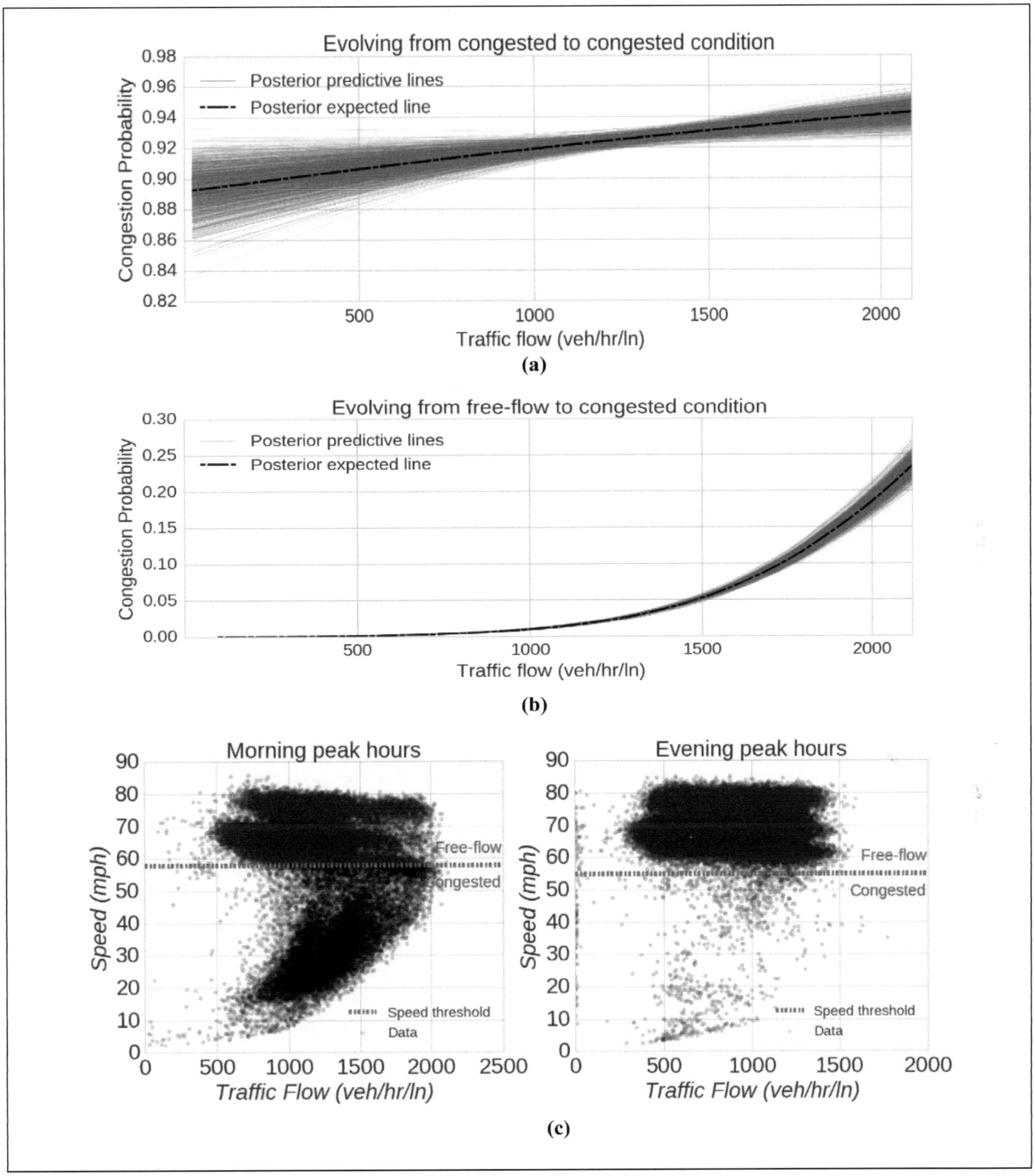

Figure 5. (*a*) Breakdown probability against traffic flow for the morning peak period, (*b*) Transition probability from free-flow to congested state against flow rate for the morning peak period, and (*c*) traffic speed–flow diagram.

congestion for morning peak period. It can be observed that the probabilities of remaining in the congested state in a current observation are higher than those of evolving from the free-flow to the congested regime. This trend is consistent with prior knowledge; once the breakdown occurs, the likelihood of a segment to remain in this state is high, especially when the observation time window is short.

From Figure 5*a*, the transition probability gradual increases with a unit increase in traffic flow. The posterior predictive lines show that at low traffic flow between 0 and 1,000 vphpl and 1,500 and 2,100 vphpl the uncertainty of estimate is high. The uncertainty is demonstrated by the posterior predictive lines in Figure 5*a*. The lines are widely spread from the mean (the posterior expected line). One could say this can be attributed to the higher variation of data in those regions than the area between 1,000 and 1,500 vphpl.

The odds of evolving from the free-flow to the congested state (probability of traffic breakdown) in Figure 5*b* indicates a slightly different pattern compared with the above-described process. The transition probability at low flow rate increases slowly, after 1,500 vphpl there is a sharp increase in breakdown probability with flow rate. Although the majority of the previous studies applied survival models to estimate the breakdown probability, the pattern estimated in Figure 5*b* is somewhat consistent with previous findings that the breakdown probability increases monotonically with the traffic flow (*7, 36*).

On the other hand, during the evening peak period, transitioning from the free-flow to the congested condition revealed a counterintuitive pattern—that is, increase in traffic flow reduces the likelihood of breakdown. The estimate was not statistically significant. Similar findings were estimated on the probability of traffic state to remain in the congested state given that the previous state was congested, suggesting that once congestion was observed in the previous state, the likelihood of being congested now decreases with a unit increase in traffic flow. These results may be associated with the fact that the evening peak period is not congested (see Figures 1*b* and 2), such that few sample data that were highly random and scattered were observed below the speed threshold estimated by the BCR model. Figure 5*c* illustrates the noisy data characteristics for the evening peak period compared with the morning peak period data. Based on this observation, the MSR model failed to detect the actual pattern of RTC during the evening peak period.

Conclusion

This study developed the stochastic model built on the Markov chains process assumption to describe the dynamic transition of RTC. The traffic flow parameters of speed, flow, and occupancy were used. The analyses were performed on historical traffic data collected from a 4.8-miles basic freeway segment. These data were aggregated on a 5-min basis for a 1-year period. To identify discrete traffic states—that is, free-flow and congested regimes—the study applied a statistical approach based on BCR. The resulting traffic regimes were then evaluated using the MSR model to describe their dynamic evolution process over time. Before developing the Markov chains model, the appropriate memory length was examined using the BIC approach.

The statistical approach used to estimate the speed threshold and that of describing the transition process between traffic regimes demonstrated their applicability to explain RTC occurrence over time. More specifically, the BCR approach used to estimate the speed threshold revealed promising estimation accuracy when verified with simulated data. The findings from the BCR suggest that congested condition on the analyzed segment occurs once speed is below 58 mph during the morning peak period, whereas during the evening peak this occurs at below 55 mph. These results were also verified through a hypothesis test, which confirmed that the pattern before and after these speed thresholds are credibly different at 95% HDI during both peak periods.

In describing the dynamic transition of recurring traffic regimes, the odds of evolving from the free-flow to the congested state (probability of traffic breakdown) and those of staying in the congested state given that the previous traffic state was congested increase monotonically with a unit increase in the traffic flow rate. These findings were observed during the morning peak period. On the other hand, the evening peak period does not get congested as much, resulting in high data variability, especially in the region below the estimated speed threshold. The MSR revealed counterintuitive results and the traffic flow was insignificant in explaining the dynamic transition process of the recurring traffic condition.

The proposed stochastic methodological framework developed by this study can be a useful resource in detecting different traffic conditions, including RTC using the speed threshold concept. The speed threshold was estimated using a statistical approach (BCR) which accounts for the noisy relationship between speed and occupancy. Using the BCR approach, uncertainty associated with the randomness of data is taken into consideration in the threshold estimate, instead of the visual estimation approach. Furthermore, the MSR model used to estimate the dynamic evolution characteristics of RTC enhance the understanding of the occurrence of congestion over time. Based on that, it can facilitate planning actions in implementing congestion-related countermeasures in urban areas.

This study assumed that the speed–occupancy relationship is linear during the BCR modeling process. It is not clear if the non-linear change point regression model will improve significantly the estimates of the parameters based on the goodness of fit statistics. Additional work is needed to evaluate the non-linear assumption and compare it with the linear assumption presented here to detect speed threshold. It will also be beneficial to test the proposed approaches to determine if similar results might be obtained in different locations and different types of bottlenecks. Furthermore, the BCR model can be modified to detect more than one change point to account for more than two traffic states, such as congested, free-flow, congestion onset/dissipation. In this case, multinomial regression analysis will be the appropriate model in the Markov structure regression for analyzing the dynamic characteristics of traffic regimes. Also, future studies can be designed to evaluate the influence of spatial location, vehicle mix, weather, and driving characteristics on the dynamic evolution of RTC.

Author Contributions

The authors confirm contribution to the paper as follows: study conception and design: Emmanuel Kidando; data collection: Emmanuel Kidando; analysis and interpretation of results: Emmanuel Kidando, Ren Moses; draft manuscript preparation: Emmanuel Kidando, Ren Moses, Thobias Sando, and Eren E. Ozguven. All authors reviewed the results and approved the final version of the manuscript.

References

1. Hojati, A. T. *Modelling the Impact of Traffic Incidents on Travel Time Reliability*. Dissertation. The University of Queensland, 2014.
2. Sofer, T., A. Polus, and S. Bekhor. A New Comprehensive, Congestion Measure for Freeways. *Advances in Transportation Studies*, Vol. 10, No. 29, 2012, pp. 45–58.
3. Bassan, S., A. Polus, and A. Faghri. Modeling of Freeway Breakdown Process with Log-Periodic Oscillations. *Canadian Journal of Civil Engineering*, Vol. 34, No. 12, 2007, pp. 1577–1586.
4. Shojaat, S., J. Geistefeldt, S. A. Parr, C. G. Wilmot, and B. Wolshon. Sustained Flow Index Stochastic Measure of Freeway Performance. *Transportation Research Record: Journal of the Transportation Research Board*, 2016. 2554: 158–165.
5. Hong, Z., H. S. Mahmassani, and Y. Chen. Empirical Analysis of Freeway Flow Breakdown and Recovery: The Effect of Snow Weather. Presented at 94th Annual Meeting of the Transportation Research Board, Washington, D.C., 2015.
6. Brilon, W., J. Geistefeldt, and M. Regler. Reliability of Freeway Traffic Flow: A Stochastic Concept of Capacity. *Proc., 16th International Symposium on Transportation and Traffic Theory*, College Park, Md., Elsevier, 2005, pp. 125–144.
7. Kim, J., H. S. Mahmassani, and J. Dong. Likelihood and Duration of Flow Breakdown: Modeling the Effect of Weather. *Transportation Research Record: Journal of the Transportation Research Board*, 2010. 2188: 19–28.
8. Dong, J., and H. S. Mahmassani. Predicting Flow Breakdown Probability and Duration in Stochastic Network Models: Impact on Travel Time Reliability. *Transportation Research Record: Journal of the Transportation Research Board*, 2009. 2124: 203–212.
9. Xu, T. D., Y. Hao, Z. R. Peng, and L. J. Sun. Modeling Probabilistic Traffic Breakdown on Congested Freeway Flow. *Canadian Journal of Civil Engineering*, Vol. 40, No. 10, 2013, pp. 999–1008.
10. Shiomi, Y., T. Yoshii, and R. Kitamura. Platoon-Based Traffic Flow Model for Estimating Breakdown Probability at Single-Lane Expressway Bottlenecks. *Procedia - Social and Behavioral Sciences*, Vol. 17, 2011, pp. 591–610.
11. Kondyli, A., B. George, L. Elefteriadou, and G. Bonyani. Defining, Measuring and Estimating Freeway Capacity. *Journal of Transportation Engineering, Part A: Systems*, Vol. 143, No. 3, 2017, pp. 1–12.
12. Lorenz, M., and L. Elefteriadou. A Probabilistic Approach to Defining Freeway Capacity and Breakdown. *Proc., 4th International Symposium on Highway Capacity,* TRB-Circular E-C018, Transportation Research Board, Washington, D.C., 2000, pp. 84–95.
13. Persaud, B., S. Yagar, and R. Brownlee. Exploration of the Breakdown Phenomenon in Freeway Traffic. *Transportation Research Record: Journal of the Transportation Research Board*, 1998. 1643: 64–69.
14. Minderhoud, M., H. Botma, and P. Bovy. Assessment of Roadway Capacity Estimation Methods. *Transportation Research Record: Journal of the Transportation Research Board*, 1997. 1572: 59–67.
15. Dehman, A. Breakdown Maturity Phenomenon at Wisconsin Freeway Bottlenecks. *Transportation Research Record: Journal of the Transportation Research Board*, 2013. 2395: 1–11.
16. Kondyli, A., L. Elefteriadou, W. Brilon, F. L. Hall, B. Persaud, and S. Washburn. Development and Evaluation of Methods for Constructing Breakdown Probability Models. *Journal of Transportation Engineering*, Vol. 139, No. 9, 2013, pp. 931–940.
17. Laflamme, E. M., and P. J. Ossenbruggen. Effect of Time-of-Day and Day-of-the-Week on Congestion Duration and Breakdown: A Case Study at a Bottleneck in Salem, NH. *Journal of Traffic and Transportation Engineering (English Edition)*, Vol. 4, No. 1, 2017, pp. 31–40.
18. Yeon, J., S. Hernandez, and L. Elefteriadou. Differences in Freeway Capacity by Day of the Week, Time of Day, and Segment Type. *Journal of Transportation Engineering*, Vol. 135, No. 7, 2009, pp. 416–426.
19. Shawky, M., and H. Nakamura. Characteristics of Breakdown Phenomenon in Merging Sections of Urban Expressways in Japan. *Transportation Research Record: Journal of the Transportation Research Board*, 2012. 2012: 11–19.
20. Wu, N. A Stochastic Model for Reliability Analysis in Freeway Networks. *Procedia-Social and Behavioral Sciences*, Vol. 96, 2013, pp. 2823–2834.
21. Geistefeldt, J., and W. Brilon. Comparative Assessment of Stochastic Capacity Estimation Methods. *Proc., 18th International Symposium on Transportation and Traffic Theory*, London, U.K. Springer, Boston, Mass., 2009, pp. 583–602.
22. Khodadadi, A., and M. Asgharian. Change-point problem and Regression: An Annotated Bibliography. *COBRA*

Preprint Series, Working Paper 44, 2008. http://biostats.bepress.com/cobra/art44.
23. Kulasekera, K. Comparison of Regression Curves Using Quasi-Residuals. *Journal of the American Statistical Association*, Vol. 90, No. 431, 1995, pp. 1085–1093.
24. Walker, E., and S. P. Wright. Comparing Curves Using Additive Models. *Journal of Quality Technology*, Vol. 34, No. 1, 2002, pp. 118–129.
25. Yang, M. H., T. T. Luong, and W. Recker. Extracting Traffic Patterns from Loop Detector Data using Multiple Changepoints Detection. Presented at 93rd Annual Meeting of the Transportation Research Board, Washington, D.C., 2014.
26. Bhagat, A., J. D. Saphores, and R. Jayakrishnan. Detecting Changes in Accident Rates using a Hierarchical Bayesian Approach: An Application to the I-710 and the Implementation of the PierPASS Program. Presented at 96th Annual Meeting of the Transportation Research Board, Washington, D.C., 2017.
27. Kidando, E., R. Moses, E. E. Ozguven, and T. Sando. Evaluating Traffic Congestion Using the Traffic Occupancy and Speed Distribution Relationship: An Application of Bayesian Dirichlet Process Mixtures of Generalized Linear Model. *Journal of Transportation Technologies*, Vol. 7, No. 3, 2017, p. 318.
28. Shamshad, A., M. Bawadi, W. W. Hussin, T. Majid, and S. Sanusi. First and Second Order Markov Chain Models for Synthetic Generation of Wind Speed Time Series. *Energy*, Vol. 30, No. 5, 2005, pp. 693–708.
29. Kwon, J., and K. Murphy. *Modeling Freeway Traffic with Coupled HMMs*. Technical Report. University of California at Berkeley, 2000.
30. Qi, Y., and S. Ishak. A Hidden Markov Model for Short Term Prediction of Traffic Conditions on Freeways. *Transportation Research Part C: Emerging Technologies*, Vol. 43, No. 1, 2014, pp. 95–111.
31. Guo, F., and Q. Li. Multi-state Travel Time Reliability Models with Skewed Component Distributions. *Transportation Research Record: Journal of the Transportation Research Board*, 2011. 2315: 47–53.
32. Kang, S. *Bayesian Change-Point Analysis in Linear Regression Model with Scale Mixtures of Normal Distributions*. Master's Thesis. Michigan Technological University, 2015.
33. Singer, P., D. Helic, B. Taraghi, and M. Strohmaier. Detecting Memory and Structure in Human Navigation Patterns Using Markov Chain Models of Varying Order. *PLos One*, Vol. 9, No. 12, 2014, e114952. https://doi.org/10.1371/journal.pone.0114952.
34. Salvatier, J., T. V. Wiecki, and C. Fonnesbeck. Probabilistic Programming in Python Using PyMC3. *PeerJ Computer Science*, Vol. 2, 2016, e55. https://doi.org/10.7717/peerj-cs.55.
35. Kruschke, J. K., and T. M. Liddell. The Bayesian New Statistics: Hypothesis Testing, Estimation, Meta-analysis, and Power Analysis from a Bayesian Perspective. *Psychonomic Bulletin & Review*, Vol. 25, No. 1, 2018, pp. 178–206.
36. Xie, K., K. Ozbay, and H. Yang. *Transportation Research Circular E-C197: The Heterogeneity of Capacity Distributions among Different Freeway Lanes* Transportation Research Board of the National Academies, Washington, D.C., 2014.

The Standing Committee on Traffic Flow Theory and Characteristics (AHB45) peer-reviewed this paper (18-06081).

TRR
JOURNAL OF THE TRANSPORTATION RESEARCH BOARD

Article

Transportation Research Record
2018, Vol. 2672(20) 75–85

DOI: 10.1177/0361198118778927
journals.sagepub.com/home/trr
SAGE

Investigating Transfer Flow between Urban Networks Based on a Macroscopic Fundamental Diagram

Sunghoon Kim[1], Sehyun Tak[2], and Hwasoo Yeo[1]

Abstract

The aim of the concept of perimeter control is to manage the transfer flow between urban road networks. However, there is a lack of description on the behavior of transfer flow between networks in relation to the macroscopic fundamental diagram (MFD). Hence, this paper suggests a description of transfer flow and investigates it with microscopic simulation experiments. The results found that the network outbound demand strongly influences the outflow regardless of spatial inhomogeneity. Another finding is that the boundary capacity influences the network outflow. The effect of restriction on outflow of a network due to limited supply level of a neighboring network was also checked. Furthermore, it was found that a network's inflow and outflow in each direction (north, south, east, and west) is proportional to the demand ratio. Finally, the suggested description of transfer flow has been confirmed, and this can be useful for various analyses on inter-network traffic dynamics.

The field of urban traffic control has been studied for several decades, and the perimeter control concept is one of the strategies recently developed for network-wide signal control solutions. The aim of perimeter control is to control the traffic demand for urban networks prior to managing the flow inside networks. In other words, the focus of the concept is to control the transfer flow between urban networks.

This concept has been developed based on a macroscopic fundamental diagram (MFD), which represents the relationship between aggregated network flow and density. The idea of utilizing macroscopic behavior was introduced by Daganzo (*1*), and since the noteworthy investigations on the relationship based on simulation studies (*2*) and real-world data (*3*) there has been growing interest in the network-wide properties of traffic. Some subsequent studies discussed the applicability of the aggregated relationship due to the effect of the network's spatial homogeneity on the shape of the MFD (*4–8*), but so far the aggregated relationship is considered to be recognizable by the research community. Studies on the hysteretic behavior of the aggregated relationship during traffic loading and unloading phases (*9–12*) and on network vulnerability in major incident cases (*13*, *14*) are some other branches of MFD studies. Another subsequent branch is that of perimeter control (*15–23*), and various strategies have been devised for different control purposes. These are: the single-protected network gating problem (*19–21*), the gating problem for multiple boundaries (*22*), the problem of one-on-one network transfer flow (*16*), the problem of transfer among multiple networks (*15*, *23*), and the mixed traffic problem (*17*). These strategies adopted control engineering-based methods such as proportional-integral control and model-predicted control.

The MFD enables the establishment of model-based signal control strategies. Hence, the utilization of this elegant concept has been the basis of perimeter control approaches. Uncertainty due to MFD scatter has always been an issue when utilizing the model, but many efforts have been made in order to reduce the effect of scatter (*16*, *18*, *24–26*). In particular, Gao and Gayah (*25*) and Mariotte et al. (*26*) are the most recent examples. Based on several considerations of the MFD, the previous studies focused on making decisions as to how many vehicles should be transferred between networks at each time step. However, when they make such decisions, there is a

[1]Korea Advanced Institute of Science and Technology, Yuseong-gu, Daejeon, Republic of Korea
[2]The Korea Transport Institute, Sejong-si, Republic of Korea

Corresponding Author:
Address correspondence to Hwasoo Yeo: hwasoo@kaist.ac.kr

lack of description on how the transfer flow behaves in relation to a network's MFD.

There are at least two major factors to be considered for transfer flow: travel demand pattern and supply level at network boundaries (boundary capacity). First, as Geroliminis and Sun (*6*) mentioned, the MFD is not universal. Since it is the representation of the given capability of a network based on infrastructural conditions (e.g., road length, network geometry, and intersection signal timings), it can be considered a fixed property as long as such conditions do not change. However, network outflow may be influenced by demand patterns, such as the proportions of demand toward each side of a network and the rate of outbound demand over total demand. Note that the term "outflow" means the flow leaving a network to neighboring networks and it excludes the flow entering the local area of the network. For instance, considering the multidirectional flow as in Figure 1*d*, there would be the case that the travel demand within a network is concentrated in one direction. In this case, the vehicles would not fully utilize the infrastructure and the outflow may not reach the maximum capability of the network. Furthermore, if a network has a role as the major traffic sink, the outbound demand would be relatively low. This may also cause reduced outflow compared with the given MFD capability. Thus, the reduced outflow effect should be examined for describing the transfer flow. In fact, there were a few noteworthy studies that investigated the effect of traffic pattern on the shape of the MFD (*27*, *28*), but these works were done under a single network perspective rather than under the view of transfer flow between two networks.

The second factor is the boundary capacity, which represents the maximum transfer rate between networks. Some of the previous studies (*16–18*) applied the maximum and minimum transfer flow rates as the constraints of the optimization process, but such constraints were not fully described in relation to the infrastructural

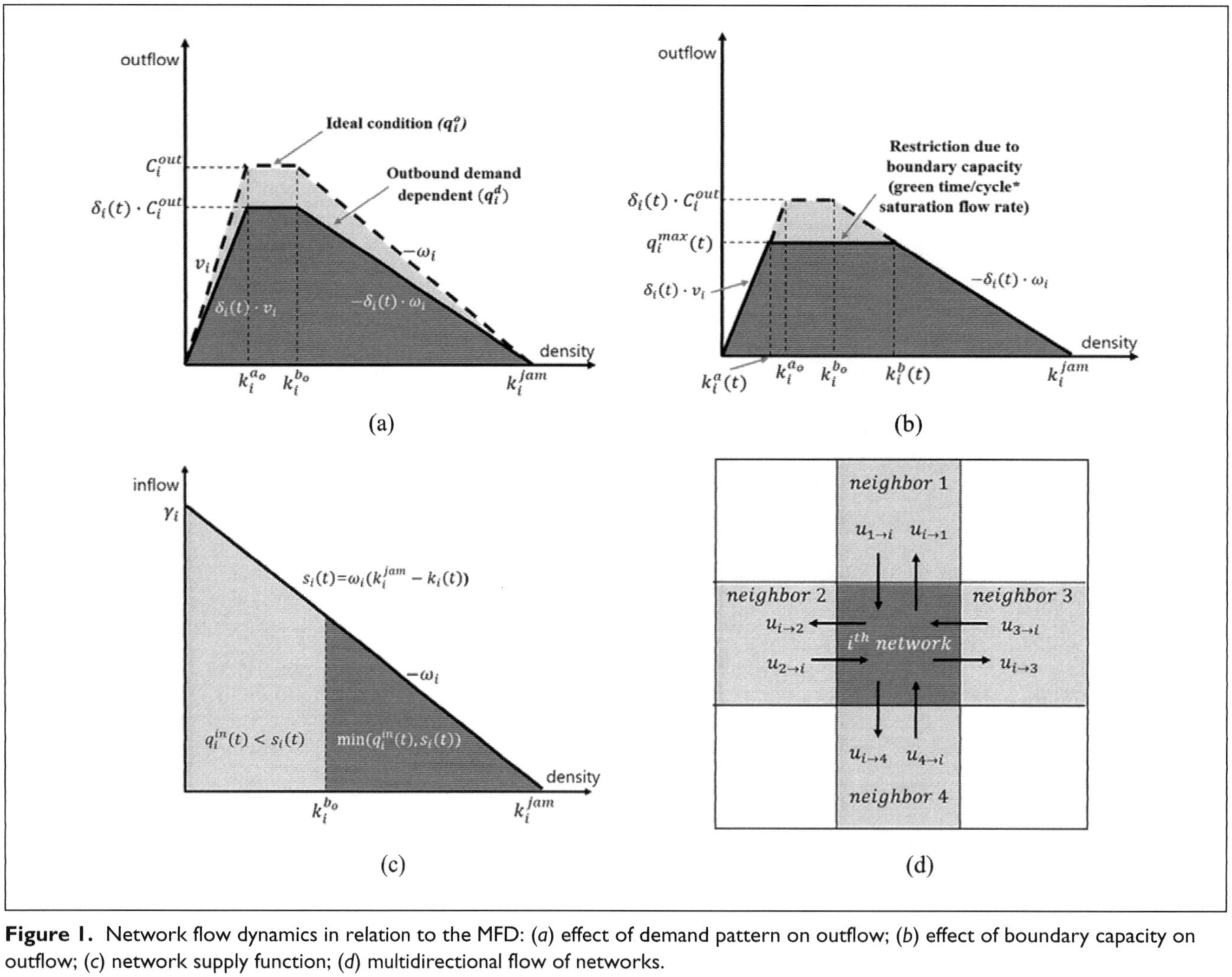

Figure 1. Network flow dynamics in relation to the MFD: (*a*) effect of demand pattern on outflow; (*b*) effect of boundary capacity on outflow; (*c*) network supply function; (*d*) multidirectional flow of networks.

conditions at the intersections of network boundaries. The boundary capacity should be described based on properties like the number of connected links, number of lanes, saturation flow rate, green time, and cycle length. Then, the effect of boundary capacity on outflow should be examined for describing the transfer flow. In addition, even though a network's outflow toward a neighboring network is decided, if the supply level (e.g., remaining vehicle spaces) of the neighboring network is low, then the transfer flow would be restricted. Hence, such effect should be examined as well for describing the transfer flow behavior.

Therefore, the behavior of the transfer flow between networks should be investigated based on several MFD-related aspects. The aim of this paper is to investigate the effect of demand pattern on network outflow. The effect of boundary capacity on outflow is also investigated, as well as the effect of restriction on transfer flow due to a network's supply level. Such investigations will lead us to better understanding of transfer flow between networks.

The rest of this paper is organized as follows. The second section provides the description of how demand and boundary capacity would influence network outflow and transfer flow. The third section presents the investigation into the effect of demand and boundary capacity with microscopic simulation experiments. The fourth section presents the effect of supply level of a neighboring network. The fifth section concludes this paper with a few considerations for further works.

Underlying Dynamics

In the following subsections, how the transfer flow can be determined is described in the relation to the MFD concept. Table 1 shows the MFD-related properties, and Figure 1 provides graphical descriptions of network outflow and supply.

Table 1. MFD-Related Properties

Denotation	Meaning
C_i^{out}	Outflow capacity of network *i*
$\delta_i(t)$	Ratio of outbound demand over total demand within network *i*
$q_i^{\max}(t)$	Maximum outflow affected by outbound demand and boundary capacity
v_i	Increase rate of outflow under unsaturated condition
ω_i	Decrease rate of outflow under oversaturated condition
k_i^{jam}	Jam density
$k_i^{a_o}$	Ideal value of lower bound of critical density
$k_i^{b_o}$	Ideal value of upper bound of critical density
$k_i^a(t)$	Situational value of lower bound of critical density
$k_i^b(t)$	Situational upper bound of critical density
γ_i	Intercept of decrease rate of outflow

Effect of Demand Pattern on Network Outflow

Network outflow may be influenced by demand pattern. Let us consider a network that has a given MFD. Then, let us assume the ideal condition where the amount of outbound demand is the same as the total travel demand (outbound demand 100%). In this condition, the outflow would be the same as the network's maximum output, and the outflow is defined as follows:

$$q_i^o(t) = \begin{cases} v_i k_i(t) & \textit{for } k_i(t) < k_i^{a_o} \\ C_i^{\text{out}} & \textit{for } k_i^{a_o} \le k_i(t) \le k_i^{b_o} \\ \gamma_i - \omega_i k_i(t) & \textit{for } k_i^{b_o} < k_i(t) \le k_i^{\text{jam}} \end{cases} \tag{1}$$

where v_i is the increase rate of outflow under unsaturated condition, C_i^{out} is the maximum outflow based on the MFD, ω_i is the decrease rate of outflow under oversaturated condition, γ_i is the intercept of the decrease rate of outflow, $k_i^{a_o}$ is the lower bound of critical density, $k_i^{b_o}$ is the upper bound of critical density, and k_i^{jam} is the jam density of network *i*.

Now, if we assume that the outbound demand is less than 100% of the total demand, the network would have to process both the outbound demand and the local area demand. Thus, the network's capability would be utilized partially for processing the outbound traffic and partially for processing the traffic toward the local area. This partial fraction is denoted with δ, and is defined as follows:

$$\delta_i(t) = \frac{\sum_{m_i}^{M_i} q_{i\to m_i}^{\text{vod}}(t)}{\sum_{m_i}^{M_i} q_{i\to m_i}^{\text{vod}}(t) + q_{i\to x_i}^{\text{vod}}(t)} \tag{2}$$

where m_i is the *m*-th neighboring network *i*, and the local area of network *i* is denoted with x_i. Vehicles on demand (VOD) is denoted with q^{vod}, and it represents vehicle demands (number of vehicles) to travel in each direction. The denominator is the total travel demand and the numerator is the outbound demand. Then, the outflow that is influenced by demand pattern is

$$q_i^d(t) = \delta_i(t) \cdot q_i^o(t). \tag{3}$$

Figure 1*a* shows the effect of demand pattern on the outflow. There are no changes in the lower and upper bound of critical density due to demand pattern, based on the linear relationship between production and outflow (*2*).

Effect of Boundary Capacity on Network Outflow

Network outflow is also influenced by boundary capacity. A few studies have dealt with the boundary capacity previously (*23*, *29*), and their common concept is that the transfer flow is the minimum value between the transfer demand and boundary capacity. However, in these studies, the maximum value of boundary capacity is considered as a constant property, even though it may be changed over time due to the split of green signal at each of the road links connecting two regions. Hence, a new way of describing the boundary capacity is necessary, and it can be defined in relation to the infrastructural condition at the intersections of network boundaries as follows:

$$C_{i\to m_i}(t) = \sum_h g^h_{i\to m_i} \cdot l_h(t) \tag{4}$$

where h is a link connecting network i and its neighbor, l_h is the number of lanes on h, and $g^h_{i\to m_i}$ is the product of green time over cycle length and saturation flow rate. The value of saturation flow rate is considered to be 1,800 veh/h/lane as in general. Then, network outflow influenced by both the boundary capacity and demand pattern is

$$q_i(t) = \begin{cases} q^d_i(t) & \text{for } k_i(t) < k^a_i(t) \text{ and } k^b_i(t) < k_i(t) \le k^{\text{jam}}_i \\ q^{\max}_i(t) & \text{for } k^a_i(t) \le k_i(t) \le k^b_i(t) \end{cases} \tag{5}$$

where

$$q^{\max}_i(t) = \max\left(\sum_{m_i}^{M_i} \min\left(q^{\text{vod}}_{i\to m_i}(t), C_{i\to m_i}(t) + \varepsilon_{i\to m_i}(t)\right)\right). \tag{6}$$

Figure 1*b* graphically shows such a description. The term $\varepsilon_{i\to m_i}(t)$ is the error boundary capacity, and this value may usually depend on the turning traffic ratio at intersections and its effect should be investigated in further works. From the above equation, if the outbound demand were less than boundary capacity, then the maximum outflow would follow the demand pattern. On the other hand, if the boundary capacity were less than outbound demand or the capacity of the MFD, the maximum outflow would be reduced. With such restriction, there would be changes in the lower and upper bounds of critical density, as follows:

$$k^a_i(t) = \begin{cases} k^{a_o}_i & \text{for } q^{\max}_i(t) = \delta_i(t) \cdot C^{\text{out}}_i \\ q^{\max}_i(t)/(\delta_i(t)\cdot v_i) & \text{for } q^{\max}_i(t) < \delta_i(t) \cdot C^{\text{out}}_i \end{cases} \tag{7}$$

$$k^b_i(t) = \begin{cases} k^{b_o}_i & \text{for } q^{\max}_i(t) = \delta_i(t) \cdot C^{\text{out}}_i \\ k^{\text{jam}}_i - q^{\max}_i(t)/(\delta_i(t)\cdot \omega_i) & \text{for } q^{\max}_i(t) < \delta_i(t) \cdot C^{\text{out}}_i \end{cases} \tag{8}$$

Network Supply and Transfer

The number of vehicles that can enter a network is determined based on the remaining space in the network, and its function can be given by the MFD as in Figure 1*c*. As in Geroliminis and Daganzo (*2*), if the network density is greater than the upper bound of critical density, then the network inflow decreases due to the supply function, which is defined as

$$s_i(t) = \omega_i\left(k^{\text{jam}}_i - k_i(t)\right). \tag{9}$$

On the other hand, if the network density is less than the upper bound, instead of following the supply function, the network inflow would be the same as the inbound traffic from neighbors because it is lower than the supply function. With this logic, the total network inflow is the minimum value between the outflow from neighbors and the supply level of the subject network. Here, we assume that a network outflow is proportionally distributed by the VOD toward each neighboring network. The remaining space (supply) of a network is also considered to be proportionally distributed by the amount of VOD from each neighbor. The transfer flow from network i to one of its neighboring networks, y, can then be defined as follows:

$$u_{i\to y}(t) = \min\left(q_i(t)\cdot\frac{q^{\text{vod}}_{i\to y}(t)}{\sum^{M_i}_{m_i} q^{\text{vod}}_{i\to m_i}(t)}, s_y(t)\cdot\frac{q^{\text{vod}}_{i\to y}(t)}{\sum^{M_y}_{m_y} q^{\text{vod}}_{m_y\to y}(t) + q^{\text{vod}}_{x_y\to y}(t)}\right) \tag{10}$$

Experiments on Single Network

Simulation Settings

To test the effect of demand pattern on the network outflow, microscopic simulation experiments were conducted. The AIMSUN program (*30*) was used for simulation experiments, and the network setting is shown in Figure 2. There is a 4×4 grid-shaped road network. The length of each road link is 500 meters and each link has two lanes. All links are bi-directional. The free flow speed is 50 km/h for all roads. The green time for each direction (for straight movement) at all intersections is set at 40 seconds hence the cycle length is 120 seconds. The centroids generate and attract the traffic, and there are 16 centroids around the network (4 in each of the cardinal directions) and eight centroids inside the network. The centroids located inside the network are considered as the local area (local streets or parking spaces). The simulation runtime is 300 minutes and the traffic data (link flow and density) are collected at every 5 minutes. The origin-destination (OD) of the traffic is given

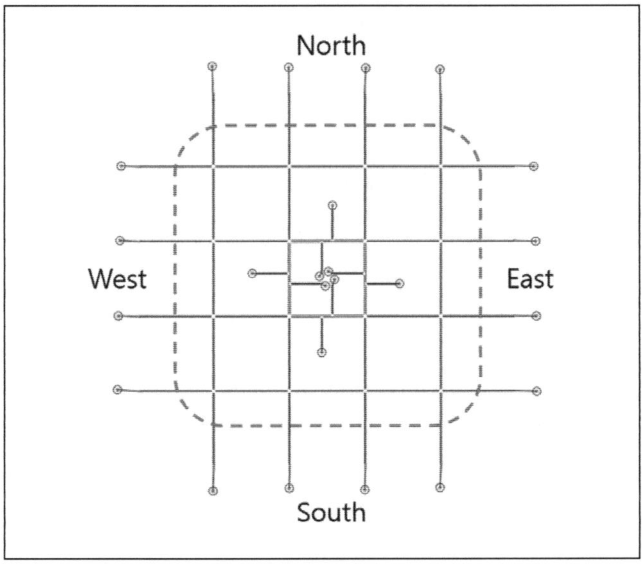

Figure 2. Single network setting in AIMSUN.

at every hour, and the volume for the set of 1-hour periods was:4,000, 10,000, 16,000, 22,000, 22,000. For homogenously loaded traffic, the traffic is evenly distributed by each OD pair so the demand proportion toward each direction is the same. The logit model that is implemented in the simulation program was used for route assignment (*30*).

Effect of Outbound Demand Level

This study ran five cases with different levels of outbound demand (δ value). The tested outbound demand levels were 100%, 90%, 80%, 70%, 60%. Figure 3 shows a comparison of the results of the five cases. As shown in Figure 3*a*, the cases show reduced outflow with respect to the outbound demand level. As the outbound demand level decreases, the reduction effect of the outflow increases. We can also see that the increase rate of the outflow is reduced by the outbound demand level. There are differences in the increase and decrease rates of the outflow of the five cases. However, Figure 3*b* shows that the spatial inhomogeneity (standard deviation of average density) for all cases is similar. This is clearly a new finding, that the outflow is strongly affected by the outbound demand within the network, regardless of the spatial inhomogeneity. Such results provide new information that the network outflow is influenced not only by the spatial condition, but also by the ratio between the outbound demand and the local area demand. Hence, the demand ratio should also be taken into account when modeling the outflow reduction. Furthermore, the δ values follow the outbound demand level, as shown in Figure 3*c*. The consistency of the values of all cases collapses soon after 240 minutes of the simulation because the network gets into a gridlock situation, but overall the results show that Equation 2 holds. Figure 3*d* shows the network production over outflow. Except for the gridlock situation, the linear relationship between the two properties is shown for all cases, but only the linearity value varies by the reduced amount of outflow.

Figure 4 shows the comparison between the cases of outbound demand level 100% and 60%. In Figure 4*a* and *b*, the black solid lines are the outbound demand in each direction. For both cases, the outflow in each direction follows the demand line during unsaturated condition. During oversaturated traffic, they both decrease far from the demand lines. In the outbound 100% case, as the network goes into saturated traffic (between 190 and 220 min), the outflow values do not reach the demand line due to the boundary capacity. With the given simulation settings and Equation 4, the estimated boundary capacity in each direction is 4,800 veh/h. Hence, the maximum outflow in this case does not reach the demand line. On the other hand, in the case of outbound demand level 60%, the outflow values reach the demand line even in saturated traffic (between 140 and 190 min) because the outbound level is lower than the boundary capacity. These results show that Equation 6 holds.

Effect of Boundary Capacity

To test the effect of boundary capacity on network outflow, other simulation cases were run with the same traffic settings except the boundary capacity in the eastbound direction. The default green time values at the eastbound intersections are 40 seconds, and the tested green time values are 30 and 20 seconds at 100% of outbound level. Figure 5*a* shows the outflow diagram of the reduced boundary capacity cases. We can see that the outflow is restricted due to the reduced boundary capacity. Unlike the results in Figure 3*a*, which shows reduced slopes during unsaturated traffic, Figure 5*a* shows no change in the slope due to the reduced boundary capacity. Only the maximum outflow is restricted. Hence, in these cases, the lower and upper bounds of critical density are changed due to such restriction, as suggested by Equations 7 and 8. In Figure 5*b*, the restriction effect of 30 seconds of green time is shown between 190 and 220 min, and the effect of the other case is shown between 140 and 190 min. For all cases, there are influences of boundary capacity on the outflow in each direction, particularly during the saturated traffic.

Heterogeneity of Demand for Each Direction

A sensitivity analysis was carried out to see the effect of different demand patterns of each direction. This is

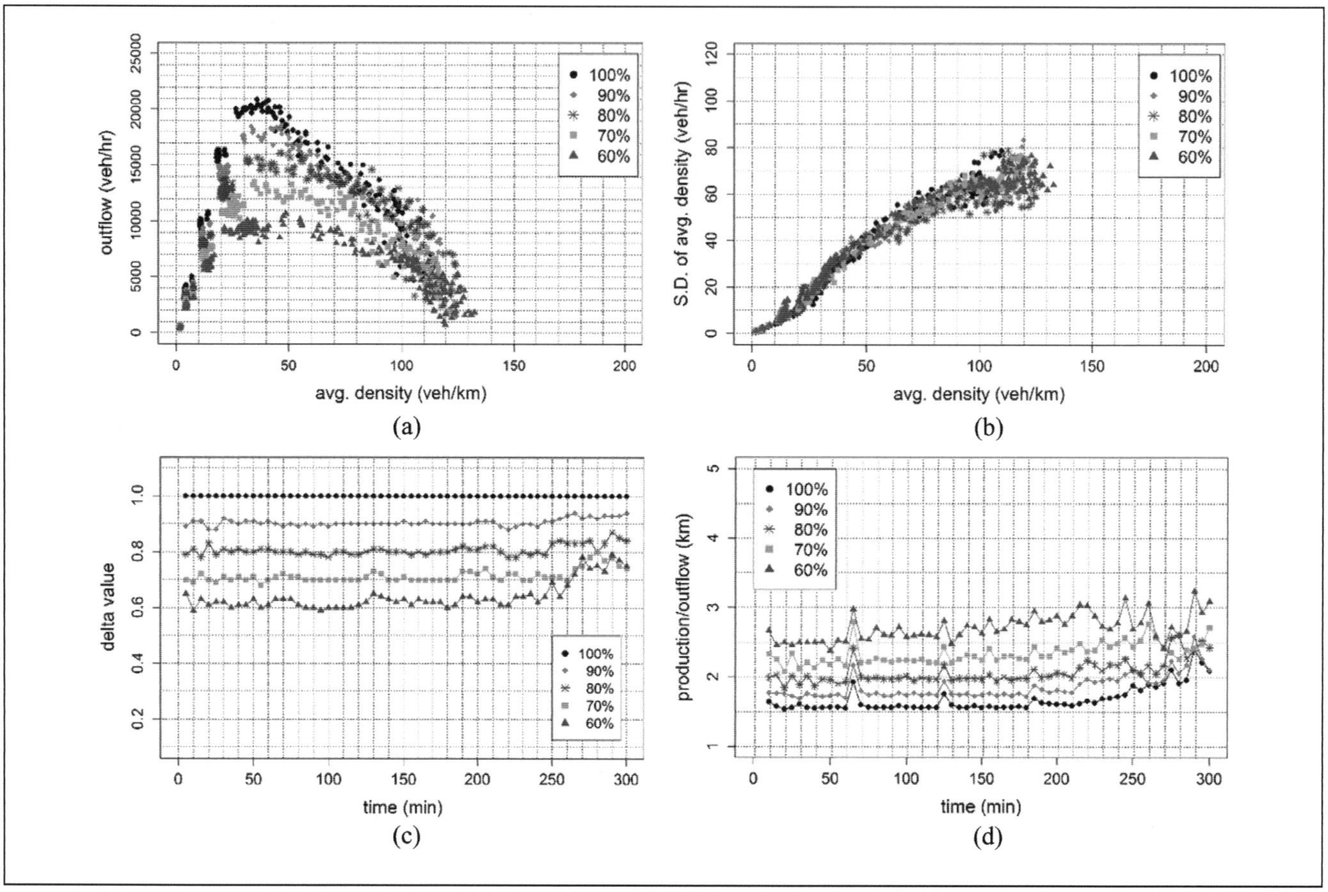

Figure 3. Effect of demand pattern on network outflow: (*a*) outflow diagram; (*b*) spatial homogeneity; (*c*) delta value; (*d*) production/outflow.

related to the first term of Equation 10, which assumes that the outflow of a direction is proportional to the demand for the direction over the entire outbound demand. In order to simply the term, let $\delta_{i\to y}(t) = q^{\text{vod}}_{i\to y}(t)/\sum_{m_i} q^{\text{vod}}_{i\to m_i}(t)$ and $f_{i\to y}(t) = q_i(t)\cdot\delta_{i\to y}$. Then, $f_{i\to y}(t)$ represents the outflow of a direction given by the demand ratio. We compared the simulated outflow and $f_{i\to y}(t)$ to see if the outflow of each direction follows the demand ratio. Figure 6*a* shows an example of such comparison. The comparison work is done by different heterogeneity levels of demand for each direction. The heterogeneity is represented by the standard deviation of the demand ratio values of each direction: northbound, eastbound, southbound, and westbound. For example, standard deviation 0.02 represents low heterogeneity, where the demand ratio values of each direction are 0.27, 0.26, 0.24, and 0.23, respectively. On the other hand, standard deviation 0.20 represents high heterogeneity, where the demand ratio values are 0.55, 0.20, 0.15, and 0.10, respectively. Ten cases with different heterogeneity levels (from 0.02 to 0.20) were tested.

Figure 6*b* shows the error values between the simulated outflow and $f_{i\to y}(t)$ for the 10 different cases. The error is expressed with the average value of the mean absolute percentage errors (MAPE) of each direction. As we can see, the error values are low (less than 10%) for heterogeneity of less than 0.12. However, the error values increase for higher heterogeneity. Hence, the assumption that the outflow of a direction is proportional to the demand ratio is reliable, as long as the heterogeneity of the demands in each direction is less than a certain condition. If the heterogeneity is high, the demand is concentrated in one or two directions. The demand concentration may have caused some delays and affected the outflow of other directions. Such an effect is definitely worth being investigated more deeply in further works.

Experiments on Two Networks

Simulation Settings

To test the effect of a network's supply level, additional microscopic simulation experiments were conducted. For this test, we set two 4×4 grid-shaped networks as in

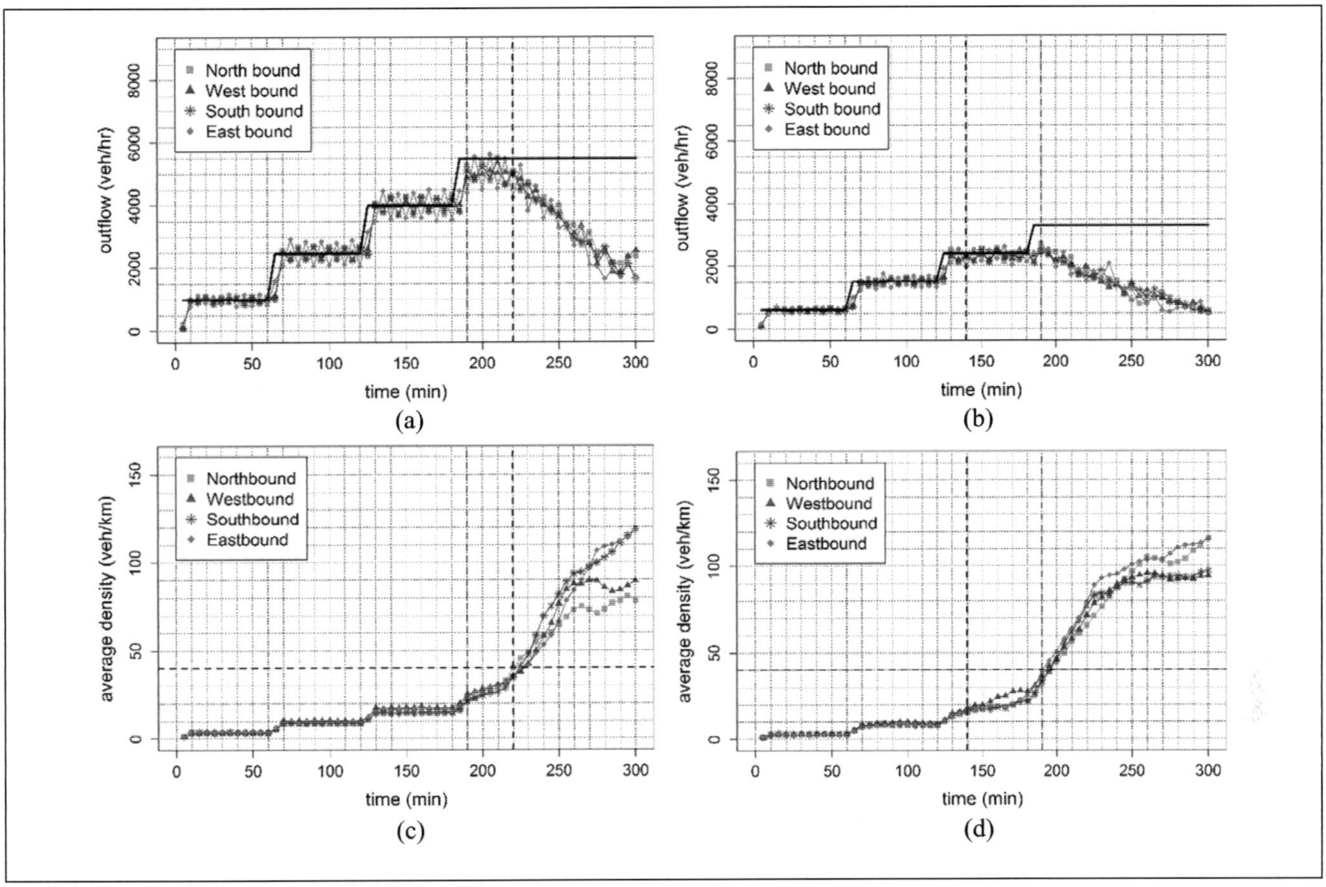

Figure 4. Comparison between the outbound demand level 100% and 60%: (*a*) outflow of each direction (100%); (*b*) outflow of each direction (60%); (*c*) average density at each boundary (100%); (*d*) average density at each boundary (60%).

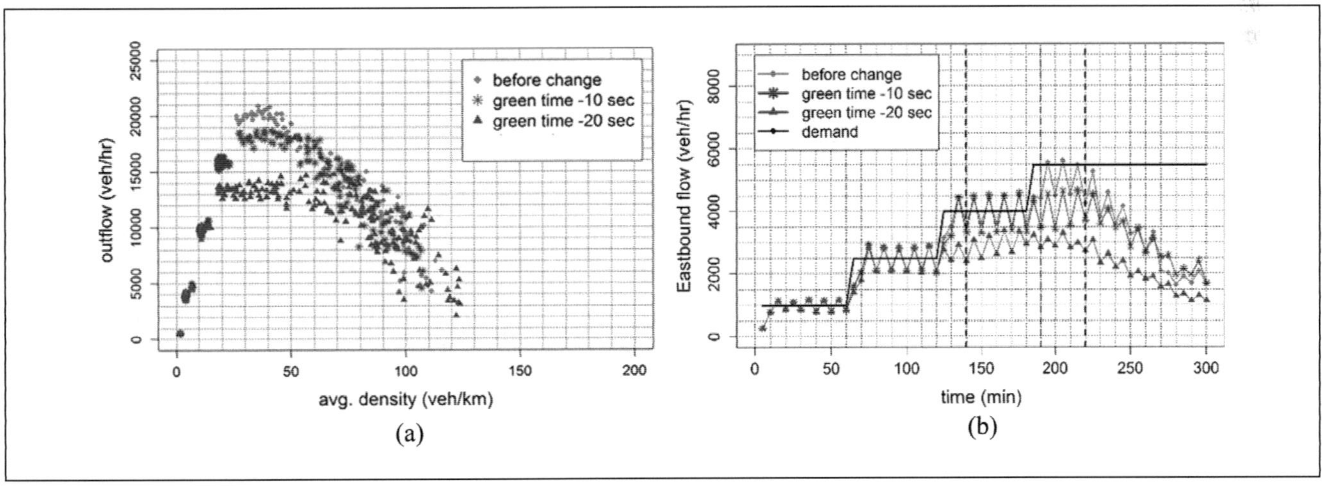

Figure 5. Effect of boundary capacity on network outflow: (*a*) outflow diagram; (*b*) eastbound flow.

Figure 7*a*. The rest of the settings are the same as the previous section, except the traffic volume. Like the case studies in the previous section, the traffic volume increases over time, but the heavy loading is given only to network 2, as shown in Figure 7*b*. Little traffic is generated from network 1, and the destination of the majority of traffic is concentrated in the local area of network 2. Such a setting makes only one network congested.

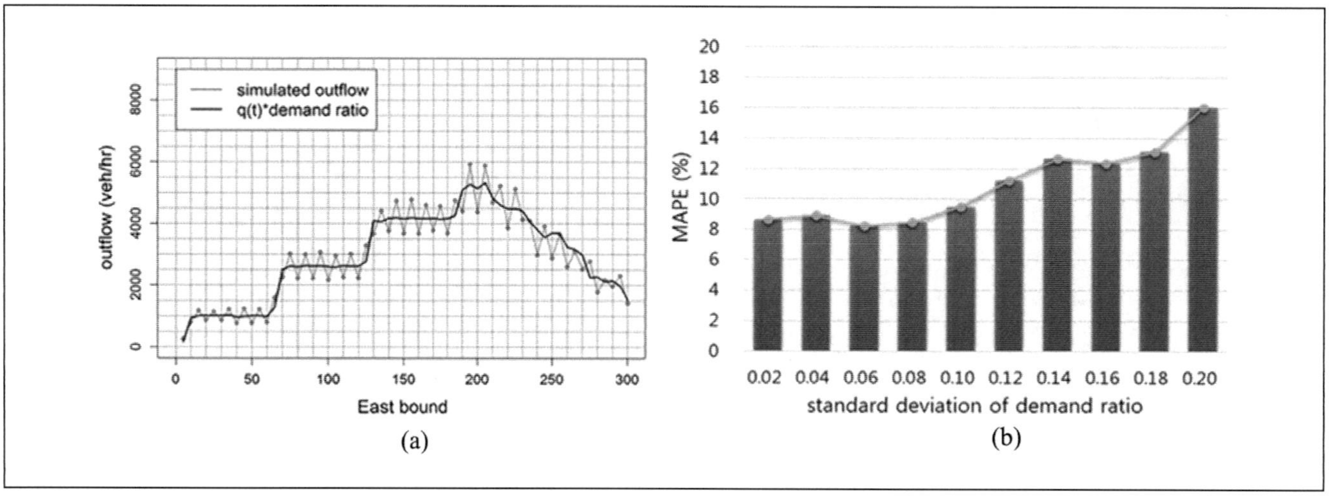

Figure 6. Comparison of the simulated outflow and $f_{i\to y}(t)$ for each direction: (*a*) comparison example for a direction (eastbound); (*b*) increasing error due to heterogeneity of demand ratio for each direction.

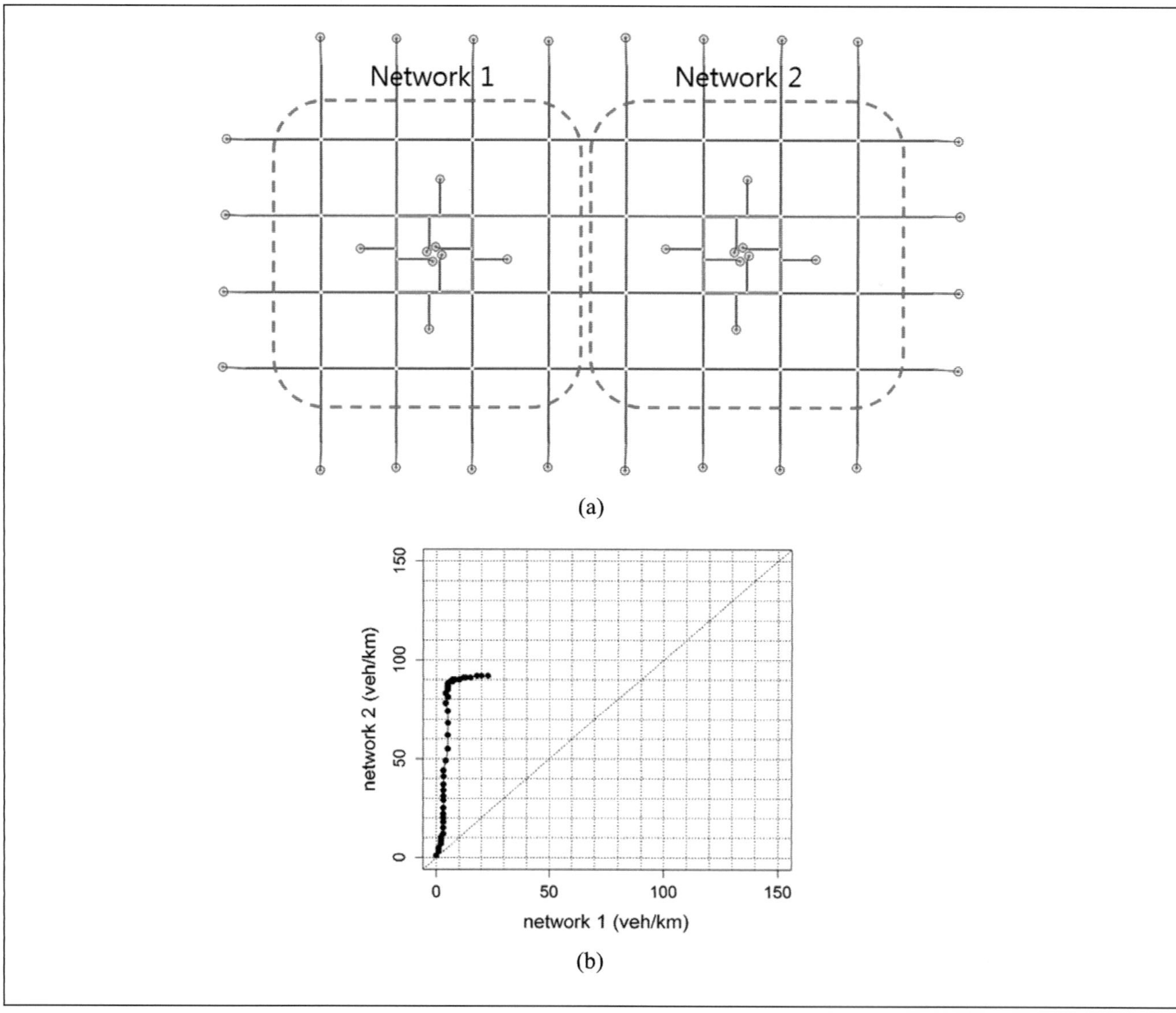

Figure 7. Simulation settings for two networks: (*a*) two-network settings in AIMSUN; (*b*) traffic loading for the two networks.

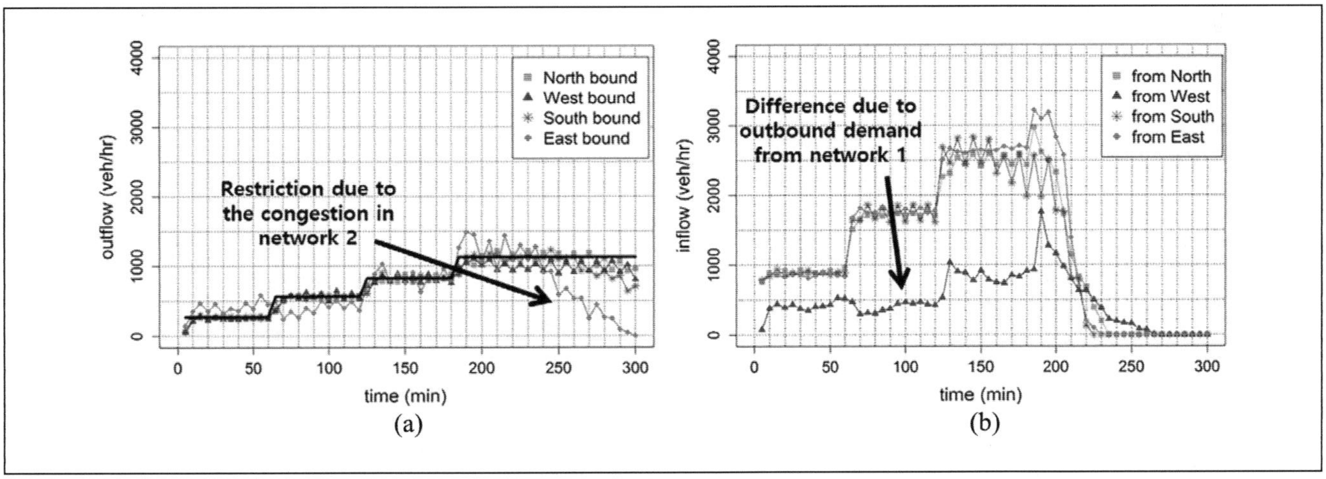

Figure 8. Transfer flow due to supply level and inbound demand: (*a*) outflow of network 1 in each direction; (*b*) inflow of network 2 in each direction.

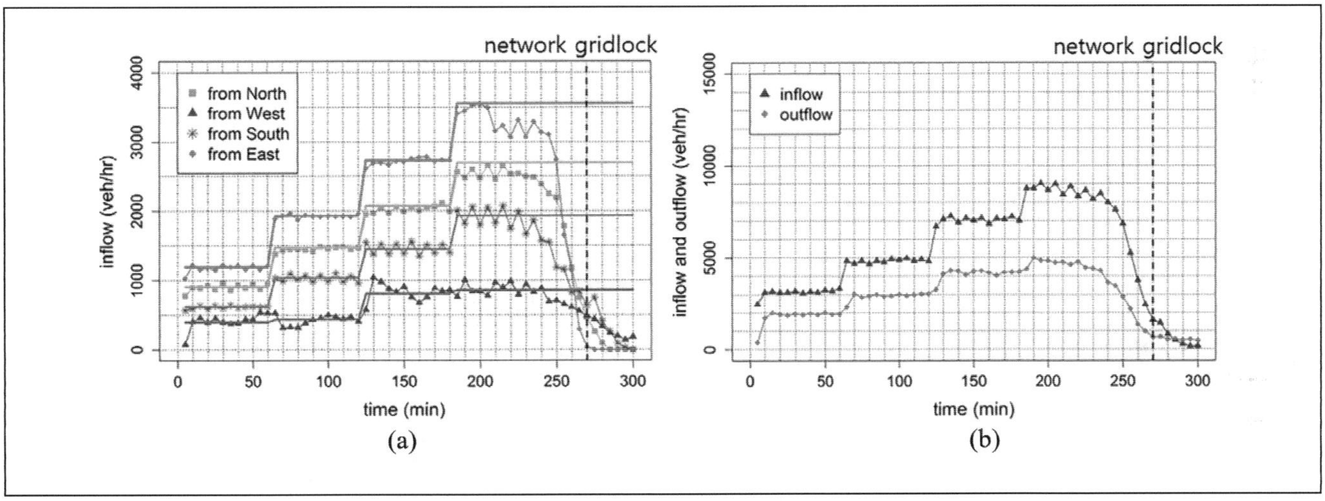

Figure 9. Transfer flow due to ratio of inbound demand: (*a*) inflow of network 2 in each direction; (*b*) inflow and outflow of network 2.

Transfer between Two Networks

Figure 8*a* shows the restriction effect of network supply on transfer flow. At no time does network 1 go into the oversaturated traffic condition. Thus, the outflow of the network in each direction does not decrease except the outflow toward network 2 (eastbound). Due to the congestion in network 2, the supply level is lower than the eastbound demand of network 1, and this restricts the flow from network 1 to 2. Figure 8*b* shows the inflow of network 2 from each direction. Since little traffic is generated from network 1 and the majority of traffic is generated from the centroids around network 2, the demand ratio from network 1 to 2 is much lower than the other directions (north, south, and east of network 2). As expected by the demand ratio, we can see that the inflow from network 1 is much lower compared with the inflow from other directions.

To see if the inflow distribution from each direction follows the demand ratio, another case was tested. In this case, we set traffic differently so that the inbound demands for network 2 from all directions had different ratios. The ratio of inbound demands from the north, south, east, and west (network 1) was approximately 3:2:4:1, respectively. As we can see in Figure 9*a*, except when the network gets into the gridlock situation (after 270 min), the inflow from each direction follows the demand ratio.

This section has shown the effect of restriction of network supply on transfer flow. It also shows that the inflow is distributed by the demand ratio from each direction. Therefore, Equation 10describeswell the transfer flow between networks.

Conclusion

In this paper, the transfer flow between road networks is investigated in relation to a macroscopic fundamental diagram (MFD). First, we found that the network outbound demand pattern strongly influences the maximum outflow rate regardless of spatial inhomogeneity. This provides new information that the ratio of travel demand toward a network's local area should be taken into account as well for modeling the reduction of network outflow. We found that the boundary capacity affects the outflow as well. We also checked the restriction effect of supply level of a neighboring network on transfer flow. Furthermore, we found that a network's inflow and outflow in each direction is proportional to the demand ratio, when the heterogeneity of demand for each direction is under a certain threshold. Through these experiments, the modeling equation of network transfer flow has been confirmed.

The transfer flow is described based on the effects of the three different factors: travel demand pattern within a network, boundary capacity, and supply level of a neighboring network. This shows the possibility of developing a new perimeter control method that considers all these factors. Particularly, the boundary capacity is the situational variable that can be adjusted for perimeter control purposes. We provide a new way of describing the boundary capacity in relation to the infrastructural condition at the intersections of network boundaries. Furthermore, when adjusting this value, a control system can take into account both the demand pattern and a neighboring network's supply level, for either maximizing the demand throughput or minimizing the congestion heterogeneity between networks. Thus, the investigated results of this study can be useful for various analyses on inter-network traffic dynamics.

Some further work can be suggested regarding transfer flow. We have not yet tested the case of outbound demand level less than 60%. Even though it may seem somewhat unrealistic in urban traffic environments, what would happen if all traffic demand was to travel toward the local area of a network? There is a certain capacity level in local areas (parking spaces and local streets), and the given situation may cause unexpected results for the network's inflow and outflow. Hence, it is worth investigating such low outbound level cases. Furthermore, it is known that the hysteretic phenomena of network flow occur during traffic loading and recovery phases (*9*, *10*). This situation may also cause unexpected behaviors of outflow and transfer flow, thus it is worth examining such cases as well. Lastly, the reason for using homogenously shaped networks in this study was to see only the effect of travel demand, rather than the effect of network topology. Since most real-world networks are not homogenously shaped, investigation of the effect of network topology shall be carried out as well, and this current paper can be considered as the base study for such further work.

Acknowledgments

This research was supported by a grant (17CTAP-C129888-01) from the Technology Advancement Research Program (TARP) funded by the Ministry of Land, Infrastructure, and Transport of the Korean government and the Basic Science Research Program through the National Research Foundation of Korea (NRF) funded by the Ministry of Science and ICT (NRF-2017R1A2B2002329).

Author Contributions

The authors confirm contribution to the paper as follows: study conception and design: Sunghoon Kim, Sehyun Tak; data collection: Sunghoon Kim; analysis and interpretation of results: Sunghoon Kim, Sehyun Tak, Hwasoo Yeo; draft manuscript preparation: Sunghoon Kim. All authors reviewed the results and approved the final version of the manuscript.

References

1. Daganzo, C. F. Urban Gridlock: Macroscopic Modeling and Mitigation Approaches. *Transportation Research Part B: Methodological*, Vol. 41, No. 1, 2007, pp. 49–62.
2. Geroliminis, N., and C. F. Daganzo. Macroscopic Modeling of Traffic in Cities. Presented at 86th Annual Meeting of the Transportation Research Board, Washington, D.C., 2007.
3. Geroliminis, N., and C. F. Daganzo. Existence of Urban-Scale Macroscopic Fundamental Diagrams: Some Experimental Findings. *Transportation Research Part B: Methodological*, Vol. 42, No. 9, 2008, pp. 759–770.
4. Buisson, C., and C. Ladier. Exploring the Impact of Homogeneity of Traffic Measurements on the Existence of Macroscopic Fundamental Diagrams. *Transportation Research Record: Journal of the Transportation Research Board*, 2009. 2124: 127–136.
5. Daganzo, C. F., V. V. Gayah, and E. J. Gonzales. Macroscopic Relations of Urban Traffic Variables: Bifurcations, Multivaluedness and Instability. *Transportation Research Part B: Methodological*, Vol. 45, No. 1, 2011, pp. 278–288.
6. Geroliminis, N., and J. Sun. Properties of a Well-Defined Macroscopic Fundamental Diagram for Urban Traffic. *Transportation Research Part B: Methodological*, Vol. 45, No. 3, 2011, pp. 605–617.
7. Knoop, V. L., H. van Lint, and S. P. Hoogendoorn. Traffic Dynamics: Its Impact on the Macroscopic Fundamental Diagram. *Physica A: Statistical Mechanics and Its Applications*, Vol. 438, 2015, pp. 236–250.
8. Mazloumian, A., N. Geroliminis, and D. Helbing. The Spatial Variability of Vehicle Densities as Determinant of

Urban Network Capacity. *Philosophical Transactions of the Royal Society A: Mathematical Physical and Engineering Sciences*, Vol. 368, No. 1928, 2010, pp. 4627–4647.

9. Gayah, V. V., and C. F. Daganzo. Clockwise Hysteresis Loops in the Macroscopic Fundamental Diagram: An Effect of Network Instability. *Transportation Research Part B: Methodological*, Vol. 45, No. 4, 2011, pp. 643–655.
10. Geroliminis, N., and J. Sun. Hysteresis Phenomena of a Macroscopic Fundamental Diagram in Freeway Networks. *Transportation Research Part A: Policy and Practice*, Vol. 45, No. 9, 2011, pp. 966–979.
11. Mahmassani, H. S., M. Saberi, and A. Zockaie. Urban Network Gridlock: Theory, Characteristics, and Dynamics. *Transportation Research Part C: Emerging Technologies*, Vol. 36, 2013, pp. 480–497.
12. Zhang, L. L., T. M. Garoni, and J. de Gier. A Comparative Study of Macroscopic Fundamental Diagrams of Arterial Road Networks Governed by Adaptive Traffic Signal Systems. *Transportation Research Part B: Methodological*, Vol. 49, 2013, pp. 1–23.
13. Kim, S., and H. Yeo. A Flow-Based Vulnerability Measure for the Resilience of Urban Road Network. *Procedia: Social and Behavioral Sciences*, Vol. 218, 2016, pp. 13–23.
14. Kim, S., and H. Yeo. Evaluating Link Criticality of Road Network Based on the Concept of Macroscopic Fundamental Diagram. *Transportmetrica A: Transport Science*, Vol. 13, No. 2, 2017, pp. 162–193.
15. Aboudolas, K., and N. Geroliminis. Perimeter and Boundary Flow Control in Multi-Reservoir Heterogeneous Networks. *Transportation Research Part B: Methodological*, Vol. 55, 2013, pp. 265–281.
16. Geroliminis, N., J. Haddad, and M. Ramezani. Optimal Perimeter Control for Two Urban Regions with Macroscopic Fundamental Diagrams: A Model Predictive Approach. *IEEE Transactions on Intelligent Transportation Systems*, Vol. 14, No. 1, 2013, pp. 348–359.
17. Haddad, J., M. Ramezani, and N. Geroliminis. Cooperative Traffic Control of a Mixed Network with Two Urban Regions and a Freeway. *Transportation Research Part B: Methodological*, Vol. 54, 2013, pp. 17–36.
18. Haddad, J., and A. Shraiber. Robust Perimeter Control Design for an Urban Region. *Transportation Research Part B: Methodological*, Vol. 68, 2014, pp. 315–332.
19. Keyvan-Ekbatani, M., A. Kouvelas, I. Papamichail, and M. Papageorgiou. Exploiting the Fundamental Diagram of Urban Networks for Feedback-Based Gating. *Transportation Research Part B: Methodological*, Vol. 46, No. 10, 2012, pp. 1393–1403.
20. Keyvan-Ekbatani, M., M. Papageorgiou, and V. L. Knoop. Controller Design for Gating Traffic Control in Presence of Time-Delay in Urban Road Networks. *Transportation Research Part C: Emerging Technologies*, Vol. 59, 2015, pp. 308–322.
21. Keyvan-Ekbatani, M., R. C. Carlson, V. L. Knoop, S. Hoogendoorn, and M. Papageorgiou. Queuing Under Perimeter Control: Analysis and Control Strategy. *Proc., 2016 IEEE 19th International Conference on Intelligent Transportation Systems (ITSC)*, Rio de Janeiro, Brazil, 2016.
22. Keyvan-Ekbatani, M., M. Yildirimoglu, N. Geroliminis, and M. Papageorgiou. Traffic Signal Perimeter Control with Multiple Boundaries for Large Urban Networks. *Proc., 16th Interntional IEEE Conference on Intelligent Transportation Systems (ITSC)*, Hague, Netherlands, 2013.
23. Ramezani, M., J. Haddad, and N. Geroliminis. Dynamics of Heterogeneity in Urban Networks: Aggregated Traffic Modeling and Hierarchical Control. *Transportation Research Part B: Methodological*, Vol. 74, 2015, pp. 1–19.
24. Haddad, J., and B. Mirkin. Adaptive Perimeter Traffic Control of Urban Road Networks Based on MFD Model with Time Delays. *International Journal of Robust and Nonlinear Control*, Vol. 26, No. 6, 2016, pp. 1267–1285.
25. Gao, X., and V. V. Gayah. An Analytical Framework to Model Uncertainty in Urban Network Dynamics Using Macroscopic Fundamental Diagrams. *Transportation Research Procedia*, Vol. 23, 2017, pp. 497–516.
26. Mariotte, G., L. Leclercqa, and J. A. Laval. Macroscopic Urban Dynamics: Analytical and Numerical Comparisons of Existing Models. *Transportation Research Part B: Methodological*, Vol. 101, 2017, pp. 245–267.
27. Leclercq, L., and N. Geroliminis. Estimating MFDs in Simple Networks with Route Choice. *Transportation Research Part B: Methodological*, Vol. 57, 2013, pp. 468–484.
28. Leclercq, L., C. Parzani, V. L. Knoop, J. Amourette, and S. P. Hoogendoorn. Macroscopic Traffic Dynamics with Heterogeneous Route Patterns. *Transportation Research Part C: Emerging Technologies*, Vol. 59, 2015, pp. 292–307.
29. Knoop, V. L., and S. P. Hoogendoorn. An Area-Aggregated Dynamic Traffic Simulation Model. *European Journal of Transport and Infrastructure Research*, Vol. 15, No. 2, 2015, pp. 226–242.
30. AIMSUN. *AIMSUN Version 7 Users Manual*. TTS-Transport Simulation Systems, Barcelona, 2011. www.aimsun.com. Accessed November 26, 2016.

The Standing Committee on Traffic Flow Theory and Characteristics (AHB45) peer-reviewed this paper (18-00597).

Article

TRR
JOURNAL OF THE TRANSPORTATION RESEARCH BOARD

Transportation Research Record
2018, Vol. 2672(20) 86–97

DOI: 10.1177/0361198118773884
journals.sagepub.com/home/trr
SAGE

Extended Desired Safety Margin Car-Following Model That Considers Variation of Historical Perceived Risk and Acceptable Risk

Junjie Zhang[1], Yunpeng Wang[1], and Guangquan Lu[1]

Abstract

In this paper, an extended desired safety margin (DSM) car-following model is proposed by accounting for the effect of the variation of historical perceived risk and acceptable risk in terms of the DSM model. By conducting gray correlation analysis, an investigation is carried out into whether the variation of historical perceived risk and acceptable risk has important effects on the acceleration or deceleration of a target vehicle based on a real-vehicle test platform. Through simulated results, the dynamic performance of an extended DSM model is investigated in comparison with the DSM model. Conclusions show that the extended DSM model can better reflect the characteristics of traffic flow compared with the DSM model, and can improve the performance of the vehicle in the start, stop, and car-following processes. Numerical simulations further demonstrate that the extended DSM model can improve traffic safety via the changes of time-to-collision and safety margin when external disturbance is introduced into the leading vehicle. Thus, historical information about target vehicles should be considered to improve the performance of car-following in adaptive cruise control (ACC) and automotive platoon driving.

Car-following is the most common phenomenon in a single lane. Simulation of car-following behavior is important to characterize confusing traffic flow conditions, which are widely reported. To capture the mechanism of stop-and-go oscillations behind the phenomena of vehicular traffic flow from the microscopic viewpoint, building various car-following models have consistently been the concern of researchers and have been under way for 60 years. Understanding the influence of car-following behaviors on traffic flow evolution is a research concern, especially in the adaptive cruise control (ACC) system and automotive platoon driving.

Many mathematical models have been developed to describe car-following behavior based on the stimulus-response framework. Well-known car-following models have been introduced based on both experimental observations and theoretical analyses, such as those by Chandler, Herman, and Montroll (CHM) (*1*); Gazis–Herman–Rothery (GHR) (*2*); Bexelius (*3*); Lenz (*4*); Addison and Low (*5*); Helly (*6*); and the intelligent driver model (IDM) (*7*). Among them, modified models based on the GHR model have been proposed to investigate the influence of certain factors, such as driver reaction time, memory function, and multi-vehicle interaction, on the stability of traffic flow (*8–10*). In addition, given the desired measures and safety distance consideration in the car-following process, some scholars have proposed desired measure models and safety distance models considering different driving behavior parameters (*11–14*). The optimized velocity approach proposed by Bando et al. in 1995 assumes that car-following is based on a driver's desire to maintain optimal velocity in a traffic stream; this is the optimal velocity (OV) model (*15*). To overcome the limitations of the OV model, other extended models, such as the generalized force (GF) (*16*), the full velocity difference (FVD) (*17*), and the asymmetric full velocity difference (AFVD) (*18*) models, have been proposed. Subsequently, the aforementioned models and their extensions have been widely used in investigating the effects of drivers' physiological and

[1]Beijing Advanced Innovation Center for Big Data and Brain Computing, Beijing Key Laboratory for Cooperative Vehicle Infrastructure Systems and Safety Control; Jiangsu Province Collaborative Innovation Center of Modern Urban Traffic Technologies, School of Transportation Science and Engineering, Beihang University, Beijing, China

Corresponding Author:
Address correspondence to Guangquan Lu: lugq@buaa.edu.cn

psychological characteristics, as well as traffic environments, on traffic flow stability (*19–45*). Among them, some scholars have in recent years investigated the influence of vehicular gap and driver's memory on the stability of traffic flow, based on the OV or FVD model. Yu and Shi proposed a new FVD that considered headway changes with memory (*33*). Peng et al. also constructed a new FVD model by accounting for the effect of OV changes using memory (*34*). Yu and Shi further constructed a new FVD model by accounting for the effect of vehicular gap (*41*). Jiang and Wu redefined the OV function to study night driving behaviors in the car-following model under periodic boundary conditions (*43*). Tang et al. developed two new car-following models that considered the driver's forecast effect (*44*) and driver's memory (*45*) to improve the stability of traffic flow. Therefore, those analysis results imply that the historical acceleration or deceleration of vehicles cannot be neglected and should be considered in modeling car-following behavior.

The car-following process is dependent not only on driver behavior and traffic environments but also on the psychological characteristics of drivers. Considering this perspective, Van Winsum proposed a new car-following model based on the linear car-following model that considered driver error and distraction (*35*). Yang and Peng proposed a stochastic car-following model that considered three major types of driver errors (perceptual limitation, time delay, and distraction) in a homogenous driving population (*36*). Andersen and Sauer modified Helly's model (*6*) by considering visual angle as the stimulus; this is the driving by visual angle (DVA) model (*37*). In general, driving behavior is largely affected by the risk perception of drivers in actual traffic. Hamdar et al. developed a driver behavior model that considered risk-taking tasks using prospect theory (*39*). Lu et al. proposed a desired safety margin (DSM) model that can simulate different drivers' physiological and psychological characteristics such as driver's response time, sensitivity factor for acceleration or deceleration, and DSM (*40*). The present study takes the view that human performance can be quantified by subjective risk perception of drivers in real traffic. In the car-following process, driver decision-making is mainly a reflection of subjective risk perception, acceleration and deceleration sensitivity, and response time. Therefore, this paper will employ the DSM model of Lu et al. (*46*) to describe the car-following process, as well as investigate the influence of variation of historical perceived risk and acceptable risk on the stability of traffic flow. The objective of this study is to answer the following questions as regards variations in historical perceived risk and acceptable risk revealed in car-following behavior:

1. Is there a strong degree of relationship between the variation of historical perceived risk and acceptable risk and acceleration of the following vehicle?
2. How significant is the effect of variations in historical perceived risk and acceptable risk on car-following performance?
3. How large is the influence of the variation of historical perceived risk and acceptable risk on traffic safety in the car-following process?

To this end, the rest of this paper is organized as follows. In Section 2, test cars and components of ACC system are introduced. The relationships between the variation of historical perceived risk and acceptable risk and the acceleration of the following vehicle are further investigated by using the field data collected from many tests of real vehicles in a test platform. In Section 3, an extended car-following model is proposed based on the DSM model that considered the effect of the variation of historical perceived risk and acceptable risk. In Section 4, the effects of the variation of historical perceived risk and acceptable risk on car-following performance are analyzed using numerical experiments. Moreover, two risk indicators are used to investigate the effect of the variation of historical perceived risk and acceptable risk on traffic safety in car-following. The concluding section summarizes the obtained results.

Experimental Data Acquisition and Gray Correlation Analysis

Experimental Vehicles and DSM Control Strategy

Two vehicles with an ACC system are used as experimental vehicles in the tests. As shown in Figure 1, the ACC system consists of information collection, speed control, communication, and human–computer interactive systems. The information collection system includes a GPS, Inertial Navigation System, and Controller Area Network (CAN) information collection device; its main function is to collect vehicle speed, acceleration, direction, and CAN information. The speed control system needs to control the vehicle drive at a certain speed to complete specific driving assistance tasks. The main component of the system is an electronic vacuum booster that can control the brake and accelerator pedals to make vehicles accelerate, decelerate, or move at a constant speed. The communication system achieves data information communication between vehicles. The communication mode is dedicated short-range communications, which enables short-range communication in real time. The human–computer interactive system consists of a horn and viewing screen that is designed to improve communication between drivers and the system.

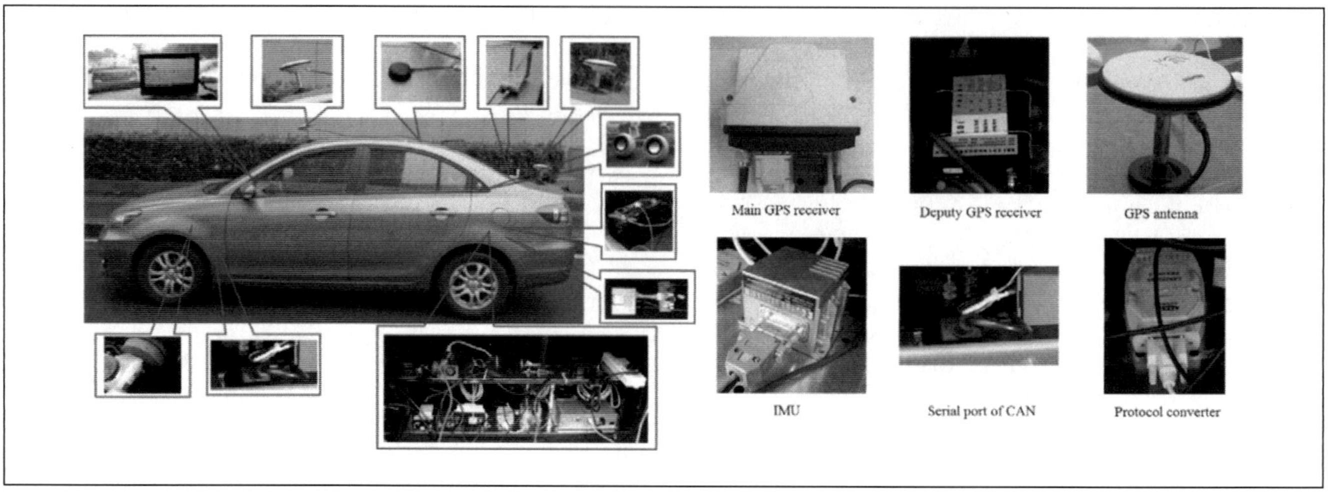

Figure 1. Test car and components of ACC system.

Under such an experimental platform, Tian Gong Road in Beijing is selected as a data collection site. The control algorithm of the ACC system in the test car is the DSM. That is, the speed control system is based on the DSM car-following model as shown in Equations 1 and 2.

The basic idea of the DSM model relies on the assumption that the driver of a following car always maintains adequate spacing to prevent collision being caused by emergency braking of the leading car. It uses a minimum safe gap, which is determined by maximum decelerations of the following and leading cars to evaluate risk level, instead of a desired minimum gap which is related to comfortable deceleration used in the Gipps model (*47*) and IDM (*7*). Therefore, the safety margin is proposed as shown in Equation 2 based on the braking process during a car-following situation (Figure 2).

The DSM of car-following is expressed as

$$a_n(t) = \begin{cases} \alpha_1(\mathrm{SM}_n(t) - \mathrm{SM}_{n\mathrm{DH}}(t)), \mathrm{SM}_n(t) > \mathrm{SM}_{n\mathrm{DH}}(t) \\ \alpha_2(\mathrm{SM}_n(t) - \mathrm{SM}_{n\mathrm{DL}}(t)), \mathrm{SM}_n(t) < \mathrm{SM}_{n\mathrm{DL}}(t) \\ 0, \text{otherwise} \end{cases} \quad (1)$$

and

$$\mathrm{SM}_n(t) = 1 - \frac{v_n(t)\cdot\tau + \left([v_n(t)]^2/2d_n(t)\right)}{x_{n-1}(t) - x_n(t) - l_{n-1}} + \frac{[v_{n-1}(t)]^2/2d_{n-1}(t)}{x_{n-1}(t) - x_n(t) - l_{n-1}} \quad (2)$$

where

$a_n(t)$ is the acceleration of a following vehicle at time t;

$\mathrm{SM}_n(t)$ is defined as the safety margin (SM);

$\mathrm{SM}_{n\mathrm{DH}}(t)$ is the upper limit of the DSM in the car-following condition;

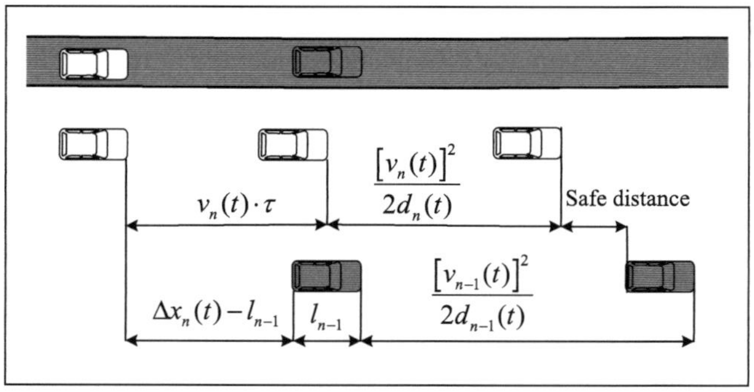

Figure 2. Car-following model with risk perception set-up.

$\mathrm{SM}_{n\mathrm{DL}}(t)$ is the lower limit of the DSM in the car-following condition;

α_1 and α_2 are the sensitivity factors for acceleration and deceleration;

τ represents the response time of the brake system;

$v_n(t)$ and $v_{n-1}(t)$ denote the velocity of the n th and $(n-1)$ th vehicle at time t, respectively;

$\Delta x_n(t) = x_{n-1}(t) - x_n(t)$ represents the space headway between the $(n-1)$ th and n th vehicles;

$d_n(t)$ and $d_{n-1}(t)$ are the deceleration of the $(n-1)$ th and n th vehicle at time t, respectively;

l_{n-1} denotes the length of the preceding vehicle.

The DSM may change because of psychological factors. However, the main considerations are the influence of the primary factors on the car-following process. Random factors are disregarded and the DSMs are considered to be fixed values. Equation 1 is simplified to

$$a_n(t) = \begin{cases} \alpha_1(\mathrm{SM}_n(t) - \mathrm{SM}_{n\mathrm{DH}}), \mathrm{SM}_n(t) > \mathrm{SM}_{n\mathrm{DH}} \\ \alpha_2(\mathrm{SM}_n(t) - \mathrm{SM}_{n\mathrm{DL}}), \mathrm{SM}_n(t) < \mathrm{SM}_{n\mathrm{DL}} \\ 0, \text{otherwise} \end{cases} \quad (3)$$

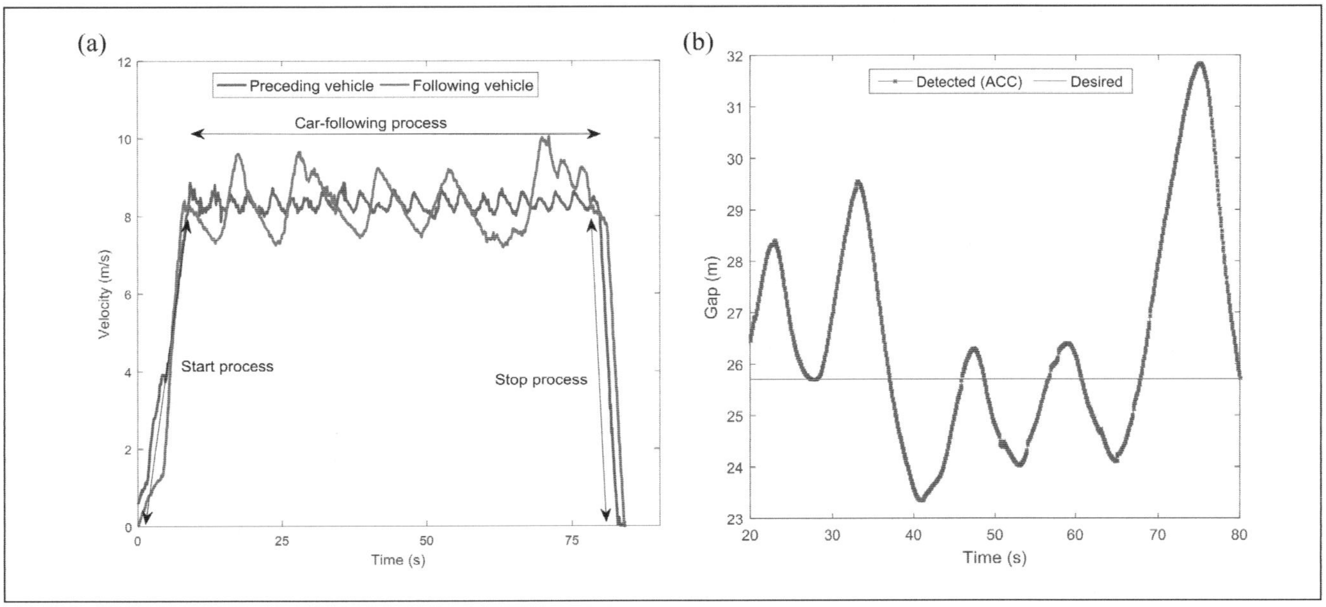

Figure 3. The car-following process in the experiment, showing (*a*) the whole process, and (*b*) the detected gap and desired gap.

In the experiment, the data structures of the two vehicles include Gauss coordinates, velocity, acceleration or deceleration, length and width of vehicle, latitude and longitude, and timestamp. In this study, to analyze the relationship between the variation in historical perceived risk and acceptable risk and the acceleration or deceleration of the following vehicle, the corresponding vehicle data is obtained from the information collection system.

As shown in Figure 3*a*, the vehicle dynamic performance in this experiment includes three traffic situations, the start, car-following, and stop processes. The figure shows that the following vehicle can maintain the consensus state and finish acceleration or deceleration quickly based on the DSM control in the car-following process. Moreover, Figure 3*b* shows that there is a certain fluctuation between the gap and the desired gap in the car-following process. Therefore, those results further imply that the DSM control strategy should be modified to improve the vehicle dynamic performance.

Gray Correlation Analysis

To improve the stability of the car-following process and dynamic performance of the start-and-stop process, the relationship between the variation of historical perceived risk and acceptable risk and the acceleration of the following vehicle is investigated. Gray correlation analysis is considered to show the linkage between the variation of historical perceived risk, acceptable risk, and dynamic performances.

Gray correlation analysis is one of the gray system analyses, which is used to determine whether or not variables are correlated and to determine the degree of their correlation. By calculation of characteristic serial curves and the degree of geometrical similarity of these curves, key factors and minor factors can be determined. Specifically, the gray correlation degree of different sequences is bigger when their curved shapes are more similar. On the contrary, that is smaller when the curves are less similar.

The basic idea of gray correlation analysis is as follows:

Let X_0 be the system characteristic series, and X_i the correlation factor series. Then,

$$\left.\begin{array}{l} X_0 = \{x_0(1), x_0(2), \cdots, x_0(n)\} \\ X_i = \{x_i(1), x_i(2), \cdots, x_i(n)\}, i = 1, 2, \cdots, m. \end{array}\right\} \quad (4)$$

where n is the number of characteristic variables, and m is the number of correlation factor.

Then, the correlation coefficient $\gamma_{0i}(k)$ at point k between system characteristic and correlation factor series can be described as

$$\gamma_{0i}(k) = \frac{1}{n}\sum_{k=1}^{n} \gamma(x_0(k), x_i(k)) \quad (5)$$

in which

$$\gamma(x_0(k), x_i(k)) = \frac{\min_i \min_k |x_0(k) - x_i(k)| + \zeta \max_i \max_k |x_0(k) - x_i(k)|}{|x_0(k) - x_i(k)| + \zeta \max_i \max_k |x_0(k) - x_i(k)|}$$

where ζ is the differentiation coefficient. The value of $\zeta = 0.5, \zeta \in [0, 1]$ is adopted in this paper.

Table 1. Results of Gray Correlation Analysis under Different Traffic Situations When $\varsigma = 0.1s$

Traffic situation	v_f(m/s)	Δx_{pf}(m)	Δv_{pf}(m/s)	PSM	ΔSM
Start process	0.7254	0.8159	0.7595	0.8121	0.7997
Car-following process	0.6287	0.6025	0.7151	0.6128	0.6160
Stop process	0.6650	0.6704	0.7091	0.6770	0.7244
Whole process	0.4442	0.8211	0.8345	0.8389	0.8551

Based on the gray correlation analysis, the degrees of relation are obtained under different traffic situations as shown in Table 1. As indicated in Table 1, a_f is acceleration of the following vehicle; Δx_{pf} is space headway between the preceding vehicle and the following vehicle; v_f is the velocity of the following vehicle; Δv_{pf} is the velocity difference between the preceding vehicle and the following vehicle; SM is the safety margin (risk perception); the effect of the variation of historical perceived risk and acceptable risk is necessary.

This study considers a platoon of n vehicles running on a straight road. All vehicles are equipped and each vehicle can receive the movement information from itself and the preceding vehicle. Thus, variation of historical perceived risk and acceptable risk is introduced into the DSM model, and the extended DSM model is proposed as

$$a_n(t) = \begin{cases} \alpha_1(\mathrm{SM}_n(t) - \mathrm{SM}_{n\mathrm{DH}}) + \lambda\alpha_1(\mathrm{SM}_n(t-1) - \mathrm{SM}_{n\mathrm{DH}}), \mathrm{SM}_n(t) > \mathrm{SM}_{n\mathrm{DH}} \\ \alpha_2(\mathrm{SM}_n(t) - \mathrm{SM}_{n\mathrm{DL}}) + \lambda\alpha_2(\mathrm{SM}_n(t-1) - \mathrm{SM}_{n\mathrm{DL}}), \mathrm{SM}_n(t) < \mathrm{SM}_{n\mathrm{DL}} \\ 0, \text{otherwise} \end{cases} \quad (6)$$

and $\Delta\mathrm{SM} = \mathrm{SM}(t-\varsigma) - \begin{cases} \mathrm{SMH}(t-\varsigma), \mathrm{SM}_n(t) > \mathrm{SM}_{n\mathrm{DH}} \\ \mathrm{SML}(t-\varsigma), \mathrm{SM}_n(t) < \mathrm{SM}_{n\mathrm{DL}} \end{cases}$ is defined as the variation of historical perceived risk and acceptable risk, which indicates the difference between SM and DSM with memory step ς.

Table 1 shows the similarities between variables and the acceleration of the following vehicle in order of importance. These are the variation of historical perceived risk and acceptable risk, perceived SM, relative velocity between the preceding vehicle and the following vehicle, space headway between the preceding vehicle and the following vehicle, and the velocity of the following vehicle under the entire process. The similarities between the variation of historical perceived risk and acceptable risk and the acceleration or deceleration of the following vehicle are different under various traffic situations in the start process. However, the car-following process, the stop process, and their relational degrees also have a relatively high value. The variation of historical perceived risk and acceptable risk has important effects on car-following behaviors. Therefore, the variation of historical perceived risk and acceptable risk should be considered in modeling car-following behaviors.

Extended DSM Model

Based on the preceding results of gray correlation analysis, the variation of historical perceived risk and acceptable risk has significant effects on car-following behaviors. Thus, developing an extended DSM model that considers

and

$$\mathrm{SM}_n(t) = 1 - \frac{v_n(t)\cdot\tau + \left([v_n(t)]^2/2d_n(t)\right)}{x_{n-1}(t) - x_n(t) - l_{n-1}} + \frac{[v_{n-1}(t)]^2/2d_{n-1}(t)}{x_{n-1}(t) - x_n(t) - l_{n-1}} \quad (7)$$

where λ is the reaction coefficient; $\mathrm{SM}_{n\mathrm{DL}}$ and $\mathrm{SM}_{n\mathrm{DH}}$ assume a fixed value for the upper limit of the DSM and the lower limit of the DSM in ACC, respectively; when $\lambda = 0$, equation (6) is the same as equation (3). As shown in reference (*40*), the upper limit of the DSM is roughly 0.94; the lower limit of the DSM is roughly 0.73; and the mean value of α_1 and α_2 are 8 and 14 m/s^2, respectively. With reference to test data in references (*48*) and (*49*), $d_n(t)$ and $d_{n-1}(t)$ are set as 0.75g, where g is the acceleration caused by gravity, and τ is set as 0.15 s under emergency situations.

Dynamic Performance

This section provides numerical experiments to investigate the dynamic performance of the extended DSM model. To illustrate the effect of the variation of historical perceived risk and acceptable risk on dynamic performance, the simulation results of the DSM model and the extended DSM model are compared under three traffic situations: start process, car-following process, and stop process.

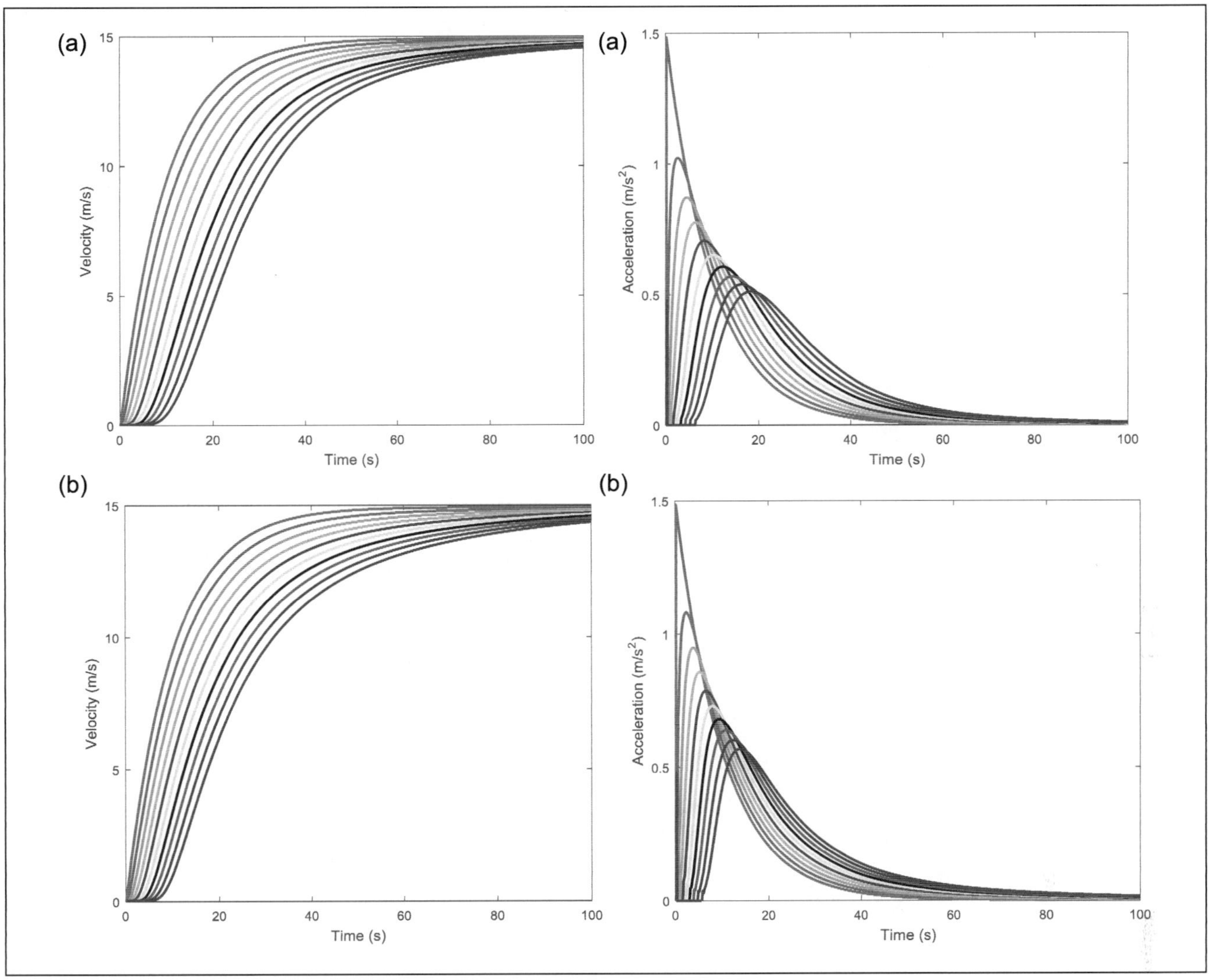

Figure 4. Velocity and acceleration profiles of vehicles 1–10 starting from a green signal, showing (*a*) DSM model and (*b*) extended DSM model. Each curve shows the velocity and acceleration of each vehicle in corresponding figures, respectively.

Start Process

In this section, the length of all vehicles is $l_n = 5\text{m}$, $n = 1, 2, \cdots, N$. Numerical experiments are provided to explore the effect of the variation of historical perceived risk and acceptable risk on the starting process through the extended DSM model.

The start process is set up as detailed in the following. Initially, the traffic signal is red, and 10 vehicles wait behind the signal with gap of 1 m. Vehicles start to run at time $t = 0$ as the signal is switched to green. Then, the leading vehicle starts to accelerate until it reaches the OV at 15 m/s. To ensure comfort of car-following in ACC, the leading vehicle starts to accelerate with typical values of comfortable acceleration at 1.5 m/s^2, as suggested by (*50*). The value of the parameter λ is 0.35. Other parameters are set as described in the above section.

In this situation, the velocity and acceleration profiles of the DSM model are depicted in Figure 4*a*. The velocity and acceleration profiles of the extended DSM model are shown in Figure 4*b*. As shown in Figure 4, *a* and *b*, all following vehicles can maintain the same manner as the preceding vehicle but with some time delay. However, the figures further show that the time delay for the extended DSM model is small compared with that of the DSM model. Furthermore, the magnitude of acceleration for the extended DSM model is the highest as shown in Figure 4*b*, followed by the DSM model in Figure 4*a*. This situation indicates that the vehicles in the extended DSM model can finish the acceleration phase quickly. Results show that the

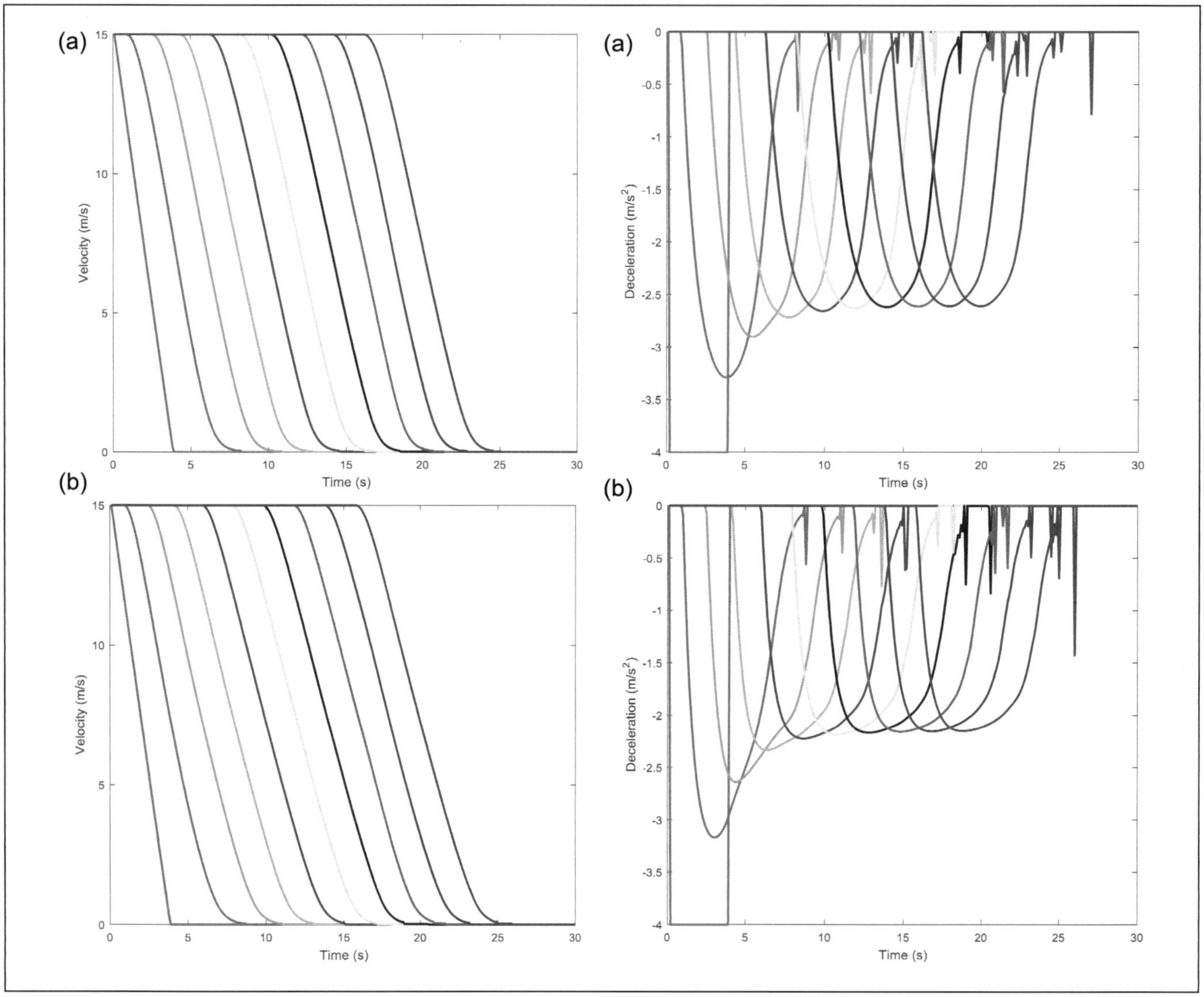

Figure 5. Velocity and acceleration profiles of vehicles 1–10 stopping at a red signal, showing (*a*) DSM model and (*b*) extended DSM model. Each curve shows the velocity and acceleration of each vehicle in corresponding figures, respectively.

vehicles with consideration of variation of historical perceived risk and acceptable risk can maintain the consensus state and finish the acceleration phase in a shorter time in the case study. Therefore, the dynamic performance in the start process of the extended DSM model is better, compared with that of the DSM model.

Stop Process

The stop process is set up as follows: initially, 10 vehicles run on a single lane with the same velocity of 20 m/s and space headway of 35 m. The leading vehicle brakes when the red signal comes up at $t = 0$. As suggested by (*51*), the leading vehicle brakes at -4 m/s^2 until it reaches a full stop. The other vehicles follow the leading vehicle to decelerate until all vehicles reach a full stop.

For comparison, the deceleration and velocity profiles of the DSM model and extended DSM model are illustrated in Figure 5, *a* and *b*, respectively. The figures demonstrate that the magnitude of deceleration for the extended DSM model is the least, followed by that of the DSM model. Figure 5 shows that all the following vehicles can duplicate the velocities of the leading vehicles but with some time delay. Furthermore, the time delay simulated by the extended DSM model is shorter than that simulated by the DSM model. Such a result implies that the extended DSM model can provide a good comfortable driving pattern compared with the DSM model. Thus, the dynamic performance in the stop process of the extended DSM model is better, compared with that of the DSM model.

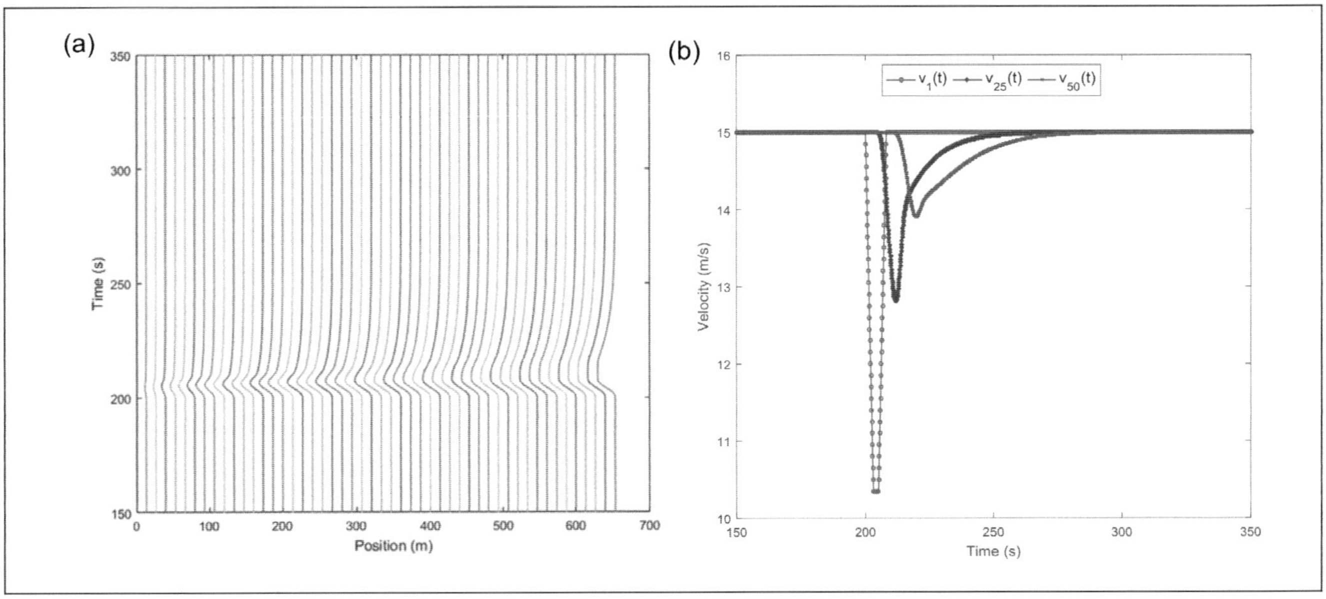

Figure 6. Numerical simulations for DSM model, showing (*a*) spatial–temporal pattern of vehicular traffic flow and (*b*) temporal velocity behavior of the 1st, 25th, and 50th vehicles.

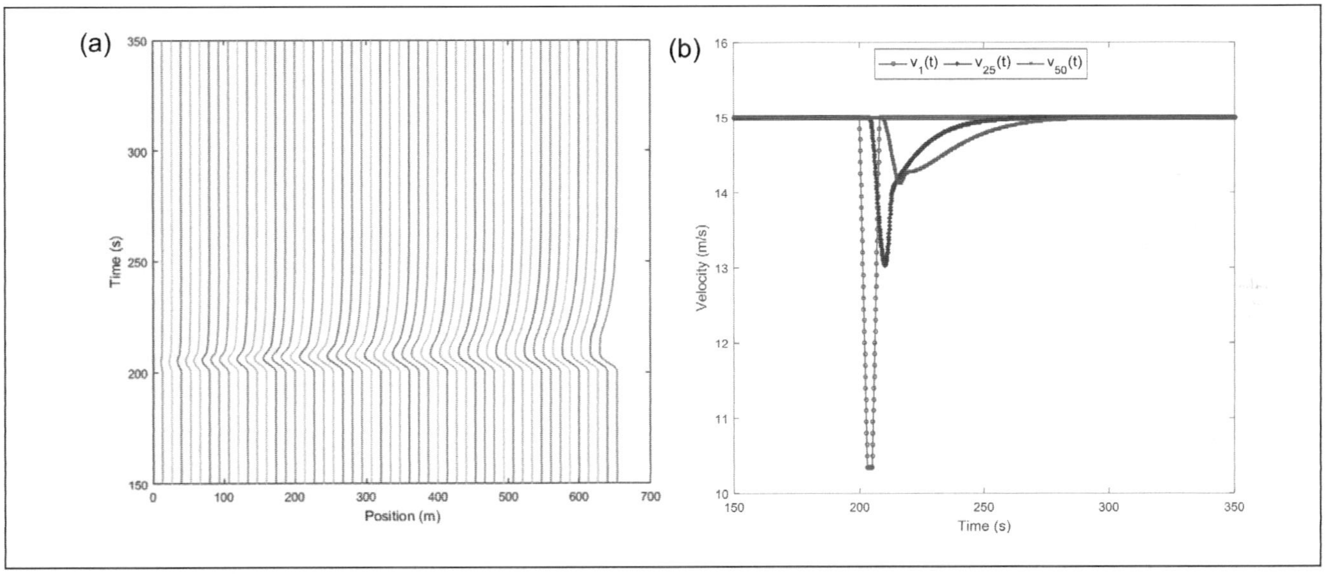

Figure 7. Numerical simulations for extended DSM model, showing (*a*) spatial–temporal pattern of vehicular traffic flow and (*b*) temporal velocity behavior of the 1st, 25th, and 50th vehicles.

Car-Following Process

In this subsection, several simulations are conducted under an open boundary condition using the extended DSM model to explore how the variation of historical perceived risk and acceptable risk affects the traffic flow evolution process. Generally, all the vehicles move on a single lane with space headway $L = 40\,\text{m}$ and vehicle number $N = 50$. It is assumed that the leading vehicle brakes at $-1.5\ \text{m/s}^2$ suddenly for a short time (200–203 s), and then the leading vehicle accelerates at $1.5\ \text{m/s}^2$ (205–208 s). The external disturbance is used to investigate the effect of the extended DSM model compared with the DSM model.

The simulation results of the DSM model and extended DSM model are shown in Figures 6 and 7, respectively. Figures 6*a* and 7*a* show the spatial–temporal pattern of vehicular traffic flow after 150 s. Figures 6*b* and 7*b* show the temporal velocity behavior

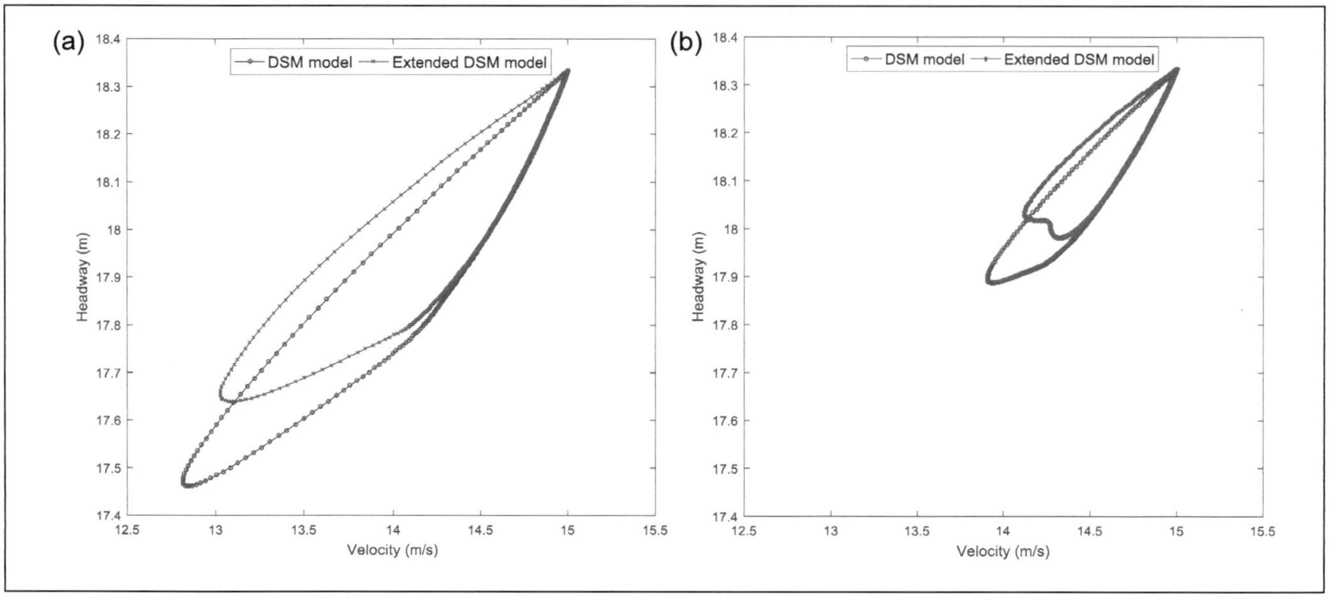

Figure 8. Hysteresis loops at both DSM model and extended DSM model from different vehicles: (*a*) 25th vehicle and (*b*) 50th vehicle.

of the 1st, 25th, and 50th vehicles. The degrees of space headway and velocity fluctuation are smaller in Figure 7, *a* and *b*, compared with those in Figure 6, *a* and *b*, respectively. This situation illustrates that the extended DSM model can improve the stability and smoothness of traffic flow in ACC and automotive platoon driving. Furthermore, the hysteresis loop of the 25th vehicle is larger in the DSM model compared with that in the extended DSM model, as shown in Figure 8*a*. Likewise, the hysteresis loop of the 50th vehicle is smaller in the extended DSM model compared with that in the DSM model, as shown in Figure 8*b*. Thus, the hysteresis loops of the vehicle further prove that the stability of traffic that considers the variation of historical perceived risk and acceptable risk flow is superior to that which does not consider this control strategy in the ACC system.

In summary, these simulations show that the variation of historical perceived risk and acceptable risk significantly affects the stability of traffic flow. This result implies that the historical information of the target vehicle has an important role in the dynamic performance. In ACC and automotive platoon driving, a similar control strategy should be considered in the speed control system to enhance the performance of car-following.

Analysis for Risk Indicators in Car-Following

The previous section Section 4 investigated the influence of the variation of historical perceived risk and acceptable risk on dynamic performance. In real traffic, smoothness and safety are the two most important problems. Therefore, the effect of the variation of historical perceived risk and acceptable risk on traffic safety will now be analyzed. In the car-following process, time-to-collision (TTC) and time headway (TH) are typically used to describe risk. Of these, TTC is the time that remains until collision occurs when both continue on the same course. TH is the time difference between the consecutive arrival instances of two vehicles that pass the measurement point on a lane. However, given that the effect of relative velocity is disregarded in the formulation of TH, risk at a large relative velocity may be underestimated. Therefore, in this study, TTC is chosen as a risk measure index, the expression of which can be computed as

$$\mathrm{TTC}_n = \frac{D_n(t)}{V_n(t) - V_{n-1}(t)}, \forall V_n(t) > V_{n-1}(t), \quad (8)$$

where

$V_n(t)$ denotes the velocity of the following vehicle;
$V_{n-1}(t)$ denotes the velocity of the preceding vehicle;
$D_n(t)$ represents the relative spacing between n and $n-1$.

Furthermore, the SM can be obtained as a risk indicator based on homeostatic risk perception in car-following. Based on the DSM model, SM is simplified as follows:

$$\mathrm{SM}_n = 1 - \frac{0.15 \cdot V_n(t) + [V_n(t)]^2/1.5g}{D_n(t)} + \frac{[V_{n-1}(t)]^2/1.5g}{D_n(t)} \quad (9)$$

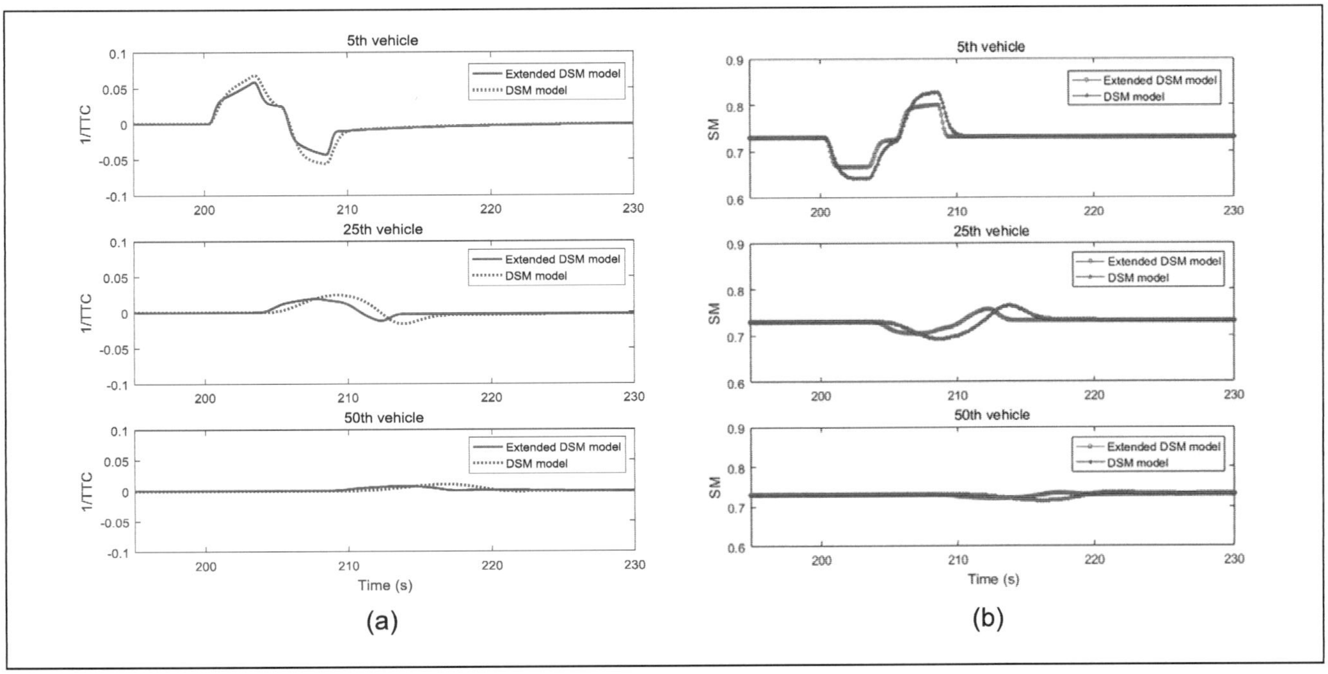

Figure 9. The evolution of two risk indicators obtained from the 5th, 25th, and 50th vehicles simulated by DSM model and extended DSM model: (*a*) 1/TTC of the vehicles and (*b*) SM of the vehicles.

This section attempts to investigate the effect of the variation of historical perceived risk and acceptable risk on two risk indicators by numerical simulation. The simulation is repeated many times to obtain the change in the two risk indicators. The initial conditions of vehicles and the external disturbance are kept the same as in the previous section.

The 5th, 25th, and 50th vehicles that were simulated by the extended DSM model and DSM model were selected as the target vehicles to investigate the difference between the two risk indicators. In the car-following process, because TTC of vehicles changes sharply, a large or even undefined TTC value may occur when the velocities of two successive vehicles in the platoon are almost equal. To avoid this condition, 1/TTC is used to represent TTC in this study. Figure 9*a* shows that 1/TTC of vehicles, obtained from the 5th, 25th, and 50th vehicles, likely fluctuates around zero. The minimum values of 1/TTC in the extended DSM model are smaller than those in the DSM model when the external disturbance is introduced into the platoon. That is, TTC of the extended DSM model is higher than that of the DSM model. Under the homeostatic risk perception indicator, Figure 9*b* shows that the minimum values of SM in the extended DSM model are larger than that in the DSM model. Thus, considering the variation of historical perceived risk and acceptable risk has a positive effect on traffic safety in the car-following process when an external disturbance occurs in the platoon.

Conclusion

In this paper, an extended DSM model is proposed, which considers the effect of variation of historical perceived risk and acceptable risk in the DSM car-following model. Findings are that the variation of historical perceived risk and acceptable risk has important effects on car-following behavior, using gray correlation analysis based on the real-vehicle test platform. The dynamic performance of the extended DSM model is investigated. Considering the simulation results, the conclusion is that the extended DSM model can better reflect the characteristics of traffic flow and improve the performance of the vehicle in three situations, namely, start process, stop process, and car-following process. Numerical simulations demonstrate that the extended DSM model can further improve traffic safety compared to the DSM model based on TTC and SM when the leading vehicle stops suddenly for a short time. Conclusions show that the historical information of the target vehicle should be considered to improve the performance of car-following in ACC and automotive platoon driving.

However, this paper has the following limitations:

(1) The reaction coefficient of the variation of historical perceived risk and acceptable risk is not calibrated by the use of experimental data.
(2) The influence of historical information of the previous moment on the driving behavior is

investigated, but the effects of the scale of historical information of the target vehicle on the driving behavior are not studied.

In view of the limitations, the real-vehicle test platform will be used to propose an exact DSM car-following model to further analyze the effects of the variation of historical perceived risk and acceptable risk on the driving behavior under different traffic situations such as bottlenecks, emergency braking, or congestion in ACC and automotive platoon driving.

Acknowledgments

This work was supported by the National Natural Science Foundation of China (Grant U1564212, U1664262).

Author Contributions

The authors confirm contribution to the paper as follows: study conception and design: Guangquan Lu, Junjie Zhang, Yunpeng Wang; data collection: Junjie Zhang; analysis and interpretation of results: Junjie Zhang, Guangquan Lu, Yunpeng Wang; draft manuscript preparation: Junjie Zhang. All authors reviewed the results and approved the final version of the manuscript.

References

1. Chandler, R. E., R. Herman, and E. W. Montroll. Traffic Dynamics: Studies in Car Following. *Operations Research*, Vol. 6, No. 2, 1958, pp. 165–184.
2. Gazis, D. C., R. Herman, and R. W. Rothery. Nonlinear Follow-the-Leader Models of Traffic Flow. *Operations Research*, Vol. 9, No. 4, 1961, pp. 545–567.
3. Bexelius, S. An Extended Model for Car-Following. *Transportation Research*, Vol. 2, No. 1, 1968, pp. 13–21.
4. Lenz, H., C. K. Wagner, and R. Sollacher. Multi-Anticipative Car-Following Model. *The European Physical Journal B*, Vol. 7, No. 2, 1999, pp. 331–335.
5. Addison, P. S., and D. J. Low. A Novel Nonlinear Car-Following Model. *Chaos: An Interdisciplinary Journal of Nonlinear Science*, Vol. 8, No. 4, 1998, pp. 791–799.
6. Helly, W. Simulation of Bottlenecks in Single-Lane Traffic Flow. In *Proc., Symposium on Theory of Traffic Flow*, Research Laboratories, General Motors. Elsevier, New York, 1959, pp. 207–238.
7. Treiber, M., A. Hennecke, and D. Helbing. Congested Traffic States in Empirical Observations and Microscopic Simulations. *Physical Review E*, Vol. 62, No. 2 Pt A, 2000, p. 1805.
8. Lee, G. A Generalization of Linear Car-Following Theory. *Operations Research*, Vol. 14, No. 4, 1966, pp. 595–606.
9. Ahmed, K. I. *Modeling Drivers' Acceleration and Lane Changing Behavior*. Massachusetts Institute of Technology, 1999.
10. Herman, R., and R. W. Rothery. Car Following and Steady State Flow. In *Proc., 2nd International Symposium on the Theory of Traffic Flow* (J. Almond, ed.), OECD, Paris, 1965.
11. Koshi, M., M. Kuwahara, and H. Akahane. Capacity of Sags and Tunnels on Japanese Motorways. *ITE Journal*, Vol. 62, No. 5, 1992, pp. 17–22.
12. Xing, J. A Parameter Identification of a Car-Following Model. In *Steps Forward*. Intelligent Transport Systems World Congress, Yokohama, Japan, 1995, pp. 1739–1745.
13. Gipps, P. G. A Behavioural Car-Following Model for Computer Simulation. *Transportation Research Part B Methodological*, Vol. 15, No. 2, 1981, pp. 105–111.
14. Newell, G. F. Nonlinear Effects in the Dynamics of Car Following. *Operations Research*, Vol. 9, No. 2, 1961, pp. 209–229.
15. Bando, M., K. Hasebe, A. Nakayama, A. Shibata, and Y. Sugiyama. Dynamical Model of Traffic Congestion and Numerical Simulation. *Physical Review E*, Vol. 51, No. 2, 1995, p. 1035.
16. Farin, G. E., and D. Hansford. Generalized Force Model of Traffic Dynamics. *Physical Review E*, Vol. 58, No. 1, 1998, pp. 133–138.
17. Jiang, R., Q. Wu, and Z. Zhu. Full Velocity Difference Model for a Car-Following Theory. *Physical Review E*, Vol. 64, No. 1 Pt 2, 2001, p. 017101.
18. Gong, H., H. Liu, and B. H. Wang. An Asymmetric Full Velocity Difference Car-Following Model. *Physica A Statistical Mechanics and Its Applications*, Vol. 387, No. 11, 2008, pp. 2595–2602.
19. Li, Y. Car-Following Model Based on the Information of Multiple Ahead & Velocity Difference. *Systems Engineering—Theory & Practice*, Vol. 30, No. 7, 2010, pp. 1326–1332.
20. Zhao, X., and Z. Gao. A New Car-Following Model: Full Velocity and Acceleration Difference Model. *The European Physical Journal B—Condensed Matter and Complex Systems*, Vol. 47, No. 1, 2005, pp. 145–150.
21. Tao, W. Multiple Velocity Difference Model and Its Stability Analysis. *Acta Physica Sinica*, Vol. 55, No. 2, 2006, pp. 634–640.
22. Li, Y., D. Sun, W. Liu, M. Zhang, M. Zhao, X. Liao, and L. Tang. Modeling and Simulation for Microscopic Traffic Flow Based on Multiple Headway, Velocity and Acceleration Difference. *Nonlinear Dynamics*, Vol. 66, No. 1–2, 2011, pp. 15–28.
23. Orosz, G., B. Krauskopf, and R. E. Wilson. Bifurcations and Multiple Traffic Jams in a Car-Following Model with Reaction-Time Delay. *Physica D: Nonlinear Phenomena*, Vol. 211, No. 3, 2005, pp. 277–293.
24. Orosz, G., and G. Stépán. Subcritical Hopf Bifurcations in a Car-Following Model with Reaction-Time Delay. *Proceedings Mathematical Physical & Engineering Sciences*, Vol. 462, No. 2073, 2006, pp. 2643–2670.
25. Low, D., and P. Addision. Chaos in a Car-Following Model with a Desired Headway Time. *Proc., 30th International Symposium on Automotive Technology & Automation Conference*, Florence, Italy. ATT/ITS Advances for Enhancing Passenger Freight and Intermodal Transportation Systems, Automotive Automation, 1997, pp. 175–182.

26. Tang, T., W. Shi, H. Shang, and Y. Wang. A New Car-Following Model with Consideration of Inter-Vehicle Communication. *Nonlinear Dynamics*, Vol. 76, No. 4, 2014, pp. 2017–2023.
27. Tang, T., Y. Wang, X. Yang, and Y. Wu. A New Car-Following Model Accounting for Varying Road Condition. *Nonlinear Dynamics*, Vol. 70, No. 2, 2012, pp. 1397–1405.
28. Tang, T. Q., L. Chen, S. C. Yang, and H. Y. Shang. An Extended Car-Following Model with Consideration of the Electric Vehicle's Driving Range. *Physica A Statistical Mechanics and Its Applications*, Vol. 430, 2015, pp. 148–155.
29. Tang, T. Q., H. J. Huang, and G. Xu. A New Macro Model with Consideration of the Traffic Interruption Probability. *Physica A Statistical Mechanics and Its Applications*, Vol. 387, No. 3, 2009, pp. 975–983.
30. Tang, T. Q., W. F. Shi, H. Y. Shang, and Y. P. Wang. An Extended Car-Following Model with Consideration of the Reliability of Inter-Vehicle Communication. *Measurement*, Vol. 58, No. 11, 2014, pp. 286–293.
31. Li, Y., L. Zhang, S. Peeta, H. Pan, T. Zheng, Y. Li, and X. He. Non-Lane-Discipline-Based Car-Following Model Considering the Effects of Two-Sided Lateral Gaps. *Nonlinear Dynamics*, Vol. 80, No. 1–2, 2015, pp. 227–238.
32. Ge, H. X., H. B. Zhu, and S. Q. Dai. Effect of Looking Backward on Traffic Flow in a Cooperative Driving Car Following Model. *European Physical Journal B—Condensed Matter and Complex Systems*, Vol. 54, No. 4, 2006, pp. 503–507.
33. Yu, S., and Z. Shi. An Improved Car-Following Model Considering Headway Changes with Memory. *Physica A Statistical Mechanics and Its Applications*, Vol. 421, 2015, pp. 1–14.
34. Peng, G., W. Lu, H. He, and Z. Gu. Nonlinear Analysis of a New Car-Following Model Accounting for the Optimal Velocity Changes with Memory. *Communications in Nonlinear Science and Numerical Simulation*, Vol. 40, 2016, pp. 197–205.
35. Winsum, W. V. The Human Element in Car Following Models. *Transportation Research Part F: Traffic Psychology and Behaviour*, Vol. 2, No. 4, 1999, pp. 207–211.
36. Yang, H. H., and H. Peng. Development of an Errorable Car-Following Driver Model. *Vehicle System Dynamics*, Vol. 48, No. 6, 2010, pp. 751–773.
37. Andersen, G. J., and C. W. Sauer. Optical Information for Car Following: The Driving by Visual Angle (Dva) Model. *Human Factors*, Vol. 49, No. 5, 2007, p. 878.
38. Jin, S., D. H. Wang, Z. Y. Huang, and P. F. Tao. Visual Angle Model for Car-Following Theory. *Physica A Statistical Mechanics and Its Applications*, Vol. 390, No. 11, 2011, pp. 1931–1940.
39. Hamdar, S. H., M. Treiber, H. S. Mahmassani, and A. Kesting. Modeling Driver Behavior as Sequential Risk-Taking Task. *Transportation Research Record: Journal of the Transportation Research Board*, 2008. 2088: 208–217.
40. Lu, G., B. Cheng, Y. Wang, and Q. Lin. A Car-Following Model Based on Quantified Homeostatic Risk Perception. *Mathematical Problems in Engineering*, Vol. 2013, No. 6, 2013, pp. 165–185.
41. Yu, S., and Z. Shi. An Extended Car-Following Model Considering Vehicular Gap Fluctuation. *Measurement*, Vol. 70, 2015, pp. 137–147.
42. Ge, H. X., S. Q. Dai, L. Y. Dong, and Y. Xue. Stabilization Effect of Traffic Flow in an Extended Car-Following Model Based on an Intelligent Transportation System Application. *Physical Review E*, Vol. 70, No. 6 Pt 2, 2004, p. 066134.
43. Jiang, R., and Q. S. Wu. The Night Driving Behavior in a Car-Following Model. Physica *A Statistical Mechanics and Its Applications*, Vol. 375, No. 1, 2007, pp. 297–306.
44. Tang, T. Q., C. Y. Li, and H. J. Huang. A New Car-Following Model with the Consideration of the Driver's Forecast Effect. *Physics Letters A*, Vol. 374, No. 38, 2010, pp. 3951–3956.
45. Tang, T. Q., H. J. Huang, S. G. Zhao, and G. Xu. An Extended Ov Model with Consideration of Driver's Memory. *International Journal of Modern Physics B*, Vol. 23, No. 5, 2009, pp. 743–752.
46. Lu, G., B. Cheng, Q. Lin, and Y. Wang. Quantitative Indicator of Homeostatic Risk Perception in Car Following. *Safety Science*, Vol. 50, No. 9, 2012, pp. 1898–1905.
47. Gipps, P. G. A Behavioural Car-Following Model for Computer Simulation. *Transportation Research Part B: Methodological*, Vol. 15, No. 2, 1981, pp. 105–111.
48. Vangi, D., and A. Virga. Evaluation of Emergency Braking Deceleration for Accident Reconstruction. *Vehicle System Dynamics*, Vol. 45, No. 10, 2007, pp. 895–910.
49. Sokolovskij, E. Experimental Investigation of the Braking Process of Automobiles. *Transport*, Vol. 20, No. 3, 2010, pp. 91–95.
50. Treiber, M., A. Kesting, and C. Thiemann. *Traffic Flow Dynamics: Data, Models and Simulation*. Springer, Heidelberg, New York, 2013.
51. *Highway Capacity Manual*. TRB, National Research Council, Washington, D.C., 2000.

The Standing Committee on Traffic Flow Theory and Characteristics (AHB45) peer-reviewed this paper (18-02547).

Article

TRR
JOURNAL OF THE TRANSPORTATION RESEARCH BOARD

Transportation Research Record
2018, Vol. 2672(20) 98–109

DOI: 10.1177/0361198118796395
journals.sagepub.com/home/trr
SAGE

Multi-Thread Optimization for the Calibration of Microscopic Traffic Simulation Model

Zenghao Hou[1] and Joyoung Lee[2]

Abstract

This paper proposes an innovative multi-thread stochastic optimization approach for the calibration of microscopic traffic simulation models. Combining Quasi-Monte Carlo (QMC) sampling and the Particle Swarm Optimization (PSO) algorithm, the proposed approach, namely the Quasi-Monte Carlo Particle Swarm (QPS) calibration method, is designed to boost the searching process without prejudice to the calibration accuracy. Given the search space constructed by the combinations of simulation parameters, the QMC sampling technique filters the searching space, followed by the multi-thread optimization through the PSO algorithm. A systematic framework for the implementation of the QPS QMC-initialized PSO method is developed and applied for a case study dealing with a large-scale simulation model covering a 6-mile stretch of Interstate Highway 66 (I-66) in Fairfax, Virginia. The case study results prove that the proposed QPS method outperforms other methods utilizing Genetic Algorithm and Latin Hypercube Sampling in achieving faster convergence to obtain an optimal calibration parameter set.

Microscopic traffic simulation has gained ever-increasing attention from both industry and academia in the field of transportation engineering. Providing users with a high-fidelity modeling environment, the use of microscopic traffic simulation has become popular as a viable tool enabling traffic engineers to conduct risk-free analysis for evaluating the wide variety of congestion mitigation strategies and cutting-edge intelligent transportation systems technologies. Relying on various mathematical models comprising numerous coefficients that directly affect the accuracy of simulation results, calibrating microscopic traffic simulation is one of the most crucial steps to ensure the highest validity of the simulation models.

A microscopic traffic simulation consists of various driving behavior models handling drivers' microscopic maneuvers for car-following, lane-changing, and route choosing to precisely describe the movement of an individual vehicle, as depicted in Equation 1:

$$P_{Sim,t} = S(\beta_1, \beta_2, \ldots \beta_i); \beta_i \in [l_i, u_i] \quad (1)$$

where, i is the total number of parameters that modelers select for the calibration and β_i denotes a set of parameters, each of which is subject to a lower (i.e., l_i) and an upper (i.e., u_i) boundaries. The function $S(.)$ indicates the combination of each driving behavior model. Depending on the size of simulation coverage, the combinations of such parameters construct a multi-dimensional complex optimization problem forming a hypercube searching space confined by the lower and upper boundaries of each dimension (i.e., parameter).

A wide variety of approaches have been proposed to deal with the problem of microscopic traffic simulation calibration (*1–11*). However, there is no commonly accepted dominant approach in the state-of-the art and practice. This paper proposes a novel multi-thread optimization method for the calibration of microscopic traffic simulation by integrating Quasi-Monte Carlo (QMC) (*12*) and Particle Swarm Optimization (PSO) (*13*), namely the Quasi-Monte Carlo Particle Swarm (QPS) calibration method. The QMC sampling method generates homogeneous parameter combinations within the hypercube searching space. With the samples obtained from QMC, the proposed algorithm conducts the distribution estimation satisfying the conservation condition: $P_{sim,t} \propto i$. In turn, PSO is implemented to perform optimization processes based on the filtered searching spaces obtained from QMC.

[1]University of California at Davis, Davis, CA
[2]John A. Reif, Jr. Department of Civil and Environmental Engineering, New Jersey Institute of Technology, Newark, NJ

Corresponding Author:
Address correspondence to Joyoung Lee: jo.y.lee@njit.edu

The rest of this paper is organized as follows. In the next section, a literature review examines the pros and cons of several existing calibration methods. The methodology section presents the overall framework of the proposed multi-thread optimization-based calibration approach. A case study is then presented to demonstrate the effectiveness of the proposed approach by conducting an actual calibration effort for a VISSIM-based (*14*) freeway simulation model. The findings and conclusions of this research are discussed in the concluding remarks section.

Literature Review

This section presents reviews for the state-of-the-art and practice of the calibration approaches for microscopic traffic simulation models. Treiber & Kesting (*15*) and Kim et al. (*16*) proposed classic statistical methods to estimate parameters. However, this paper emphasizes solving the calibration problem by using heuristic methodologies.

Gardes et al. (*2*) and Gomes et al. (*3*) conducted the calibrations of PARAMICS (*17*) and VISSIM models, respectively, by applying sequential trial-and-error-based methods. In their methods, a simulation modeler arbitrarily chooses one parameter and conducts manual tweaking for the selected parameter while fixing other parameters. This approach is straightforward and easy to implement, particularly for a small-scale network. However, the entire calibration process heavily relies on the modeler's subjective judgments, thereby resulting in unreliable calibration results.

Park and Qi (*4*) and Park and Schneeberger (*5*) proposed systematic frameworks for the calibration of VISSIM simulation model by integrating genetic algorithm (GA) (*18*) and Latin hypercube sampling (LHS) (*19*). LHS is employed to generate unbiased sample sets out of all possible combinations of parameter values. GA demonstrated promising performance to find near-global optimal solutions that can make the calibration more reliable. Despite the promising performance of the procedure, the authors had to conduct multiple runs of LHS and GA optimizations as a result of the stochastic nature of LHS, which resulted in excessive calibration time (*4*, *5*).

Lee et al. (*6*) simplified the calibration procedure in Park and Qi (*4*) by eliminating the feasibility test. To this end, the authors proposed enlarging the LHS sample sizes to cover wider ranges of parameter values. The addressed multi-step calibrations in Park and Qi (*4*) become unnecessary by allowing bigger sample size that would likely contain feasible candidate parameter sets. Using 5000 samples obtained from LHS that included several feasible candidate sets, the authors proved that the LHS-incorporated procedure outperformed the GA-only method proposed in Park and Qi (*4*) in terms of saving total calibration time for a large-scale simulation network. Yet, the authors did not clearly address how to determine the proper sample sizes to deal with the variety of simulation scales.

Lee and Ozbay (*7*), Ma et al. (*8*), and Paz et al. (*9*) proposed a simultaneous perturbation stochastic approximation (SPSA) (*20*) technique for microscopic simulation calibration. In Paz et al. (*9*), the authors demonstrated that the SPSA technique outperformed the trial-and-error methods and GA. They also proved that SPSA needs fewer rounds of evaluations to produce the same level or improved calibration performance than GA and the trial-and-error methods. In addition, Paz et al. (*9*) proposed an enhanced SPSA methodology to reduce the number of evaluations.

Abdalhaq and Abu Baker (*10*) reviewed GA, PSO, SPSA, and Tabu Search (TS) (*21*) and conducted numerical studies for the calibration of microscopic traffic simulation by implementing these algorithms. The result of this study demonstrated that PSO, as an easy parallel computing algorithm, is capable of simultaneously calibrating microscopic traffic simulation models. However, the dimension for calibration is often greater than 4 in practice, and it is challenging to ensure that PSO has the promising accuracy with high-dimensional calibration.

In summary, the trial-and-error methods (*2*, *3*) and heuristic algorithms, such as GA (*4*, *5*), SPSA (*7–9*), and PSO (*10*), have all been proposed to perform the calibration of microscopic traffic simulation models. In addition, research (*5*, *6*) has proposed LHS to filter out the searching spaces to increase the probability of finding the best solution. Particularly in Lee et al. (*6*), LHS is solitarily employed to search for the best parameter set for the calibration without optimization methods. However, thus far, none of the aforementioned methods has become dominant owing to the following challenges. First, the trial-and-error method requires subjective judgments of modelers and consumes a tremendous amount of time and effort for the calibration. Second, although GA is capable of seeking out the best solution, its slow convergence speed in high-dimensional problems undermines the performance in the case of the calibration of large-scale simulation models. Third, SPSA also has limitations on parallelization in that it is unsuitable for calibrating a large-scale simulation model. Furthermore, LHS utilizing a large sample size needs multiple runs for searching for the best solution because of the stochastic nature of its sampling approach. Last, although PSO appears promising, its accuracy became a critical challenge, as shown in Abdalhaq and Abu Baker (*10*). Through reviewing the literature, it is discovered that developing a novel approach is necessary to improve the efficiency, accuracy, and robustness of microscopic traffic simulation model calibration.

Methodology

With the flexibility suitable for the wide spectrum of optimization problems, heuristic algorithms, such as GA, PSO, and SPSA, have been adopted for the calibration of microscopic traffic simulation models. Relying on randomly selected initial solutions, however, heuristic algorithms likely miss a chance to reach the global optimum from the first iteration. Repeatedly implementing the heuristic algorithm with various initial solutions might prevent missing the chance to find an optimal solution. Yet, depending on the network size, this is often impractical for the calibration of a traffic simulation model, as it requires significant time and computational efforts to complete the optimization task. However, incorporating sampling techniques into the heuristic optimization can dramatically reduce the searching space for the optimization process. In this paper, as a robust sampling technique producing low-discrepancy samples, QMC is selected to be integrated into the implementation of the PSO algorithm. The details of advantages and disadvantages of the PSO are presented in the following section by a comparison with GA, and the details of the QMC and the forming of the QPS are presented after it.

Comparison between Generic Algorithm and Particle Swarm Optimization

Genetic Algorithm. GA is a population-based stochastic optimization algorithm, which has been proposed as a calibration framework elsewhere (*4*, *5*). It is also popular in the state-of-the practice of optimization applications because of its flexibility, robustness, and easy-to-implement structure. Instead of being best performance driven, GA has no dominant chromosomes guiding the tendency discovered at each generation; many better-performing chromosomes (a.k.a., elites) are brought to the next generation without crossover or mutation. Therefore, such chromosomes move in crowd sweeping through searching spaces.

Particle Swarm Optimization. PSO was inspired by the foraging behavior of birds. A flock of birds looks for food and shares the location information of food while they are flying. Once a food source is found, the information is propagated to birds in the flock and, in turn, the flock approaches the food by constantly adjusting their velocity and position to approach the food. From the perspective of heuristic optimization, such a biomimetic process can be divided into two stages: (1) initialization (i.e., initialization of a particle's position); and (2) iteration. The iteration stage consists of the following three steps: (1) evaluate fitness value; (2) update particles' position and velocity; and (3) update best values and positions in the swarm and particle.

PSO randomly samples the initial point for initialization using:

$$x_0^i = \mathrm{rand}(R), R \in [l_i, u_i], l_i, u_i \text{are lower and upper bound of i} \tag{2}$$

The iteration stage is conducted with the following equations:

$$v_{t+1}^i = \omega v_t^i + c_1 \frac{\mathrm{rand}\left(p_t^{i,lb} - x_t^i\right)}{\Delta t} + c_2 \frac{\mathrm{rand}\left(p_t^{i,sb} - x_t^i\right)}{\Delta t} \tag{3}$$

$$x_{t+1}^i = x_t^i + v_{t+1}^i \Delta t \tag{4}$$

$$p_{t+1}^{i,lb} = \begin{cases} x_{t+1}^i, \mathbf{V}(x_{t+1}^i) - \mathbf{V}(p_t^{i,lb}) > 0 \\ p_t^{i,lb}, \mathbf{V}(x_{t+1}^i) - V\left(p_t^{i,lb}\right) < 0 \end{cases} \tag{5}$$

$$p_{t+1}^{sb} = \max_{i \in P}\left\{p_{t+1}^{i,lb}\right\} \tag{6}$$

where, $v_{t+1}^i, x_{t+1}^i,$ denote the particle i's velocity and location at the time interval *t + 1*, respectively. $p_t^{i,lb}, p_t^{i,sb}$ are the best position of the particles (local optimal) and the best of best positions in its history (global optimal) at time t, respectively. V(.) denotes the vector of the parameters. ω denotes the strength of inertia from the previous speed status. c_1, c_2 are coefficients (i.e., weight) that determine the next movement for each particle by examining the performance of the local optima $p_t^{i,lb}$ and the global optima $p_t^{i,sb}$ of the swarm. Generally, $c_1 \in [1.5, 2], c_2 \in [2, 2.5]$.

The speed of particle denoted by v_{t+1}^i balances the time expense and accuracy performance, and it is determined by the inertia of previous speed status, ωv_t^i, the affection of local best position, $c_1 \frac{\mathrm{rand}\left(p_t^{i,lb} - x_t^i\right)}{\Delta t}$, and the affection of global best position, $c_2 \frac{\mathrm{rand}\left(p_t^{i,sb} - x_t^i\right)}{\Delta t}$. As shown in Figure 1*a*, $\beta_{i,t}$ is the parameter set that can be selected in the hypercube space at the time interval *t*. The velocity of $\beta_{i,t}$ is determined by Equation 3 using the global best (p_t^{sb}), the local best ($p_t^{i,lb}$), and its own inertia (ωv_t^i) as illustrated by the dashed lines and circles for affecting position with distinct colors in Figure 1, *a* and *b*. As $\beta_{i+2,t}$ is the both global and local best spot at time *t*, it was affected by only its inertia as depicted in Figure 1*b*. $\beta_{i+1,t}$ is determined by local and global optimums at time *t* as shown in Figure 1*c*. Then, the whole process of swarm's moving finishes: calculated by Equations 4–6,

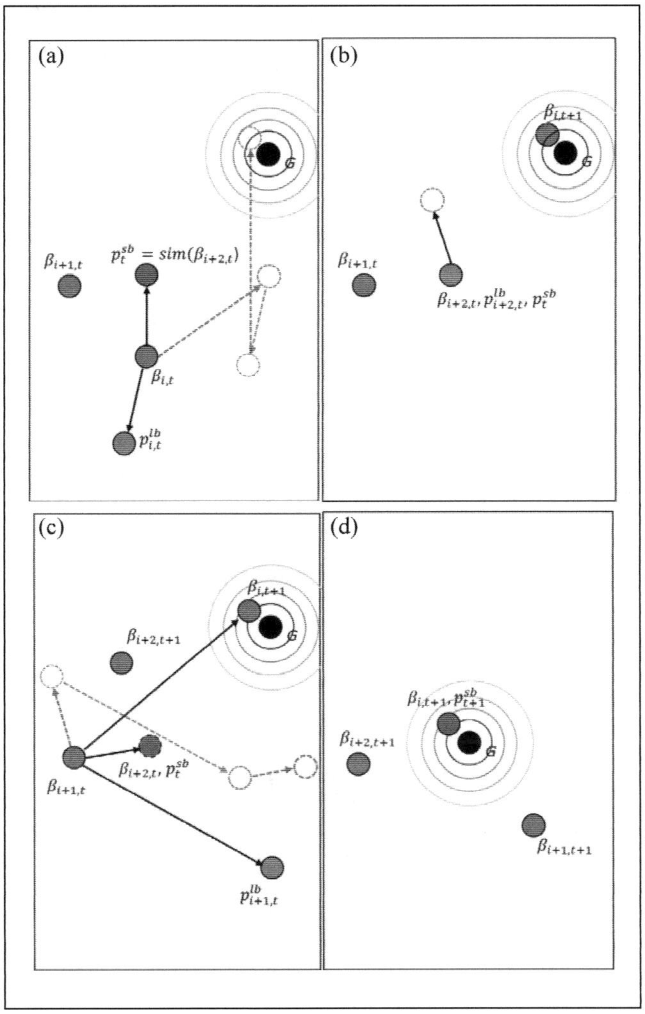

Figure 1. Particles' optimization steps: (*a*) Step 1; (*b*) Step 2; (*c*) Step 3; (*d*) Step 4.

the parameter set $\beta_{i,t+1}\left(p_{t+1}^{i,lb}, p_{t+1}^{i,sb}\right)$ becomes the both local and global optimal leading the next round of swarm moving, as shown in Figure 1*d*.

The Comparison. PSO has two dominant factors, that is, local best position from each generation $p_t^{i,lb}$ and global best performance position $p_t^{i,sb}$, that lead the tendency of the swarm moving, which makes the searching process fast. That is, PSO performs a cooperative and parallelized optimization process compared with GA. As a result, it was proved that PSO outperforms GA in terms of computation efficiency under an unconstrained nonlinear problem with continuous variables (*22*). As almost all the parameters in traffic simulation model calibration are continuous variables, PSO would therefore be better than GA to solve the calibration problem.

However, the weakness of PSO cannot be ignored. Although random factors are added to Equation 2 to avoid early convergence, local optimum "traps" for PSO are unavoidable. There exist various viable solutions to lower the possibility of being trapped by a local optimum; for instance, (1) a mutation process in GA that helps a chromosome jump out of early convergence and searches for better performances, and (2) a TS (*21*) algorithm that leaves one seed at the optimum and spreads other seeds to search for better optimums. However, such methods have both pros and cons at the same time, meaning that engineers need to balance the time expense and the performance of optimization results. To tackle this issue, this paper adopts a QMC-based sampling technique to filter in the potential best solution candidates within the searching space.

The Methodology of QPS

Quasi-Monte Carlo Sampling. In the real-world practices of microscopic traffic simulation model calibration, simulation modelers are often unable to conduct in-depth calibration because of a lack of time and resources (e.g., software, hardware). In that sense, an efficient and robust method is needed for modelers to successfully screen out redundant samples to find the global optimum in a timely manner. The QMC sampling method is receiving ongoing attention from a wide variety of application areas. It has been used in numerical finance (*23*), electrical engineering (*24*), and civil engineering (*25*) because of its proven performance in generating low-discrepancy samples. With this proven performance, this paper employed the QMC sampling technique for filtering the potential solution areas in the searching space.

QMC is a deterministic sampling method (*26*) in nature. QMC starts from base conversion (i.e., binary, decimal, and hexadecimal) to create a deterministic low-discrepancy sequence (LDS) (*27*), and combines the sequences, which are constructed by different bases, to conduct multi-dimension sampling. Depending on base-converting methods, QMC is categorized into two sequencing methods: (1) Halton sequence (*28*, *29*) and (2) Sobol sequence (*30*).

Both sequencing methods are initiated by a one-dimensional LDS: Van der Corput sequence (*28*). A hexadecimal representation shall be calculated by:

$$\alpha(i) = \sum_{\gamma=0}^{I} a_\gamma(i) x b^\gamma \quad (7)$$

where, $\alpha(i)$ is the hexadecimal representation of *i* and b^γ is the hexadecimal-based exponent. The summation of each digit is the hexadecimal representation of *i*. For the Van der Corput sequence, the number of digits of $\alpha(i)$ will be mirrored to the other side of the decimal point, and converted back to hexadecimal representation, $\phi_b(i)$:

$$\phi_b(i) = a_1(i) \cdot b^{-1} \ldots a_I(i) \cdot b^{-I} = \sum_{\gamma=0}^{I} \frac{a_\gamma(i)}{b^{\gamma+1}} \quad (8)$$

Taking $b = 2$, for example, $\alpha(i) = \boxed{1}*2^1 + \boxed{0}*2^0 \rightarrow 10_2$ (binary); then it mirrors the binary to the decimal point, for example, 10.0 to 0.01. Then, it converts binary 0.01 to decimal $\phi_2(i) = 0*2^0 + 0*2^{-1} + 1*2^{-2} = \frac{1}{4}$. When converting the decimal sequence {1, 2, 3, 4...}, the binary Van der Corput sequence becomes:

$$\left.\frac{1}{2}, \frac{1}{4}, \frac{3}{4}, \frac{1}{8}, \frac{5}{8}, \frac{3}{8}, \frac{7}{8}, \frac{1}{16}, \frac{9}{16}, \frac{5}{16}, \cdots\right\} \quad (9)$$

Then, ternary Van der Corput sequence is:

$$\left.\frac{1}{3}, \frac{2}{3}, \frac{1}{9}, \frac{4}{9}, \frac{7}{9}, \frac{2}{9}, \frac{5}{9}, \frac{8}{9}, \frac{1}{27}, \frac{10}{27}, \cdots\right\} \quad (10)$$

Thus, in the Van der Corput sequence, the new element in the set lies in the area that was not seized by previous elements.

The Halton sequence pairs the Van der Corput sequence with different bases combined together to construct two or higher dimensional samples, as expressed in Equation 11:

$$\left\{\left(\frac{1}{2}, \frac{1}{3}\right), \left(\frac{1}{4}, \frac{2}{3}\right), \left(\frac{3}{4}, \frac{1}{9}\right), \left(\frac{1}{8}, \frac{4}{9}\right), \left(\frac{5}{8}, \frac{7}{9}\right) \cdots\right\} \quad (11)$$

The deterministic sequence generated from Equation 11 assures samples of uniform distribution. It is necessary to note that using the Halton sequence it is often challenging to produce high-dimensional uniform samples because of a sequence correlation issue caused by higher base numbers (*31*). The Sobol sequence, on the other hand, is less sensitive to the number of dimensions. Constructed from the Van der Corput sequence with base 2 for the first dimension, the Sobol sequence builds higher dimension sequences by using Equation 12:

$$x_n = a_1(i)v_1 \oplus a_2(i)v_2 \oplus \ldots \oplus a_m(i)v_m, n \geq 0 \quad (12)$$

where, $a_1(i)$ is the hexadecimal representation of I and v_i is the number that help the first dimension deform into other dimensions. $v_i = \frac{m_i}{2^i}, 0 < m_i < 2^i$, are odd integers. $\oplus$ denotes a bit-by-bit exclusive-or-operation (XOR). Compared with LHS, a stratified sampling technique that randomly chooses a sample from an evenly divided strata in a hypercube $(0, 1)^s$, Asmussen and Glynn (*24*) showed that the Sobol sequence achieves faster convergence with higher dimension cases.

Figure 2 graphically shows that QMC outperforms LHS in terms of creating uniform hypercube distribution. The red circles in Figure 2*a* indicate the biased area that LHS would likely generate, which are aggregated by random sampling from each strata. Using the same sample data in Figure 2, *a* and *b*, Figure 2, *c* and *d* illustrate mesh charts created by filling out the affection area within the 0.04 radius from each data point. As clearly appeared in both Figure 2, *c* and *d*, samples from LHS produce more uncovered areas (i.e., shown as white area) than Halton QMC. This experimental observation clearly confirms that QMC outperforms LHS in terms of generating unbiased samples, which also confirms the results revealed by Wang and Fang (*31*).

An additional effort to investigate the performance of QMC is conducted by calculating Euclidean distances (EDs) in Equation 13 with samples randomly selected by QMC-Sobol, QMC-Holton, and LHS:

$$\mathrm{ED} = \sqrt{\sum_{i=1}^{n} (s_i - o_i)^2} \quad (13)$$

where, n, s_i, o_i denote the number of dimension, the sample at ith dimension, and the objective point at ith dimension, respectively. The samples are generated based on the different dimensions and sample size. Figure 3*a* shows ED changes using 100 random samples generated by changing the number of dimensions from 10 to 1,000. Contrarily, Figure 3*b* also shows the changes of ED with the different number of samples from 10 to 1,000 obtained given 100 dimensions. The blue dashed lines in Figure 3 are the average distances that were converted into ED. Every experiment is executed on the premise of statistical significance. The test results in Figure 3*a* show that both Sobol and Halton samplings outperform LHS when the number of dimension is less than 100. With the growth of number dimension, Sobol increases the advantage and Halton falls behind LHS and Sobol in terms of ED. In brief, Sobol has outstanding performance in multi-dimensional sampling. Furthermore, it is also discovered that QMC (i.e., both QMC-Sobol and QMC-Halton) outperforms LHS when the number of samples is higher than 100 samples while fixing the number of dimension in Figure 3*b*, and the Sobol sampling appears less sensitive to the sample sizes than LHS.

With the findings regarding QMC in this paper, a QMC-based filtering was adopted for the proposed calibration framework. It is worth noting that the Halton sequence is recommended for 10 to 15 dimensional samplings compared with Sobol sequence; for higher dimension samplings, the Sobol sequence is suitable. The current Sobol sequence sampling can cover 21,201 dimensions (*32*), which is sufficient enough to cover most traffic simulation calibration cases.

QPS Algorithm Development. In the QPS algorithm, QMC performs a filtering function to narrow down the searching space to increase the probability of finding potential

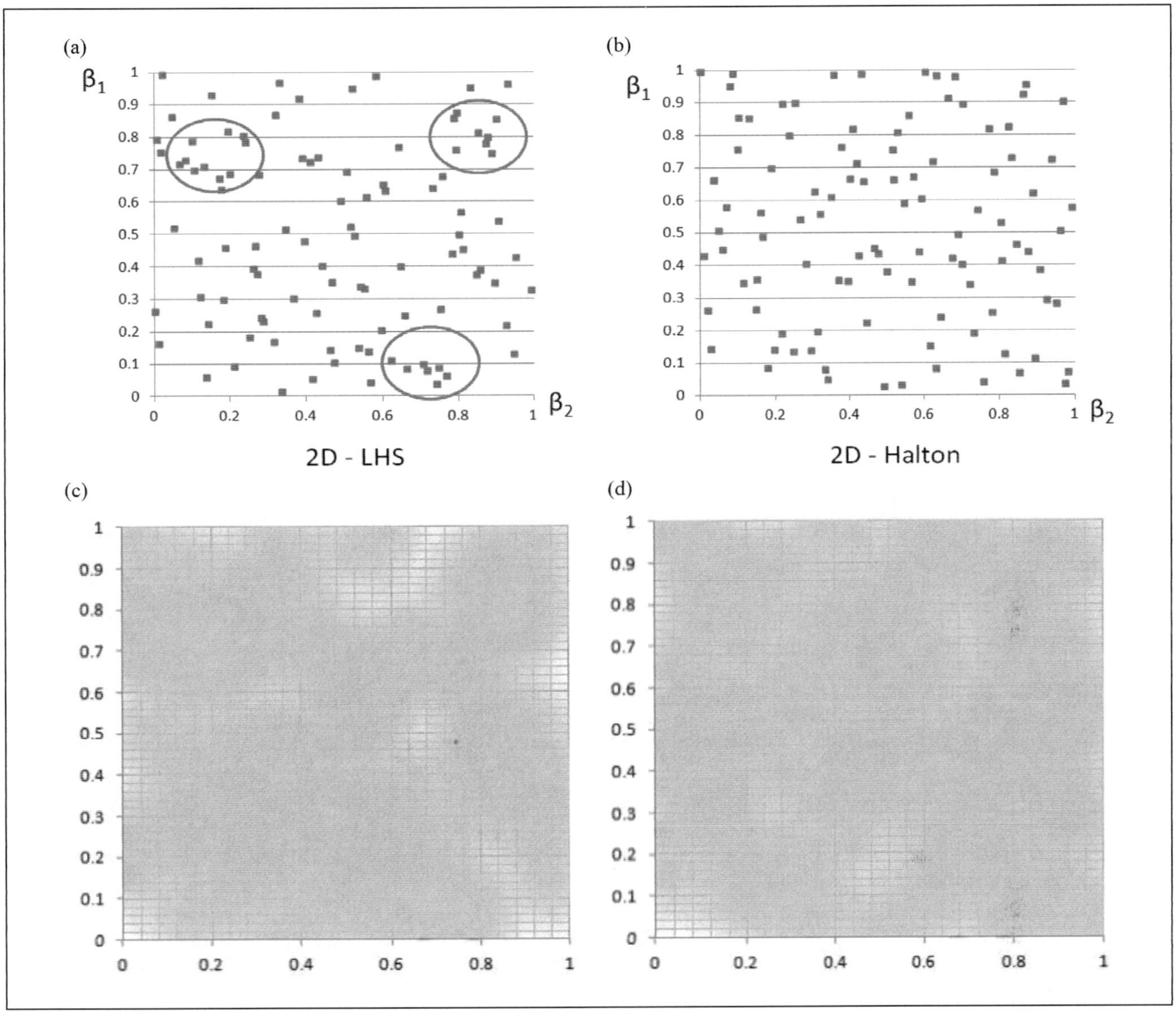

Figure 2. Comparison between 2D LHS and 2D Halton sequence: (*a*) LHS samples (dots); (*b*) QMC samples (dots); (*c*) LHS samples (mesh); (*d*) QMC samples (mesh).

optimal solutions. Based on the evaluation of samples selected by QMC, PSO repeatedly conducts the optimization process until the optimization results converge. Combining QMC enables PSO to quickly converge, thereby resulting in dramatic savings in calibration time and effort. Figure 4 shows the overall framework of the QPS algorithm applied for the calibration of a microscopic traffic simulation.

The QPS algorithm comprises the following step-wise procedure, as illustrated in Figure 5. In Step 1, QMC sampling is conducted using Equation 14:

$$\{\beta_{\mathrm{mn}}\} = \mathrm{QMC}(\beta_1, \beta_2, \ldots \beta_n); \beta_n \in [l_n, u_n]; \tag{14}$$

$$\begin{bmatrix} \beta_{11} & \cdots & \beta_{1n} \\ \vdots & \ddots & \vdots \\ \beta_{m1} & \cdots & \beta_{mn} \end{bmatrix} \overset{yields}{\longrightarrow} \begin{bmatrix} P_{sim,1,t} \\ \vdots \\ P_{sim,m,t} \end{bmatrix} \tag{15}$$

where, n denotes the number of parameters; m denotes the number of samples; t is the aggregated time intervals (i.e., $t = 1, 2, \ldots$). The generated sequence will be multiplied by the differences of upper bound u_n and lower bound l_n for each dimension to generate parameters. Depending on the field data for the calibration, the interval could be aggregated as much as 5 min or even longer. In Step 2, the samples obtained in Step 1 are evaluated using Equation 16:

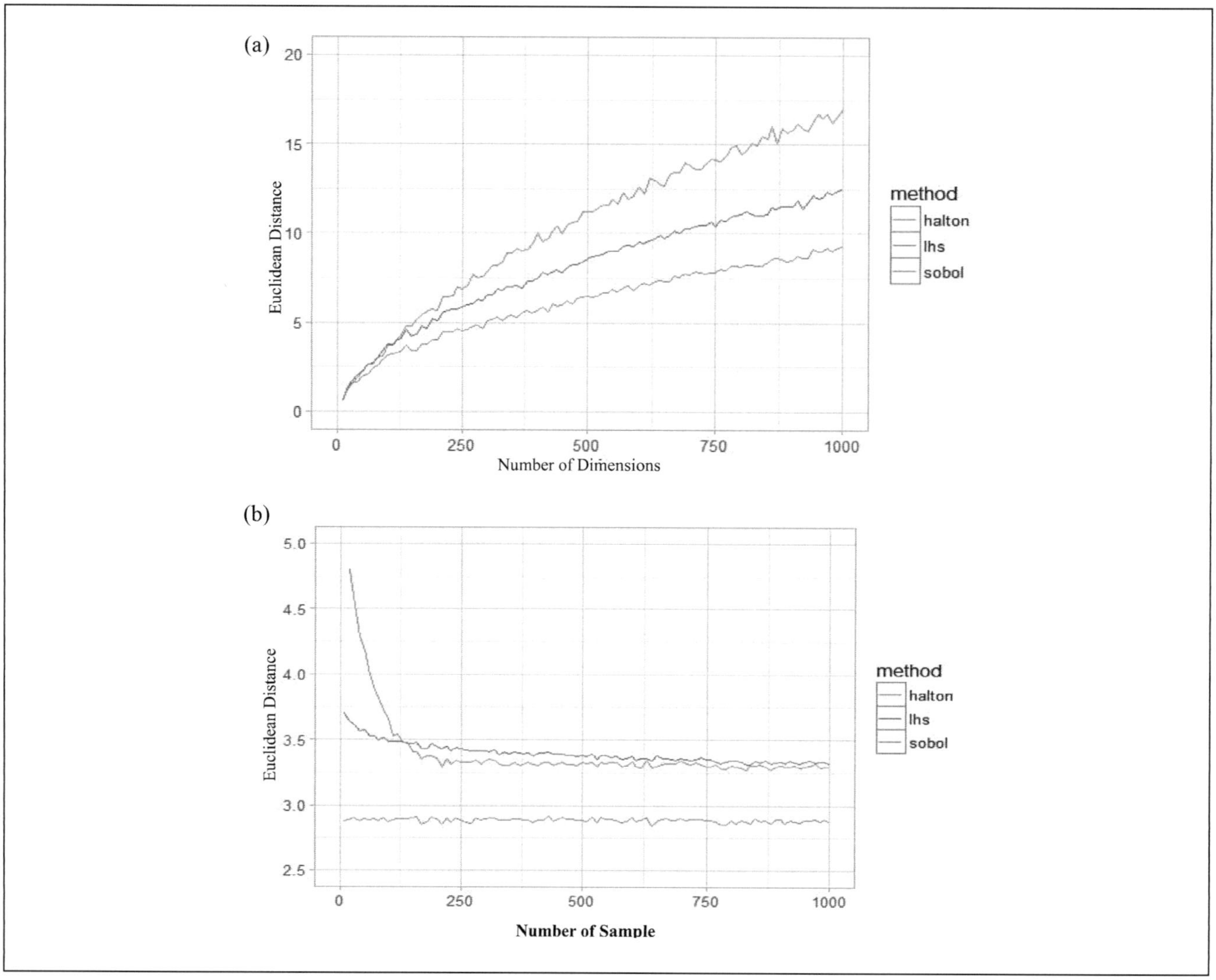

Figure 3. QMC and LHS performance comparison: (*a*) performance of 100 samples by different dimensions(10–1,000); (*b*) performance of 100 dimensions in samples from 10 to 1,000.

$$Obj.: min\ z = f(P_{Field,t} - P_{Sim,t}), P_{Sim,t} \in \{P_{Sim,t,m}\} \quad (16)$$

In Step 3, the number of better performed samples, denoted by δ, in$\{\beta_{mn}\}$ are chosen for the initial optimization points in the searching space. Where, $\delta < m$. Then, a PSO optimization is conducted using the fitness values defined in both Equations 17 and 18. The initial velocities and positions are assigned from δ to conduct the minimization process by repeatedly calculating Equations 3–6:

$$APM_{(i,j,k)} = \frac{\sum_{j=1}^{N}\left(\frac{\frac{\sum_{i=1}^{M} sim(i,j)}{M} - \frac{\sum_{k=1}^{P} field(k,j)}{P}}{\frac{\sum_{k=1}^{P} field(k,j)}{P}}\right)}{N} \quad (17)$$

$$Fitness\ Value = \frac{TT_{APM_{(i,j,k)}} + TF_{APM_{(i,j,k)}}}{2} \quad (18)$$

where

$TT_{APM_{(i,j,k)}}, TF_{APM_{(i,j,k)}}$ = average of performance measure for travel time (TT), and traffic flow (TF);
$sim(i,j)$ = jth segment performance measure obtained at ith replication;
$sim(k,j)$ = data collected from kth day for jth segment;
APM = average of performance measure;
APM_{Field} = average of performance measure of multi-days obtained from the field;
M = the number of replications;
N = the number of measured segments; and
P = the number of days.

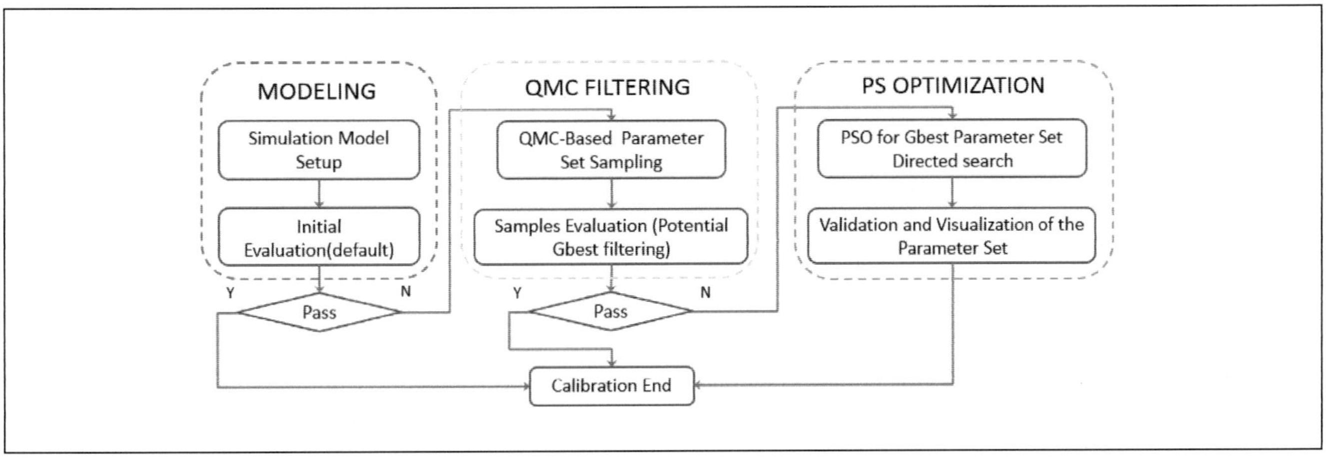

Figure 4. Overall framework of QPS algorithm.

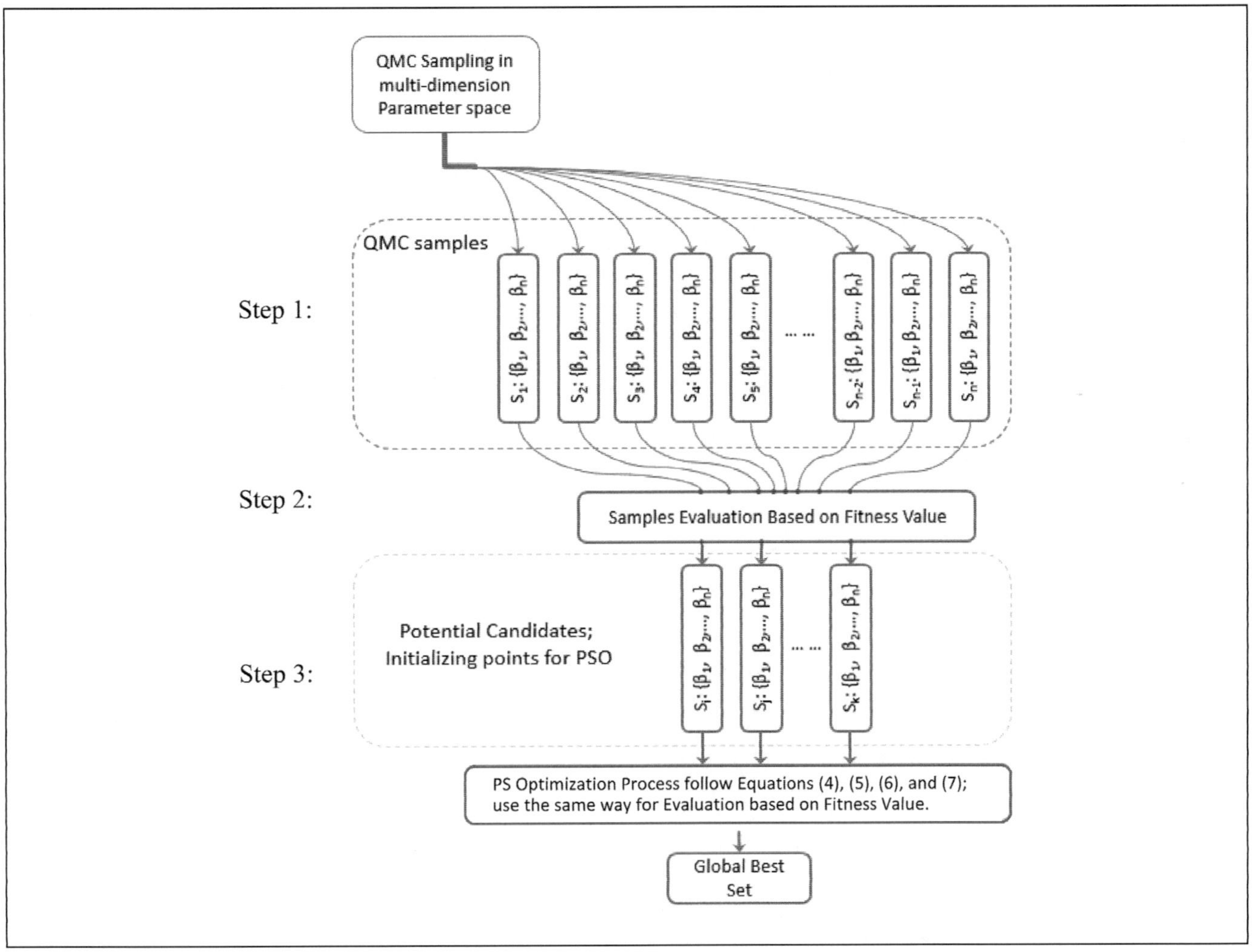

Figure 5. QPS algorithm workflow.

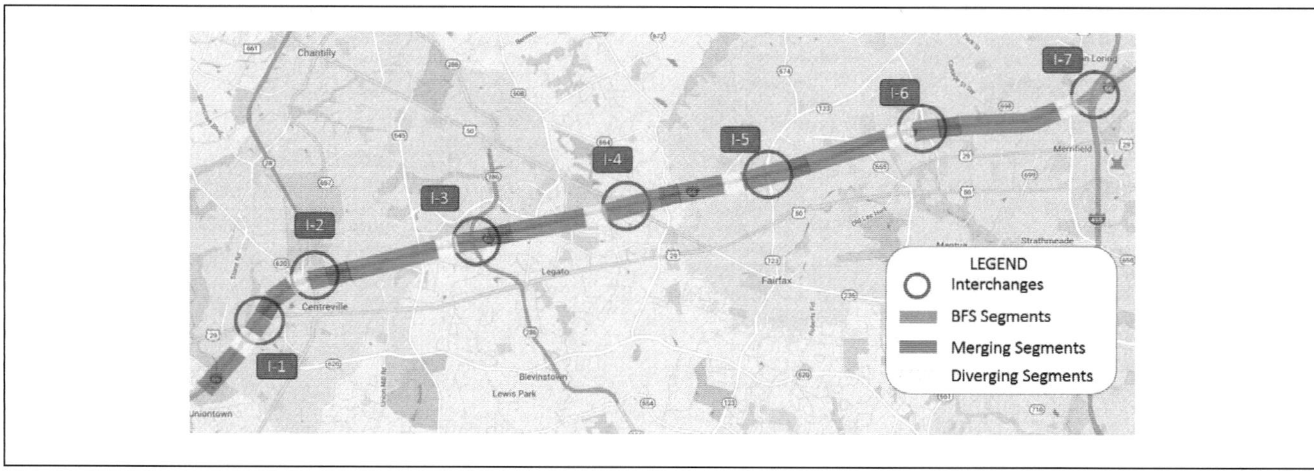

Figure 6. Case study corridor.

Case Study

This section presents a case study to examine the performance of the QPS calibration algorithm by using a VISSIM-based microscopic simulation model. This paper conducts multiple comparative experiments to test the efficiency of the QPS (i.e., QMC + PSO) algorithm by comparing it with other calibration methods such as PSO, GA, LHS + GA, and QMC + GA. The effectiveness and robustness of the QPS calibration algorithm are investigated using 100 samples.

Case Study Site

A 13-mile-long freeway stretch on the Interstate Highway 66 (a.k.a., I-66) Eastbound in Fairfax, Virginia, has been selected for the case study site, as shown in Figure 6. The case study site consists of diverse functional characteristics categorized into basic freeway segment (BFS), merging, diverging, and on- and off-ramp sections, which increase the complexity of simulation calibration. The mainline of the case study site has four lanes with one or two additional lanes depending on merging and diverging conditions. The simulation model of the case study site covers the morning peak period from 6:30 AM to 9:30 AM to consider recurrent traffic congestion.

Probe-based travel time data were used as the ground truth data to compute the fitness values of the calibrations in Equations 17 and 18. The ground truth travel time data used for the case study were obtained from INRIX (*33*) by using consecutive traffic message channels defined along with the selected I-66 section in the study area. In addition, traffic count data in and around the case study site were collected via Virginia ADMS (*34*), a database portal managing traffic count data from permanent data collection stations in Virginia. Both travel time and traffic count data were captured every Tuesday, Wednesday, and Thursday from 6:30 AM to 9:30 AM in September and October 2013.

Parameter Selection

VISSIM's Wiedemann-99 model (*14*) is a car-following model widely used in the simulations of uninterrupted traffic flow. Table 1 summarizes the list of parameters used for the Wiedemann-99 model for the case study. It must be noted that all the parameters were selected by assuming unique driving behaviors depending on BFS, merging, diverging, and interchange (i.e., on- and off-ramp) sections. Furthermore, to take into consideration drivers' mandatory lane-changing behaviors for their travels to an exit ramp, two lane-changing model parameters, emergency stop distance and lane-changing distance, were included in the calibration parameters list. The minimum and maximum range of each parameter set is presented along with its default value in Table 1. Categorized into five functional characteristic categories (i.e., BFS, merging, diverging, on-ramp, and off-ramp), the parameter set in Table 1 is independently applied for each category, thereby resulting in a total of 85 calibration dimensions (i.e., $17 \times 5 = 85$). Therefore, this paper adopts the Sobol sequence for the QMC sampling.

Results

Figure 7 demonstrates the changes of fitness values over the generation, obtained from (1) QPS (i.e., QMC + PSO); (2) PSO-Only; (3) GA-Only; (4) LHS + GA; and (5) QMC + GA, respectively. 100 samples for each generation, equivalent to particles for PSO-Only and populations for GA-Only, were chosen for the calibration of the case study site. For PSO, inertia, c_1,

Table 1. Calibration Parameters and Ranges

No.	Parameters name	Unit	Default	Min	Max
1	Maximum look-ahead distance	meter	250	150	350
2	Number of observed preceding vehicles	veh	2	1	4
3	CC0 average standstill distance	meter	1.5	0.5	1.5
4	CC1 headway time	sec	0.9	0.5	1.5
5	CC2 "following" variation	meter	4.0	0.0	15.0
6	CC3 threshold for entering "following"	sec	−8.0	−30	0
7	CC4 negative "following" threshold	m/s	−0.35	−1.0	0
8	CC5 positive "following" threshold	m/s	0.35	0.0	1.0
9	CC6 speed dependency of oscillation	Rad/sec	11.44	0.0	25.0
10	CC7 oscillation acceleration	m/s^2	0.25	0.0	1.0
11	CC8 standstill acceleration	m/s^2	3.50	1.0	8.0
12	CC9 acceleration at 80 km/h	m/s^2	1.50	0.5	3.0
13	Maximum deceleration	m/s^2	−4	−5	−1
14	Accepted deceleration	m/s^2	−1	−3	−0.2
15	Minimum headway	meter	0.5	0.1	1
16	Emergency stop distance	meter	5	5	30
17	Lane-changing distance	meter	200	430	1370

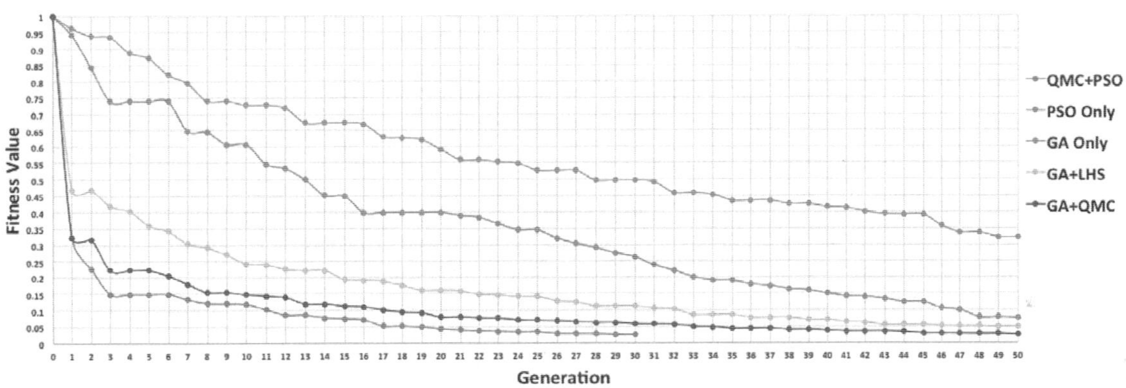

Figure 7. Fitness value changes by QPS.

and c_2 are designated to be 1.5, 1.5, and 1.5, respectively. For GA, the elitist and mutation rates are 20% and 20%, respectively.

This case study employed the fitness value (FV) defined in Equation 18 to measure the difference between simulation data and observed data. The FV is a dimensionless number between 0 and 1. That is, 0 means the simulated driving behaviors completely match the driving behaviors from study corridors; on the contrary, 1 means a bad speculation of driving behaviors in simulation. Put simply, the smaller value, close to 0, demonstrates a better driving behavior parameter set.

Figure 7 shows the change of each calibration method over the generations. As is clearly shown in Figure 7, QPS arrives at the FV of 0.05 at the 17th generation whereas GA + QMC, LHS + GA, GA, and PSO achieve 0.1, 0.18, 0.68, and 0.4 FVs, respectively, at the same generation. By combining QMC with GA, QPS still has 0.05 more accurate result for FV at the 17th generation. Compared with the group of "without filtering" methods (GA-Only and PSO-Only), the "with filtering" methods (QPS and LHS-GA) have notable superiority in the optimization process in terms of convergence speed. The combined LHS and QMC methods drop at 0.32 and 0.46 FV, respectively, at the very first sampling generation. According to the fast-convergence speed of PSO, it outperformed GA for more than 10 generations faster under the premise of using QMC for filtering. The changes of FVs obtained at every five generations by each calibration approach is tabulated in Table 2.

Table 2. Change of Fitness Values

Generation	Fitness value				
	QPS	QMC + GA	PSO-Only	GA-Only	GA + LHS
0	1	1	1	1	1
5	0.15	0.22	0.74	0.87	0.36
10	0.12	0.15	0.61	0.73	0.24
15	0.07	0.11	0.45	0.67	0.19
20	0.04	0.08	0.40	0.59	0.16
25	0.04	0.07	0.35	0.53	0.14
30	0.03	0.06	0.26	0.50	0.11
35	–	0.05	0.19	0.44	0.09
40	–	0.04	0.15	0.41	0.07
45	–	0.03	0.12	0.40	0.05
50	–	0.03	0.08	0.33	0.05

Conclusion

It is discovered that incorporating a sampling technique into a heuristic optimization method generates great benefits to improve the efficiency and robustness of a large-scale optimization process. Creating a complex hyper-dimensional optimization problem, heuristic optimization integrating a sampling technique appears suitable for the calibration of microscopic traffic simulation models.

This paper proposed a multi-thread optimization method to build a solid calibration algorithm for microscopic traffic simulation models by integrating QMC and PSO. As also proven by Hassan et al. (*22*) and Singhee and Rutenbar (*23*), PSO demonstrated outstanding performance in an optimization problem compared with GA in this paper. QMC has proved its promising performance as a tool to assist PSO to expedite the convergence in optimization problems, compared with Monte Carlo and LHS.

A VISSIM-based microscopic simulation calibration case study dealing with an 85-dimensional optimization problem was conducted to examine the performance of the proposed QPS algorithm. The simulation model for the case study contains distinctive highway segments (e.g., BFS, on-ramp, off-ramp, merge, and diverge). The case study results showed that the QPS algorithm outperforms LHS + GA, QMC + GA, GA-Only, and PSO-Only by dramatically reducing the number of generations. As demonstrated in the case study, the QPS algorithm achieved a more straightforward and maneuverable calibration of microscopic traffic simulation, thereby enabling seamless application for real-world calibration practices.

Author Contributions

The authors confirm contribution to the paper as follows: study conception and design: JL and ZH; calibration framework development: ZH and JL; draft manuscript preparation: ZH and JL; data analysis: ZH, JL. All authors reviewed the results and approved the final version of the manuscript.

References

1. Park, B., and J. Won. *Microscopic Simulation Model Calibration and Validation Handbook*, Virginia DOT, Richmond, VA, 2006. http://www.virginiadot.org/vtrc/main/online_reports/pdf/07-cr6.pdf. Accessed June 27, 2016.
2. Gardes, Y., A. D. May, J. Dahlgren, and A. Skabardonis Freeway Calibration and Application of the PARAMICS Model. Presented at 81st Annual Meeting of the Transportation Research Board, Washington, D.C., 2002.
3. Gomes, G., A. May, and R. Horowitz. Congested Freeway Microsimulation Model using VISSIM. *Transportation Research Record: Journal of the Transportation Research Board*, 2004. 1876: 71–81.
4. Park, B., and H. Qi. Development and Evaluation of a Procedure for the Calibration of Simulation Models. *Transportation Research Record: Journal of the Transportation Research Board*, 2005. 1934: 208–217.
5. Park, B. B., and J. D. Schneeberger. Microscopic Simulation Model Calibration and Validation: Case Study of VISSIM Simulation Model for a Coordinated Actuated Signal System. *Transportation Research Record: Journal of the Transportation Research Board*, 2003. 1856: 185–192.
6. Lee, J., B. Park, J. Won, and I. Yun. A Simplified Procedure for Calibrating Microscopic Traffic Simulation Models. Presented at 92nd Annual Meeting of the Transportation Research Board, Washington, D.C., 2013.
7. Lee, J.-B., and K. Ozbay. New Calibration Methodology for Microscopic Traffic Simulation using Enhanced Simultaneous Perturbation Stochastic Approximation Approach. *Transportation Research Record: Journal of the Transportation Research Board*, 2009. 2124: 233–240.
8. Ma, J., H. Dong, and H. Michael Zhang. Calibration of Microsimulation with Heuristic Optimization Methods. *Transportation Research Record: Journal of the Transportation Research Board*, 2007. 1999: 208–217.

9. Paz, A., V. Molano, and A. Khan. Calibration of Micro-Simulation Traffic Flow Models considering All Parameters Simultaneously. Presented at 93rd Annual Meeting of the Transportation Research Board, Washington, D.C., 2014.
10. Abdalhaq, B. K., and M. I. Abu Baker. Using Meta Heuristic Algorithms to Improve Traffic Simulation. *Journal of Algorithms*, Vol. 2, No. 4, 2014, pp. 110–128.
11. Dowling, R., J. Holland, and A. Huang. *Guidelines for Applying Traffic Micro Simulation Modeling Software*. Dowling Associates, Oakland, CA, 2002.
12. Joy, C., P. P. Boyle, and K. S. Tan. Quasi-Monte Carlo Methods in Numerical Finance. *Management Science*, Vol. 42, No. 6, 1996, pp. 926–938.
13. Kennedy, J. *Particle Swarm Optimization Encyclopedia of Machine Learning*. Springer, New York, NY, 2011, pp. 760–766.
14. PTV. *VISSIM Traffic Simulation User Manual: Version 6.0*, 2013.
15. Treiber, M., and A. Kesting. Microscopic Calibration and Validation of Car-Following Models—A Systematic Approach. *Procedia-Social and Behavioral Sciences*, Vol. 80, 2013, pp. 922–939.
16. Kim, S.-J., W. Kim, and L. Rilett. Calibration of Microsimulation Models using Nonparametric Statistical Techniques. *Transportation Research Record: Journal of the Transportation Research Board*, 2005. 1935: 111–119.
17. PARAMICS. http://www.sias.com/2013/sp/sparamicshome.htm. Accessed July 31, 2017.
18. Lucasius, C. B., and G. Kateman. Understanding and Using Genetic Algorithms Part 1. Concepts, Properties and Context. *Chemometrics and Intelligent Laboratory Systems*, Vol. 19, No. 1, 1993, pp. 1–33.
19. McKay, M. D., R. J. Beckman, and W. J. Conover. Comparison of Three Methods for Selecting Values of Input Variables in the Analysis of Output from a Computer Code. *Technometrics*, Vol. 21, No. 2, 1979, pp. 239–245.
20. Spall, J. C. Implementation of the Simultaneous Perturbation Algorithm for Stochastic Optimization. *IEEE Transactions on Aerospace and Electronic Systems*, Vol. 34, No. 3, pp. 817–823.
21. Glover, F. Future Paths for Integer Programming and Links to Artificial Intelligence. *Computers & Operations Research*, Vol. 13, No. 5, 1986, pp. 533–549.
22. Hassan, R., R. Cohanim, and O. de Weck. A Comparison of Particle Swarm Optimization and the Genetic Algorithm. *Proc., 46th AIAA/ASME/ASCE/AHS/ASC Structures, Structural Dynamics and Materials Conference*, Austin, TX, 2005.
23. Singhee, A., and R. A. Rutenbar. Why Quasi-Monte Carlo Is Better than Monte Carlo or Latin Hypercube Sampling for Statistical Circuit Analysis. *IEEE Transactions on Computer-Aided Design of Integrated Circuits and Systems*, Vol. 29, No. 11, 2010, pp. 1763–1776.
24. Asmussen, S., and P. W. Glynn. *Stochastic Simulation: Algorithms and Analysis* Springer-Verlag, New York, NY, 2007.
25. Glasserman, P. *Monte Carlo Methods in Financial Engineering*. Springer-Verlag, New York, NY, 2004.
26. Halton, J. H. On the Efficiency of Certain Quasi-Random Sequences of Points in Evaluating Multi-Dimensional Integrals. *Numerische Mathematik*, Vol. 2, No. 1, 1960, pp. 84–90.
27. Halton, J. H. Algorithm 247: Radical-Inverse Quasi-Random Point Sequence. *Communications of the ACM*, Vol. 7, No. 12, 1964, pp. 701–702.
28. Sobol, I. M. On Quasi-Monte Carlo Integrations. *Mathematics and Computers in Simulation*, Vol. 47, No. 2, 1998, pp. 103–112.
29. Hess, S., and J. Polak. *An Alternative Method to the Scrambled Halton Sequence for Removing Correlation between Standard Halton Sequences in High Dimensions*. ERSA Conference Paper. ersa03p406. European Regional Science Association, Louvain-la-Neuve, Belgium, 2003.
30. Wang, X., and I. H. Sloan. Low Discrepancy Sequences in High Dimensions: How Well Are Their Projections Distributed? *Journal of Computational and Applied Mathematics*, Vol. 213, No. 2, 2008, pp. 366–386.
31. Wang, X., and K.-T. Fang. The Effective Dimension and Quasi-Monte Carlo Integration. *Journal of Complexity*, Vol. 19, No. 2, 2003, pp. 101–124.
32. Joe, S., and F. Y. Kuo. Remark on Algorithm 659: Implementing Sobol's Quasirandom Sequence Generator. *ACM Transactions on Mathematical Software (TOMS)*, Vol. 29, 2003, pp. 49–57.
33. INRIX. www.inrix.com. Accessed July 30, 2017.
34. ADMS Virginia. http://openroadsconsulting.com/pdf/paper_admsvirginia.pdf. Accessed July 30, 2017.

The Standing Committee on Traffic Flow Theory and Characteristics (AHB45) peer-reviewed this paper (18-03527).

Article

TRR
JOURNAL OF THE TRANSPORTATION RESEARCH BOARD

Transportation Research Record
2018, Vol. 2672(20) 110–120

DOI: 10.1177/0361198118786607
journals.sagepub.com/home/trr
SAGE

Perimeter Control as an Alternative to Dedicated Bus Lanes: A Case Study

Nicolas Chiabaut[1], Michael Küng[2], Monica Menendez[2,3,4], and Ludovic Leclercq[1]

Abstract

Dedicated bus lanes (DBLs) are a common traffic management strategy in cities as they improve the efficiency of the transit system by preventing buses from getting trapped in traffic jams. Nevertheless, DBLs also have certain disadvantages: they consume space, reduce available capacity for general traffic, and can thus lead to even more congested car traffic situations. It is appealing to find more efficient alternatives that maintain a sufficient network supply for general traffic while guaranteeing high commercial speeds for the bus system. This paper investigates whether perimeter control (gating) could be such an alternative to DBL strategies. This solution aims at controlling the traffic conditions of a given area by monitoring vehicle accumulations and adapting traffic signal parameters to reach the targeted conditions. If free-flow states can be maintained within the zone, then DBLs become superfluous. This hypothesis is examined through a simulation case study with an urban arterial acting as the targeted area. A dual-objective control approach was applied to allow for not only the vehicle accumulation inside the area but the queue lengths at its perimeter, thereby addressing one of the main issues associated with gating schemes. Due to the gating strategy, traffic performance in the arterial, measured through vehicle accumulation plus mean speed and density, improved significantly. Moreover, results showed that bus operations reach almost the same efficiency level when DBLs are replaced by perimeter control. Furthermore, the availability of an additional lane for general traffic in the control case significantly increased the arterial capacity for cars.

Traffic congestion and its negative impacts is an ongoing topic of discussion among city authorities all over the world, and will continue to remain on the agenda as urban economic development (*1*) and number of vehicles are increasing (*2*). Some regions (e.g., city centers) are predestined for bottlenecks due to their functions as junctions and economic hubs. Once congestion arises in these sensitive zones, congested traffic states start to propagate in the surrounding network sections and may lead to major gridlocks. Therefore, it becomes crucial to monitor and control in real-time urban traffic conditions in order to utilize the existing network as efficiently as possible (*3*) and minimize congestion hours in such critical regions.

Urban congestion has direct effects on both travelers (increased travel time, reduced speed, decrease in time reliability, greater fuel consumption, higher wear, etc.) and the environment (increased pollution and noise). In addition, traffic jams also affect other transportation modes, including public transport. In particular, bus lines might experience increases in both average travel times and unreliability for passengers (*4*). As a result, the transit system loses appeal and the shift of transportation mode from individual vehicles to massive transportation modes as emphasized in Amirgholy et al. (*5*) can decrease. Local authorities and operators are therefore compelled to deploy strategies to improve performance and increase attractiveness of the public transport system, paying special attention to bus operations.

A common measure deployed worldwide to reduce bus delays is the use of dedicated bus lanes (DBLs), which protect bus operations from general traffic by reserving a special lane. Among others, Balke et al. and Zhu emphasize various advantages of DBL, in particular the fact that they prevent buses from getting trapped in traffic queues at signalized intersections, thereby significantly improving the bus system performance (*6,7*).

[1]University of Lyon, ENTPE, IFSTTAR, LICIT, F-69518 Lyon, France
[2]Traffic Engineering Group, Institute for Transport Planning and Systems, ETH Zurich, Switzerland
[3]Division of Engineering, New York University, Abu Dhabi, United Arab Emirates
[4]Tandon School of Engineering, New York University, NY

Corresponding Author:
Address correspondence to Nicolas Chiabaut: nicolas.chiabaut@entpe.fr

Nevertheless, DBLs still remain controversial and not always well accepted. Chiabaut et al. especially point to the unavailability of the dedicated lane for individual vehicles as a main drawback, whereby capacity for cars is decreased (*8*). As a consequence DBLs are only appropriate for low traffic flows from a car user perspective (*9, 10*), a disadvantage that must be taken into account. Research is ongoing in the field of DBLs. Pre-signals, as proposed by Guler and Menendez, can be implemented to better use the intersection capacity even when bus flows are low; hence keeping the bus priority and reducing the negative effects on cars (*11*). Intermittent bus lanes (IBLs), as proposed by Viegas and Lu, are a further development: DBL are opened intermittently to general traffic to mitigate the negative effects of DBLs on car traffic, while keeping bus operations optimal (*12*). However, although IBLs have been analytically studied (*8, 9*), attempts at real-world deployment remain very rare. On a larger scale, decisions for DBL deployment can also serve different purposes like promoting the use of public transport by reducing space for cars, which makes it a matter of political intention.

The current paper proposes an alternative to DBLs by applying the concept of perimeter control, also known as gating, to a sensitive region, and at the same time replacing the bus lanes in the area with general purpose lanes. More precisely, the paper aims to provide proof of concept that perimeter control could be an alternative to DBL. To this end, this approach is tested on as generic an urban arterial as possible to ensure that results are transferable, recognizing that the methodology may need to be adapted to the specific conditions of a city. Perimeter control as described by Bretherton et al. addresses the above-mentioned point of increasing traffic efficiency in a existing network (*13*). Basically, the idea of this control strategy is to restrict the input flow to a predefined area, to maintain non-congested traffic states inside that region (*14*) and thereby improve overall traffic performance. Thus, congestion is relieved at the network level. More precisely, the incoming flow is controlled through the traffic signals at the boundaries of the "protected" area. One advantage of this approach is that it can be implemented in real time through real traffic data. There are various examples of perimeter control strategies in literature with positive results (*13–19*). The strategy is successful when the saved delays in the controlled region are higher than the additionally generated delays by the controller at the entries. Perimeter control schemes are already applied in some cities, for example in Chania and Zurich (*3,20*). The main drawback of gating is that long queues can occur at the perimeter entries, which potentially lead to spillbacks that affect the surrounding network [for details see (*14, 16*)]. Only a few papers focus more deeply on this issue. One approach can be found in Haddad, in which the queue length at the perimeter boundary is considered and controlled in the model (*19*). Yang et al. recently developed a so called multi-scale controller that takes the queue lengths at perimeter entries into account by integrating the occurring delay in the queues directly into the objective function (*21*). Traffic performance is therefore improved globally by considering both the network scale and the local level.

If a high level of service can be obtained through perimeter control, the original purpose of DBLs to guarantee free-flow conditions for buses, potentially becomes redundant, meaning DBLs could be eliminated. This hypothesis is examined in the following by applying a perimeter controller as an alternative to DBLs in a simplified simulation case study. In contrast to former studies, a multi-lane urban arterial is considered as the controlled perimeter instead of a region. Similar to a city center, an arterial is often a traffic hot spot and a critical section for congestion emergence. The control scheme is tested specifically in the context of bus operations, as the travel times of buses are compared for cases with perimeter control and with DBL. Because of the worldwide presence of DBLs, the current case study is of relevance and could provide decision makers with new perspectives, all the more so as it strikes a new path through the complete replacement of DBLs in contrast to the optimization of DBLs examined in former research.

The paper is organized as follows. After describing the simulation and controller framework in the first section, findings are presented. This section starts with an evaluation of the controller to validate that the framework developed for large-scale traffic flow models works with a detailed description of traffic dynamics, that is, in micro-simulation. Then, the controller can be safely applied to study the opportunity of replacing a DBL with a gating strategy. The last section is devoted to the conclusions.

Methodology

The concept of the controller applied in this work originates from Yang et al., but the original methodology has been adapted to address different research questions and objectives (*21*). Specifically, this method has been adapted to the finest level of detail, that is, micro-simulation, whereas it was originally developed for large-scale models as the biggest part of existing papers in the field of transportation network controller. Yang et al. provide an accurate framework to implement and test, in simulation, a control that aims to maintain a protected area in free-flow conditions while simultaneously limiting spillbacks at the entries of the protected area (*21*). Different scenarios, presented in the next section, will be simulated.

Case Study

All the simulations for the present work were executed with the traffic simulation platform SymuVia. SymuVia is a dynamic microscopic traffic simulator that provides a very high level of detail. It is based on the kinematic wave model (KW) described by Lighthill and Whitham (*22*) and Richards (*23*) with a Lagrangian resolution introduced in Leclercq et al. (*24*). Multiple components to reproduce urban networks (intersections, lane-changing, multi-class, etc.) are embedded in the simulator and the positions of all vehicles are calculated every second giving access to their full trajectories. Furthermore, SymuVia uses a dynamic traffic assignment based on users' equilibrium.

Network. Figure 1 gives an overview of the network setup and of the controlled intersections in the case study. The network consists of a 14 × 14 street grid, also known as a *Manhattan-type* network. This layout is large enough to represent roughly a district of a city with its interlinked traffic dynamics. In the center of the network, an arterial road crosses the whole district. All links are 200 m long. The arterial has three lanes and is unidirectional from west to east. The rest of the links are bidirectional with one lane in each direction. The speed limit for the arterial is 50 km/h, and 30 km/h for the rest of the links. We attempted to use as general a network as possible because this paper aims to provide proof of concept that perimeter control can be an alternative to DBL. For further practical deployment, more research would be needed to allow for the specific conditions of a city.

Two regions are denoted: the arterial is considered a center region A, whose traffic states should be controlled, the side streets represent the outer region B, see Figure 1. The border of these two regions is composed of signalized intersections in which vehicles can change from one region to another.

Vehicles can enter and leave the arterial, that is, region A, at any of its intersections. Only vehicles in region A in the far-left lane can turn into a side street on the left-hand side (see the generic geometry intersection 1). In the same vein, only vehicles on the far-right lane can turn right into region B. Vehicles entering the arterial from the side streets, that is, region B, have to take the nearest lane (i.e., far-left or far-right lane). Once cars move onto the arterial, they can switch lanes without restriction.

Signal settings are as follows. On every intersection, the total cycle length c of the signals is 60 s, and there is neither intermediate time nor a yellow phase. Each cycle contains two phases, in which the two opposite approaches belong to the first phase and the arterial to the second. The phase of the arterial gets assigned 40 s of green time, and the side-street entries 20 s of green time for the cases without control. Due to the higher demand, the arterial also has a higher green-time ratio, which is assumed to be realistic and consistent with the hierarchy of real urban networks. In total there are 14 intersections linking the arterial with the outside region, on which the controller is applied via the traffic signals, that is, green times are updated for each control period.

Finally, in region B, links are also connected with signalized intersections. We assume that the green time g_i is split evenly 30 s/30 s for both phases.

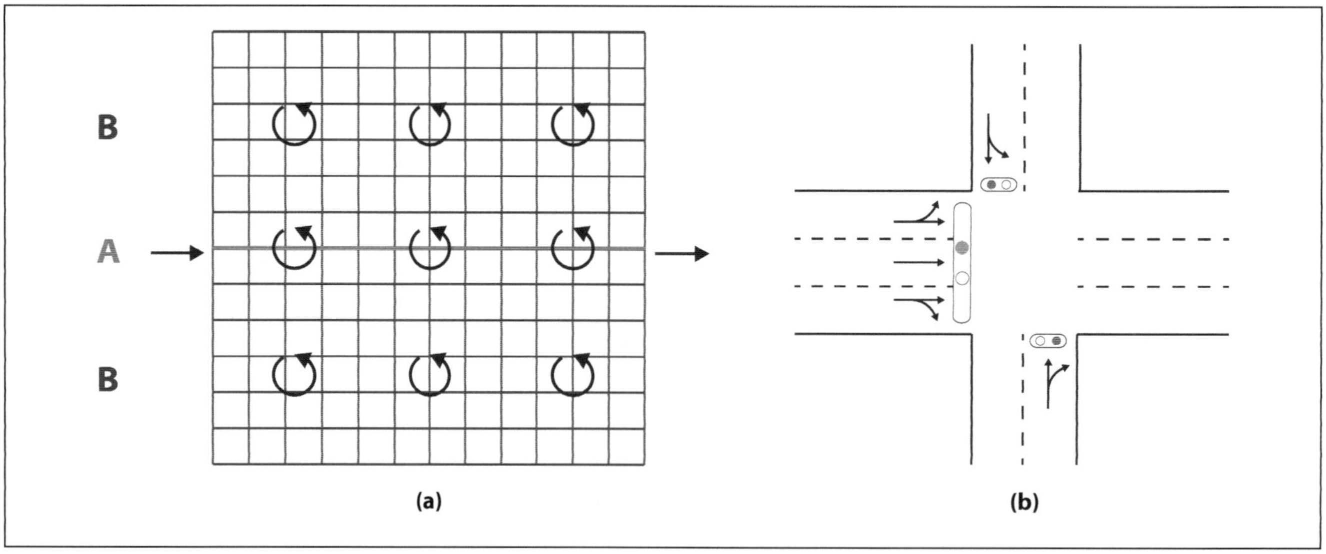

Figure 1. (*a*) Manhattan-type network with an arterial as the controlled region *A* and the outer region *B*. Arrows represent inner and outer demand; (*b*) Schematic intersection arterial: side street with controlled traffic signals and traffic streams.

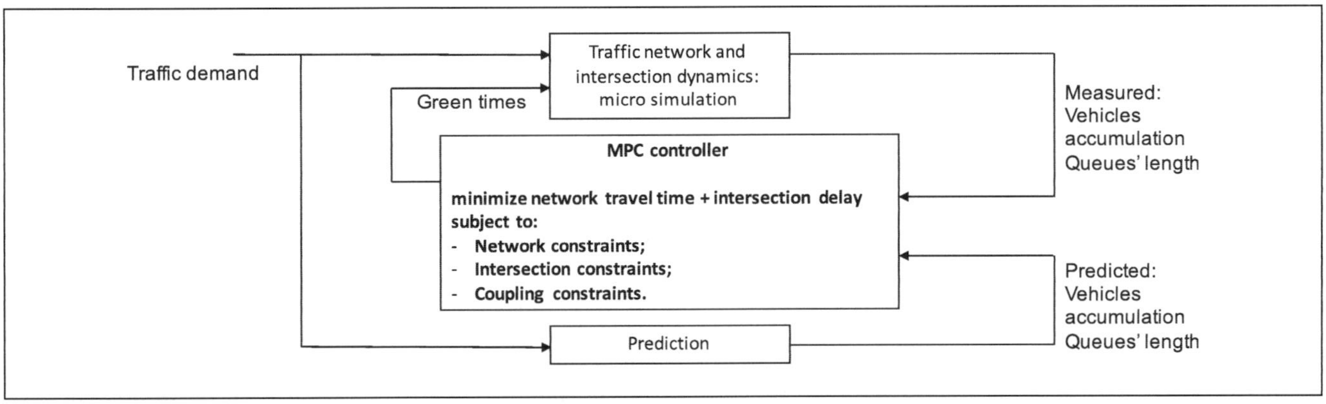

Figure 2. Control diagram.

Demand. There are two types of demand. First, there is an outer demand entering and leaving the network through the arterial and passing the network from west to east. Second, there is an inner demand with both origin and destination within the network. To generate this demand, the network is divided into nine zones of equal size and a demand level is defined for the number of vehicles traveling from each zone to each other zone. The arterial (outer) demand is fixed at 2,600 vph, whereas the zone-to-zone (inner) demand varies over time with the intention to simulate a typical peak period – 300 vph during the first 30 min, 600 vph during the second 30 min, and 300 vph again during the last 30 min. A simulation run lasts 90 min.

Feedback Control Approach

The controller is feedback-based, which means that the control action is dependent on the previous output. It is worth noting that the proposed methodology can be used with other control schemes. Traffic detectors (sensors) indirectly measure the objective function J_D (total time spent), whereas the traffic signal parameters (actuator) influence the control variable (green-time ratios for every perimeter intersection). The detectors are placed on every street entering and leaving the arterial. As already mentioned, the challenge of the control algorithm used in the paper is the two competing control objectives, which need to be coupled into an integrated control scheme (*21*). Traffic performance should be optimized simultaneously at the network (i.e., arterial) and local level (i.e., entries into the arterial). As the indicator for the global traffic state in the protected perimeter, we used the total vehicle accumulation in region A. For the local level, the total delay, which is directly related to the queue lengths at the entry intersections, is considered an indicator. Traffic conditions are monitored through virtual detectors that give access to the total vehicle accumulation of the arterial as well as to queue lengths at the entries. Note that accumulation is directly connected to the total time vehicles spend inside the arterial. The applied controller calculates the optimal green-time ratios for every entry intersection based on the measured data, to optimize the two objectives mentioned above. The new green-time ratios are updated and the traffic signal parameters are specifically modified at each intersection every K seconds. This implies that one control period lasts K seconds. To change signal timings more often did not seem realistic in this context as the cycle length on intersections in our case study was 60 s.

The optimization problem was formulated in the same vein as in Yang et al. and solved with the same algorithm (*21*). Figure 2 shows the controller diagram. In contrast to Yang et al., we used micro-simulation to reproduce traffic system dynamics instead of macroscopic modeling (*21*). The optimization problem can now be formulated and explained by the following equations, where Equation 1 is the objective function and equations 2 to 9 are the constraints.

$$min\ J_D = \sum_{p=1}^{L} \left(K^*(n_{AA}(p) + n_{AB}(p)) + \sum_{i \in I} x_i(p)^* k_x^* \frac{c - g_i(p)}{2} * \frac{K}{c} \right) \quad (1)$$

$$s.t.n_{AA}(p+1) = n_{AA}(p) + \lambda^{in}_{Arterial}(p)^*K - \lambda^{out}_{Arterial}(p)^*K \quad (2)$$

$$n_{AB}(p+1) = n_{AB}(p) + \lambda_{BA}(p)^*K - \lambda_{AB}(p)^*K \quad (3)$$

$$\lambda_{BA}(p) = \sum_{i \in I} \mu_i(p) \quad (4)$$

$$\lambda_{AB}(p) = \sum_{i \in I} q_i^{out}(p) \quad (5)$$

$$\mu_i(p) = \min\{q_i(p), \frac{g_i(p)}{c} * q_{max}\} \quad (6)$$

$$x_i(p) = \frac{u^* w}{u + w} * (c - g_i(p)) * \frac{q_i(p)}{(q_{max} - q_i(p))} \quad (7)$$

$$q_{max} = \frac{u^* w}{u + w} * k_x \quad (8)$$

$$\gamma^{min} \leq \frac{g_i(p)}{c} \leq \gamma^{max} \quad (9)$$

In the objective function (Equation 1), the total travel costs J_D are minimized. These travel costs are composed of (i) the total time spent by vehicles in region A (arterial) during one control period, that is, the first two terms multiplied by the length K of the control period, and (ii) the total delay at the different entry intersections. We assume that traffic obeys KW theory with a triangular fundamental diagram with free-flow speed u, backward wave speed w and jam density k_x. This model is known to accurately reproduced urban traffic flow dynamics and especially the queue length at signalized intersections. Moreover, we assume that vehicle arrivals at intersections are uniformly distributed. At a macroscopic scale, this assumption is valid to accurately reproduce the average delay (*25*). According to KW theory, this delay is calculated as the product of the queue length x_i with $k_x * \frac{c-g_i}{2} * \frac{K}{c}$
where

g_i is the green time at intersection i,

c is the cycle duration, and

$\frac{K}{c}$ is the number of traffic signal cycles during one control period.

The queue lengths in the outflowing direction are not taken into consideration here, as they are accounted for when computing the accumulations in the protected region. Note that the two objective criteria have the same weight, which can be adjusted for different purposes, according to the corresponding main aim, see also Xue et al. (*26*).

Constraints (Equation 2 and Equation 3) stand for the vehicle accumulation in region A, which is basically inflow minus outflow. The predefined flow entering $\lambda^{in}_{Arterial}$ and leaving $\lambda^{out}_{Arterial}$ region A through the arterial is simple to measure with loop detectors. Meanwhile the flows that enter and exit region A through the side streets are represented by Equation 4 and Equation 5, respectively. This last flow must be the minimum of the measured demand on the upstream loop detectors and the capacity of the entry (Equation 6), whereas the flow leaving region can be directly measured.

The capacity, however, depends on the allocated green time for the entry and changes, therefore, with every control cycle. Equation 7 shows how mean queue lengths during one period are estimated based on a KW model, taking into account the measured arrival flow, the green time, and the link capacity expressed in Equation 8. Contrary to Yang et al., we express the length of the queue in terms of distance (*21*). Because the controller is based on a detailed description of traffic dynamics, it makes it possible to precisely track the propagation of the queue. To this end, virtual loop detectors have been emulated at the entry of the approach links, giving access to the average demand q_i during one control period. This value is then used to derive the average length of the queue for the control period. If the queue cannot disappear during the control period, Equation 7 will underestimate the length of the queue until the queue spills back on the detector located at the entry of the approach link. Consequently, queue lengths are approximated during the transition phase. Moreover, it is important to note that, in the case of strong spillbacks, Equation 7 strongly underestimates the length of the queues and the associated delays in the objective function (Equation 1). However, in such a case, a dynamic traffic assignment (DTA) process will be required because congestion spreading into region B will have an impact on route choices and traffic assignments. Consequently, the controller proposed here is valid only when the queues can be maintained within the approach links. Validated by visual inspection of the simulated results, the assumption seems appropriate and this analytical formulation gives an accurate average estimation of the queues' lengths.

Furthermore, a minimum and maximum allowed green-time ratio for the entries is defined by Equation 9. These boundaries were defined to emulate realistic scenarios. It does not seem plausible to have green durations below 6 s, that is, 10% of the cycle's length. The same holds for the maximum boundary. If the red sequence for the high-demand arterial becomes too long, this immediately leads to spillbacks.

Now the problem is well formulated to be solved according to the model predictive control (MPC) presented in Yang et al. (*21*). Basically, the constraints can be linearized and the problem can be reformulated into a linear programming problem. We used a commercial solver, Cplex, to calculate the solution and derive the optimized values of the control variables. Note that, for an overview, Table 1 lists the used variables.

Input Data. As is evident from the objective equation (Equation 1), the required data inputs for the controller are: traffic accumulation in region A and queue lengths at entry intersections. The outputs are the green ratios, which are then used in the computations for the next period. For calculating other values, real-time simulated measurements are used such as flow at entry and exit of the arterial and the different approach links. Approximation through extrapolation would be a possibility in the case of a lack of detectors. Besides that,

Table 1. Nomenclature

Variable	Description	Value (if constant)
J_D	Total travel costs	
K	Control period (s)	300
p	Index of the current control period	
$n_{AA(k)}$	Calculated accumulation of vehicles in Region *A* with destination in Region *A* at the end of period *k* (veh)	
$n_{AB(k)}$	Calculated accumulation of vehicles in Region *A* with destination in Region *B* at the end of period *k* (veh)	
I	Set of intersections	
L	Prediction horizon (number of control periods)	18
$x_i(p)$	Calculated mean queue length at intersection *i* during period *k*, $i \in I$ (veh)	
$\lambda^{in}_{Arterial}$	Predefined flow entering arterial	2,600 (vph)
$\lambda^{out}_{Arterial}$	Measured arterial flow leaving the network (vph)	
λ_{AB}	Calculated total flow from region *A* into region *B* (vph)	
λ_{BA}	Calculated total flow from region *B* into region *A* (vph)	
$\mu_i(p)$	Calculated departure flow at intersection *i* in period *p*, $i \in I$ (vph)	
$q_i(p)$	Measured arrival flow at intersection *i* in period *p*, $i \in I$ (vph)	
$q_i^{out}(p)$	Calculated leaving flow in direction of region *B* at intersection *i* in period *p*, $i \in I$ (vph)	
$g_i(p)$	Green time at intersection *i* for cycle *C* during period *p*, $i \in I$	
u	Free-flow speed on a link corresponding to the speed limit (m/s)	13.89 (*A*); 8.34 (*B*)
w	Shock wave speed of vehicles (m/s)	–5.8
c	Cycle length (s)	60
q_{max}	Capacity-flow of one lane (vph)	2,200 (*A*); 1,860 (*B*)
k_x	Jam density of one lane (vpm)	0.15
γ^{min}	Minimum allowed green-time ratio at intersection *i*, $i \in I$	0.1
γ^{max}	Maximum allowed green-time ratio at intersection *i*, $i \in I$	0.67

Table 2. Scenarios

		Dedicated bus lane	
Controller		Yes	No
	Yes	1	2
	No	3	4

connected vehicles as suggested by Yang et al., GPS data from mobile phones, navigation devices, or a combination of sources, could also provide the required input data (*21,27,28*).

Results

In the following, simulation results are interpreted and discussed. As previously explained, the efficiency of the controller was first evaluated for the no DBL case. The objective was to validate that the framework developed for large-scale traffic flow models could be adapted to micro-simulation and be consistent for a detailed description of traffic dynamics. The controller could then be safely applied to bus operations. Specifically, we studied the opportunity of replacing a DBL with a gating strategy while maintaining a high commercial speed for the bus lines. To this end, four different scenarios were identified, see Table 2. Obviously, comparison of Scenario 2 with Scenario 3 would be the most insightful approach to examine if gating is an alternative to DBL. In addition, Scenario 4 was used to evaluate the performance of the controller, as mentioned earlier. Finally, Scenario 1 was set aside because it would probably have been too extreme and would therefore not make sense to apply it in the real world.

Efficiency of the Control Scheme

Primarily, the consequences of applying the presented controller to the urban arterial are illustrated by comparing Scenario 2 with Scenario 4. Traffic in the arterial is expected to move faster owing to the control scheme that optimizes the signal plans.

Traffic Conditions. Here we define two variables as performance indicators: vehicle accumulation and mean speed in the arterial. These parameters are common as traffic performance indicators, especially against the background of the Macroscopic Fundamental Diagram (MFD) concept (*29*). The results are described in Figure 3 below. The left column holds for the traffic performance in case Scenario 4 in contrast to case Scenario 2 on the right-hand side.

As expected under constant demand, vehicle accumulation in the arterial stabilized over time, although on a lower level when the control scheme was implemented, see Figure 3, *a* and *b*. This is because the controller extends green phases for the arterial if there is no queue

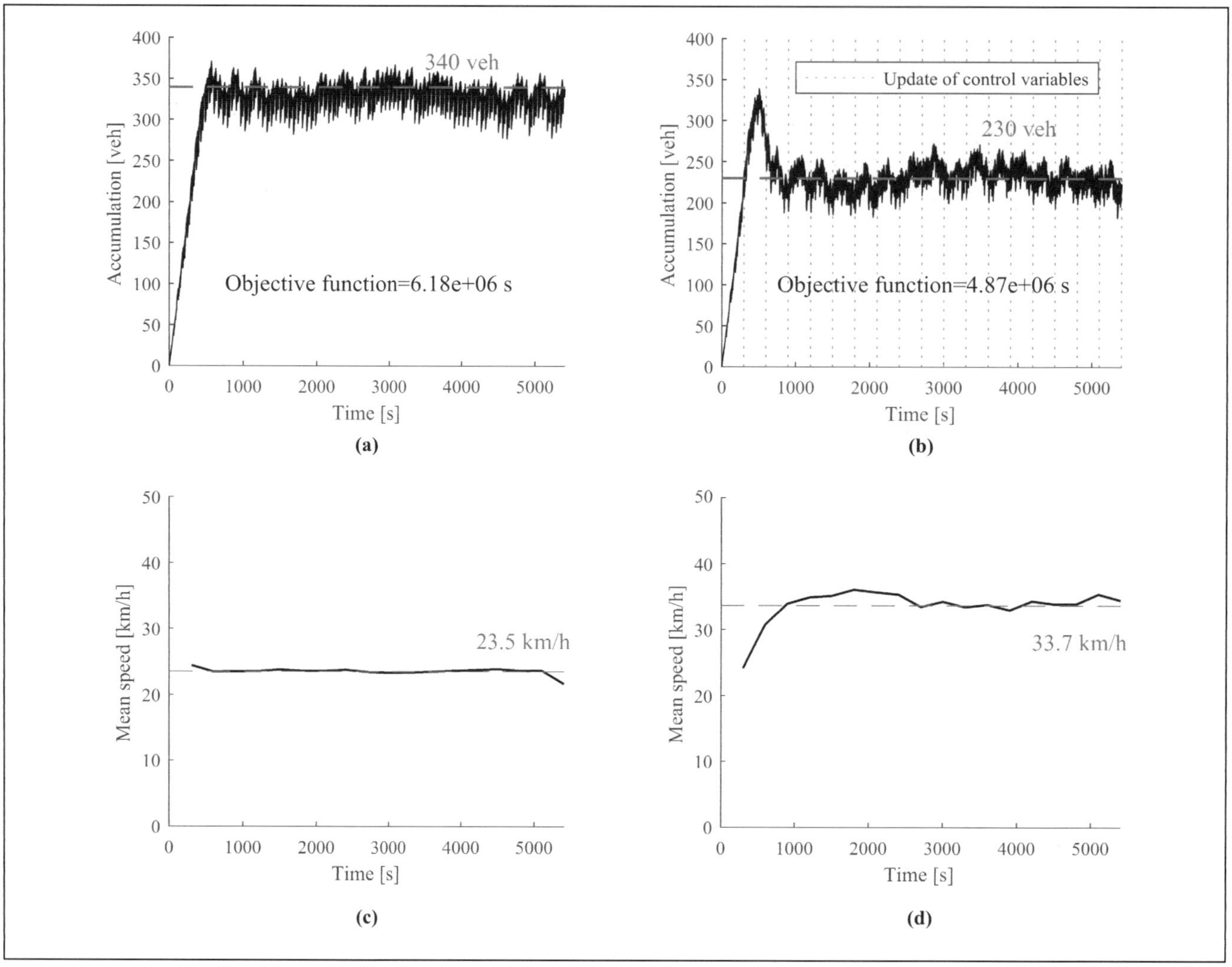

Figure 3. Traffic performance in the urban arterial illustrated by vehicle accumulation (*a* and *b*) and mean speed (*c* and *d*). Left column: Scenario 4; right column: Scenario 2.

or a small queue at the related entry. On average, there were approximatively 100 vehicles less in the arterial for case *B*. The accumulation curve in Figure 3*b* peaked at the beginning as new signal plans were allocated to the traffic lights for the first time after 5 min. Afterwards, the level dropped when the controller started operating. The controller interventions are illustrated in Figure 3*b* through the vertical dashed lines. With the controller, vehicles can drive at a significantly higher mean speed in the arterial, see Figure 3, *c* and *d*. The difference between case Scenario 2 and case Scenario 4 was 10 km/h on average. Overall, the objective function was lower for Scenario 2 than Scenario 4, indicating that the controller indeed improves the combined operations at the arterial and its entries. In other words, the proposed controller improves traffic performance – as represented by the parameters named above – strongly and continuously over the whole simulated period. Differences became even clearer when demand in the arterial increased up to 3,000 vph and higher. In that case, heavy congestion was caused in the arterial for Scenario 4 and results became less illustrative for the comparisons. Another indicator for improved traffic performance is the total arterial output, both at the main arterial exit and at side-street junctions. During the whole simulation run, nearly 300 vehicles more were measured passing the arterial in Scenario 2 compared with Scenario 4.

Application for Bus Operation

In urban arteries or boulevards, DBLs are common in many cities as they promise various advantages, as described for example by Basso et al. (*30*). Primarily, bus times and service reliability are improved through DBLs.

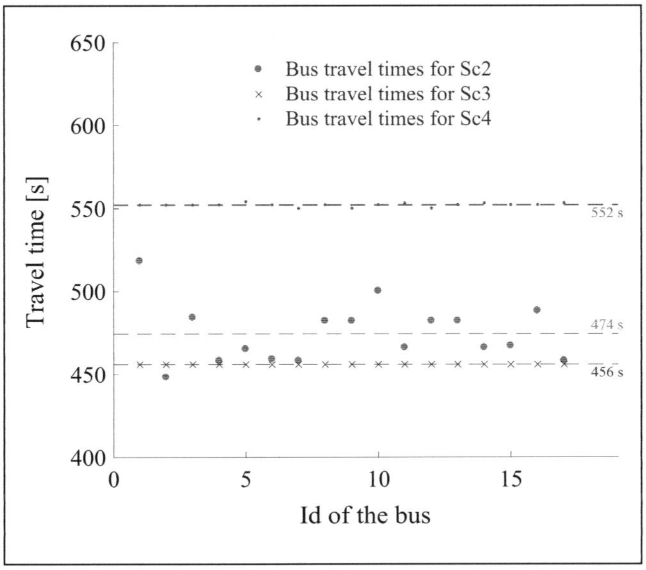

Figure 4. Travel time of buses through the urban arterial for scenarios: Scenario 2, Scenario 3 and Scenario 4. Every point corresponds to one bus trip.

As shown in the previous section, the proposed controller enhanced traffic performance in the arterial considerably, as a consequence traffic moved faster. From this point of view, one could argue that DBLs become superfluous as their original purpose is now redundant. In addition, DBLs have the disadvantage of reducing the arterial capacity devoted to general traffic (*8*). When seeking to optimize the transportation system at a global level, such a strategy could be counterproductive (*10*, *31*). Therefore, the potential DBL replacement by the perimeter control scheme was tested in the following by comparing the travel times of buses in the arterial for three cases and in the described simulation environment.

As previously explained, we will now focus on the comparison of Scenario 2 with Scenario 3 in which a DBL is deployed on the rightmost lane. Buses ran at a frequency of 5 min through the arterial. There are five stations, and dwell times were fixed at 10 s for each station. These conditions held for all three cases. Note that dynamic dwell times, that is, dwell times that depend on the effective time headways between two successive buses, were tested. Such an approach led to similar results, as the bus system was very regular for the different test cases. Figure 4 provides the graphical results of the comparisons. The travel times of the last two buses in the simulation were disregarded because the simulation ceased before these buses arrived at their destination.

Due to the constant dwell time and fixed traffic signal timing, the variations in travel time were relatively small for cases Scenario 4 and Scenario 3. Scenario 2, on the other hand, showed more variations because traffic signal timings changed from one bus trip to the next (*32*). However, variations remained very limited. On the whole, buses passed significantly faster through the arterial when perimeter control was in operation, in contrast to reference case Scenario 4. This occurred as every signal plan at the 14 intersections was optimized according to the current traffic situation. Due to the controller, almost 1.5 min per bus was saved on average compared with the reference scenario without traffic management. This corresponds to a reduction in bus travel time of 15% for the 3 km-long arterial. It was anticipated that this difference in favor of case Scenario 2 would grow with increasing distances traveled. Moreover, the advantages of the control scheme regarding bus operations would become even more noticeable were arterial demand further increased. Note also, that the higher inner demand during the middle section of the simulation had no visible effect on bus travel times. Even when queue lengths were taken into account, the controller still ensured good traffic conditions in the arterial. In terms of bus operations, the option with DBLs (Scenario 3) was still the most efficient option. Buses could travel even faster than in Scenario 2 as they did not encounter any queues at all. However, DBL gains compared with the controller were not significant (20 s or 4% per bus), considering the drawbacks of DBLs described before.

Concerning the impact on traffic flow, Scenario 2 was definitely the best option. With DBLs in operation, the assigned arterial demand in the simulation could not be fully satisfied. The number of lanes for general traffic was reduced from three to two, which resulted in heavy congestion outside and inside of the arterial. Thus, the available capacity for general traffic in Scenario 3 was lower than in Scenario 2. This resulted in spillbacks at the different intersections but also upstream of the arterial. It turned out that Scenario 2 and Scenario 3 were too dissimilar to perform a detailed comparison of the traffic conditions.

However, queue lengths could be analyzed for Scenario 2. Indeed, in addition to the traffic performance advantages provided by the controller, the common perimeter control problem of a long delay caused at the perimeter intersections was mitigated owing to the dual-objective controller. Figure 5 shows the queue lengths at the arterial entries for Scenario 2. Because comparing the results with those of Scenario 3 is not a consistent approach, as previously explained, we proposed comparing Scenario 2 with Scenario 4, the situation without control. Our expectation was that Scenario 4 would underestimate the Scenario 3 queue length. Even if no control was applied in either case, the arterial was a two-lane road in Scenario 3, whereas three lanes were simulated in Scenario 4. Note that the mean queue length over the whole simulation period is plotted.

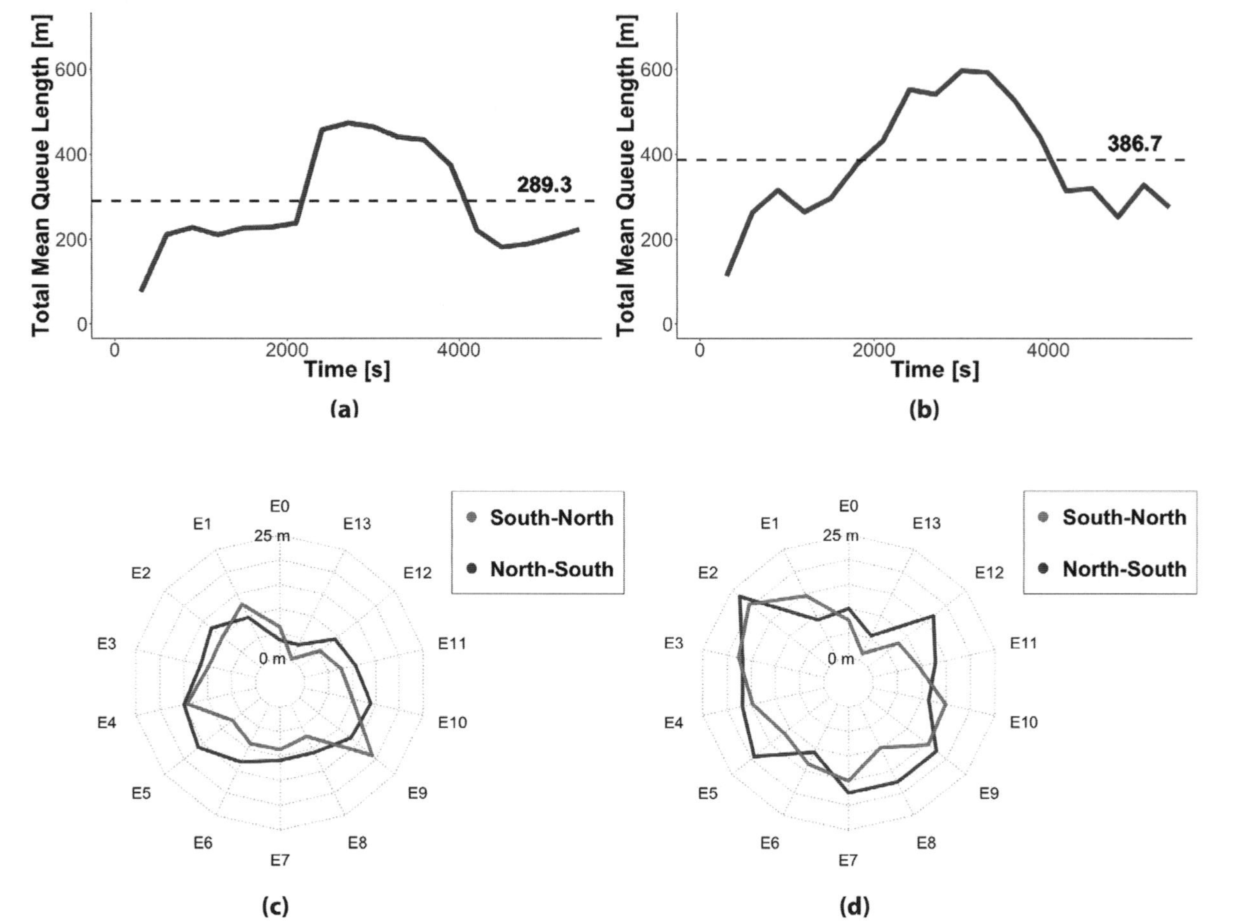

Figure 5. Mean queue lengths at perimeter entries cumulated over all intersections (*a* and *b*) and split up by every single intersection (*c* and *d*). Left column: Scenario 4; right column: Scenario 2.

Both diagrams for Scenario 4 and Scenario 2 show a comprehensible peak of queue lengths after half of the simulation time, when inner demand from zone-to-zone in the network is increased. On average, the entry queues were indeed longer when the controller was running, but the difference remained within reasonable limits and might be much smaller in comparison to a bang-bang type of control algorithm, as pointed out by Yang et al. (*21*). The kiviat charts (Figure 5, *c* and *d*) highlight that queues are quite evenly distributed among the intersections. The exceptions are entries 0 and 13, where queues are clearly below average because these intersections are located right on the edge of the network. The equitable distribution is a consequence of a DTA applied in the simulation. Moreover, the graphs indicate that queues on north–south approaches are somewhat longer on average than for the opposite direction. This is caused through the priority given to vehicles coming from the south over oncoming vehicles. Finally, because Scenario 4 gave only an underestimation of the queue length in Scenario 3, Scenario 2 did not have a significant effect on traffic conditions even at the intersections. These different results, that is, decrease in bus travel times and limited effects on traffic conditions, make the replacement of DBLs by perimeter control very promising.

Discussion and Conclusion

In this paper, we investigated, through simulation, the potential for using perimeter control to improve bus system efficiency. This strategy aimed at maintaining free-flow conditions in the vicinity of a major bus line, preventing buses from queuing together with cars at the intersections, thus, obtaining similar benefits to those from DBL, but without drastically reducing the capacity available for general traffic. To prevent spillbacks outside the protected area, we resorted to a perimeter and queues' length at intersections controller, that was then tested with the novelty of using an urban arterial as the controlled area. The control algorithm assigned the optimal green time to each of the 14 perimeter intersections, where side streets enter into the arterial, to minimize total travel costs expressed as a function of vehicle accumulations. Necessary data about traffic accumulation in the arterial and queue lengths at the entering links were provided through virtual detectors. Simulation results show that traffic performance, as indicated by vehicle accumulation and mean speed in the arterial, is significantly improved with the control approach applied compared with the reference case without a controller.

Because traffic conditions remain in free-flow when activating the controller, replacing DBLs in the arterial with the proposed control scheme is suggested. As explained before, the disadvantages of DBLs could be eliminated as a result of this measure. The hypothesis was confirmed by simulation, which shows that perimeter control is a possible alternative to DBL. Bus travel times with the perimeter control algorithm are almost as fast as bus travel times when using a DBL. At the same time, the available capacity for general traffic is higher, and this leads to reductions in overall delays. As a result, perimeter control leads to better traffic conditions both inside and outside the protected area. Overcoming the various drawbacks of DBLs, perimeter control could therefore be a promising alternative traffic management strategy in urban arterials.

To further pursue this research direction, the controller could still be modified and extended. For instance, the two objective criteria for the proposed controller were weighted the same in the present study. The weighting could be adapted for other policy goals. Focus could be directed more toward queue lengths, for instance. Additionally, other data sources besides loop detectors could be used for the measurements. This could lead to an increase in the accuracy of the estimations as these alternative technologies become more reliable, especially for high-demand situations, during which loop detectors are possibly covered by vehicles and, therefore, imprecise. Furthermore, the introduced feedback controller could be extended to an MPC, also including a prediction of traffic conditions. Another option would be to shorten the control period down to 1 or 2 min instead of 5 min, and possibly increase the efficiency of the controller.

Further comparisons could be made taking transit signal priority together with DBLs into consideration, which would probably result in additional time savings for buses traveling on DBLs. Another applicable strategy could be bus lanes with intermittent priority, as they represent a dynamic solution that offers greater capacity to general traffic compared with DBLs. In the same vein, whether bus lanes could also be replaced when the controlled region is larger and includes a city center could be investigated. Finally, the fact, that the simulated environment is simplified should be borne in mind when results are interpreted. On a larger scale, the enhanced traffic performance through perimeter control increases the attractiveness for motorized individual traffic, which results in higher demand and does not necessarily reduce congestion in the long-term view, as described by Duranton and Turner (*33*). Altogether, the proposed perimeter control scheme is very beneficial, because it improves the traffic conditions in a certain area and prevents this area from exceeding demand.

Acknowledgments

This research was partly supported by the European Research Council under the European Union's Horizon 2020 research and innovation program (grant no. 646592 — MAGnUM project). The authors thank the IT team of the traffic engineering laboratory in Licit Lyon-France for the assistance with the traffic simulator SymuVia in particular. Sincere gratitude is also expressed to the SVT Group in Zurich, Switzerland, for providing us with the preliminary work.

Author Contributions

The authors confirm contribution to the paper as follows: study conception and design: NC, LL, MM; data collection: MK, NC; analysis and interpretation of results: MK, NC; draft manuscript preparation: MC, NC, MM. All authors reviewed the results and approved the final version of the manuscript.

References

1. Buhaug, H., and H. Urdal. An Urbanization Bomb? Population Growth and Social Disorder in Cities. *Global Environmental Change*, Vol. 23, No. 1, 2013, pp. 1–10.
2. Dargay, J., D. Gately, and M. Sommer. Vehicle Ownership and Income Growth, Worldwide: 1960-2030. *The Energy Journal*, Vol. 28, No. 4, 2007, pp. 143–170.
3. Keyvan-Ekbatani, M., M. Papageorgiou, and I. Papamichail. Perimeter Traffic Control via Remote Feedback Gating. *Procedia-Social and Behavioral Sciences*, Vol. 111, 2014, pp. 645–653.
4. Rietveld, P., F. Bruinsma, and D. van Vuuren. Coping with Unreliability in Public Transport Chains: A Case Study for Netherlands. *Transportation Research Part A: Policy and Practice*, Vol. 35, No. 6, 2001, pp. 539–559.
5. Amirgholy, M., M. Shahabi, and H. O. Gao. Optimal Design of Sustainable Transit Systems in Congested Urban Networks: A Macroscopic Approach. *Transportation Research Part E: Logistics and Transportation Review*, Vol. 103, 2017, pp. 261–285.
6. Balke, K., C. Dudek, and T. Urbanik, II. Development and Evaluation of Intelligent Bus Priority Concept. *Transportation Research Record: Journal of the Transportation Research Board*, 2000. 1727: 12–19.

7. Zhu, H. Numerical Study of Urban Traffic Flow with Dedicated Bus Lane and Intermittent Bus Lane. *Physica A: Statistical Mechanics and its Applications*, Vol. 389, No. 16, 2010, pp. 3134–3139.
8. Chiabaut, N., X. Xie, and L. Leclercq. Road Capacity and Travel Times with Bus Lanes and Intermittent Priority Activation. *Transportation Research Record: Journal of the Transportation Research Board*, 2012. 2315: 182–190.
9. Eichler, M., and C. F. Daganzo. Bus Lanes with Intermittent Priority: Strategy Formulae and an Evaluation. *Transportation Research Part B: Methodological*, Vol. 40, No. 9, 2006, pp. 731–744.
10. Chiabaut, N. Evaluation of a Multimodal Urban Arterial: The Passenger Macroscopic Fundamental Diagram. *Transportation Research Part B: Methodological*, Vol. 81, 2015, pp. 410–420.
11. Guler, S. I., and M. Menendez. Analytical Formulation and Empirical Evaluation of Pre-Signals for Bus Priority. *Transportation Research Part B: Methodological*, Vol. 64, 2014, pp. 41–53.
12. Viegas, J., and B. Lu. Widening the Scope for Bus Priority with Intermittent Bus Lanes. *Transportation Planning and Technology*, Vol. 24, No. 2, 2001, pp. 87–110.
13. Bretherton, D., G. Bowen, and K. Wood. Effective Urban Traffic Management and Control: Recent Developments in SCOOT. Presented at 82nd Annual Meeting of the Transportation Research Board, Washington, D.C., 2003.
14. Aboudolas, K., and N. Geroliminis. Perimeter and Boundary Flow Control in Multi-Reservoir Heterogeneous Networks. *Transportation Research Part B: Methodological*, Vol. 55, 2013, pp. 265–281.
15. Daganzo, C. F. Urban Gridlock: Macroscopic Modeling and Mitigation Approaches. *Transportation Research Part B: Methodological*, Vol. 41, No. 1, 2007, pp. 49–62.
16. Keyvan-Ekbatani, M., A. Kouvelas, I. Papamichail, and M. Papageorgiou. Exploiting the Fundamental Diagram of Urban Networks for Feedback-Based Gating. *Transportation Research Part B: Methodological*, Vol. 46, No. 10, 2012, pp. 1393–1403.
17. Geroliminis, N., J. Haddad, and M. Ramezani. Optimal Perimeter Control for Two Urban Regions With Macroscopic Fundamental Diagrams: A Model Predictive Approach. *IEEE Transactions on Intelligent Transportation Systems*, Vol. 14, No. 1, 2013, pp. 348–359.
18. Haddad, J., and A. Shraiber. Robust Perimeter Control Design for an Urban Region. *Transportation Research Part B: Methodological*, Vol. 68, 2014, pp. 315–332.
19. Haddad, J. Optimal Perimeter Control Synthesis for Two Urban Regions with Aggregate Boundary Queue Dynamics. *Transportation Research Part B: Methodological*, Vol. 96, 2017, pp. 1–25.
20. Ortigosa, J., M. Menendez, and H. Tapia. Study on the Number and Location of Measurement Points for an MFD Perimeter Control Scheme: A Case Study of Zurich. *EURO Journal on Transportation and Logistics*, Vol. 3, No. 3, 2014, pp. 245–266.
21. Yang, K., N. Zheng, and M. Menendez. Two-level Perimeter Control Approach in a Connected-Vehicle Environment. *arXiv preprint arXiv:1612.05910*, 2016.
22. Lighthill, M. J., and G. B. Whitham. On Kinematic Waves. II. A Theory of Traffic Flow on Long Crowded Roads. In *Proceedings of the Royal Society of London A: Mathematical, Physical and Engineering Sciences*, Vol. 229, The Royal Society, 1955, pp. 317–345.
23. Richards, P. I. Shock Waves on the Highway. *Operations Research*, Vol. 4, No. 1, 1956, pp. 42–51.
24. Leclercq, L., J. A. Laval, and E. Chevallier. The Lagrangian Coordinates and What it Means for First Order Traffic Flow Models. In *Transportation and Traffic Theory 2007. Papers Selected for Presentation at ISTTT17*, London, 2007, pp. 735–753.
25. Chevallier, E., and L. Leclercq. A Macroscopic Theory for Unsignalized Intersections. *Transportation Research Part B: Methodological*, Vol. 41, No. 10, 2007, pp.1139–1150.
26. Xue, Z., N. Chiabaut, and L. Leclercq. Effect of Traffic Modeling on Control of Traffic Networks. *Transportation Research Record: Journal of the Transportation Research Board*, 2016. 2560: 47–56.
27. Herrera, J. C., D. B. Work, R. Herring, X. J. Ban, Q. Jacobson, and A. M. Bayen. Evaluation of Traffic Data Obtained via GPS-Enabled Mobile Phones: The Mobile Century Field Experiment. *Transportation Research Part C: Emerging Technologies*, Vol. 18, No. 4, 2010, pp. 568–583.
28. Leduc, G. *Traffic Data: Collection Methods and Applications.* European Commission, Joint Research Centre, Institute for Prospective Technological Studies, Luxembourg, 2008.
29. Daganzo, C. F., and N. Geroliminis. An Analytical Approximation for the Macroscopic Fundamental Diagram of Urban Traffic. *Transportation Research Part B: Methodological*, Vol. 42, No. 9, 2008, pp. 771–781.
30. Basso, L. J., C. A. Guevara, A. Gschwender, and M. Fuster. Congestion Pricing, Transit Subsidies and Dedicated Bus Lanes: Efficient and Practical Solutions to Congestion. *Transport Policy*, Vol. 18, No. 5, 2011, pp. 676–684.
31. He, H., M. Menendez, and I. Guler. Analytical Evaluation of Flexible Sharing Strategies on Multi-Modal Arterials. *Transportation Research Procedia*, Vol. 23, 2017, pp. 980–999.
32. Hans, E., N. Chiabaut, and L. Leclercq. Clustering Approach for Assessing the Travel Time Variability of Arterials. *Transportation Research Record: Journal of the Transportation Research Board*, 2014. 2422: 42–49.
33. Duranton, G., and M. A. Turner. The Fundamental Law of Road Congestion: Evidence from US Cities. *American Economic Review*, Vol. 101, No. 6, 2011, pp. 2616–2652.

The Standing Committee on Traffic Flow Theory and Characteristics (AHB45) peer-reviewed this paper (18-00029).

TRR
JOURNAL OF THE TRANSPORTATION RESEARCH BOARD

Article

Transportation Research Record
2018, Vol. 2672(20) 121–132

Article reuse guidelines:
sagepub.com/journals-permissions
DOI: 10.1177/0361198118796714
journals.sagepub.com/home/trr
SAGE

Traffic Flow of Merging Pedestrian Crowds: How Architectural Design Affects Collective Movement Efficiency

Zahra Shahhoseini[1] and Majid Sarvi[1]

Abstract

The need for developing reliable and rigorous models that can replicate and make predictions of pedestrian crowd evacuations has necessitated an understanding of the impact of architecture on individuals' interactions with their surroundings and the behavioral rules that govern their movements. Due to the challenges of providing such behavioral data from natural evacuations and previous crowd incidents, simulation-based and laboratory-based evacuation experiments have recently been employed as innovative data-provision approaches to study crowd behavior notably under emergency conditions. This study explores pioneer experiments of emergency escape with a view to investigating the relationship between spatial constraints and collective behavior of human crowds. Here, we make use of two types of empirical and analytical data obtained from a large number of well-controlled laboratory and evacuation simulation experiments. This study presents findings corresponding to how and to what extent the presence of conflicting layouts in egress areas, particularly merging corridors, affect the collective motion of pedestrians. The focus of attention will be on measures of performance at macroscopic level derived from both observations. Our results suggested that the movement patterns observed in both types of experiments are sensitive to the angle between the two merging streams and the symmetry/asymmetry of the merging layouts, with symmetric layouts almost invariably outperforming the asymmetric counterparts. Also, within each symmetry/asymmetry structural type, the angle at which the flows combined with each other affected the efficiency of discharge. Our findings provide further evidence as to the significant role of the architectural structure of the movement area in facilitating the traffic flow of heavy crowds of pedestrians.

A great deal of attempt has been made recently to understand how humans behave when fleeing a confined, crowded space. The answer to this question has important implications for planning safe and efficient evacuations as part of the risk assessment process essential for any mass event (*1*). The risk assessment conception can apply to a wide range of applications such as organizing sports events, music concerts, to name a few. In the simplest form, one might raise these basic but vital questions as to the time that it is going to take for the entire crowd to be safely evacuated in case of an emergency and if there is anything we can do to expedite this process. This question has guided a rapidly growing area of research focusing on developing prediction modeling frameworks (*2–27*) based on laboratory experiment (human and non-human subjects) and simulation experiments that have proved to be practical for assessing the emergency performance of built environments.

A major challenge that analysts face towards providing a reliable answer to such questions is that the impact of the design of escape areas as important determinants of crowd dynamics in evacuation scenarios are poorly understood. As is suggested by Pinter-Wollman, Fiore and Theraulaz (*28*), the architecture of built environments, especially the layout of egress areas, strongly influences the collective behavior of crowds especially at mass gatherings. Since human interactions with each other and their interaction with the physical features of escape areas can considerably impede the collective movements of the panicked crowd and result in dangerous bottlenecks and loss of time, it might culminate in injuries or loss of life. In social science, Kabo et al. (*29*) have found that the design of buildings, particularly

[1]Centre for Disaster Management and Public Safety, Department of Infrastructure Engineering, The University of Melbourne, Australia

Corresponding Author:
Address correspondence to Zahra Shahhoseini:
zahra.shahhoseini@unimelb.edu.au

confining environment, affects interactions and formation of collaborations among occupants. Despite the obvious impact of space on pedestrians' interactions and the importance of quantifying this impact, spatial constraints and conflicting geometrical layouts are rarely considered in the studies of crowd behavior at both macroscopic and microscopic levels. Therefore, the literature lacks an extensive understanding of the effect of architectural features on the pattern of pedestrians' movements. Merging streams of the pedestrian flow is one of the frequently observed features of mass events and public infrastructures which can restrict the efficiency of evacuation when normal, or emergency evacuation arises. In fact, the description of congestion caused by heavy flows of merging pedestrians and their conflicting movements must be accounted for in the efficient design of new facilities (such as public buildings, train stations, airports, or intersections of urban streets) and the daily operations of these facilities (*30*). This has so far been mainly challenged by the fact that we scarcely have access to observations and information from naturally occurring settings observed in past incidents in a form that can be used for scientific analysis purposes. As a result, in an attempt to tackle this problem, many researchers have resorted to data obtained from laboratory-based virtual-evacuation experiments and simulation-based virtual experiments for gleaning insights into human behavior and developing their modeling tools. Considering the inherent complexity of human behavior and the extra complexities that an emergency scenario can add in providing reliable data in safe experimental setups, simulated experiments have received a clear boost in attention in recent years.

Based on the investigation of the literature related to the notion of geometric designs, particularly the impact of conflicting geometries on crowd dynamic, a number of studies (*11*, *31*, *32*) explored the effect of corridor configurations in merging passage using non-human subjects. These experiments have been considered as a proximate model of the scale-free behavior of humans. This type of experiment allows a range of phenomena to be explored which is very hard to duplicate in experiments with human subjects. So far, utilizing ants and mice as subjects of experiments has been the most common which has proved to be practical for gaining insights into aggregate measures of movements. Dias et al. (*31*) investigate two groups of ants escaping through merging pathways. Their observation indicates that transition of clogging between two merging streams as well as oscillating flows happen when two streams of ants meet each other in crossing corridors. However, it should be noted that they did not consider different designs for merging corridors in their experiments. In another laboratory study (*11*), the impact of different merging designs on emergency performance is explored utilizing two groups of panicking ants passing through a shared pathway. A wide range of experiments with ants was conducted under various angles and geometric characteristics to examine the underlying geometric factors that influence the dynamics of two merging streams of humans. They observed significant dependency between geometrical configurations of merging corridors and traffic flow characteristics of the merged flow.

Zhang et al. (*16*) experimentally precisely explored merging 180° also known as T-junction and constructed fundamental diagrams for this specific configuration. They demonstrated that the fundamental diagrams of two regions at upstream of merging point match, however, the velocities associated with a merged corridor (downstream of merging point) are higher than those of upstream corridors at same densities. Shi et al. (*33*) investigated the dependency of outflow of merged corridors on the angle between two merging streams. Their findings suggest that the angle between two merging branches exerted a major influence on fluctuations of flows. Another study (*34*) used experimental data for estimating the impact of merging configurations on motion pattern of pedestrians. Their findings showed that the velocities of pedestrians decrease in the merging regions resulting in lower output rate especially for the larger angle between two merging corridors. Shahhoseini et al. (*35*) performed controlled experiments in order to explore the impact of merging angle and angle configuration on the efficiency of pedestrians' movements. Their findings demonstrated that symmetric setups considerably outperform their asymmetric setups (same angles with a different design) in terms of discharge rate and velocity distribution. However, unlike the study of Shiwakoti et al. (*34*), their results do not support the notion of decreasing relationship between merging angle and discharge rate. Their findings indicate inexistence of a monotonic trend between the throughput and the merging angle.

In another laboratory experiment, Lian et al. (*36*) performed a series of controlled experiments of merging flows of pedestrians in which one stream joined a major stream. In these experiments, they particularly looked at the effect of the width of passage which imposes itself to the straight stream on the efficiency of flow. Their findings suggest increasing the width of deviated branch results in a reduction in discharge rate at downstream of junction area.

A number of studies have approached this problem using crowd simulation models. They utilized simulated environments in view of the opportunity they provide in terms of observing a range of phenomena which is hard to reproduce in experiments with humans without compromising participants' safety. Peng and Chou (*37*) examined the conformation of congestion in a "T" intersection proposing a multi-floor fields cellular automata procedure. They observed that the occurrence of phase transition is irrespective of the competitive or cooperative behavior of pedestrians that has an effect on average first jamming time. Chen, Li and Zhang (*38*) studied the movements of pedestrians evacuating from the T-shaped intersection by utilizing a force-driving cellular automata simulation model. They found out that the most serious blockage happened at the entrance of the emergency staircase which is a typical kind of T-shaped channel.

As Haghani and Sarvi (*39*) recently discussed in a review of all experimental methods, the current literature demands a deep understanding of the impact of architecture especially merging corridors which result in conflicting maneuvers and complex interactions of humans on the collective behavior of pedestrian crowds. Insufficient behavioral data has prevented making solid inferences from the existing literature to understand how building architecture (i.e., physical attributes of merging corridors) influences human interactions. In this study, we staged mock evacuation trials in various merging layouts with human participants and gather empirical evidence from controlled laboratory experiments. Then we mimic those scenarios in a simulation environment, as well as a wide range of other merging configurations. Utilizing a simulation tool allows us to observe i) the impact of architectural features of merging passages on movement efficiency of pedestrian crowds; ii) whether or not the influence of configuration of merging corridors dramatically differs from one design to another design. Making use of two data sets obtained from mock evacuation and simulated experiments, we attempt to quantify architectural features for which we require to examine the macroscopic measure of performance. To our knowledge, none of the studies that have made use of simulation tools for examining the impact of constraints imposed by the merging corridors has provided mock evacuation trials of their simulated scenarios.

Methodology

A series of high-speed, high-density experiments took place on 23 February 2015 at a basketball court. For these experiments, 150 participants were hired to conduct trials and perform experimental tasks. The experimental procedure was approved by Monash University Human Research Ethics Committee. These experiments replicate emergency escape of pedestrian crowds through various merging passages. Therefore, we prepared a number of artificial, temporary merging corridors using panels. Each run of the trials consisted of the combination of two factors: the merging angle and the symmetry/asymmetry of the experimental setup, with these factors determining the physical layout at each trial. The participants were asked to gather at a designated place in the environment and stay in the waiting areas (at the beginning of each merging corridors) and were instructed to escape through the corridors as soon as the experiment started. They were instructed to run, compete with other participants, and leave the built environment as soon as possible. Participants' behavior and the track their movements were recorded by a camera which was mounted 9 m from the floor and an individual data-extraction approach, Petrack software (*40*), was practiced in order to provide a required dataset for analysis. In total 30 scenarios with two repetitions for each scenario were exercised through which factors such as the angle between two merging passages and the configurations of merging branches were varied. Two main structures of paths were considered for designing experimental setups in this study. One type of layout includes setups in which one stream imposes itself on a straight branch (asymmetric structures). The second configuration comprises merging designs in which two streams join each other symmetrically (symmetric structures). The imitation of those laboratory-based experiments was constructed in a bi-layer, force-based simulation model using A-star algorithm for wayfinding layer (*41*). The evacuation model that was employed in the simulation we used consists of two layers. At the operation level, the simulated agents move based on the notion of social force model introduced initially by Helbing and Molnár (*42*).

At the tactical level, the directional-path choices of simulated agents are generated using a probabilistic econometric model. The outcomes produced from the tactical level of the model are considered as input by the operational level in order to provide simulated motions at the operational level. Each step of each simulated subject is determined by the net outcome of three forces which are the desired force of each agent driving him or her towards his or her desirable direction, the force from any architectural barrier that makes agents avoid them to prevent collision, and the social interaction forces from other simulated agents. In the relevant equations, the body mass of each individual, the velocity vector for each individual, and the body radius of each simulated pedestrian was estimated using a dataset of individuals.

Snapshots of raw footage of the experiments associated with different scenarios are shown in Figure 1. Here we utilize a subset of our dataset related to the scenarios in which participants were asked to run as fast as they can and evacuate through merging corridors in a way they escape from imminent danger (since we practiced three different speed regimes in our experiments).

In order to elucidate the role of geometrical design of merging paths in shaping the scale-free behavior of pedestrian crowds, a wide range of physical attributes

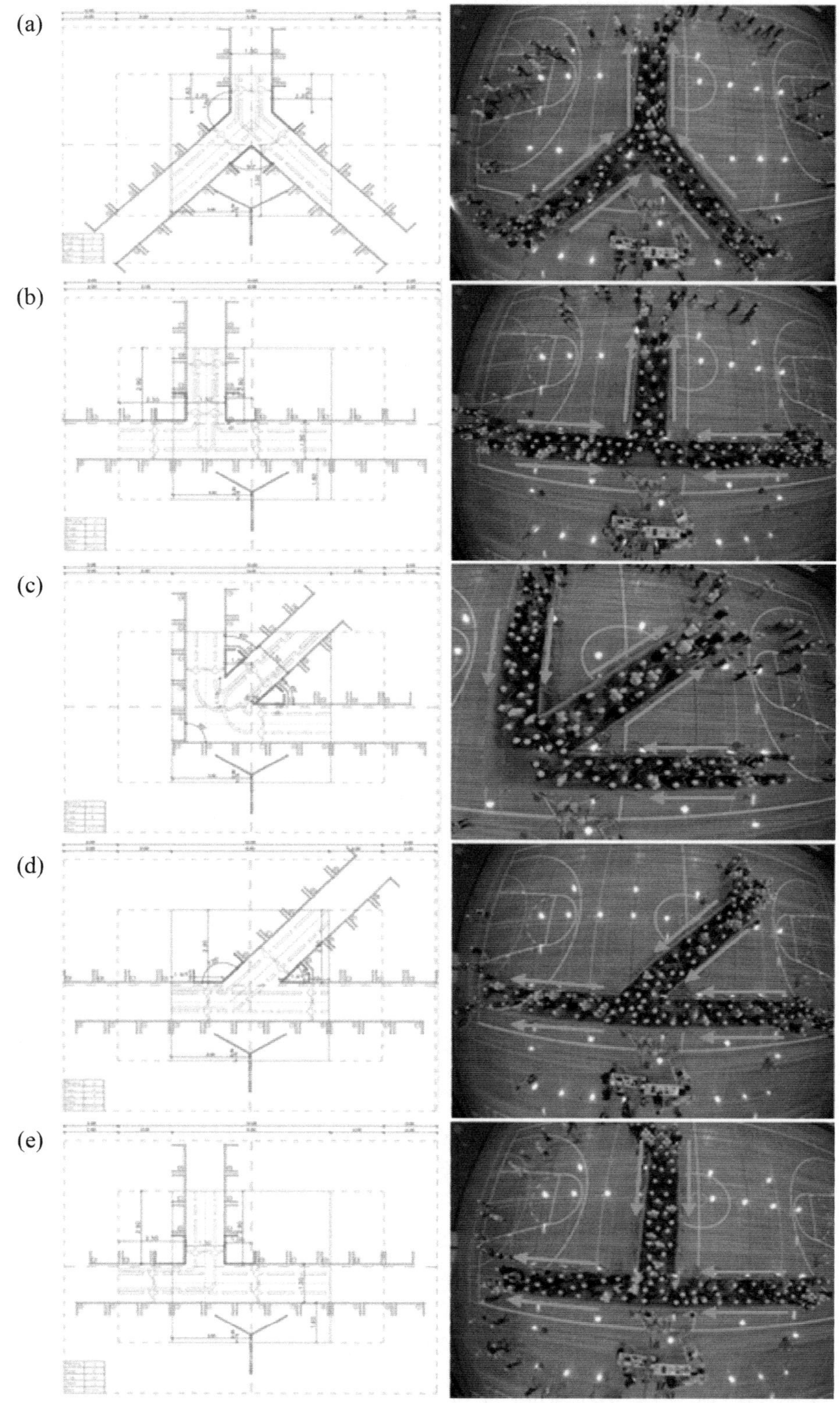

Figure 1. The blueprint of experimental setups (on the left) and snapshots from the raw footage of the experiments (on the right): (*a*) symmetric 90°; (*b*) symmetric 180°; (*c*) symmetric 270°; (*d*) asymmetric 45°; (*e*) asymmetric 90°.

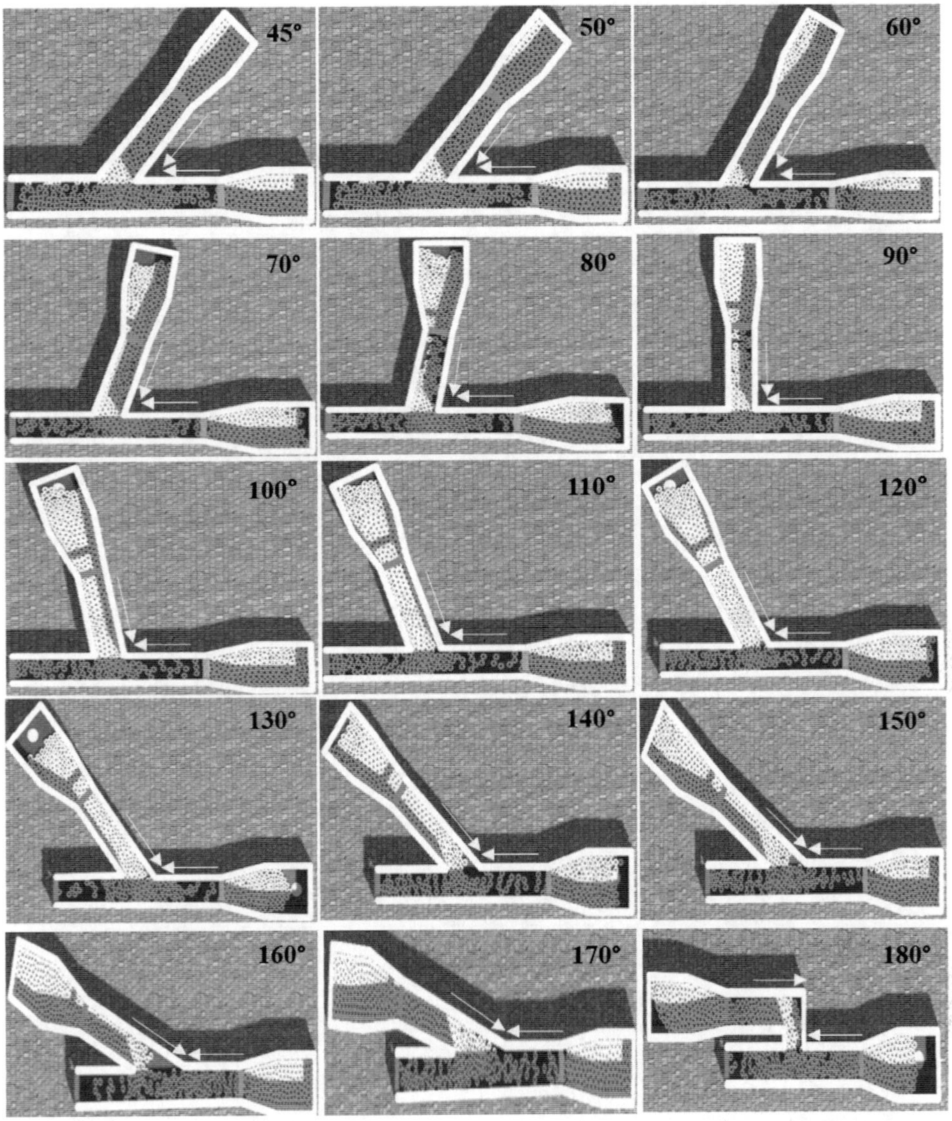

Figure 2. The architectural setups used for simulation process. This figure illustrates asymmetrical merging configurations in which deviated stream of agents needs to impose itself on the straight stream. The angles between the two streams represent the general direction of movement in each branch.

were considered for defining different merging layouts in a force-based simulation context. The architectural setups illustrated in Figures 2 and 3 are used for this analysis. Thirty scenarios and 100 repetitions for each scenario were processed in a simulation environment over which symmetry and the angle between two joining corridors varied. We made sure that the traversed distance and desired speed for all agents in all scenarios were the same to make our simulated outcomes comparable. The simulated scenarios were treated as emergency evacuation conditions in which the desired speed of agents was defined as the maximum speed of human participants in the laboratory evacuation experiments.

Results

Average evacuation time of individuals and total evacuation time were measured for each laboratory-based scenario as well as each simulation-based scenario over 100 repetitions since having access to accurate estimates of the evacuation times has become a major concern related to the safety of crowded spaces.

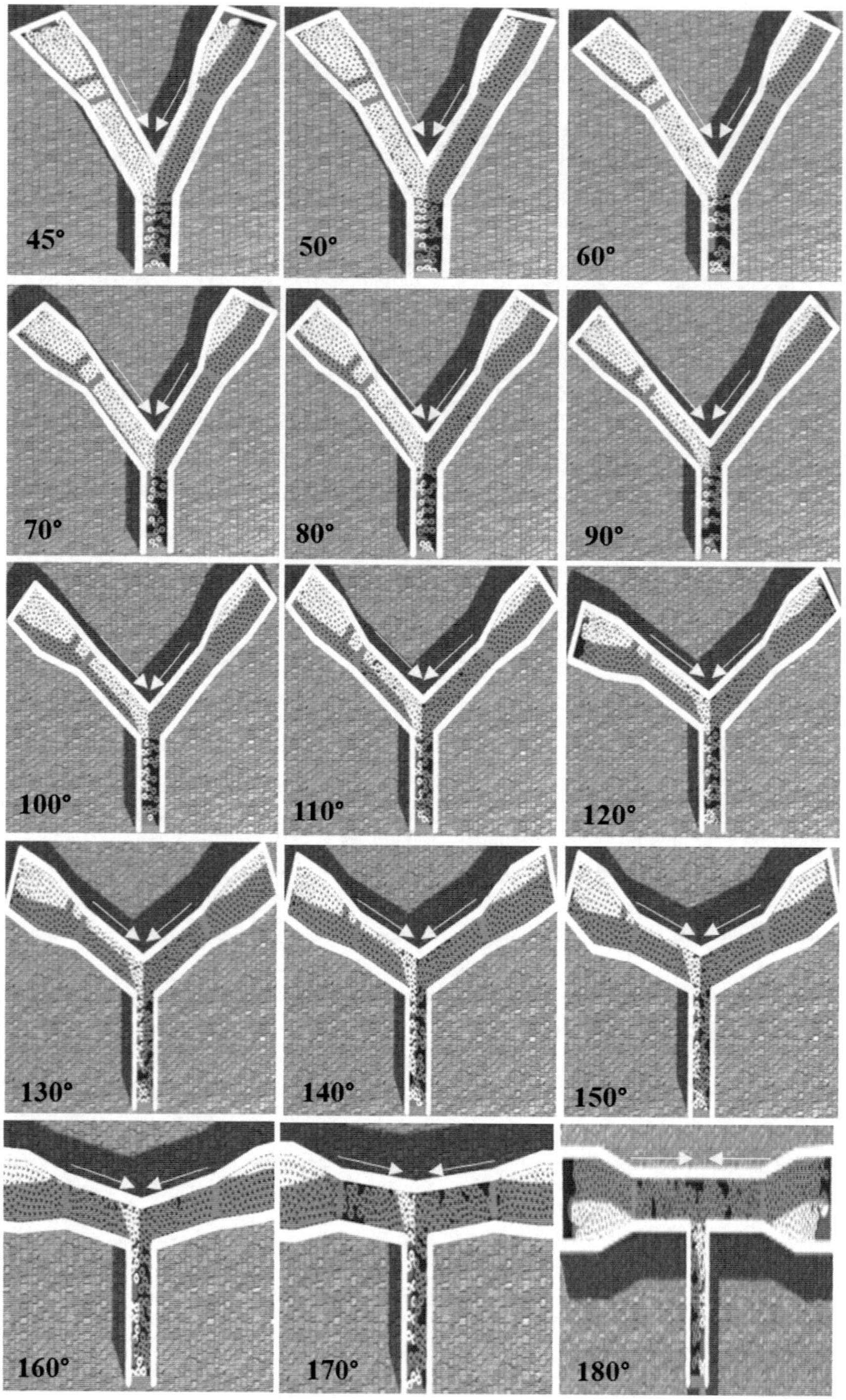

Figure 3. The architectural setups used for simulation process. This figure illustrates symmetrical merging configurations in which two streams of agents meet each other symmetrically. The angles between the two streams represent the general direction of movement in each branch.

The distribution of these total evacuation times observed in the experiments associated with each particular experimental setup and their simulation-based imitations is depicted in Figure 4. The plots visualize the average of total evacuation time for the laboratory experiments as well as minimum, first quartile, average,

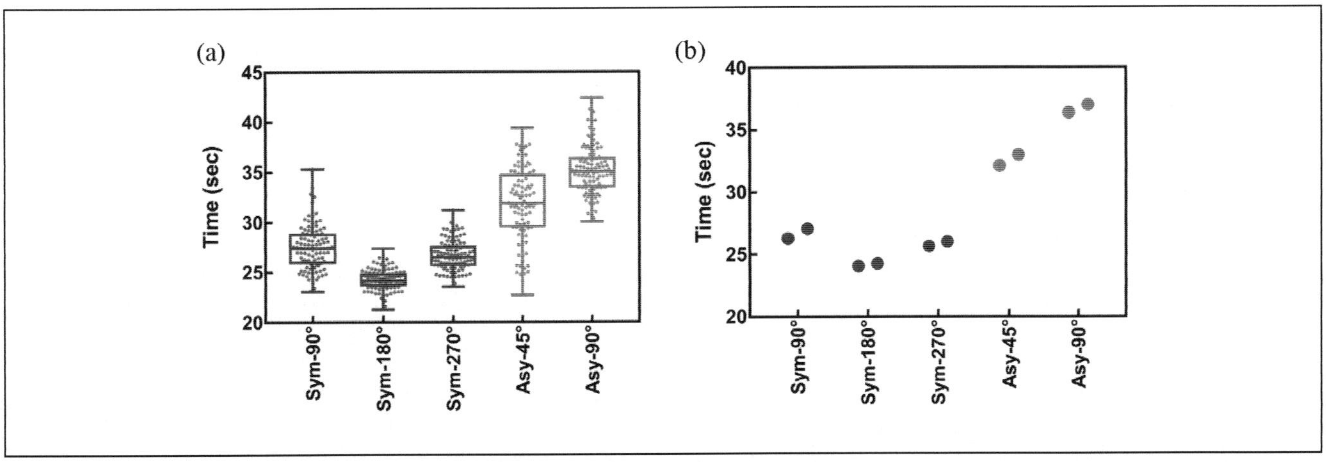

Figure 4. (*a*) Box-and-whisker plots for the minimum, 25th, 50th (median), 75th, and maximum percentiles of the total evacuation times calculated over 100 times of simulation repetitions, corresponding with the architectural setups in Figure 1 based on the simulation models and laboratory experiments; (*b*) total evacuation times associated with experimental setups in Figure 1.

third quartile, and maximum of the observed evacuation time for each imitated scenario in simulation environment over 100 repeats. The fundamental question that we are examining in this study is how the architectural features of merging corridors would influence on collective outcomes.

As can be observed, both plots reveal a degree of pattern resemblance between the obtained outcomes from laboratory-based experiments and evacuation simulation experiments. They both demonstrate that asymmetric merging configurations deliver poor performance in terms of evacuation time in comparison with merging setups in which two branches meet each other symmetrically. Analyzing movements of pedestrians in laboratory experiments and agents in simulation environments, it can be observed that conflict of trajectories is much higher in asymmetrical setups suggesting merging layouts in which one flow of pedestrians requires to impose itself on straight flow.

The simulated experiments over which we intended to examine the sensitivity of system-wise aggregate measures of performance were performed while the value of other parameters was kept constant (i.e., length and width of corridors, the number of agents, agents' speed, agents' size etc.) and the value of the angle between two joining branches was gradually increased while recording the total evacuation time and individual evacuation time. In other words, for each merging angle, we measured the time that it takes for all agents to vacate the experimental setups over 100 repetitions performed for that structure. The time lapse between the egress of the first individual and the last individual was considered to estimate average escape rate. The results of these analyses are illustrated in Figure 5 where the average of total evacuation time and the average of individual evacuation time over 100 repetitions of the simulation are visualized for each merging configuration in the form of a line chart. The standard deviations of total evacuation times are visualized as error envelopes. In other words, Figure 5 presents the comparison of the average total evacuation and the contrast of the average of average individual evacuation times observed in simulation repetitions associated with each experimental setup.

One clear observation based on these plots is the consistency between the two aggregate measures of performance that we considered (total evacuation time and an average of individual evacuation time). As can be seen in these graphs, these two measures exhibit the same pattern in response to the physical attributes of merging configurations.

Another conclusive observation based on these plots is the larger scatter of evacuation times measured for asymmetrical setups compared to those of symmetrical configurations. This could be attributed to the fact that the creation of clogging areas is more probable in the asymmetrical shapes which have reflected in significant variations of predicted total evacuation times.

Another observation based on Figure 5, total evacuation time, is a significant dependency between the time that it takes for a setup to be fully evacuated and the angle between two merging streams. This statement is suggested to be valid for both types of configurations. However, the dependency of total evacuation time on merging angle is proposed to be non-monotonic indicating the specified range of angles results in best performance of joining layouts. The average of the average individual evacuation times also seems to convey a

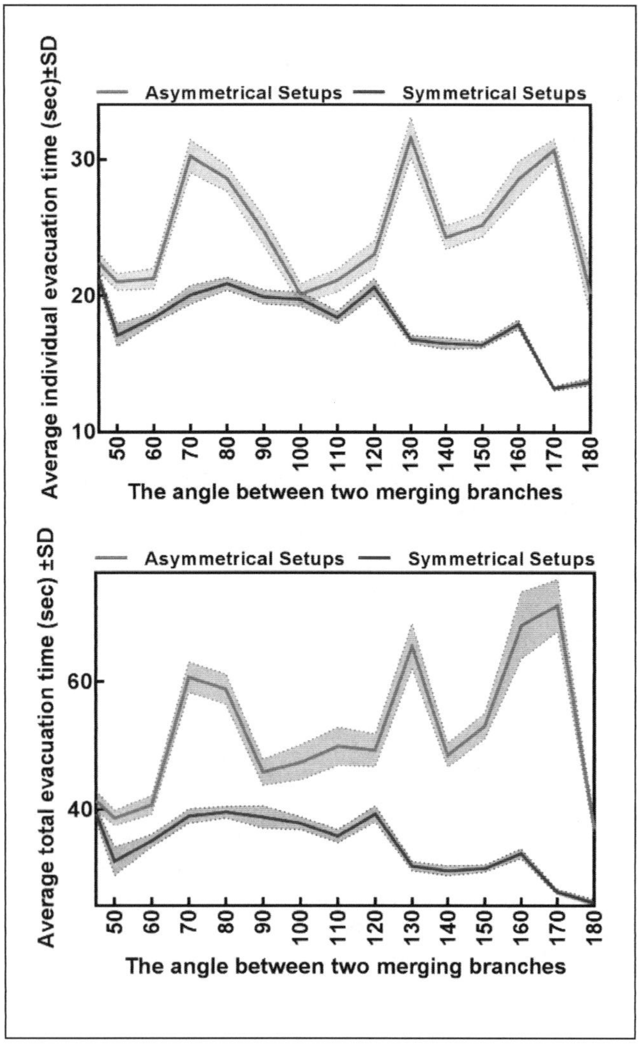

Figure 5. The variation of total evacuation times and average individual evacuation times associated with different designs of paths and different merging angles. Each point on both graphs is obtained by averaging 100 simulation runs. The error bands represent the standard deviation of the simulated observations.

similar indication. Although the extent of the difference in evacuation time associated with each experimental setup appears to be smaller when the average of individual evacuation times is regarded as the primary measure.

In order to quantify the sensitivity of the outcomes of simulation to the value of merging angle, we conducted *t*-test between the observed total evacuation time of experimental setups corresponding with 100 repetitions for symmetrical and asymmetrical configurations separately. The relative percentage of the difference between the averages of total evacuation times observed for each architectural setup in simulation repetitions was calculated. Our results suggest that the difference between the levels of sensitivity of aggregate measures of performance that we applied to the merging angle varies from architectural setup to architectural setup. For example, as far as the total evacuation times are concerned (as the most conventional system-wise measure of evacuation performance), the difference has varied from nearly 1.5%–26% for symmetrical setups and 9%–51% for asymmetrical configurations depending on which architectural setup is considered.

As reflected in plots of Figures 4 and 5, both experimental and simulated observations agree upon a critical finding suggesting that the impact of the merging angle on the efficiency of movement is not monotonic as our outcomes do not exhibit a continually decreasing or increasing pattern between the average of total evacuation time and the merging angle. This non-monotonic pattern suggests that there could be a specific range of angles that could result in better system performance in terms of the aggregate measures of evacuation.

Another common observation based on this figure is the significant imbalance between the aggregate performance measure (total evacuation time) for symmetrical and the asymmetric counterparts. It is suggested that symmetrical merging layouts, in general, deliver a better performance in terms of emergency response compared to their equivalent asymmetrical setups. Therefore, with a view to quantifying the impact of symmetry (or asymmetry) of the design on the efficiency of collective movements, we put into contrast the total evacuation time estimated for symmetrical and asymmetrical designs for each merging angle over repeated simulation. In addition to the percentage of difference, we conducted *t*-test between the observed total evacuation times of two designs of each angle corresponding to 100 repetitions. The pairwise comparison of the evacuation time of symmetrical and asymmetrical configurations for each merging angle is illustrated in Figure 6 separately. The graphs in this figure have visualized the values of evacuation time and its average for each angle value associated with 100 repetitions of the simulation process. The connecting lines on top of two corresponding columns show the relative percentage of the difference between the averages of simulated total evacuation times represented by the two columns. The star under the connecting lines indicates a highly significant difference ($p < 0.001$) between the total evacuation time of the symmetrical configurations and the asymmetrical layouts of same merging angle. Our simulated results reveal the impact of asymmetry or symmetry of design differs from 5%–149% depending on which merging angle is considered as far as total evacuation time is concerned. Based on these plots, it appears that the average differences in the predicted evacuation time could magnify when the merging angle is greater than 120°.

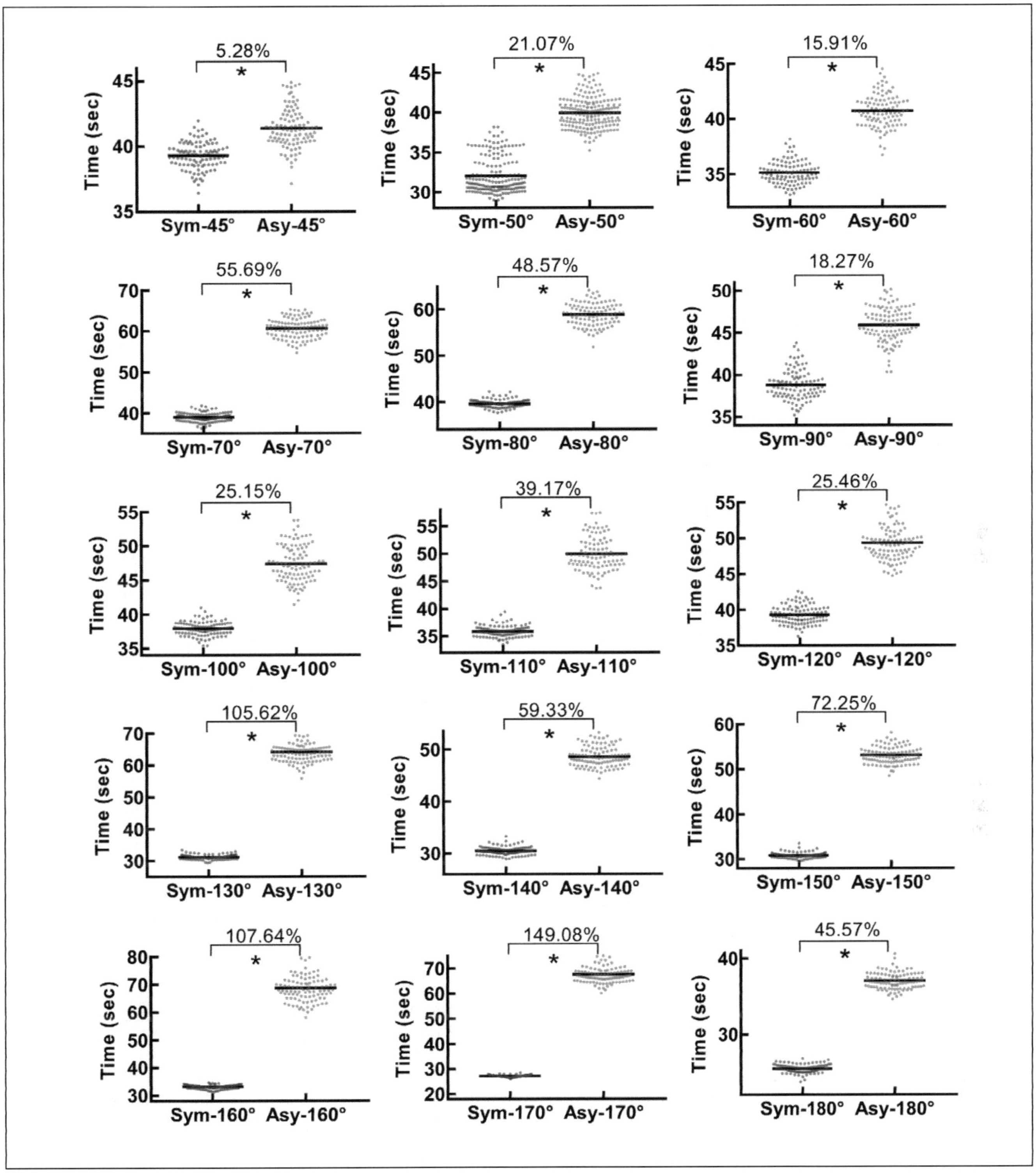

Figure 6. Comparison of total evacuation times for merging setup pairs with the same angle and different symmetry layout (i.e., symmetric and asymmetric). The connecting lines on the top show the relative percentage of the difference between the two averages represented by data and the star displays the statistical significance of this difference.

Discussion

Researchers in a variety of disciplines, notably in transport, have attempted to study the impact of architecture on collective behavior and micro-level interactions of humans with their surroundings during the evacuation of confined built environments, and incorporate their findings into planning and designing spaces tools. These studies mostly aim at developing reliable mathematical or simulation models that can capture and predict crowd movement through conducting virtual evacuations (*42, 43*). Such forecast tools will enable planners to identify the worst-case scenarios and potential dangers thereby improving the safety of human crowds through control, management, or architectural design.

Here, we report on a large number of laboratory-type and simulation-type experiments to study how structural constraints of merging corridors influence on behavioral mechanisms that govern collective movements of humans as well as identifying the optimum designs that facilitate the most efficient collective outcome. In other words, we investigated how the macro-scale predictions would differ by defining different geometrical features for merging corridors based on two data sources. As we discussed previously, virtual simulation environments have proved to offer more flexibility compared with field experiment in terms of an opportunity to take the control of all parameters, collect very large samples within a reasonable amount of time, stretch attributes of scenarios (i.e., agent speed or number of agents) to extreme level, and provide a safe environment to explore a wide range of potentially dangerous conditions. Furthermore, simulation-based experiments provide us an ethical environment to rapidly explore a wide array of merging maneuvers and outcomes, thus expediting the process of finding optimum architectural designs that deliver the best performance.

Our laboratory and simulated experiments quantify the sensitivity for predicted evacuation times to the architectural features of merging corridors. This highlights the crucial necessity for exploring the scale-free behavior of pedestrians escaping from densely crowded environments through joining passages when emergency evacuation arises. In other words, the outcome of the simulation is sensitive to the angle between two merging streams and the symmetry (or asymmetry) of the design. In addition, it is suggested by this study that the asymmetrical merging layouts deliver poorer performance in terms of total escape time compared with symmetrical setups. Also, within the symmetric designs, certain angles delivered a faster throughput and higher discharge rate. In addition, it also observed that the variations of total evacuation time and symmetry structure of merging setups correlates suggesting that outcomes of asymmetrical setups show more considerable scatter in comparison. Furthermore, our findings reveal that both types of merging layouts exhibit sensitivity to the angle between two joining branches, although the asymmetric arrangements clearly display more sensitivity to the merging angle.

These findings can offer important behavioral insights on the causative relationships between specific geometrical features and collective outcome and how conflicting geometries affect the mass motion of pedestrians as well as how architectural design can improve the efficiency with which pedestrians move. Lessons from laboratory and simulated scenarios may help formalize the quantification of spaces to inspire architects to design spaces that can facilitate the scale-free behavior of human crowds.

Acknowledgments

The financial support of Public Transport Victoria (PTV) and Department of Economic Development, Jobs, Transport and Resources (DEDJTR) for this research (Arc Linkage project LP120200361) is highly acknowledged.

Author Contributions

The authors confirm contribution to the paper as follows: Study conception and design: ZS and MS; data collection: ZS and MS; analysis and interpretation of results: ZS; draft manuscript preparation: ZS and MS. All authors reviewed the results and approved the final version of the manuscript.

References

1. Still, G. K. *Introduction to Crowd Science*. CRC Press, Boca Raton, FL, 2014.
2. Helbing, D., I. J. Farkas, P. Molnar, and T. Vicsek. Simulation of Pedestrian Crowds in Normal and Evacuation Situations. *Pedestrian and Evacuation Dynamics*, Vol. 21, 2002, pp. 21–58.
3. Moussaïd, M., D. Helbing, S. Garnier, A. Johansson, M. Combe, and G. Theraulaz. Experimental Study of the Behavioural Mechanisms Underlying Self-Organization in Human Crowds. *Proceedings of the Royal Society of London B: Biological Sciences*, May 13, 2009. https://doi.org/10.1098/rspb.2009.0405.
4. Haghani, M., and M. Sarvi. Stated and Revealed Exit Choices of Pedestrian Crowd Evacuees. *Transportation Research Part B: Methodological*, Vol. 95, 2017, pp. 238–259.
5. Dogbe, C. On the Modelling of Crowd Dynamics by Generalized Kinetic Models. *Journal of Mathematical Analysis and Applications*, Vol. 387, No. 2, 2012, pp. 512–532.
6. Haghani, M., and M. Sarvi. How Perception of Peer Behaviour Influences Escape Decision Making: The Role of Individual Differences. *Journal of Environmental Psychology*, Vol. 51, 2017, pp. 141–157.
7. Do, T., M. Haghani, and M. Sarvi. Group and Single Pedestrian Behavior in Crowd Dynamics. *Transportation Research Record: Journal of the Transportation Research Board*, 2016. 2540: 13–19.

8. Haghani, M., and M. Sarvi. Following the Crowd or Avoiding it? Empirical Investigation of Imitative Behaviour in Emergency Escape of Human Crowds. *Animal Behaviour*, Vol. 124, 2017, pp. 47–56.
9. Abdelghany, A., K. Abdelghany, and H. Mahmassani. A Hybrid Simulation-Assignment Modeling Framework for Crowd Dynamics in Large-Scale Pedestrian Facilities. *Transportation Research Part A: Policy and Practice*, Vol. 86, 2016, pp. 159–176.
10. Haghani, M., M. Sarvi, Z. Shahhoseini, and M. Boltes. How Simple Hypothetical-Choice Experiments Can Be Utilized to Learn Humans' Navigational Escape Decisions in Emergencies. *PLoS One*, Vol. 11, No. 11, 2016, pp. 1–24.
11. Shahhoseini, Z., M. Sarvi, and M. Saberi. Insights Toward Characteristics of Merging Streams of Pedestrian Crowds Based on Experiments with Panicked Ants. *Transportation Research Record: Journal of the Transportation Research Board*, 2016. 2561: 81–88.
12. Haghani, M., and M. Sarvi. Social Dynamics in Emergency Evacuations: Disentangling Crowd's Attraction and Repulsion Effects. *Physica A: Statistical Mechanics and its Applications*, Vol. 475, 2017, pp. 24–34.
13. Schadschneider, A., W. Klingsch, H. Klüpfel, T. Kretz, C. Rogsch, and A. Seyfried. Evacuation Dynamics: Empirical Results, Modeling and Applications. *Encyclopedia of Complexity and Systems Science*, 2009, pp. 3142–3176.
14. Haghani, M., and M. Sarvi. Pedestrian Crowd Tactical-Level Decision Making during Emergency Evacuations. *Journal of Advanced Transportation*, Vol. 50, No. 8, 2016, pp. 1870–1895.
15. Chraibi, M., M. Freialdenhoven, A. Schadschneider, and A. Seyfried. Modeling the Desired Direction in a Force-Based Model for Pedestrian Dynamics. *Traffic and Granular Flow '11*, 2013, pp. 263–275.
16. Zhang, J., W. Klingsch, A. Schadschneider, and A. Seyfried. Experimental Study of Pedestrian Flow Through a T-junction. *Traffic and Granular Flow '11*, 2013, pp. 241–249.
17. Haghani, M., M. Sarvi, and Z. Shahhoseini. Accommodating Taste Heterogeneity and Desired Substitution Pattern in Exit Choices of Pedestrian Crowd Evacuees Using a Mixed Nested Logit Model. *Journal of Choice Modelling*, Vol. 16, 2015, pp. 58–68.
18. Shahhoseini, Z., M. Sarvi, and M. Saberi. Pedestrian Crowd Dynamics in Merging Sections: Revisiting the "Faster-Is-Slower" Phenomenon. *Physica A: Statistical Mechanics and its Applications*, Vol. 491, 2018, pp. 101–111.
19. Saberi, M., K. Aghabayk, and A. Sobhani. Spatial Fluctuations of Pedestrian Velocities in Bidirectional Streams: Exploring the Effects of Self-organization. *Physica A: Statistical Mechanics and its Applications*, Vol. 434, 2015, pp. 120–128.
20. Haghani, M., and M. Sarvi. Human Exit Choice in Crowded Built Environments: Investigating Underlying Behavioural Differences Between Normal Egress and Emergency Evacuations. *Fire Safety Journal*, Vol. 85, 2016, pp. 1–9.
21. Moussaid, M., E. G. Guillot, M. Moreau, J. Fehrenbach, O. Chabiron, S. Lemercier, J. Pettré, C. Appert-Rolland, P. Degond, and G. Theraulaz. Traffic Instabilities in Self-Organized Pedestrian Crowds. *PLoS Computational Biology*, Vol. 8, No. 3, 2012, p. e1002442.
22. Moussaïd, M., and M. Trauernicht. Patterns of Cooperation During Collective Emergencies in the Help-or-Escape Social Dilemma. *Scientific Reports*, Vol. 6, 2016.
23. Haghani, M., M. Sarvi, O. Ejtemai, M. Burd, and A. Sobhani. Modeling Pedestrian Crowd Exit Choice Through Combining Sources of Stated Preference Data. *Transportation Research Record: Journal of the Transportation Research Board*, 2015. 2490: 84–93.
24. Seyfried, A., M. Boltes, J. Kähler, W. Klingsch, A. Portz, T. Rupprecht, A. Schadschneider, B. Steffen, and A. Winkens. Enhanced Empirical Data for the Fundamental Diagram and the Flow Through Bottlenecks. *Pedestrian and Evacuation Dynamics 2008*, 2010, pp. 145–156.
25. Haghani, M., and M. Sarvi. Identifying Latent Classes of Pedestrian Crowd Evacuees. *Transportation Research Record: Journal of the Transportation Research Board*, 2016. 2560: 67–74.
26. Aghabayk, K., M. Sarvi, O. Ejtemai, and A. Sobhani. Impacts of Different Angles and Speeds on Behavior of Pedestrian Crowd Merging. *Transportation Research Record: Journal of the Transportation Research Board*, 2015. 2490: 76–83.
27. Shahhoseini, Z., M. Haghani, and M. Sarvi. Estimation and Application of a Multi-class Multi-criteria Mixed Paired Combinatorial Logit Model for Transport Networks Analysis. *Transportmetrica B: Transport Dynamics*, Vol. 3, No. 1, 2014, pp. 59–78.
28. Pinter-Wollman, N., S. M. Fiore, and G. Theraulaz. The Impact of Architecture on Collective Behaviour. *Nature Ecology & Evolution*, Vol. 1, 2017, p. 0111.
29. Kabo, F. W., N. Cotton-Nessler, Y. Hwang, M. C. Levenstein, and J. Owen-Smith. Proximity Effects on the Dynamics and Outcomes of Scientific Collaborations. *Research Policy*, Vol. 43, No. 9, 2014, pp. 1469–1485.
30. Timmermans, H. J. *Pedestrian Behavior: Models, Data Collection and Applications: Models, Data Collection and Applications*. Emerald Group Publishing, Bingley, UK, 2009.
31. Dias, C., M. Sarvi, N. Shiwakoti, O. Ejtemai, and M. Burd. Investigating Collective Escape Behaviours in Complex Situations. *Safety Science*, Vol. 60, 2013, pp. 87–94.
32. Shahhoseini, Z., and M. Sarvi. Collective Movements of Pedestrians: How We Can Learn from Simple Experiments with Non-Human (Ant) Crowds. *PLoS One*, Vol. 12, No. 8, 2017, p. e0182913.
33. Shi, X., Z. Ye, N. Shiwakoti, D. Tang, C. Wang, and W. Wang. Empirical Investigation on Safety Constraints of Merging Pedestrian Crowd through Macroscopic and Microscopic Analysis. *Accident Analysis & Prevention*, Vol. 95, 2015, pp. 4015–416.
34. Shiwakoti, N., Y. Gong, X. Shi, and Z. Ye. Examining Influence of Merging Architectural Features on Pedestrian Crowd Movement. *Safety Science*, Vol. 75, 2015, pp. 15–22.
35. Shahhoseini, Z., M. Sarvi, M. Saberi, and M. Haghani. Pedestrian Crowd Dynamics Observed at Merging Sections: Impact of Designs on Movement Efficiency.

Transportation Research Record: Journal of the Transportation Research Board, 2017. 2622: 48–57.
36. Lian, L., X. Mai, W. Song, Y. K. K. Richard, Y. Rui, and S. Jin Pedestrian Merging Behavior Analysis: An Experimental Study. *Fire Safety Journal*, Vol. 91, 2017, pp. 918–925.
37. Peng, Y.-C., and C.-I. Chou. Simulation of Pedestrian Flow Through a "T" Intersection: A Multi-floor Field Cellular Automata Approach. *Computer Physics Communications*, Vol. 182, No. 1, 2011, pp. 205–208.
38. Chen, C.-K., J. Li, and D. Zhang. Study on Evacuation Behaviors at a T-shaped Intersection by a Force-driving Cellular Automata Model. *Physica A: Statistical Mechanics and its Applications*, Vol. 391, No. 7, 2012, pp. 2408–2420.
39. Haghani, M., and M. Sarvi. Crowd Behaviour and Motion: Empirical Methods. *Transportation Research Part B: Methodological*, Vol. 107, 2018, pp. 253–294.
40. Boltes, M., and A. Seyfried. Collecting Pedestrian Trajectories. *Neurocomputing*, Vol. 100, 2013, pp. 127–133.
41. AlShawi, I. S., L. Yan, W. Pan, and B. Luo. Lifetime Enhancement in Wireless Sensor Networks using Fuzzy Approach and A-Star Algorithm. *IET Conference Proceedings*, London, UK, 2012, p. 82. https://doi.org/10.1049/cp.2012.0611.
42. Helbing, D., and P. Molnár. Social Force Model for Pedestrian Dynamics. *Physical Review E*, Vol. 51, No. 5, 1995, pp. 4282–4286.
43. Helbing, D., I. Farkas, and T. Vicsek. Simulating Dynamical Features of Escape Panic. *Nature*, Vol. 407, No. 6803, 2000, pp. 487–490.

The Standing Committee on Traffic Flow Theory and Characteristics (AHB45) peer-reviewed this paper (18-03961).

Article

TRR
JOURNAL OF THE TRANSPORTATION RESEARCH BOARD

Transportation Research Record
2018, Vol. 2672(20) 133–143

DOI: 10.1177/0361198118796927
journals.sagepub.com/home/trr
SAGE

Control of Connected and Autonomous Vehicles with Cut-in Movement using Spring Mass Damper System

Soohyuk Bang[1] **and Soyoung Ahn**[1]

Abstract

This paper proposes a theoretical framework to control a platoon of connected and autonomous vehicles (CAVs) in the presence of cut-in movements. The control method is developed based on the spring–mass–damper (SMD) system concept and aims to improve the platoon efficiency and stability after a cut-in movement (e.g., lane change and merging from on-ramp). The method seeks to resolve a disturbance created by a cut-in vehicle by systematically setting two control parameters, spring constant and damping coefficient, of the SMD-based control model based on the prevailing traffic conditions. The control method is evaluated through a simulation based on the changes in speed and spacing, recovery time to reach the desired speed, disturbance propagation, and platoon flow. The simulation result shows that the control method can effectively reduce the disturbance caused by a cut-in movement and improve platoon flow.

Autonomous vehicle (AV) technologies have the enormous potential to drastically improve safety and efficiency of traffic flow. In particular, AV technologies such as adaptive cruise control (ACC) (without communication functions) and cooperative adaptive cruise control (CACC) (with communication among vehicles) promise high performance with regards to capacity and stability as demonstrated by several pioneering field tests (*1–4*). For example, VanderWerf et al. (*5*) showed through simulation that traffic capacity increased significantly when traffic was composed of 100% ACC or CACC vehicles owing to shorter time gaps. In particular, traffic capacity more than doubled with CACC vehicles. Shladover et al. (*2*) showed a similar result: as the penetration rate of CACC vehicles increased, the lane capacity increased rapidly. They showed, however, that ACC vehicles did not significantly increase the lane capacity.

The basic concept of ACC and CACC is to maintain a constant time gap among vehicles in a platoon. For instance, Shladover et al. and Milanés et al. developed two different control algorithms of ACC and CACC by traffic condition on freeways: one for light traffic and the other for heavier traffic, in which vehicles are in car-following mode (*2*, *3*). Milanés et al. also showed that these algorithms can achieve string stability (*3*). To validate the control logic by Milanés et al., Milanés and Shladover conducted several field tests using 10 vehicles equipped with their ACC and CACC (*3*, *4*). They showed that, owing to its communication capability, the CACC platoon attained the most improved traffic flow stability and efficiency, as compared to their ACC and the intelligent driver model (IDM), which is one of the widely used ACC control models.In IDM, vehicles in a low density state accelerate until they reach the desired speed. When they are under control, they decelerate at a rate less than the comfort deceleration rate while maintaining the desired time gap to guarantee collision-free control (*6*). Gong et al. developed new car-following logic for connected and autonomous vehicles (CAVs) based on an optimization framework to minimize traffic disturbances and maintain a stable gap between vehicles (*7*). This new logic showed better performance than the existing CACC algorithm of Schakel et al. for stability (*8*). Although the control algorithms cited above are shown to improve traffic flow efficiency and stability, most of them focused only on longitudinal control. In addition, they considered a limited range of traffic conditions (e.g., free-flow or congested) without vehicle cut-in and cut-out maneuvers.

Several studies developed control algorithms considering vehicle cut-in maneuvers. Contet et al. developed a

[1]Department of Civil and Environmental Engineering, University of Wisconsin-Madison, Madison, WI

Corresponding Author:
Address correspondence to Soyoung Ahn: sue.ahn@wisc.edu

"multi-agent system model" based on the spring–mass–damper (SMD) system concept (*9*). In this model, a vehicle platoon was considered as a virtual train so that cut-in vehicles could join the platoon only at the end. With this arrangement, however, some cut-in vehicles may have to wait awhile, particularly with a long platoon. Milanés et al. (*3*) demonstrated that their CACC algorithm worked properly for cut-in and cut-out vehicles on the freeway: a platoon of two CACC vehicles recovered the desired time gap when a vehicle merged into and left the platoon. Milanés and Shladover considered more cut-in cases on the freeway with regard to the existence of vehicle-to-vehicle (V2V) communication (*10*). In particular, if a cut-in vehicle had no V2V capability, the first follower of the cut-in vehicle would be an ACC vehicle because it cannot communicate with the cut-in vehicle. When a V2V-equipped vehicle merged into the platoon of CACC vehicles, the platoon showed better stability. Although these studies evaluated the performance of the existing CACC control algorithm in the presence of cut-in and cut-out movements, they did not explore different strategies that could be more stable or efficient.

A review of the previous studies on CAV platooning in the presence of cut-in movement reveals some major shortcomings. One is that a constraint is imposed on where a cut-in vehicle can join a platoon (e.g., only at the end), which may not be feasible in certain conditions. Furthermore, previous studies simply verified string stability after a cut-in movement without additional CAV control. Therefore, an opportunity exists to minimize the impact of cut-in maneuvers (e.g., speed disturbances) more proactively. To this end, this research aims to develop a method to specifically control a CAV platoon after a vehicle cut-in maneuver with the assumption that a cut-in vehicle can join a platoon at any time and location. Our aim for CAV control is to develop a simple and flexible control framework that improves platoon efficiency (i.e., higher flow) and stability during the transition and that can be applied in a wide range of traffic conditions. Therefore, we built on our previous research which developed a SMD-based model to describe platoon formation and develop an optimal strategy for platooning in light to near-capacity traffic states (*11*). The strategy considers a balance between freedom for cut-in and cut-out maneuvers, and efficient and rapid platooning. This model enables simple dynamic control via dynamic setting of two control parameters. The present paper develops a systematic way of setting these control parameters to minimize the impact of cut-in maneuvers on platoon efficiency and stability. Testing through simulation shows that a CAV platoon can be controlled effectively with the proposed control method.

SMD System

In our previous study (*11*), a SMD system was applied to control CAVs, analogous to the self-organization behavior of fish and birds (*12*). In this section, we briefly summarize the SMD system model described in our previous publication, including the formulation, characteristics of parameters, and damping types of the system (*11*). We extended this basic control framework to control CAVs in a platoon in the presence of vehicle cut-ins.

Notation

n:	the nth vehicle of the platoon
m:	mass of a vehicle (kg)
c:	coefficient indicating how rapidly a vehicle reaches the desired speed (kg/s)
v_d:	desired speed (m/s)
x:	position of a vehicle (m)
$\dot{x}$:	speed of a vehicle (m/s)
$\ddot{x}$:	acceleration of a vehicle (m/s^2)
k:	spring constant (kg/s^2)
b:	damping coefficient (kg/s)
l:	critical spacing (m)
S_o:	minimum spacing (=vehicle length + minimum space gap, m)
τ:	response time (or desired time gap, sec)

n *Body* ***n*** *− 1 Spring and Damper System*

The formulation of the SMD system for a CAV platoon is described below. Equation 1 shows the force needed to achieve the desired speed, v_d, for the first leader of the platoon. The two forces that govern the controlled motion of the nth CAV in the platoon are expressed in Equation 2. The first term represents the spring force to reach the critical spacing, l, represented by Newell's simplified car-following model (*13*) in Equation 3; and the second term represents the damping force that maintains the same speed as the vehicle immediately ahead. The parameters in these equations are defined above under Notation. These formulations are designed so there is no influence from a vehicle behind.

$$m_1\ddot{x}_1 = c(v_d - \dot{x}_1), \tag{1}$$

$$\begin{aligned} m_n\ddot{x}_n &= k_n(x_{n-1} - x_n - l) + b_n(\dot{x}_{n-1} - \dot{x}_n) \\ &= k_n\Delta x_n + b_n, \Delta\dot{x}_n, \end{aligned} \tag{2}$$

$$l = S_o + \tau^*\dot{x}_n. \tag{3}$$

Using Equation 1, the upper-bound of c is obtained based on the cases of maximum acceleration (to the free-flow speed) and maximum deceleration (to the speed of zero), whichever is smaller. Equation 4 shows the feasible range of c (*11*). a_{max} and d_{max} represent the maximum acceleration and deceleration rates, respectively.

$$0 < c \leq min\left(\frac{m_1 a_{max}}{v_d}, \frac{m_1 d_{max}}{\dot{x}_1}\right) \tag{4}$$

In the spring force term in Equation 2, the spring constant, k, represents the stiffness of spring: it would take a shorter time to reach l with higher k. The possible range of k is expressed as follows (*11*):

$$0 < k \leq \frac{1}{\Delta x} m a_{max} \tag{5}$$

In the damping force term, the damping coefficient, b, represents how quickly a follower adopts the speed of the leading vehicle and maintains it. The setting of b determines three types of damping—critical damping, over-damping, and under-damping—and thus string stability. Specifically, Equation 6, which is derived from Yanakiev and Kanaellakopoulos (*14*), describes the conditions that achieve string stability, in which the equality condition represents critical damping (b_{crit}) and strict inequality condition represents over-damping:

$$b \geq max\left(\frac{m}{\tau}, \sqrt{km}\right) \tag{6}$$

In our previous study we have shown that with critical damping, the system reaches an equilibrium state, in which all vehicles have the same speed and spacing, fastest without oscillations (Figure 1*a*) (*11*). With over-damping, it takes longer to reach an equilibrium state, but without oscillations (Figure 1*b*). The system shows oscillations with under-damping (Figure 1*c*). Therefore, considering safety and efficiency, we targeted critical and over-damping in our control framework.

CAV Control with Cut-in

In this section, we describe how the above control framework is extended to control CAVs in a platoon when it is disturbed by a cut-in vehicle. It should be noted that this study does not focus on controlling the cut-in movement, but CAVs in a platoon, including the cut-in vehicle once it joins the platoon.

In our previous research, we developed a method for CAV platooning in light traffic, in which CAVs maintain large spacing to better accommodate cut-in and cut-out maneuvers (*11*). If traffic volume is sufficiently high, some impact would be inevitable without alternative control. Specifically, if the cut-in spacing is less than $2l$, the control algorithm proposed in this study should be used to minimize the impact. We assume that (i) all vehicles including the cut-in vehicle are CAVs; (ii) spacing between vehicles before the cut-in maneuver is less than $2l$; and (iii) the cut-in vehicle inserts itself anywhere in the available spacing; see Figure 2. Therefore, the cut-in maneuver would result in spacings (ahead and/or behind) of less than l and instigate a speed disturbance for the vehicle(s) upstream. This study aims to minimize the disturbance via systematic parameter setting of the lower k and higher b, as will be described in detail below.

CAV Control Setting

Cut-in vehicles near a merge can create voids ahead (spacing greater than l) (*15*), which can lead to inefficient use of capacity, known as "capacity drop" (*16*). A void can occur, for example, when the insertion speed is lower than the mainline speed and the vehicle has a finite acceleration rate. It can also occur if the spacing of a cut-in vehicle is less than l at insertion, and the vehicle decelerates excessively (over-compensation) while trying to

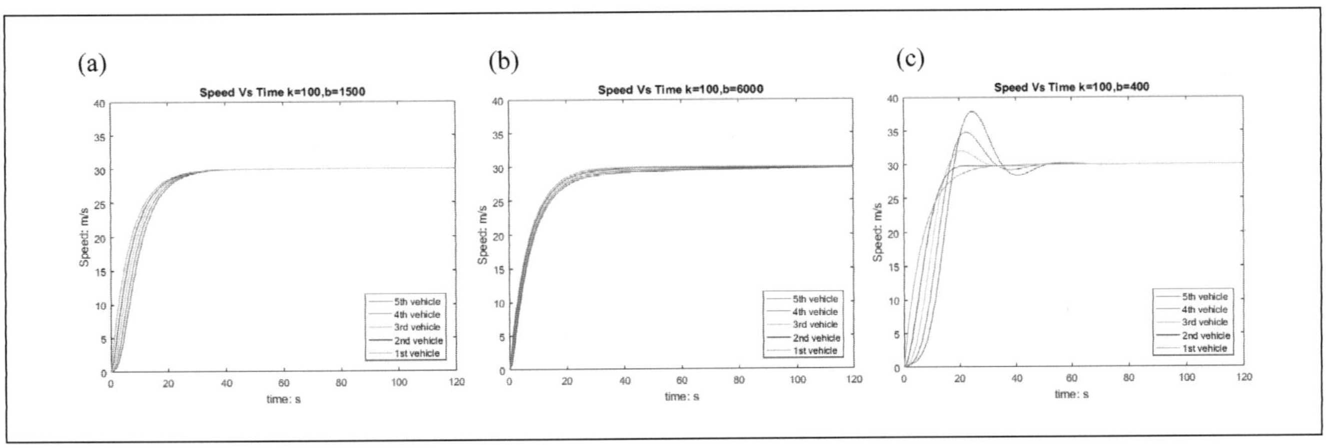

Figure 1. Types of damping (*a*) critical-damping; (*b*) over-damping; and (*c*) under-damping.

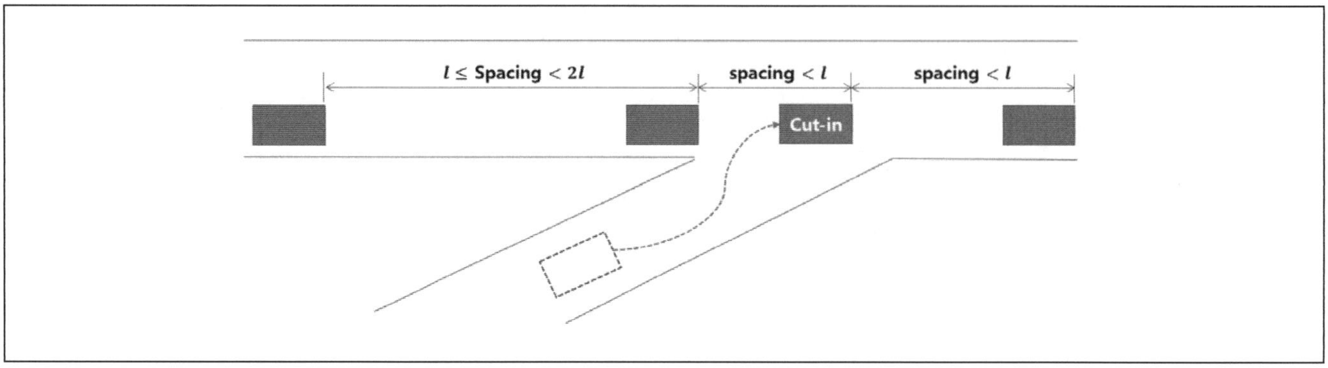

Figure 2. Hypothetical example of cut-in movement.

reach the critical spacing. Insertions in the above examples can also trigger speed disturbances (*17*, *18*), which could propagate upstream in the platoon. Therefore, we aim to minimize the time for the cut-in vehicle to reach the platoon speed to prevent void creation and control the CAV(s) upstream to prevent propagation of a speed disturbance. We measured the impact of cut-in with regard to the average speed change, average spacing change, and time to recover the desired speed of a platoon (recovery time).

To achieve these, we set lower spring constants, k, for the cut-in vehicle and the immediate follower, so that they maintain spacings shorter than l (at least temporarily). Under this control, the recovery time for the cut-in vehicle would be much less, particularly if the insertion speed is close to the platoon speed. Furthermore, we set a higher damping coefficient, b (over-damping), so that the cut-in vehicle would quickly accelerate to the platoon speed if the insertion speed is lower than the platoon speed. A higher damping coefficient would also reduce speed disturbances as vehicles would match the speed of the leader more quickly. Thus, by systematically setting lower k and higher b, the platoon can accommodate cut-in vehicles while maintaining a high, stable flow. Specifically, we set k on insertion for vehicle n, $k_{in,n}$, to be proportional to the spacing after the insertion (S_n) as in:

$$k_{in,n} = \left(\frac{S_n}{\alpha}\right)^{\beta} * k_0 \tag{7}$$

In which k_0 represents the spring constant before an insertion, and α and β represent scaling constants. An exponential relationship is assumed here so that the spring constant decreases as the spacing decreases. The scaling constants are used to regulate the control sensitivity. It should be noted that k_{in} is determined for each vehicle, including the cut-in vehicle, based on the respective spacing. The setting of k_0 in uncongested traffic is provided in our previous paper (*11*). In summary, k_0 can be determined by three relationships in Equations 8 to 10, depending on the traffic condition. Coefficients e_1 and e_2 are determined so that k_0 at capacity is equal to 121 kg/s^2.

$$\text{Cubic relationship for light traffic}: k_0 = e_1 * flow^3, \tag{8}$$

$$\text{Quadratic relationship for intermediate trafiic}: k_0 = e_2 * flow^2, \tag{9}$$

$$\text{Maximum relationship for near capacity}: k_0 = \frac{1}{\Delta x} m a_{max}. \tag{10}$$

In Equation 7, we set $\alpha > S_n$ and $\beta > 0$, so that $k_{in,n} < k_0$. Furthermore, we can decrease $k_{in,n}$ by increasing α and β.

We further investigated how k_{in} and thus spacing, changes with respect to the parameters α and β. For this, we assume that the platoon is moving at capacity with the spacing of $l = 37$ m and that a vehicle inserts in the middle, that is, $S_n = 18.5$ m. As illustrated in Figure 3*a*, k_{in} decreases exponentially with α for various values of β. However, when $\alpha > 150$ and $\beta = 1$, k_{in} decreases marginally. For given α, the relationship between k_{in} and β shows a similar trend, as expected as the coefficient of k_0 is the βth power of $\frac{S_n}{\alpha}$. When $\beta > 2$, k_{in} decreases marginally. We further calculated the force and acceleration using Equation 2, and based on these values, we computed the rate of spacing increase per second; see Figure 3*b* for the result. We assume that when the rate of spacing increase is less than 0.01 m/s, a vehicle essentially keeps its spacing. Based on the result of this parameter analysis, α and β are set to be 250 and 2, respectively, so that the rate is less than 0.01 m/s. As we discuss later, a higher k value can be restored once the effect of cut-in maneuver subsides, and the platoon is stable.

We also set the damping coefficient on insertion, b_{in}, to be proportional to the speed difference between the cut-in vehicle and its follower in the target lane at the time of insertion:

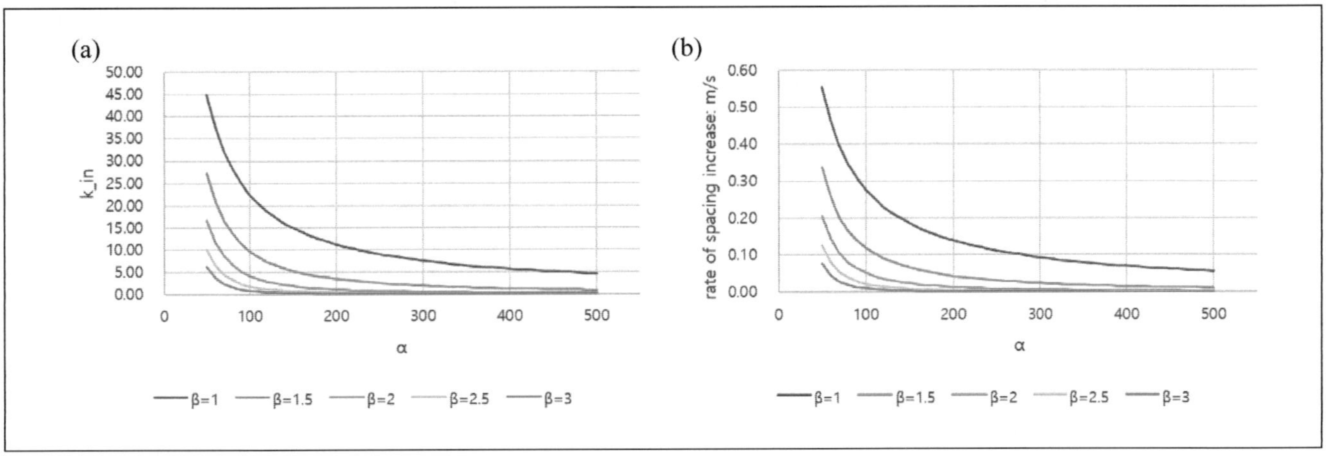

Figure 3. Parameter analysis of k_{in} (*a*) k_{in} versus α with various β and (*b*) spacing increase versus α with various β.

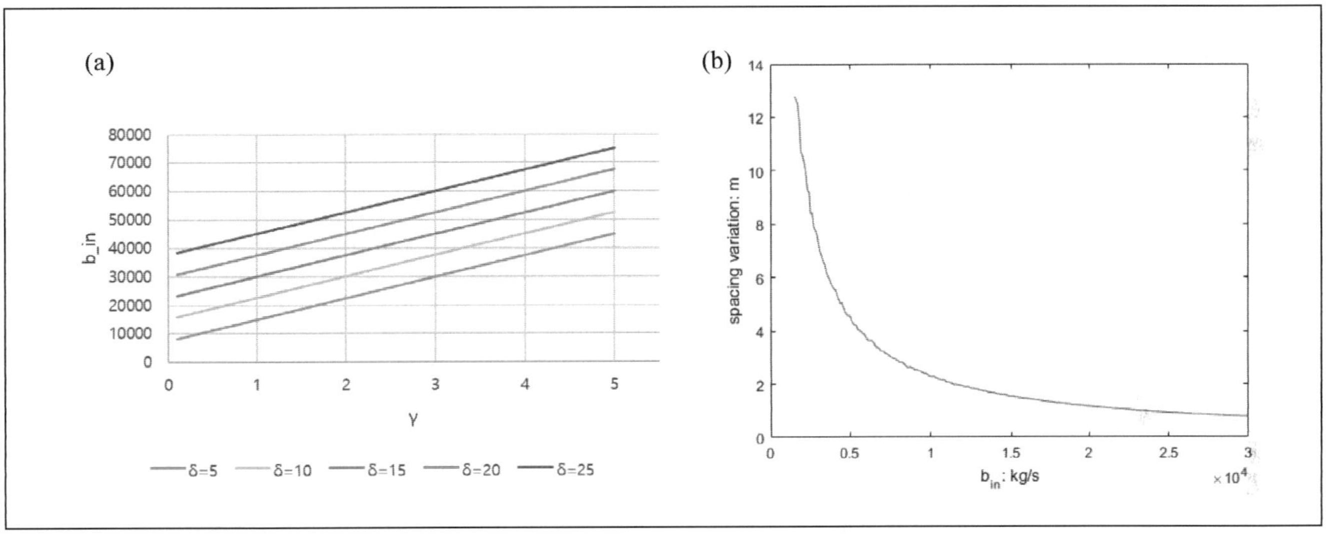

Figure 4. Parameter analysis for b_{in} (*a*) b_{in} versus γ and δ and (*b*) spacing variation versus b_{in}.

$$b_{in} = \left(\gamma^*\left(\left|v_p - v_{in}\right|\right) + \delta\right)^* b_{crit}, \tag{11}$$

In which v_{in} and v_p represent the speeds of the cut-in vehicle and its follower, respectively, and γ and δ represent the scaling constants for the coefficient of b_{crit}. We used the absolute value of $v_p - v_{in}$ considering that a cut-in vehicle could be slower or faster than the platoon. To achieve over-damping, the scaling constants should be set as $\gamma>0$ and $\delta>1$. With this setting, the platoon, including the cut-in vehicle, would move safely as one solid body with nearly constant spacing that is smaller than the critical spacing even with a dramatic speed change. After $k_{in,n}$ and b_{in} are applied, the spacing of the cut-in vehicle and the vehicle behind it increase to the critical spacing very slowly. We can restore the original CAV control with k_0 and b for stable traffic when one of three conditions is satisfied: (i) the spacing eventually reaches the critical spacing after some time; (ii) the cut-in vehicle leaves the platoon so that the critical spacing is recovered; or (iii) the vehicle in front of or behind the cut-in vehicle leaves the platoon so that the vehicle recovers the critical spacing.

We also performed a parameter analysis for γ and δ to investigate their effects on b_{in} and the rate of spacing change and to determine proper values. In this analysis, the relative speed is set at 5 m/s and b_{crit} at 1500 kg/s. As suggested by Equation 11, b_{in} has linear relationships with δ and γ. These linear relationships are illustrated in Figure 4*a* for various parameter values. To set proper γ and δ, b_{in} should be such that spacing does not change by more than the minimum/stand-still gap (2 m) to prevent a collision. For example, we consider an extreme case for conservative parameter setting, where the initial spacing is 7 m, and the leader abruptly brakes ($-4\,m/s^2$) for 5 s from the desired speed (30 m/s). Figure 4*b* shows that the spacing variation decreases exponentially as b_{in} increases and reaches below $2m$ when b_{in} is greater than 11,500 kg/s, so the minimum b_{in} should be 11,500 kg/s. Then we set δ at 7.67 to attain the minimum b_{in} even when the relative speed is zero, and set γ at 0.2 to be conservative.

To summarize, when a vehicle cuts in between vehicles in a platoon so that the spacings of the cut-in vehicle and the vehicle behind it are less than the critical spacing, we let these vehicles keep their short spacings with low k values and set higher b so that the cut-in vehicle follows the speed of the platoon. As a result, the platoon can maintain high flow (without a void) without a disturbance propagating through the platoon. Furthermore, the CAVs in the platoon, including the cut-in vehicle, move safely and stably despite the short spacings caused by nearly fixed spacing attained by high b.

Simulation

Parameter Setting. We evaluated the proposed control by simulation. In this simulation, the specifications of a 2016 Chevrolet Malibu 2.0T were used for the performance of CAVs (*19*). For this vehicle, m is 1,500 kg; the vehicle length is 5 m; and a_{max} and b_{max} are calculated as 4.43 and $9.42\,m/s^2$, respectively. Here S_o is 7 m as we assume that the minimum gap is $2m$. We also set v_d at 30 m/s, and thus the maximum value of c is equal to 221.5 kg/s. To calculate the maximum value of k, we used the 85th percentile spacing for Δx based on the Pearson type III distribution, which has been shown to represent the distribution of real traffic well (*20*). Maximum, quadratic and cubic relationships between k and flow were used for k_0 as in our previous research (*11*). For b, τ is set conservatively at 1 s as CACC vehicles are able to achieve time gaps of 0.5–1 s (*2, 21*). Finally, as previously described, we set α, β, γ, and δ at 250, 2, 0.2, and 7.67, respectively.

Simulation Set-Up. We assumed nine vehicles in a platoon and one cut-in vehicle between the first and second vehicles at 200 s. For a significant impact of a cut-in maneuver, we assumed that the platoon moves with the critical spacing of 37–42 m and the desired speed of 30–35 m/s. After the cut-in, the cut-in vehicle had 25–30 m of spacing with the first vehicle, and the follower had 12 m of spacing. Two types of cut-in maneuvers were considered: (i) cut-in from another mainline lane and (ii) cut-in because of merging from an on-ramp. We assumed that the cut-in vehicle travels at the same speed as the platoon for the first type (v_{in} = 30–35 m/s) and at the 5 m/s lower speed than the platoon (v_{in} = 25–30 m/s) for the second type. We applied $k_{in,n}$ and b_{in} only for the cut-in vehicle and the second vehicle (the immediate follower). For performance, we computed: (1) the average speed and spacing changes of all vehicles during the simulation duration of 600 s to measure the magnitude of disturbance; (2) the time taken to recover the desired speed; (3) the number of vehicles that the disturbance from an insertion propagates; and (4) flow to measure the overall platoon efficiency. For (2) and (3), we set the tolerance level at 1 m/s for speed and 0.01 m for the rate of spacing variation as defined previously.

We consider four scenarios: (i) basic control with $v_p - v_{in} = 0$ m/s, (ii) basic control with $v_p - v_{in} = 5$ m/s, (iii) proposed control with $\boldsymbol{v_p - v_{in} = 0m/s}$, and (iv) proposed control with $\boldsymbol{v_p - v_{in} = 5m/s}$. The basic control refers to the one used in our previous research without the consideration of cut-in vehicles (*11*). The values of the k and b parameters used for each scenario are presented in Table 1. For both scenarios of basic control, the maximum k value ($121.3\,kg/s^2$) and critical damping (b_{crit} = 1,500 kg/s) were used to maximize efficiency. For the control scenarios, k_{in} and b_{in} are calculated from Equations 7 and 11 for each vehicle based on its spacing and $v_p - v_{in}$, respectively. We further set b_{in} at the minimum value (11,500 kg/s) for control scenario 1 and at 13,000 kg/s for control scenario 2, as established by the parameter analysis in the previous section.

Result. Figure 5 presents the simulation result. In both cases of v_{in}, the average speed and spacing changes decrease considerably with the proposed control, suggesting smaller disturbances. As a result, vehicles recover the

Table 1. Value of the Parameters for Each Scenario

	Basic control ($v_p - v_{in}$ = 0 or 5 m/s)			Control scenario 1 ($v_p - v_{in}$ = 0 m/s)			Control scenario 2 ($v_p - v_{in}$ = 5 m/s)		
Parameter	Cut-in vehicle	2nd vehicle	Other vehicles	Cut-in vehicle	2nd vehicle	Other vehicles	Cut-in vehicle	2nd vehicle	Other vehicles
k (kg/s^2)	121.3	121.3	121.3	1.21–1.75	0.28	121.3	1.21–1.75	0.28	121.3
b (kg/s)	1,500	1,500	1,500	11,500	11,500	1,500	13,000	13,000	1,500

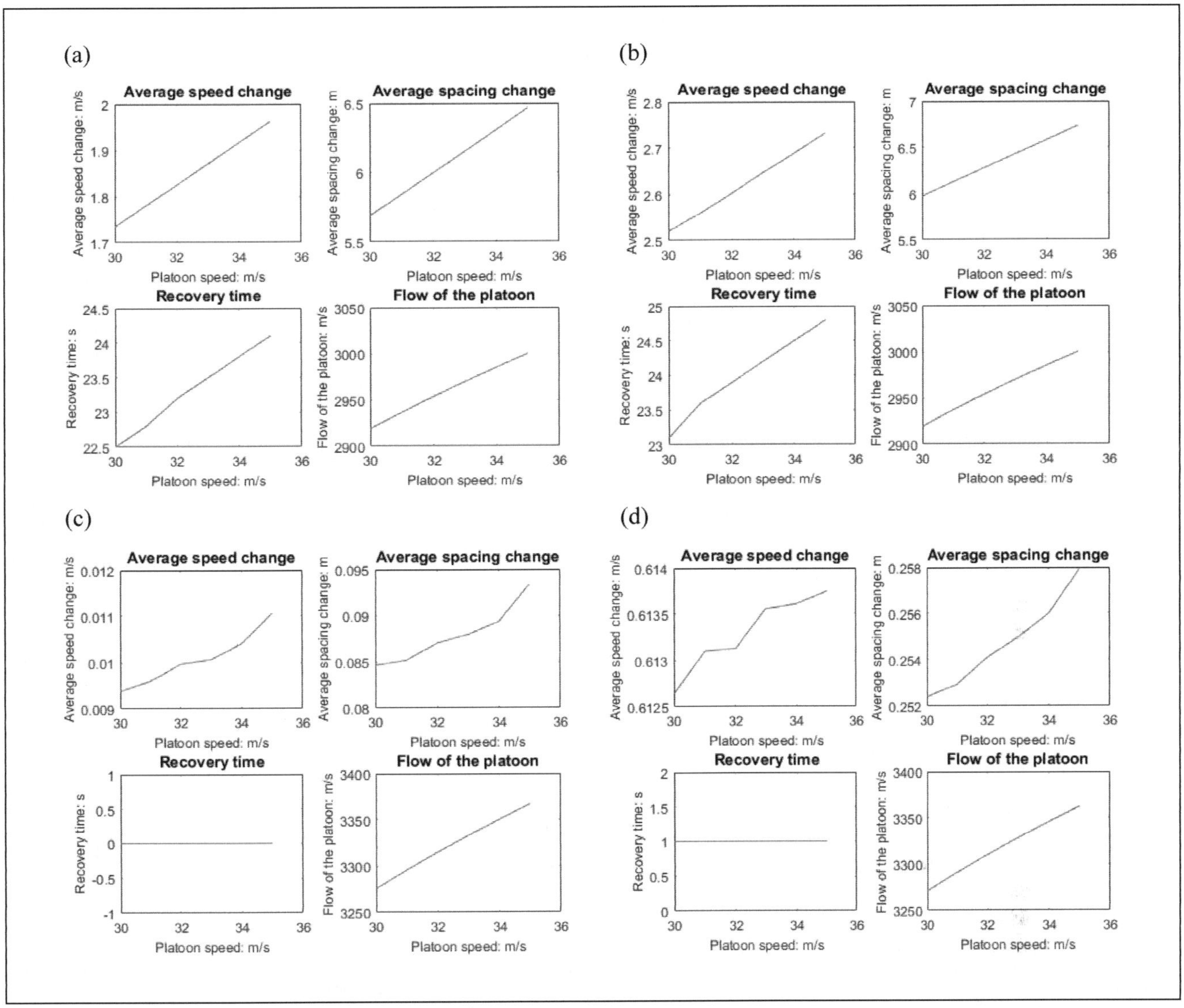

Figure 5. Simulation results: (*a*) basic control with $v_p - v_{in} = 0m/s$; (*b*) basic control with $v_p - v_{in} = 5m/s$; (*c*) proposed control with $v_p - v_{in} = 0m/s$; and (*d*) proposed control with $v_p - v_{in} = 5m/s$.

desired speed almost instantly. Figure 6 shows the vehicle trajectories and speed under the basic control (Figure 6*a*, *b*) and the proposed control (Figure 6*c*, *d*) for $v_{in} = v_p = 30\,m/s$. Comparing the two controls, all vehicles maintain their spacing after the proposed control, and there is no deceleration to reach the critical spacing. Therefore, the recovery time also reduces to 0 s. Furthermore, the disturbance propagates to the last vehicle of the platoon with basic control, as opposed to zero with the proposed control. When $v_{in} = v_p = 30\,m/s$, the platoon flow is 2,919 veh/hr with basic control as all vehicles eventually reach the critical spacing, whereas it increases to 3,275 veh/hr with the proposed control because the cut-in vehicle and its immediate follower maintain shorter spacings than the critical spacing with the desired speed. The results are similar for $v_p = 30\,m/s$ and $v_{in} = 25\,m/s$ as shown in Figure 7.

One of the consequences of the proposed control is that the controlled CAVs keep spacings that are smaller than the critical spacing, thereby "absorbing" all of the impact of a cut-in maneuver. In manual traffic, a short gap is acceptable temporarily for a short period (≈ 20 s), during which vehicles gradually gain more acceptable spacing. This phenomenon is called the "relaxation" phenomenon (*22*, *23*). Thus, readjustment of spacing would be necessary to recover spacing that is more desirable to the driver. There might also be other reasons, such as to accept more vehicles to join the platoon, to redistribute

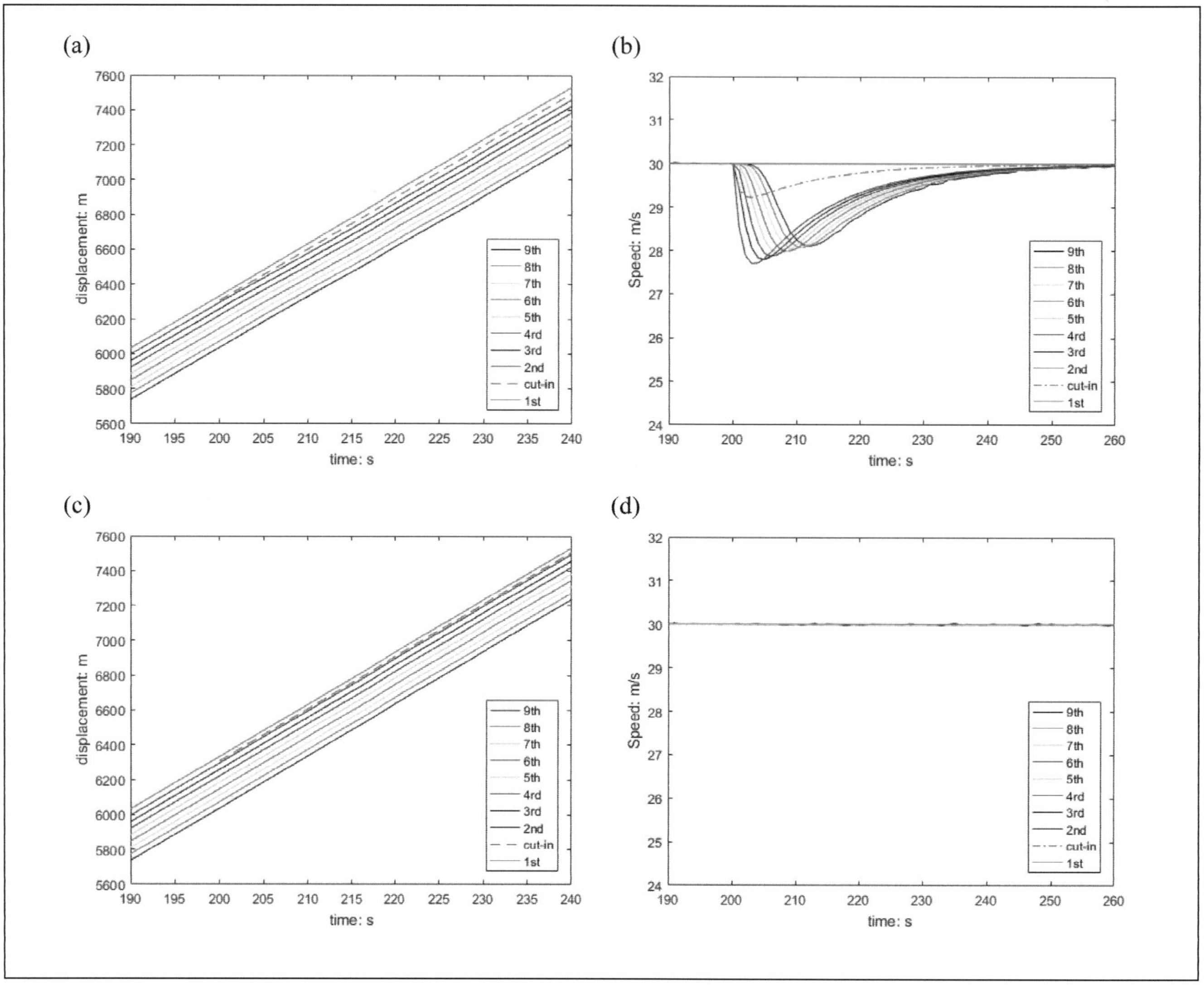

Figure 6. Vehicle trajectories and speed profiles for $v_{in} = v_p = 30$ m/s: (*a*) trajectories with basic control; (*b*) speed profile with basic control; (*c*) trajectories with the proposed control; and (*d*) speed profile with the proposed control.

spacings more evenly within the platoon as more cut-in and cut-out maneuvers occur, and simply to recover spacing that is more desirable to the driver. This can be achieved in two ways through parameter setting. One is to set the minimum value for k_{in} (while maintaining over-damping) which corresponds to spacing greater than the standstill spacing that the driver desires and/or that would guarantee convergence to the critical spacing within a specified time period. Another way is to dynamically change the k value while maintaining over-damping: k can be increased after a specified period following an insertion. An example is presented in Figure 8. In this example, $k_{in} = 0.28\,\text{kg/s}^2$ and $b = 11{,}500\,\text{kg/s}$ for the second vehicle (the one immediately behind the cut-in vehicle) when an insertion happens at 100 *s*. With this parameter setting, the vehicle maintains a very short spacing with a very low rate of spacing recovery (0.0007 m/s), as shown in Figure 8*a*. After 200 s, we change the parameter setting to $k = 121.3\,\text{kg/s}^2$ and $b = 11{,}500\,\text{kg/s}$, so that the vehicle spacing increases by 0.06 m/s. With this setting, the vehicle almost recovers its critical spacing (36.62 m) at around 600 s; see Figure 8*b*. During this transition, the speeds of the second vehicle and all the followers are reduced slightly as shown in Figure 8*c*. Notably, the speed reduction is much less severe compared to the basic control (see Figure 6*b*).

Conclusion

This study presented a control method for CAVs to minimize the impact of vehicle cut-in maneuvers, building on the SMD-based model developed in our previous study

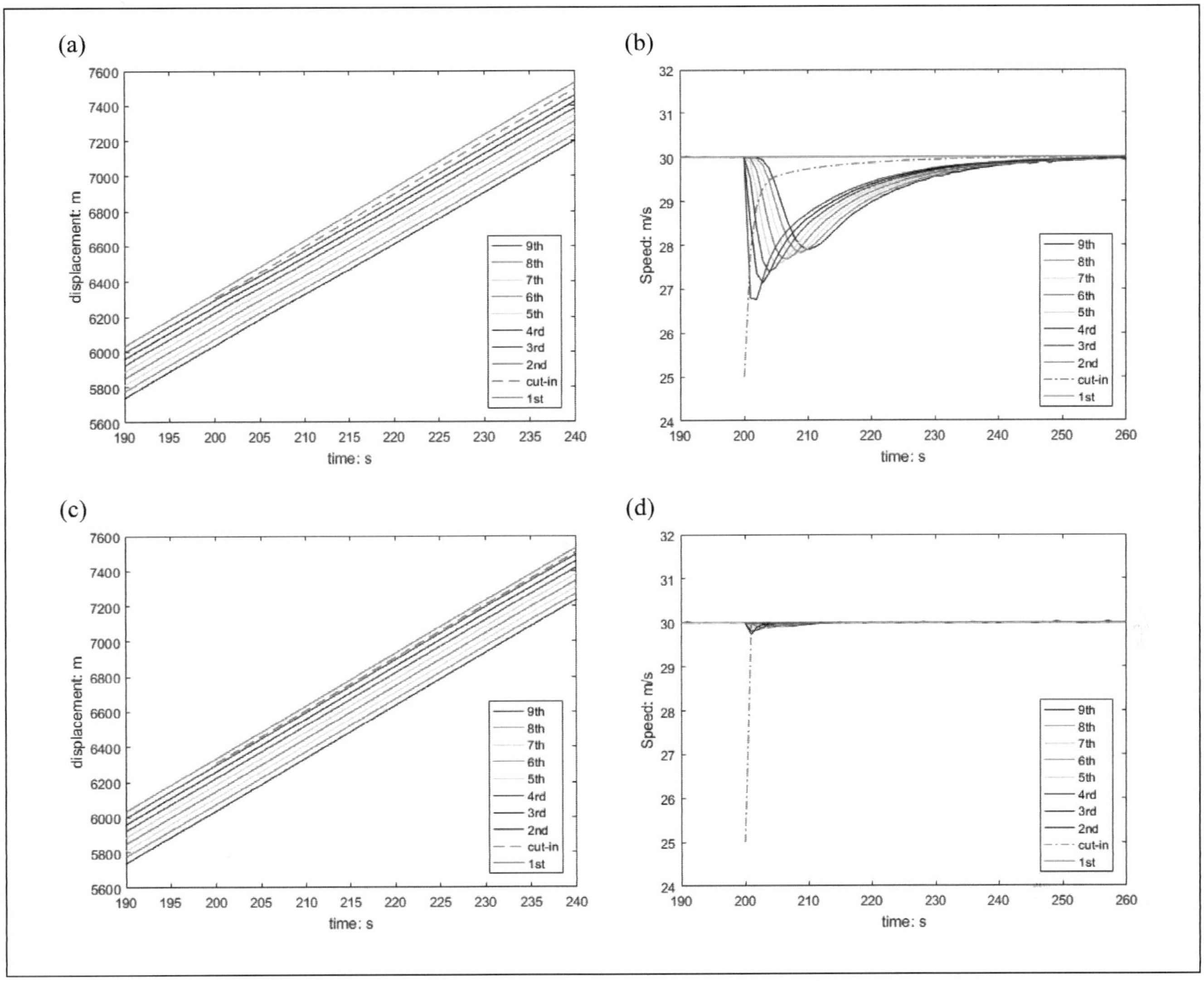

Figure 7. Vehicle trajectories and speed profiles for $v_p = 30$ m and $v_{in} = 25$ m/s: (*a*) trajectories with basic control; (*b*) speed profile with basic control; (*c*) trajectories with the proposed control; and (*d*) speed profile with the proposed control.

(*11*). Based on the characteristics of model parameters, k and b, and their effects on vehicle movement, we developed formulations to systematically set k and b on a vehicle cut-in maneuver. Specifically, a low k value is suggested for the cut-in vehicle and its immediate follower while maintaining over-damping to maintain short spacings, thereby minimizing the speed disturbance and preventing it from propagating through the platoon. We performed a simulation to evaluate the control method, which showed that the speed and spacing variation, and recovery time (to reach the desired speed) diminished significantly with the proposed control. In addition, we offer ways to smoothly recover spacing to accommodate further cut-ins.

Nevertheless, some future studies are desired. In this paper, we developed a theoretical framework for CAV control to proactively minimize the impact of cut-in maneuvers assuming perfect conditions for CAV infrastructure, communication, and sensing. Field testing using sensor-equipped vehicles is desired in the future to calibrate and validate the proposed model with real data. For real-world implementation, several pragmatic issues should be considered. For example, short spacing under the proposed control may raise safety concerns in the event of system failure caused by, for example, communication failure, sensor malfunction, and so forth. Thus, backup control should be considered for safety assurance. Furthermore, our model was developed for central control in a fully connected and automated environment with vehicle to infrastructure (V2I) communication. Specifically, we assumed that CAVs can be controlled through a traffic management center that monitors

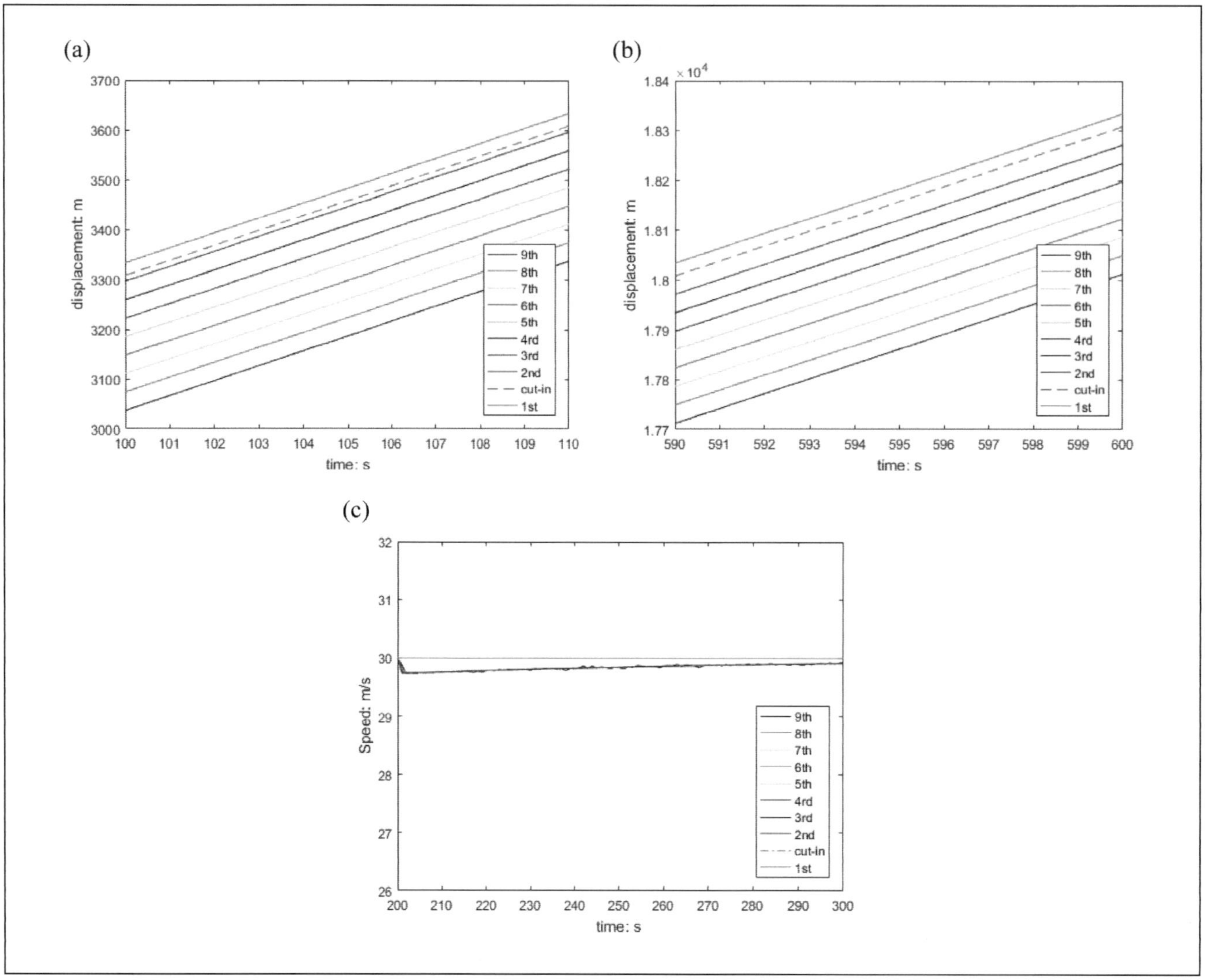

Figure 8. Spacing evolution with cut-in maneuver (*a*) vehicle trajectories with k_{in}; (*b*) vehicle trajectories with original *k*; and (*c*) speed profile.

traffic conditions in a target area and determines proper control. This type of central control system would require more sophisticated infrastructure and advanced communication and computing capabilities with negligible delays. Therefore, a distributed control scheme is desirable that requires less communication and computational capabilities, albeit with reduced benefits. For this strategy, we also assumed that all CAVs were of a single type. Thus, this strategy should be modified to accommodate heterogenous CAVs (e.g., passenger cars and trucks) and manual vehicles. Finally, if merging or weaving sections are closely spaced, where many lane changes are expected to occur, the dynamic setting of the spring constant for the lane changer and its immediate follower may still not yield adequate recovery to the critical spacing before the next merging/weaving section. One possible remedy is to extend the basic framework in this paper and distribute the impact of a cut-in maneuver to several followers. The number of controlled vehicles can be determined based on the operational and geometric characteristics of the traffic (e.g., merge, diverge, and weave sections).

Acknowledgments

This research was sponsored by the National Science Foundation through Award CMMI 1536599 and the University of Wisconsin-Madison.

Author Contributions

The authors confirm contribution to the paper as follows: study conception and design: Soohyuk Bang, Soyoung Ahn; data collection: Soohyuk Bang; analysis and interpretation of results: Soohyuk Bang, Soyoung Ahn; draft manuscript preparation: Soohyuk Bang, Soyoung Ahn. All authors reviewed the results and approved the final version of the manuscript.

References

1. Shladover, S. E., C. Nowakowski, J. O'Connell, and D. Cody. Cooperative Adaptive Cruise Control: Driver Selection of Car-Following Gaps. *Proc., ITS World Congress*, Busan, South Korea, 2010.
2. Shladover, S. E., D. Su, and X. Y. Lu. Impacts of Cooperative Adaptive Cruise Control on Freeway Traffic Flow. *Transportation Research Record: Journal of the Transportation Research Board*, 2012. 2324: 63–70.
3. Milanes, V., S. E. Shladover, J. Spring, C. Nowakowski, H. Kawazoe, and M. Nakamura. Cooperative Adaptive Cruise Control in Real Traffic Situations. *IEEE Transactions on Intelligent Transportation Systems*, Vol. 15, No. 1, 2014, pp. 296–305.
4. Milanés, V., and S. E. Shladover. Modeling Cooperative and Autonomous Adaptive Cruise Control Dynamic Responses Using Experimental Data. *Transportation Research Part C: Emerging Technologies*, Vol. 48, 2014, pp. 285–300.
5. VanderWerf, J., S. E. Shladover, M. A. Miller, and N. Kourjanskaia. Effects of Adaptive Cruise Control Systems on Highway Traffic Flow Capacity. *Transportation Research Record: Journal of the Transportation Research Board*, Vol. 1800, No. 1, 2002, pp. 78–84.
6. Treiber, M., A. Hennecke, and D. Helbing. Congested Traffic States in Empirical Observations and Microscopic Simulations. *Physical Review E*, Vol. 62, No. 2, 2000, pp. 1805–1824.
7. Gong, S., J. Shen, and L. Du. Constrained Optimization and Distributed Computation Based Car Following Control of a Connected and Autonomous Vehicle Platoon. *Transportation Research Part B: Methodological*, Vol. 94, 2016, pp. 314–334.
8. Schakel, W. J., B. Van Arem, and B. D. Netten. Effects of Cooperative Adaptive Cruise Control on Traffic Flow Stability. *Proc., 2010 13th International IEEE Conference on Intelligent Transportation Systems (ITSC)*, No. Idm, Funchal, Portugal, 2010, pp. 759–764.
9. Contet, J., F. Gechter, P. Gruer, and A. Koukam. Multiagent System Model for Vehicle Platooning with Merge and Split Capabilities. *Proc., 3rd International Conference on Autonomous Robots and Agents (ICARA)* Palmerston North, New Zealand, 2006.
10. Milanés, V., and S. E. Shladover. Handling Cut-In Vehicles in Strings of Cooperative Adaptive Cruise Control Vehicles. *Journal of Intelligent Transportation Systems*, Vol. 20, No. 2, 2016, pp. 178–191.
11. Bang, S., and S. Ahn. Platooning Strategy for Connected and Autonomous Vehicles: Transition from Light Traffic. Presented at 95th Annual Meeting of the Transportation Research Board, Washington, D.C., 2016.
12. Reynolds, C. W. Flocks, Herds, and Schools: A Distributed Behavioral Model. *Computer Graphics*, Vol. 21, No. 4, 1987, pp. 25–34.
13. Newell, G. F. A Simplified Car-Following Theory: A Lower Order Model. *Transportation Research Part B: Methodological*, Vol. 36, No. 3, 2002, pp. 195–205.
14. Yanakiev, D., and I. Kanellakopoulos. A Simplified Framework for String Stability Analysis of Automated Vehicles. *Vehicle System Dynamics*, Vol. 30, 1998, pp. 375–405.
15. Laval, J. A., and C. F. Daganzo. Lane-Changing in Traffic Streams. *Transportation Research Part B: Methodological*, Vol. 40, No. 3, 2006, pp. 251–264.
16. Leclercq, L., J. A. Laval, and N. Chiabaut. Capacity Drops at Merges: An Endogenous Model. *Transportation Research Part B: Methodological*, Vol. 45, No. 9, 2011, pp. 1302–1313.
17. Ahn, S., and M. J. Cassidy. Freeway Traffic Oscillations and Vehicle Lane-Change Maneuvers, *Transportation and Traffic Theory*, Vol. 2007, 2007, pp. 691–710.
18. Zheng, Z., S. Ahn, D. Chen, and J. Laval. Applications of Wavelet Transform for Analysis of Freeway Traffic: Bottlenecks, Transient Traffic, and Traffic Oscillations. *Transportation Research Part B: Methodological*, Vol. 45, No. 2, 2011, pp. 372–384.
19. Car and Driver. http://www.caranddriver.com/reviews/2016-chevrolet-malibu-20t-test-review. Accessed July 30, 2016.
20. May, A. D. *Traffic Flow Fundamentals*. Prentice Hall, Inc.. New Jersey, 1990.
21. Nowakowski, C., S. E. Shladover, D. Cody, F. Bu, J. O'Connell, J. Spring, S. Dickey, and D. Nelson. *Cooperative Adaptive Cruise Control: Testing Drivers' Choices of Following Distances*. Research ReportUCB-ITS-PRR-2011-01. PATH, University of California, 2011.
22. Leclercq, L., N. Chiabaut, J. Laval, and C. Buisson. Relaxation Phenomenon After Lane Changing: Experimental Validation with NGSIM Data Set. *Transportation Research Record: Journal of the Transportation Research Board*, 2007. 1999: 79–85.
23. Zheng, Z., S. Ahn, D. Chen, and J. Laval. The Impact of Lane-Changing on the Immediate Follower: Anticipation, Relaxation, and Behavioral Change. *Transportation Research Part C: Emerging Technologies*, Vol. 26, No. 1, 2013, pp. 367–379.

The Standing Committee on Traffic Flow Theory and Characteristics (AHB45) peer-reviewed this paper (18-04291).

Article

TRR
JOURNAL OF THE TRANSPORTATION RESEARCH BOARD

Transportation Research Record
2018, Vol. 2672(20) 144–153

DOI: 10.1177/0361198118792982
journals.sagepub.com/home/trr
SAGE

Modeling Driver Merging Behavior: A Repeated Game Theoretical Approach

Kyungwon Kang[1] and Hesham A. Rakha[1]

Abstract

Various lane-changing models have been developed for use within microscopic traffic simulation software to replicate driver merging behavior. An understanding of human driving behavior, which can be gained through such modeling, will be critical in harmonizing emerging advanced vehicle technology, such as connected automated vehicles, with human drivers. Therefore, it is important to ensure that lane-changing models are clearly understood, appropriately designed, and carefully calibrated. An earlier study by Kang and Rakha proposed and developed a decision-making model for merging maneuvers using a game theoretical approach considering two drivers: the driver of the subject vehicle (DS) in an acceleration lane and the driver of the following lag vehicle (DL) in the target lane. The previous model assumed that the DS and DL decide on an action at the first point only, where the subject and lag vehicles are identified. The current study extends the Kang and Rakha model by introducing the concept of a repeated game, assuming that a lane change decision is made repeatedly to adjust to changes in surrounding conditions. For example, drivers often decide to change their initial decision as a result of conflicts with other drivers. A repeated game helps the proposed model produce more realistic decision-making in the lane-changing process. To evaluate the model, driver decisions at a certain stage, along with accumulated historical decision data, were extracted from Next Generation SIMulation (NGSIM) data. The validation results reveal that the proposed repeated game model produces considerable prediction accuracy (above 75%).

Because human driving behavior is critical to the safety and throughput of the transportation system, various lane-changing models have been developed to simulate driver decisions at the microscopic level. (*1*). An understanding of human driving behavior, which can be gained through such modeling, will become critical in harmonizing human drivers with emerging advanced vehicle technologies, such as connected automated vehicles (CAVs). Consequently, modeling of human driver decision-making within the traditional transportation system should be as accurate as possible. This paper focuses on decision-making for mandatory merging lane-changing behavior, which is one of the most significant human driving maneuvers.

Capturing real-life lane-changing scenarios greatly increases the difficulty of model specification (*2*). The accuracy of microscopic traffic simulation models is highly dependent on the lane-changing model and, therefore, it is important to ensure that lane-changing models are clearly understood, appropriately designed, and carefully calibrated (*1*). Rahman et al. categorized microscopic lane-changing models into four groups: discrete choice based, artificial intelligence, incentive based, and rule based (*1*).

Ahmed proposed a dynamic discrete choice model to capture the heterogeneity in driving characteristics across the driving population and considered explanatory variables affecting driver behaviors (*1, 3*). His model defines a trigger to change lanes, a target lane choice, and a rule to accept a gap based on the discrete choice framework (*3*). Artificial neural network models process information using functional architecture and mathematical models that are similar to the neuron structure of the human brain (*4*). These models learn human behaviors from training and are capable of demonstrating those human behaviors in a new situation. The minimizing overall braking induced by lane change (MOBIL) model, modeled by Kesting et al., is based on two criteria: incentive and safety. The incentive criterion measures the attractiveness of a given lane based on its utility, and the safety criterion measures the risk associated with lane changing (*1, 5*). As one of the rule-based models, Gipps's model

[1]The Charles E. Via, Jr. Department of Civil and Environmental Engineering, Virginia Polytechnic Institute and State University, Blacksburg, VA

Corresponding Author:
Address correspondence to Hesham A. Rakha: hrakha@vtti.vt.edu

attributes decisions to change lanes and the execution of lane changes on freeways and urban streets to three factors: possibility of changing lanes, necessity for changing lanes, and desirability of changing lanes (*6*). Game theory models are based on the giveway behavior in a merging situation when a traffic conflict arises between through and merging vehicles, where these vehicles try to influence each other.

Various lane-changing models, with the exception of game theory models, deal with lane-changing maneuvers from the merging vehicle's point of view only. There is no consideration of maneuvers of surrounding vehicles in the target lane. A following vehicle driver in a target lane reacts once they perceive a merging vehicle's intention to change lanes. The following driver decelerates to make a safe gap as an acceptance of lane change or accelerates to disallow a lane change. Kita modeled on-ramp merging behavior using a discrete choice model, estimating the probability of giving way based on a game theoretical approach (*7, 8*). Liu et al. modeled merging and yielding behavior using assumptions of the vehicles' objectives (*9*), as the development of advanced communication technologies such as vehicle-to-vehicle communication and autonomous vehicles has led recent research efforts to focus on the interaction between vehicles (*10, 11*).

The objective of this study was to enhance the merging decision-making model proposed by Kang and Rakha using an advanced game theoretical approach (*12*). In this paper, the concept of a repeated game is introduced to depict realistic human lane-changing behavior at freeway merging sections. Action strategies and the corresponding payoff functions for the drivers of the subject and lag vehicle are defined following previous studies (*12, 13*). Next, the model is calibrated using datasets extracted from the Next Generation Simulation (NGSIM) dataset (*14*). Finally, model validation is presented before concluding with summary remarks and recommendations for further research.

Development of a Decision-Making Model for Merging

Game Design using a Repeated Game

As mentioned previously, this study adopted the decision-making model for merging maneuvers using a game theoretical approach proposed and developed by Kang and Rakha (*12, 13*). The model describes a merging decision-making process involving three vehicles: (1) the subject vehicle (i.e., merging vehicle), which makes a lane change to merge onto the freeway; (2) the lag vehicle, which is the closest following vehicle in the target lane; and (3) the lead vehicle, which is the nearest preceding vehicle in the target lane.

Table 1. Structure of the Merging Decision-Making Game in Normal Form (*12*)

Player and actions		Driver of lag vehicle (DL): Yield, $l_1(q_1)$[b]	Driver of lag vehicle (DL): Block, $l_2(q_2)$
Driver of subject vehicle (DS)	Change, $s_1(p_1)$[a]	(P_{11}, Q_{11})[c]	(P_{12}, Q_{12})
	Wait, $s_2(p_2)$	(P_{21}, Q_{21})	(P_{22}, Q_{22})
	Overtake, $s_3(p_3)$	(P_{31}, Q_{31})	(P_{32}, Q_{32})

[a]p_i in parentheses denotes the probability assigned to the DS's pure strategy s_i; $\sum_{i=1}^{3} p_i = 1$.
[b]q_j in parentheses denotes the probability assigned to the DL's pure strategy l_j; $\sum_{j=1}^{2} q_j = 1$.
[c]P and Q denote the payoff for DS and DL, respectively.

The merging decision-making game consists of the number of players, action strategies of each player, the characteristics of the game, and corresponding payoff functions to describe the outcome for each player throughout the game. In the model structure as shown in Table 1, DS and DL are players participating in the game. Moreover, the DS has three action strategies: change a lane for merging (s_1), wait for the lag vehicle's overtaking maneuver in the acceleration lane (s_2), or overtake the lead vehicle to use a forward gap to merge (s_3). The DL has two action strategies: yield to allow the subject vehicle's lane change maneuver (l_1) or block the subject vehicle's merging maneuver by decreasing the spacing to the subject vehicle (l_2) (*1, 2*). Let $S = \{s_1, s_2, s_3\}$ and $L = \{l_1, l_2\}$ denote the set of pure strategies for the DS and DL, respectively. In addition, $a = (s_i, l_j)$ denotes a set of actions ($a \in S \times L$) where $i$and j indicate the index of action strategies of the DS and DL (i.e., $i = 1, 2, 3$ and $j = 1, 2$). Thus, there are in total six sets of action strategies in the proposed game.

In previous work, Kang and Rakha assumed that the DS and DL each decided on an individual action only at the initial point where the subject, lead, and lag vehicles were identified (*12*). In other words, the merging decision-making game started and concluded at the beginning point of an acceleration lane (see Figure 1*a*). If the DS showed an overtaking or waiting action, additional games were formed and played with fresh opponents (see Figure 1*b* and *c*). This followed an assumption that drivers made their decision at the initial point of each game and then just performed it without any changes to their decision. In reality, however, drivers continuously make decisions, and may adapt to changes in the surrounding environment (e.g., other drivers' behavior; see Figure 1). In other words, their original decision may change at subsequent decision-making

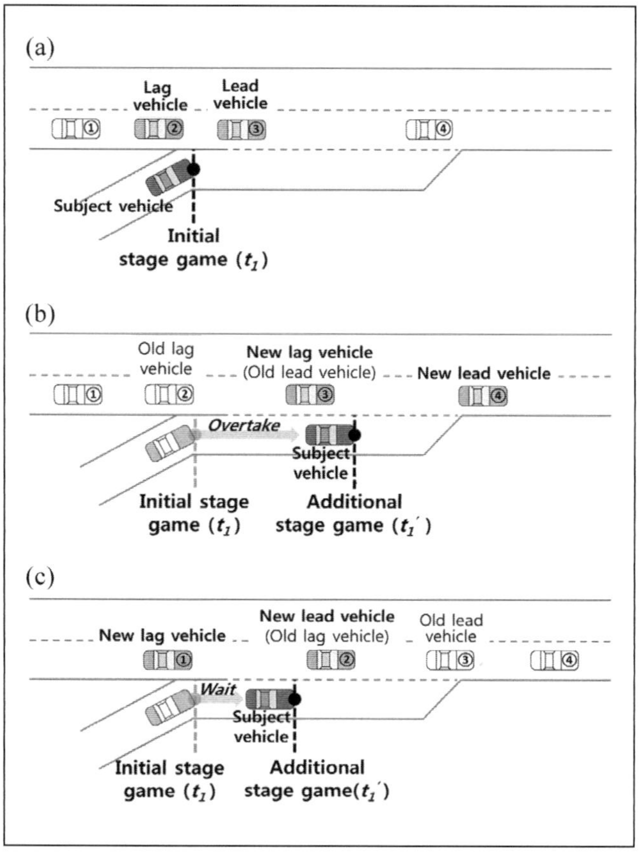

Figure 1. Merging decision-making game formation (*13*): (*a*) initial game at the beginning point of the acceleration lane, (*b*) additional game produced by the DS's overtaking action, (*c*) additional game produced by the DS's waiting action.

points in consideration of conflicts derived from competitive action strategy sets, e.g., (s_1, l_2) or (s_2, l_1) as denoted in Table 1. If the DS and DL decide on *merging* and *blocking* actions at the first decision-making point, for example, the DS initially increases his or her speed to ensure acceptable spacing for entering a target lane and the DL is unwilling to give way to the merging vehicle, then the decision might change. Specifically, after becoming aware of the opponent's behavior, one of them will change their initial decision in order to avoid a conflict. This study applies successive decision-making into Kang and Rakha's model to enhance and make it more realistic. The type of game proposed here is a *two-person non-zero-sum non-cooperative repeated game under complete information*.

In game theory, the repeated game consists of a number of repetitions of a stage game. When players interact by playing a similar stage game numerous times, the game is called a dynamic, or repeated, game. As shown in Figure 2, this study assumes that the identical base decision-making game is conducted periodically and repeatedly over discrete time periods $T \in [t_1, t_n]$ until the DS chooses to execute a lane change. Furthermore, if the identity of the players changes, the base stage game is reset as a new stage game between the DS and fresh DL at time step t_1. Each driver makes a decision at a given time step t in consideration of the entire history of play in previous time steps. Drivers do not know what the opponent will choose without any communication system at a given time step t, but recognize the other driver's decision made in previous time steps by observing the other vehicle's relative speed or acceleration level. Thus, this game is also regarded as an *imperfect information game* with *perfect recall* (or *monitoring*). It allows a player to condition their current action in consideration of their opponent's earlier actions.

This study introduces the concept of cumulative payoffs of a player in a repeated game as the sum of payoffs across time period T. It means that former decisions affect the next game. In other words, the decision-makers take previous game results into account. The cumulative payoffs prevent the drivers' decisions from fluctuating by instantaneous decision factor, such as acceleration level. Moreover, this concept concurs with human driving behavior whereby drivers intend to follow their initial decision. A common assumption in the repeated game is that the player wants to maximize a weighted sum of

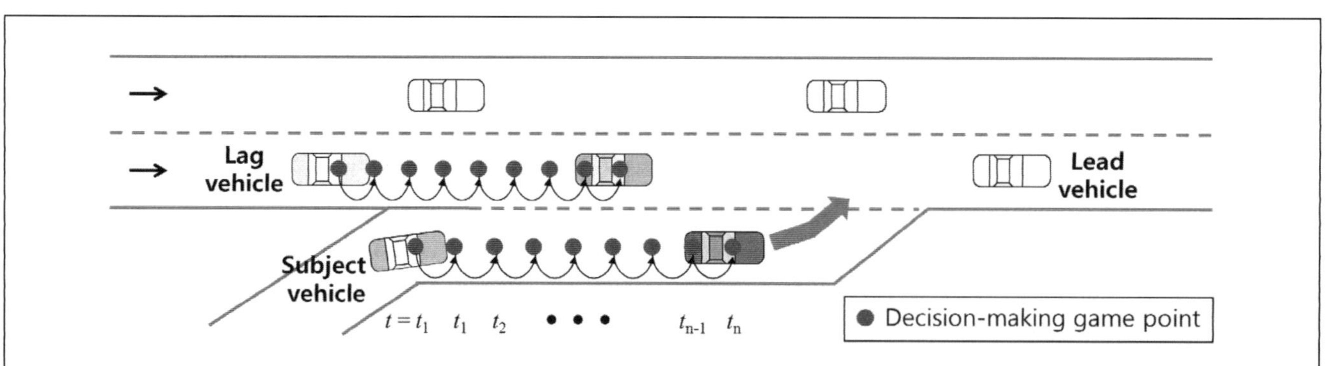

Figure 2. Conceptual diagram of a repeated game for merging decision-making.

payoffs in all periods, where the later periods weigh less than earlier periods. Regarding decision-making for driving behavior, however, the present decision-making might be more important than decisions at previous time steps. Thus, the study models time preferences by assuming that future payoffs are weighted proportionately at some rate δ, called the rate factor. The rate factor $\delta = 0$ implies that the present has no value (*15*). Like the assumption in the previous study, in other words, players follow the action strategy decided at the initial game point without change of their former decisions. Otherwise $\delta = 1$ means that the present is just as important as the past (*15*). In addition, $0<\delta<1$ indicates that the past is more valuable than the present; $\delta>1$ denotes that current payoffs are more meaningful than past payoffs. Drivers' cumulative payoffs, i.e., $U^d = P$ or Q, are presented in Equation 1.

$$U_{ij}^d(n) = \sum_{t=1}^{n} \delta^{t-1} u_{ij}^d(t) \tag{1}$$

where

$u_{ij}^d(t)$ = the utility of a driver d for an action strategies set (s_i, l_j) at time step t;

n = the number of decision-making time steps; and

d = a driver, i.e., player in a game, DS or DL.

Payoff Functions in the Merging Decision-Making Game

Payoff functions in game models reveal a players' outcome (or reward) from a base game at a time step t by choosing a specific action strategy set a. Existing lane-changing models based on a game theoretical approach define a game structure and formulated payoff functions. Kita assumed that each player's behavior goal is to minimize their risk of collision (*7, 8*). Liu et al. used the acceleration level and the time spent in an acceleration lane subject to safety constraints for the payoff functions of the DL and the DS, respectively (*9*). Talebpour et al. chose safety and speed gain as the main decision factors for a driver's decision to change lanes (*10*). Kang and Rakha initially considered five decision factors for the DS and DL (*12*). In consideration of the limitation that there was a small number of observations required to estimate more than 50 parameters, they developed the model by using simplified payoff functions and it showed better prediction accuracy than one with complicated functions (*13*). The study uses the simplified payoff functions for both the DS and DL, as will be described shortly.

Payoff Functions for the DS. In the payoff functions for the DS, the main decision factor is the *expected gap* based on the anticipated acceleration levels of the two vehicles in each set of action strategies; the *forced merging factor* is also considered. It is assumed that the DS predicts whether there are available lag and lead gaps to merge onto a freeway at the next time step. If available gaps are difficult to assure, it should be assumed that the DS intends to either *overtake* or *wait*. In addition, the third term in Equations 2 and 3 is the forced merging factor for considering the remaining distance in the decision-making process. If the DS is close to the end of the acceleration lane, they should merge onto the freeway due to the remarkable increase of the changed action's payoffs. In other words, the model outcome should result in the DS deciding to merge if there is not sufficient distance in the current lane for an overtaking or waiting action. This third term was formulated using the *hardwall* concept proposed by Rakha and Zhang as the exponential function (*16*). The *hardwall* is an imaginary lane-changing wall for mandatory lane-changing, and is located at 10 times the vehicle spacing at standstill (i.e., jam density spacing), x_j (*16*). Moreover, the last term in Equations 2 to 7 is an error term included to capture the unobserved utility. The payoff functions of the DS, denoted by P_{ij} (i = 1,2,3 and j = 1,2), are specified as:

$$P_{11} = \alpha_{11}^1 \Delta x_{11}^{S,F}(t+\Delta t) + \alpha_{11}^2 \Delta x_{11}^{P,S}(t+\Delta t) + \alpha_{11}^3 \exp(x_0/x_{RD}) + \varepsilon_{11}^S \tag{2}$$

$$P_{12} = \alpha_{12}^1 \Delta x_{12}^{S,F}(t+\Delta t) + \alpha_{12}^2 \Delta x_{12}^{P,S}(t+\Delta t) + \alpha_{12}^3 \exp(x_0/x_{RD}) + \varepsilon_{12}^S \tag{3}$$

$$P_{21} = \alpha_{21}^1 \Delta x_{21}^{S,F}(t+\Delta t) + \varepsilon_{21}^S \tag{4}$$

$$P_{22} = \alpha_{22}^1 \Delta x_{22}^{S,F}(t+\Delta t) + \varepsilon_{22}^S \tag{5}$$

$$P_{31} = \alpha_{31}^1 \Delta x_{31}^{P,S}(t+\Delta t) + \varepsilon_{31}^S \tag{6}$$

$$P_{32} = \alpha_{32}^1 \Delta x_{32}^{P,S}(t+\Delta t) + \varepsilon_{32}^S \tag{7}$$

where

$\Delta x_{ij}^{S,F}(t+\Delta t)$ = the expected lag gap between the subject and lag (i.e, following) vehicles in the action strategies set (s_i, l_j) at time $t+\Delta t$ (in m);

$\Delta x_{ij}^{P,S}(t+\Delta t)$ = the expected lead gap between the lead (i.e, proceding) and subject vehicles in the action strategies set (s_i, l_j) at time $t+\Delta t$ (in m);

x_0 = the distance threshold for forced lane-changing (= 10 Δx_j) (in m);

x_{RD} = the remaining distance in the acceleration lane for the subject vehicle (in m);

ε_{ij}^S = an error term to capture the unobserved payoffs; and

$\alpha_{11}^1, \alpha_{11}^2, \alpha_{11}^3, \alpha_{12}^1, \alpha_{12}^2, \alpha_{12}^3, \alpha_{21}^1, \alpha_{22}^1, \alpha_{31}^1, \alpha_{32}^1$ = parameters to be estimated. Note that it is expected that the DS will have different weights on variables for each action strategy.

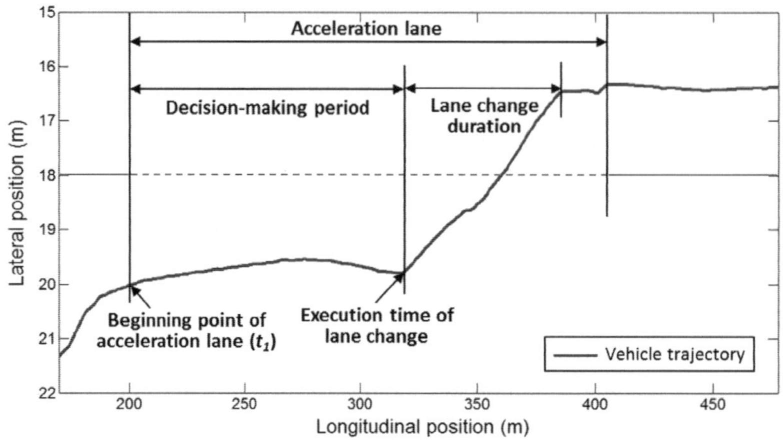

Figure 3. Definition of decision-making period and a point of lane change execution (subject vehicle 1302 in the 8:20 a.m. ~ 8:30 a.m. data).

Payoff functions for the DL. For the DL's payoff function, it is assumed that the DL evaluates the future driving condition according to the speed difference between the subject and lag vehicles. The *expected relative speed* is used for the DL's decision factor in this study. The DL wants to maintain the car-following status and also perceives the DS's intention to merge when there is available spacing. Therefore, it is assumed that the yielding action has priority for the DL if there is no considerable decrease of speed expected. In other words, if the DL is not affected by a DS's merging action, the DL generally chooses a yielding action.

If the relative speed between two vehicles is positive or close to zero, the DL will likely choose a *yielding* action. Otherwise, the DL will likely try to block the DS's merging due to an anticipated decrease of speed. The DL's payoff functions denoted by Q_{ij} are specified as:

$$Q_{ij} = \beta^1_{ij}\Delta\dot{x}^{S,F}_{ij}(t + \Delta t) + \varepsilon^F_{ij}, \forall i = 1, 2, 3, j = 1, 2 \quad (8)$$

where

$\Delta\dot{x}^{S,F}_{ij}(t + \Delta t)$ = the expected relative speed between the subject and following vehicle in the action strategies set (s_i, l_j) at time $t + \Delta t$ (in m/s);

ε^F_{ij} = an error term to capture the unobserved payoffs; and

β^1_{ij} = a parameter to be estimated.

Model Evaluation

Observation Dataset and Classification of Driver's Actions

The NGSIM data were used for calibration of the developed merging behavior model. This study used vehicle trajectory data on a segment of U.S. Highway 101 (Hollywood Freeway) in Los Angeles, California, collected every 0.1 s between 7:50 a.m. and 8:35 a.m. on June 15, 2005 (*14*). The study area is approximately 640 m in length with five mainline lanes (*14*). An auxiliary lane is present through a portion of the corridor between the on-ramp at Ventura Boulevard and the off-ramp at Cahuenga Boulevard, and the lane length is approximately 212 m (*12*, *14*).

It is important to extract observations properly from field data. To evaluate the proposed model, all game points were selected between the beginning point of the acceleration lane and the lane change execution points every 0.5 s (see Figure 2). Specific observed cases where the initial lag gap at time step t_1 was greater than twice the jam density spacing, $2x_j$, were eliminated from the dataset. In these cases, it was assumed that the merging action did not affect the acceleration level of the lag vehicles because of sufficient gap size. To apply a conservative length of lag gaps in model evaluation, therefore, this study used a constant threshold ($2x_j$) by considering a length of average accepted lag gaps in the NGSIM data (*14*). Moreover, entire game points were considered as individual observations within a historical sequence. This study assumed that the merging decision-making game was terminated at the execution point in each lane change. Figure 3 demonstrates how the point of lane change execution time was defined by the lateral movement of the subject vehicle in the vehicle trajectory data.

Reasonable classification of the action strategies chosen by the DS and DL is also a critical issue, as it is directly related to the game model results. Following an approach proposed by Kang and Rakha, for classification of the DS's actions observed in the field, this study used the types of gap: forward, original (i.e., current gap), or backward (*12*, *13*). This approach, however, is limited in defining the DS's initial decision and the exact point where the initial (or previous) decision was changed, as it only focuses on the type of the gap in the final game result (i.e., final choice of action). In the data, the vehicle speeds on the main lane varied from 32.18 to 49.88 km/h during the 45 minutes of data collection, while the average speed of the on-ramp merging vehicles when entering the auxiliary lane was around 50.01 km/h (*17*). To take the NGSIM data features into consideration, therefore, this study assumed that the DS has a priority to preserve the initial decision and the DL loses in the competitive game. This allowed for use of the merging gap types for classification of the DS's observed actions.

For classification of the DL's actions, we used the spacing between the subject and lag vehicles. It is assumed that an increase in the lag spacing occurs as a result of a *yielding* action and otherwise the lag spacing decreases. Fluctuation of the lag spacing caused by instantaneous vehicle acceleration and relative speed affects the quality of the dataset in evaluation of the model. Consequently, this study considered instantaneous fluctuations below a certain range as noise when classifying the DL's actions. The noise-cancellation range ± 1.0 *m* was used here; any variation smaller than the defined range was ignored and treated as if the DL's action did not change. Following this method, only 50 cases in the entire dataset were observed to show change of the DL's initial decision. To evaluate the proposed model, a suitable number of observations in each combination of action strategies were needed. Due to the constraints on data collection time in the NGSIM dataset, there were some limitations to data composition (e.g., cases showing a DS performing a waiting action were rare since the NGSIM data were collected during a peak hour.)

Figure 4 shows the sets of action strategies identified by the proposed approach; 4*a* and *b* show the cooperative action strategies sets (s_1, l_1) and (s_2, l_2), respectively. These cases reflect no conflict between the DS and DL. The former set (s_1, l_1) shows that the DS drove at a higher speed than the DL, and the DL decreased speed slightly and then maintained the original car-following speed. In Figure 4*b*, there was a gap smaller than x_j and the DL had a higher speed than the DS, thus the DS chose the waiting action and decreased its speed. Otherwise 4*c* and *d* show non-cooperative cases, including the action strategies sets (s_2, l_1) and (s_1, l_2), respectively. These cases show instances where there was competition between two players at an early decision-making period, which culminated in cooperative action strategies through a consecutive decision-making process. Case 4*c* reflects a case where both the DS and DL decreased their speed through a waiting and yielding decision (s_2, l_1), and then the DS revised its initial decision to change lanes in recognition of the DL's yielding action. In Figure 4*d*, the DS and DL accelerated initially. After a time of competition between drivers, the DL decelerated to yield. On the other hand, the DS maintained his/her initial decision to change lane. A total of 1,504 similar cases were extracted within the time parameters to play the merging decision-making game.

Model Calibration

Calibration Approach. In the game model, each player decides on one of the pure strategies to achieve the goal of the game (*11*). To find the best strategies for both drivers, the Nash equilibrium was used in this study. The Nash equilibrium is a solution where neither player has motivation to change their own strategy, as no player can improve their situation by individually switching to another strategy (*11*).

If the Nash equilibrium exists, players use *pure strategies*. This means that each player will choose the strategy that maximizes their own payoff while considering an opponent who also wants to maximize their payoff. The Nash equilibrium defines a *pure strategy* as:

$$\begin{cases} P(s^*, l^*) \geq P(s_i, l^*), \forall s_i \in S, i = 1, 2, 3 \\ Q(s^*, l^*) \geq Q(s^*, l_j), \forall l_j \in L, j = 1, 2 \end{cases} \tag{9}$$

where s^* and l^* indicate the DS and DL's equilibrium action strategy, respectively.

If the Nash equilibrium is absent, however, mixed strategies are considered. A probability for each player's strategy is assigned with consideration of each player's expected payoff from the different strategies (*11*). In the defined game, each driver's best response is the set of probabilities that maximizes their own payoff according to the set of probabilities chosen by the other driver (*10*). For this model, we used the MATLAB function *NPG* developed by Chatterjee to solve a two-player finite noncooperative game (*18*). Chatterjee's algorithm solves the game by computing the Nash equilibrium in mixed strategies based on the estimated parameters and expected payoffs (i.e., P_{ij} and Q_{ij}). The algorithm provides probabilities of choice of pure action strategies for each driver (i.e., p_i and q_j) in each observation. The parameters are estimated to minimize the system total deviation of probabilities of choosing actual observed actions using the following function:

$$\min \sum_{k=1}^{N} (1 - p_{a^k} \cdot q_{a^k}) \tag{10}$$

where

k = the index of observations;

N = the number of observations;

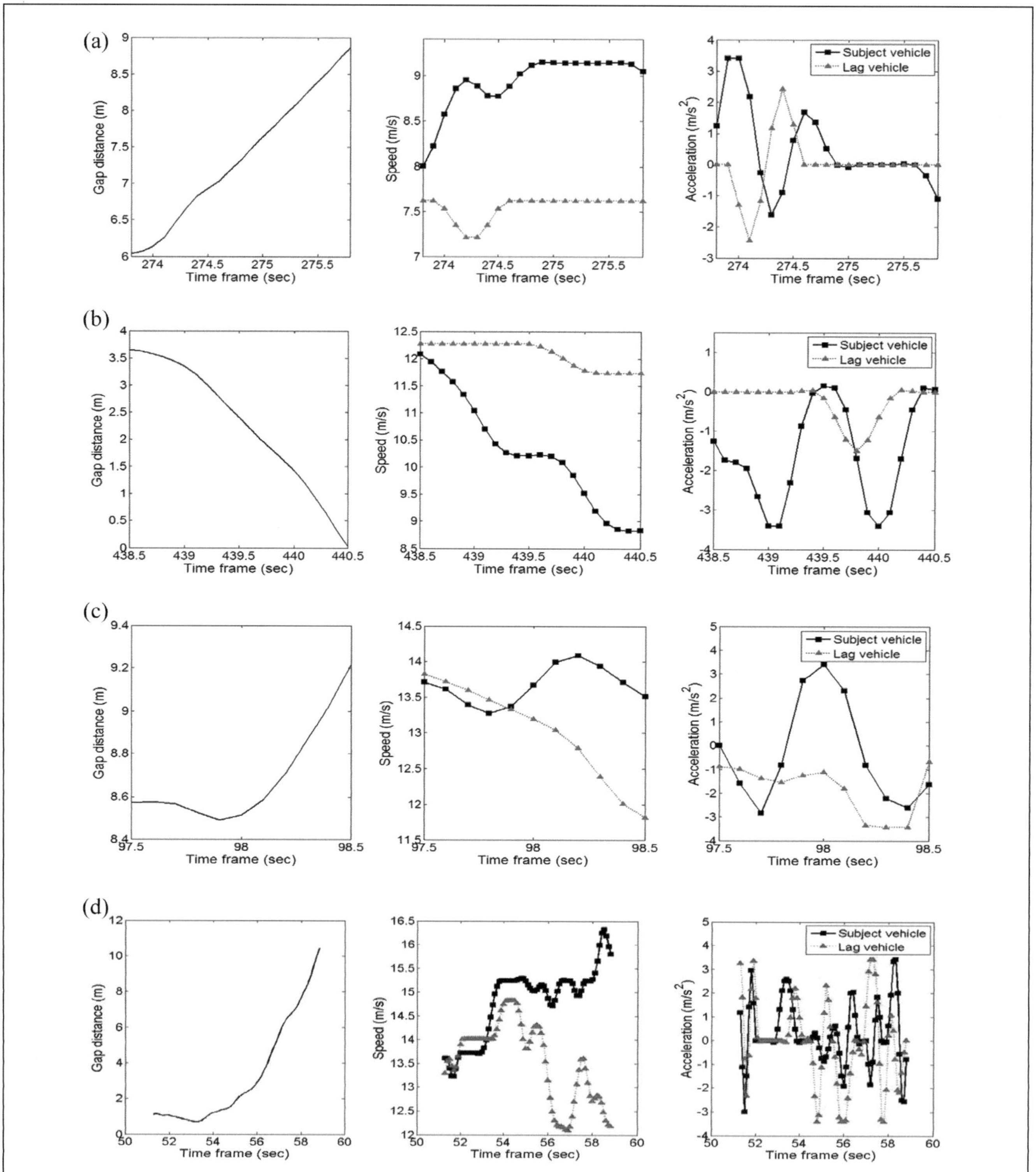

Figure 4. Classification of DS's and DL's action strategies: (*a*) cooperative case: *change* (s_1,) and *yield* (l_1) (subject vehicle 559 in the 8:05 a.m. ~ 8:20 a.m. data), (*b*) cooperative case: *wait* (s_2,) and *block* (l_2) (subject vehicle 879 in the 8:20 a.m. ~ 8:35 a.m. data), (*c*) competitive case: *wait* (s_2,) and *yield* (l_1) (subject vehicle 157 in the 8:20 a.m. ~ 8:35 a.m. data), (*d*) competitive case: *change* (s_1,) and *block* (l_2) (subject vehicle 35 in the 8:20 a.m. ~ 8:35 a.m. data).

Table 2. Estimated Parameters for the DS's and DL's Payoff Functions

Payoff functions	Parameters	Calibrated values of the model parameters					
		Model 0 (δ = 0.0)	Model 1 (δ = 0.6)	Model 2 (δ = 0.8)	Model 3 (δ = 1.0)	Model 4 (δ = 1.2)	Model 5 (δ = 1.4)
P_{11}	α_{11}^{1}	−0.589	−0.416	−0.425	−0.023	−0.680	−0.927
	α_{11}^{2}	0.330	0.522	0.409	0.434	0.497	0.632
	α_{11}^{3}	0.436	0.887	2.075	0.590	1.916	0.721
	ε_{11}^{S}	−2.307	−1.926	−1.236	−1.582	−2.364	−1.998
P_{12}	α_{12}^{1}	0.426	1.186	−0.641	0.090	−0.577	−0.248
	α_{12}^{2}	1.303	0.471	1.727	0.811	1.265	0.758
	α_{12}^{3}	1.157	1.007	0.834	0.802	0.849	1.299
	ε_{12}^{S}	3.142	0.736	2.735	−0.679	0.684	0.424
P_{21}	α_{21}^{1}	−0.191	−0.863	−0.915	−0.386	−1.572	−1.201
	ε_{21}^{S}	−2.487	−2.253	−1.875	−0.287	2.065	0.206
P_{22}	α_{22}^{1}	−0.548	−2.092	−1.243	−1.014	−1.110	−0.223
	ε_{22}^{S}	0.269	1.692	−0.817	−1.849	1.218	−1.000
P_{31}	α_{31}^{1}	0.015	0.377	−0.003	0.129	−0.209	−0.160
	ε_{31}^{S}	3.121	1.052	4.814	2.212	4.164	1.983
P_{32}	α_{32}^{1}	−1.035	−0.660	−0.629	−0.624	−1.237	−0.796
	ε_{32}^{S}	−2.274	0.514	−2.484	−0.422	−1.320	1.005
Q_{11}	β_{11}^{1}	−1.013	−0.289	0.239	−0.089	−1.733	−0.855
	ε_{11}^{F}	0.235	0.474	−2.457	−0.279	−0.549	1.166
Q_{12}	β_{12}^{1}	1.496	0.862	2.135	0.981	0.791	0.544
	ε_{12}^{F}	0.344	0.525	−4.342	−0.903	1.292	1.685
Q_{21}	β_{21}^{1}	−0.199	−0.973	−2.400	−1.172	−0.659	−1.007
	ε_{21}^{F}	0.746	0.120	3.192	0.401	−1.113	−0.404
Q_{22}	β_{22}^{1}	1.196	1.310	0.866	0.360	0.486	0.940
	ε_{22}^{F}	−0.842	0.591	1.041	1.081	−1.298	−1.135
Q_{31}	β_{31}^{1}	−0.167	−1.192	0.903	−0.036	−0.396	−1.214
	ε_{31}^{F}	1.554	−1.335	−1.377	0.271	3.459	−1.232
Q_{32}	β_{32}^{1}	1.503	0.568	2.758	1.091	0.933	0.535
	ε_{32}^{F}	0.988	−0.039	0.914	0.370	2.803	1.965

a^k = the observed action strategies set (s_i^k, l_j^k) in observation k; and

p_{a^k} and q_{a^k} = the probabilities to choose the observed action in a^k for the DS and the DL, respectively.

Calibration Results. To calibrate the model, macroscopic parameters (i.e., jam density spacing (x_j), free-flow speed (v_d), and speed at capacity (v_c)) were first calibrated. Those parameters were used to calculate the expected acceleration levels determined by the Rakha-Pasumarthy-Adjerid (RPA) car-following model (*19*). The expected acceleration level of vehicles in each action strategy set affected payoffs in the decision-making game. The RPA model used the Van Aerde steady state traffic stream model first proposed in a nonlinear, single-regime functional form (*20*). In this study, those macroscopic parameters were calibrated using a heuristic automated tool (SPD-CAL) developed by Rakha and Arafeh (*21*) as follows: x_j = 6.35 m, v_d = 80.4 km/h, and v_c = 40.6 km/h (*12, 21*).

The proposed repeated game model was calibrated to estimate parameters depending on the rate factor: $\delta = \{0.6, 0.8, 1.0, 1.2, 1.4\}$. To prove the performance of the repeated game model, the game model developed by Kang and Rakha in the previous study, i.e., model 0 with δ = 0.0, was also calibrated and validated (*13*). A total of 685 cases (out of 1,504) collected between 7:50 a.m. and 8:20 a.m. were used for model calibration. Table 2 shows the estimated parameters for the payoff functions of the DS and DL in all alternatives of the model.

The mean absolute error (MAE) was calculated using Equation 11 as follows:

$$MAE = \frac{1}{N}\sum_{k=1}^{N}\left|1 - 1\left(\hat{a}^k - a^k\right)\right| \qquad (11)$$

where $\hat{a}^k$ and a^k denote model prediction and actual observation, respectively. Note that $1\left(\hat{a}^k - a^k\right)$ is equal to one if $\hat{a}^k = a^k$, and zero otherwise. The model prediction $\hat{a}^k$ was estimated by probabilities calculated using Chatterjee's algorithm (*18*). Table 3 reveals the calibration results of the proposed models based on the estimated parameters. The MAE of all repeated games models except model 0 (i.e., δ = 0.0) was less than 0.25, implying that the proposed model captures more than 75% of the drivers' decision-making results. A detailed

Table 3. Model Evaluation Results

Models	Model 0	Model 1	Model 2	Model 3	Model 4	Model 5
Rate factor, δ	0.0	0.6	0.8	1.0	1.2	1.4
Calibration results[a]	0.2686 (73.14%)	0.2102 (78.98%)	0.2175 (78.25%)	0.2321 (76.79%)	0.2204 (77.96%)	0.2146 (78.54%)
Validation results[a]	0.2808 (71.92%)	0.2454 (75.46%)	0.2491 (75.09%)	0.2271 (77.29%)	0.2357 (76.43%)	0.2418 (75.82%)

[a]The number in parentheses indicates prediction accuracy.

explanation of the model evaluation is provided in the following section.

Model Validation

The rest of the data, a total of 819 cases (out of 1,504) collected between 8:20 a.m. and 8:35 a.m., were used for model validation purposes. Table 3 includes the model validation results. First, models 1 to 5, which considered the rates from 0.6 to 1.4, explained 75.09% to 77.29% of the observation dataset. On the other hand, model 0 showed a prediction accuracy of 71.92%. In model 0 a rate factor of 0.0 was used implying that drivers chose an action strategy based solely on the initial payoff value. It followed the assumption in previous studies by Kang and Rakha that drivers make a decision once and just follow their initial decisions (*12*, *13*). Consequently, the proposed repeated game in this study showed better prediction ability than the previous game model. In addition, model 0 does not capture changes in the driver's decision-making. If there were more competitive cases in the dataset, model 0 would have lower prediction ability. This showed why using the repeated game has an advantage.

Furthermore, a microscopic traffic simulator integrated the previous game model and demonstrated that it is challenging to represent stable car-following status and cooperative decision-making. Even though it provides quite high prediction accuracy as a result of games played, the prediction accuracy of model 0 and model 4 regarding an initial game point in each stage game was 67.71% and 74.39%, respectively. These results demonstrate that the previous model, which shows lower performance accuracy at the initial game points, can induce improper vehicle speed profiles. This might cause unnatural car-following and driving behaviors in a microscopic simulation model.

In conclusion, the repeated game models 1 to 5 show considerable prediction accuracy of more than 75%. Minor differences between models using the various rates were observed due to issues with the data. Specifically, because of lack of cases with changes in driver initial decisions, the estimated parameters could ignore these cases. Thus, the rate factor did not affect model validation results. Nevertheless, these results reveal that decisions made at previous time steps impact future decision-making, and that the repeated game model is able to predict realistic driver merging behavior along an on-ramp section.

Conclusions and Recommendations for Further Research

An understanding of human driving behavior is necessary for harmonization between emerging advanced automotive technology and human drivers. Since human driving behavior affects the safety of lane-changing, which is often seen as a major source of congestion and collision, human decision-making should be understood before integration with any advanced automotive technology system. As merging is one of the most critical human driving maneuvers, this study focused on the development of the decision-making model for mandatory lane-changing maneuvers. To expand the model proposed and developed by Kang and Rakha and incorporate more realistic decision-making processes, a repeated game was introduced (*12*, *13*). Accumulated payoffs using the rate factor were proposed to reflect the fact that drivers know the history of their decision-making and their previous decisions continuously affect vehicle control. If there are obvious conflicts between drivers, their non-cooperative set of actions will change to cooperative actions to avoid collisions. In model evaluation, the proposed model captured realistic driver merging behavior with a prediction accuracy of more than 75%, indicating that the repeated game model can describe realistic driver lane-changing behavior in a microscopic simulation model.

Further work is needed to expand the proposed model for discretionary lane-changing behavior. To verify model performance under various traffic conditions, the developed model should be tested using other observational data collected in both free-flow traffic conditions and heavy traffic conditions. Moreover, the model should be integrated with microscopic simulation models, including car-following and lane-changing execution models. In simulation models, the rate factor in cumulative payoffs for individual driver can be determined variously to show each driver's sensitivity for reacting against the opponent's action. Lastly, in consideration of the emerging transportation system, the model will be expanded to consider advanced vehicle environments in which CAVs coexist on the roadway, taking into account the impacts of CAV interaction driving behaviors. This model, based on a game theoretical approach, is anticipated to represent lane-changing behavior in the future transportation environment.

Acknowledgments

This research was funded partially by the Mid-Atlantic University Transportation Center (MAUTC) and the Urban Mobility and Equity Center (UMEC).

Author Contributions

The authors confirm contribution to the paper as follows: study conception and design: K. Kang and H. A. Rakha; data collection and analysis: K. Kang; interpretation of results: K. Kang and H. A. Rakha; draft manuscript preparation: K. Kang and H. A. Rakha. All authors reviewed the results and approved the final version of the manuscript.

References

1. Rahman, M., M. Chowdhury, Y. Xie, and Y. He. Review of Microscopic Lane-Changing Models and Future Research Opportunities. *IEEE Transactions on Intelligent Transportation Systems*, Vol. 14, No. 4, December 2013, pp. 1942–1956.
2. Laval, J. A. and C. F. Daganzo. Lane-changing in traffic streams, *Transportation Research Part B: Methodological*, Vol. 40, No. 3, 2006, pp. 251–264.
3. Ahmed, K. I. *Modeling Drivers' Acceleration and Lane-Changing Behavior*. PhD dissertation. Department of Civil and Environmental Engineering, Massachusetts Institute of Technology, Cambridge, Mass., February 1999.
4. Hunt, J. G., and G. D. Lyons. Modeling Dual Carriageway Lane-Changing Using Neural Networks. *Transportation Research Part C*, Vol. 2, No. 4, December 1994, pp. 231–245.
5. Kesting, A., M. Treiber, and D. Helbing. General Lane-Changing Model MOBIL for Car-Following Models. *Transportation Research Record: Journal of the Transportation Research Board*, 2007. 1999: 86–94.
6. Gipps, P. G. A Model for the Structure of Lane-Changing Decisions. *Transportation Research Part B: Methodological*, Vol. 20, No. 5, October 1986, pp. 403–414.
7. Kita, H. A Merging–Giveway Interaction Model of Cars in a Merging Section: A Game Theoretic Analysis. *Transportation Research Part A: Policy and Practice*, Vol. 33, Issues 3–4, 1999, pp. 305–312.
8. Kita, H., T. Keishi, and K. Fukuyama. A Game Theoretical Analysis of Merging-Giveway Interaction: A Joint Estimation Model. In *Transportation and Traffic Theory in the 21st Century* (Michael A. P. Taylor, ed.), Adelaide, Australia, 2002, pp. 503–518.
9. Liu, H. X., W. Xin, Z. M. Adam, and X. Ban, A game theoretical approach for modeling merging and yielding behavior at freeway on-ramp section. In *Transportation and Traffic Theory* (R.E. Allsop, M.G.H. Bell, B.G. Heydecker, ed.), Elsevier, Amsterdam, Netherlands, 2007, pp. 197–211.
10. Talebpour, A., H. S. Mahmassania, and S. H. Hamdar. Modeling Lane-Changing Behavior in a Connected Environment: A Game Theory Approach. *Transportation Research Part C: Emerging Technologies*, Vol. 59, October 2015, pp. 216–232.
11. Kim, C., and R. Langari. Game Theory Based Autonomous Vehicles Operation. *International Journal of Vehicle Design*, Vol. 65, No. 4, 2014, pp. 360–383.
12. Kang, K., and H. A. Rakha. A Game Theoretical Approach to Model Decision Making for Merging Maneuvers at Freeway On-Ramps. *Transportation Research Record: Journal of the Transportation Research Board*, 2017. 2623: 19–28.
13. Kang, K., and H. A. Rakha. Development of a Decision-Making Model for Merging Maneuvers: A Game Theoretical Approach. Presented at 12th Conference on Traffic and Granular Flow, Washington, D.C., 2017.
14. *Next Generation Simulation: US101 Freeway Dataset*. Federal Highway Administration, 2007. http://ops.fhwa.dot.gov/trafficanalysistools/ngsim.htm. Accessed January 25, 2016.
15. Roy, Gardner. *Games for Business and Economics*, 2nd ed. John Wiley & Sons, Inc., N.J., 2003.
16. Rakha, H. A., and Y. Zhang. INTEGRATION 2.30 Framework for Modeling Lane-Changing Behavior in Weaving Sections. *Transportation Research Record: Journal of the Transportation Research Board*, 2004. 1883: 140–149.
17. Wan, X., P. J. Jin, L. Zheng, Y. Cheng, and B. Ran. Speed Synchronization Process of Merging Vehicles from the Entrance Ramp. *Transportation Research Record: Journal of the Transportation Research Board*, 2013. 2391: 11–21.
18. Chatterjee, B. An Optimization Formulation to Compute Nash Equilibrium in Finite Games. *Proc., International Conference on Methods and Models in Computer Science 2009*, Delhi, India, IEEE, 2009, pp. 1–5.
19. Sangster, J., H. A. Rakha, and J. Du. Application of Naturalistic Driving Data to Modeling of Driver Car-Following Behavior. *Transportation Research Record: Journal of the Transportation Research Board*, 2013. 2390: 20–33.
20. Van Aerde, M. Single Regime Speed-Flow-Density Relationship for Congested and Uncongested Highways. Presented at 74th Annual Meeting of the Transportation Research Board, Washington, D.C.,1995.
21. Rakha, H. A., and M. Arafeh. Calibrating Steady-State Traffic Stream and Car-Following Models Using Loop Detector Data. *Transportation Science*, Vol. 44, No. 2, April 2010, pp. 151–168.

The Standing Committee on Traffic Flow Theory and Characteristics (AHB45) peer-reviewed this paper (18-05037).

TRR
JOURNAL OF THE TRANSPORTATION RESEARCH BOARD

Article

Transportation Research Record
2018, Vol. 2672(20) 154–161

DOI: 10.1177/0361198118784176
journals.sagepub.com/home/trr
SAGE

A Platooning Strategy for Automated Vehicles in the Presence of Speed Limit Fluctuations

Sina Arefizadeh[1] and Alireza Talebpour[1]

Abstract

Platooning is expected to enhance the efficiency of operating automated vehicles. The positive impacts of platooning on travel time reliability, congestion, emissions, and energy consumption have been shown for homogenous roadway segments. However, the transportation system consists of inhomogeneous segments, and understanding the full impacts of platooning requires investigation in a realistic setup. One of the main reasons for inhomogeneity is speed limit fluctuations. Speed limit changes frequently throughout the transportation network, due to safety-related considerations (e.g., changes in geometry and workzone operations) or congestion management schemes (e.g., speed harmonization systems). In the current transportation systems with human-driven vehicles, these speed drops can potentially result in shockwave formation, which can cause travel time unreliability. Automated vehicles, however, have the potential to prevent shockwave formation and propagation and, therefore, enhance travel time reliability. Accordingly, this study presents a constant time headway strategy for automated vehicle platooning to ensure accurate tracking of any velocity profile in the presence of speed limit fluctuations. The performance of the presented platooning strategy is compared with Gipps' car-following model and intelligent driver model, as representatives of regular non-automated vehicles. Simulation results show that implementing a fully autonomous system prevents shockwave formation and propagation, and enhances travel time reliability by accurately tracking the desired velocity profile. Moreover, the performance of platoons of regular and automated vehicles is investigated in the presence of a speed drop. The results show that as the market penetration rate of automated vehicles increases, the platoon can track the velocity profile more accurately.

The problem of transporting passengers and freight has long been a topic of interest among traffic engineers. Accordingly, safety and efficiency (e.g., travel time reliability and emissions) of transportation systems have been widely investigated. Travel time reliability is considered one of the most critical measures of efficiency in any transportation system, as any unreliability and unexpected fluctuations in travel time can potentially result in user frustration, reduction in travel comfort, and safety issues (*1*).

In general, transportation system users (including human drivers and automated vehicles) tend to minimize their travel time, while ensuring safety (e.g., avoiding collisions for vehicles). To achieve these objectives, from the operational decision-making standpoint, drivers and automated vehicles tend to drive as close as possible to the free-flow speed, while keeping a safe distance from other vehicles in the surrounding area (specifically, the immediate leader and follower). From human drivers' perspective, several factors (including visibility issues and distractions) can cause uncertainties in the underlying decision-making processes and potentially result in safety issues and an increase in congestion (i.e., shockwave formation and propagation). Consequently, variability between expected travel time and the actual travel time can increase, which results in reduction in travel time reliability. Automated vehicles, however, promise improvements in efficiency (e.g., travel time reliability and congestion reduction) and safety (i.e., keeping a safe distance between successive vehicles) by removing the human element from driving-related decisions. Moreover, the efficiency of automated vehicles can be further increased by replacing the operation of isolated automated vehicles (e.g., Google's automated vehicle, a

[1]Zachry Department of Civil Engineering, Texas A&M University, College Station, TX

Corresponding Author:
Address correspondence to Alireza Talebpour: atalebpour@tamu.edu

self-driving car without connectivity with other vehicles on the road) with platooning strategies.

The travel time of each vehicle is proportional to the inverse of its velocity. Therefore, accurately tracking a certain velocity profile can guarantee a fixed amount of travel time for all vehicles and perfect travel time reliability. On the other hand, failure to track a specific velocity profile can result in shockwave formation, congestion, and travel time unreliability, as can be seen in the current transportation systems on a daily basis. One of the key factors that contributes to travel time unreliability in the current transportation systems is the network inhomogeneity. One of the main reasons for inhomogeneity is speed limit fluctuations. In general, speed limits (i.e., velocity profiles) are based on infrastructure and behavioral factors, including roadway geometry, type of land use, and weather conditions. For example, in workzone locations, vehicles should decrease their velocity to maximize their responsiveness in dealing with unexpected situations. In many instances, drivers are not capable of following a specific speed profile and changes in speed limit around the capacity point can create shockwaves and lead to congestion. The key question is whether automated vehicles are capable of following a specific profile without sacrificing safety or efficiency.

Several physical and behavioral models have been introduced to capture human drivers' longitudinal driving behavior. These models consider two states for driving behavior: car-following and free-flow. Various models have different approaches to determine the driving state. However, as vehicles get closer to each other, their behavior can be better represented by car-following models rather than free-flow models (*2*). One of the main assumptions of most of these models is that the driver is an active, predictable control element in the system, even though some probabilistic models exist (*2*). A wide range of car-following models, such as Gazis–Herman–Rothery' model, collision avoidance models, linear models, action point models, and fuzzy logic-based models have been introduced to capture various aspects of drivers' decision making (*3–8*). In general, the structure of the driving task in such models consists of perception, decision making, and reaction (*9*) and the outcome is vehicle acceleration and speed. Therefore, if these models accurately capture the human drivers' operational decision-making and longitudinal driving behavior, they can be utilized to study travel time reliability in the presence of a speed drop.

Several platooning strategies have also been proposed for automated vehicles. In order to ensure safety, two different platooning policies were considered in the literature: constant spacing policy and constant headway policy (*10*, *11*). The constant spacing policy is focused on keeping a constant space between the vehicles (e.g., adaptive cruise control systems), whereas the constant time headway policy is focused on keeping a fixed time headway among vehicles (*12*, *13*). Several studies investigated various aspects of platooning, including safety and travel time reliability, based on these platooning strategies (*14*, *15*). Fixed time gap is another novel approach that has been considered in a few recent studies (*16*, *17*). To be able to adjust the so-called time gap in those papers (time delay) between two successive vehicles, the time that the vehicle ahead passes any particular location is required for the following vehicle, thus, it should be given for each particular location. Therefore, the full trajectory of the vehicle ahead is needed for the following vehicle. As a result, the strategy may need a sufficient amount of memory to save the trajectory of all vehicles. This can be an effective strategy, however, having access to and saving such an extensive amount of data may raise several problems in a real-world implementation. However, a fixed time headway may not need full trajectory information as it is based on the Lagrangian point of view, and the spacing between successive vehicles might be measured by on-board sensors and can be controlled at any time.

We showed in an earlier work that constant spacing policy cannot provide a control logic base on which all vehicles accurately follow a pre-defined speed profile (*18*). Therefore, constant spacing policy is not capable of improving travel time reliability in the presence of a speed drop (i.e., changes in speed limit). Accordingly, the main objective of this study is to investigate the possibility of designing a safe and string-stable platooning strategy based on the constant headway policy that is capable of following specific speed profiles. Such a controller can prevent shockwave formation and potentially delay or eliminate congestion. Following the requirements for such a controller, the present study proposes a novel controller based on the constant headway policy capable of tracking any velocity profile in the transportation system, while keeping a safe distance from the immediate leader. The performance of the proposed controller in a platoon of automated vehicles is compared with the performance of platoons of human-driven vehicles.

As this study is concerned with keeping a safe distance and preventing collisions, while dealing with inhomogeneity in the roadway segment, collision avoidance models were selected to model human drivers. Gipps' car-following model is one of the early comprehensive car-following models that takes many behavioral factors into account. The model determines the response of the vehicle to the following vehicle, based on the assumption that drivers set certain limits on their desired acceleration or deceleration rates, or on both acceleration and deceleration rates (*19*). In the same fashion, the intelligent driver model (IDM) is a time continuous car-following model that has improved upon the results of Gipps' and other car-following models (*20*). These models are very well established crash-free models. Accordingly, this study utilizes both Gipps' model with parameter values set to the calibrated values employed by Vasconcelos et al. (*21*), and IDM with typical parameter values to model human-driven vehicles (*20*).

The remainder of this paper is organized as follows: the next section presents the formal problem statement. This is followed by an introduction of the automated vehicle platooning strategy along with the formulation of Gipps' model and IDM. The next section presents the simulation results and discussion on the findings of this study. Finally, the study concludes with some summary remarks and future research directions.

Problem Statement

In the transportation networks, speed limit varies due to various factors, including geometric inhomogeneity and workzones. Thus, tracking a velocity profile, which is a function of the location, is fairly common. Consider the given desired velocity profile of Figure 1. This figure shows a gradual drop followed by an increase in the desired velocity. All vehicles are supposed to follow this velocity profile, where v_0 is the roadway speed limit before and after the speed drop location, and pv_0 is the change in speed limit ($0<p<1$). The combination of these speed limit changes results in a velocity profile as shown in Figure 1. Such a setup is common in a workzone location along a highway segment. The research objective is to design a controller for automated vehicle platooning, capable of accurately tracking the velocity profile shown in Figure 1, while ensuring safety. The next section introduces the automated vehicle controller and discusses its structure, as well as the human driver car-following models.

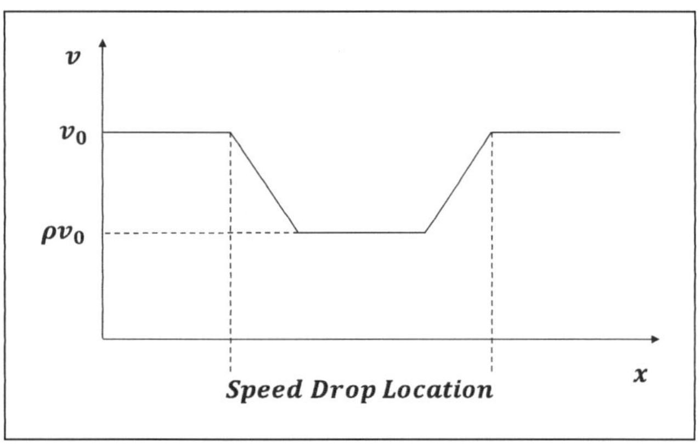

Figure 1. Desired velocity profile.

Model Formulation

As discussed previously, the main objective of this study is to design a constant headway policy-based controller for automated vehicles capable of following specific speed profiles. In this section, the details of the proposed automated vehicle controller are discussed. Moreover, the formulations of Gipps' model and IDM are presented.

Automated Vehicle Controller

In order to control a platoon of automated vehicles, let's consider the following control law:

$$\begin{cases} \ddot{x}_i(t) = \dot{x}_i(t) \cdot v'_d(x_i(t)) - \dot{x}_i(t) + v_d(x_i(t)) \\ for\, |\dot{x}_i(t) - v_d(x_i(t))| \geq |x_{i-1}(t) - x_i(t) - T \cdot \dot{x}_i(t)| \\ \ddot{x}_i(t) = \frac{1}{T}(x_{i-1}(t) - x_i(t) - T \cdot \dot{x}_i(t) + \dot{x}_{i-1}(t) - \dot{x}_i(t)) \\ Otherwise \end{cases}, \quad (1)$$

where

$x_i(t)$ and $\dot{x}_i(t)$ denote the actual location and spot speed of vehicle i at time t

$\ddot{x}_i(t)$ is the acceleration/deceleration command for vehicle i

$v'_d(x_i(t))$ denotes the slope of the desired velocity profile

T denotes the constant time headway between vehicles

This model is obtained by assuming an intelligent vehicle simultaneously adjusts acceleration considering the safe time headway and maintains the desired velocity. If the error in velocity with respect to the target velocity is greater than the error in relative distance, the vehicle prioritizes decreasing the velocity. The vehicle starts to decrease the velocity; however, the amount of decrease depends on the error and slope of the velocity profile. When the velocity error is less than the error in relative distance, the vehicle may increase the velocity to compensate for the error with respect to the current headway and relative velocity to avoid collision. An extensive mathematical analysis of the string stability of the car-following model (*1*) is presented in Arefizadeh et al. (*22*).

Figure 2 shows the behavior of the model for different values of speed and relative speed, when $v_d(x_i(t)) = 30^m/_s, T = 0.5s$, and $\dot{x}_{i-1}(t) - \dot{x}_i(t) = 0$. Figure 2*a* illustrates that when the vehicles are far enough from each other, the controller only focuses on the desired velocity (the behavior of the model does not depend on the relative speed between vehicles). Once the vehicles get closer to each other, as shown by Figure 2*b*, the controller starts to react to the situation to ensure safety. Note that the accelerations and decelerations are not physically

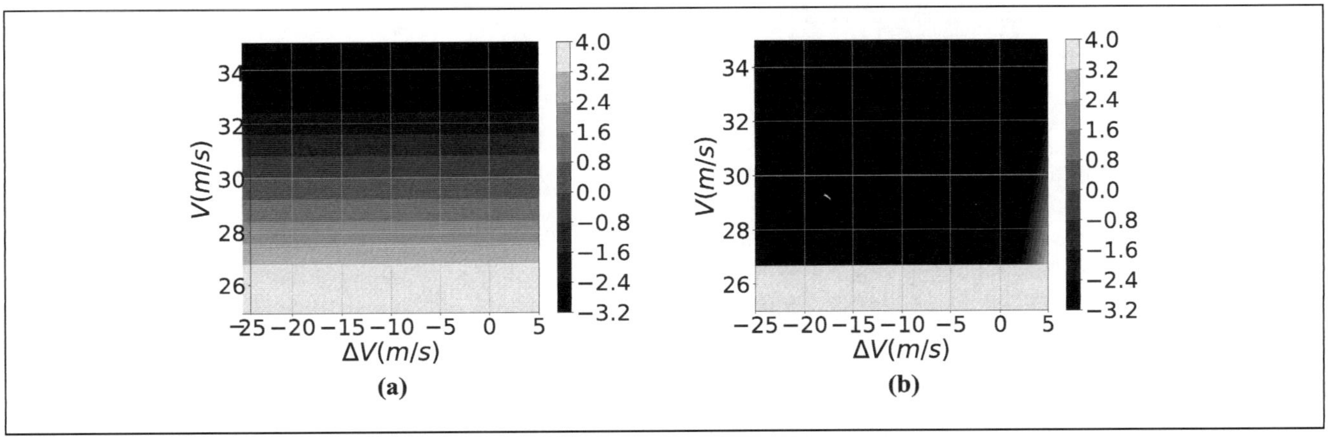

Figure 2. Acceleration profile based on the presented automated vehicle controller for different values of speed and relative speed, when $v_d(x_i(t)) = 30 \, m/s$ (a constant velocity profile), $T = 0.5$s, and $\dot{x}_{i-1}(t) - \dot{x}_i(t) = 0$. Note that a constant velocity profile in which the time headway of the successive vehicles is 0.5 s leads to 15-m spacing between vehicles: (*a*) gap between vehicles is set to the steady state constant spacing of 15 m; (*b*) the initial spacing of successive vehicles deviates from the ideal steady state for 15 m and is set to 10 m.

limited, but the figures are plotted for a specific range (−3.2 m/s^2 to 4 m/s^2). As a result, higher accelerations and decelerations are not displayed in the figures. Therefore, the extended black and white regions are shown in the figures.

Human Driver Model

Gipps' model: For the human driver system, according to Gipps' model we have the following behavior:

$$v_n(t+\tau) = \min \left\{ \begin{array}{l} v_n(t) + 2.5 a_n \tau \left(1 - \frac{v_n(t)}{v_n^d}\right) \sqrt{0.025 + \frac{v_n(t)}{v_n^d}}, \\ -\tau d_n + \sqrt{\tau^2 d_n^2 + d_n \left\{ 2[x_{n-1}(t) - x_n(t) - s_{n-1}] - \tau v_n(t) - \frac{v_{n-1}^2(t)}{d'_{n-1}} \right\}} \end{array} \right\}, \quad (2)$$

where

τ is the reaction time
$v_n(t)$ is the speed of vehicles n at time t
$v_{n-1}(t)$ is the speed of vehicle $n-1$ at time t
v_n^d is the desired velocity of vehicle n
a_n is the maximum speed of vehicle n

$d_n > 0$ and $d'_{n-1} > 0$ represent the most severe braking that the follower wishes to undertake, and the automated vehicle's estimate of the leader's most severe braking capability, respectively. $x_{n-1}(t)$ and $x_n(t)$ are respectively the leader's and the follower's longitudinal positions at time t, and s_{n-1} is the leader's effective length, that is, the leader's real length, l_{n-1}, added to the follower's desired inter-vehicle spacing at stop (between front and rear bumpers) (*21*). Considering the nominal value of parameters a, d, d', and τ for all vehicles as 1.5, 5, 5, and 1, respectively, Gipps' model is implemented to represent human driver behavior.

Intelligent driver model (IDM): This model is one the most reliable car-following models and has the following form (*20*):

$$\dot{v}_n = a \left[1 - \left(\frac{v_n}{v_d} \right)^{\delta} - \left(\frac{\max\left\{0, v_n T + \frac{v_n \Delta v_n}{2\sqrt{ab}}\right\}}{x_{n-1} - x_n} \right) \right], \quad (3)$$

where

a = maximum acceleration
b = comfortable braking deceleration
v_n = velocity of vehicle n
δ = velocity exponent parameters
v_d = desired velocity (which is a function of location)
Δv_n = velocity difference between vehicles $n-1$ and n
T = time headway
x_n = position of vehicle n

Simulation Results

In order to simulate a long string of vehicles in the current study, 100 vehicles were simulated moving successively in a straight line in a one-lane highway segment of infinite length. It is noteworthy that vehicles are dimensionless, thus the only criterion to prevent collisions is safe distance between vehicles. In addition, the run time was set to 1,000 s to provide sufficient time for all vehicles to pass through the speed drop in the velocity profile. A symmetric speed drop (see Figure 1) was considered with the total length of 3,750 m and 2,750 m of constant velocity region. Additionally, the upstream and downstream velocities and the velocity at the drop location were set to 20 m/s, 20 m/s, and 10 m/s, respectively.

Simulation results for the Gipps' model with various initial conditions are illustrated in Figure 3. The initial time headways before the speed drop were considered as 3, 2.7, and 2.5 in Figure 2, *a, b*, and *c*, respectively. Simulation results show that for time headways larger than 2.7 s, all vehicles are able to follow the velocity profile. For any time headway less than 2.7, there is a shockwave propagating upstream and changing the traffic flow regime. Figure 4 illustrates the simulation results for IDM with typical values of 1, 1.5, 4, and 2 for a,b,δ, and T, respectively. The simulation results indicate that if human drivers follow the IDM logic while keeping a time headway of 2 s, tracking the velocity profile is impossible. Increasing the number of vehicles will result in more severe shockwaves and the vehicles' speed will eventually go to zero. This also indicates a severe shockwave moving upstream in traffic. Note that Talebpour et al. showed that IDM is string stable in the presence of small perturbations (*23*, *24*).

In the simulations of automated vehicles, time headway was set to 1 s (much lower than the previous simulations). The rest of the parameters were the same as previously defined. The simulation results of the presented control logic for 100 vehicles are illustrated in Figure 5. Comparing figures 3 through 5, in the velocity profile related to the homogenous system of human drivers, a backward shockwave exists that could potentially decreases travel time reliability. However, in the autonomous system, violations of the desired velocity profile are asymptotically damped. As we can see in Figure 5*b*, time headways vary between 0.097 and 1.03 seconds and the velocity profile is accurately tracked by all vehicles.

In addition to the results of fully automated and regular vehicles, system performance was investigated for a mixed environment with both regular and automated vehicles. Three different market penetration rates of automated vehicles (i.e., 25%, 50%, and 75%) were simulated. Figure 6 illustrates the velocity profile of the mixed string of vehicles. In this case, different combinations of IDM and the proposed controller with the previously described parameter values were considered. Figure 6 shows that increasing the penetration rate of automated vehicles results in better tracking of the velocity profile. However, it seems that in a mixed environment, unlike the fully automated case, it is hard to see any evidence of string stability (compare Figure 5*a* with Figure 6).

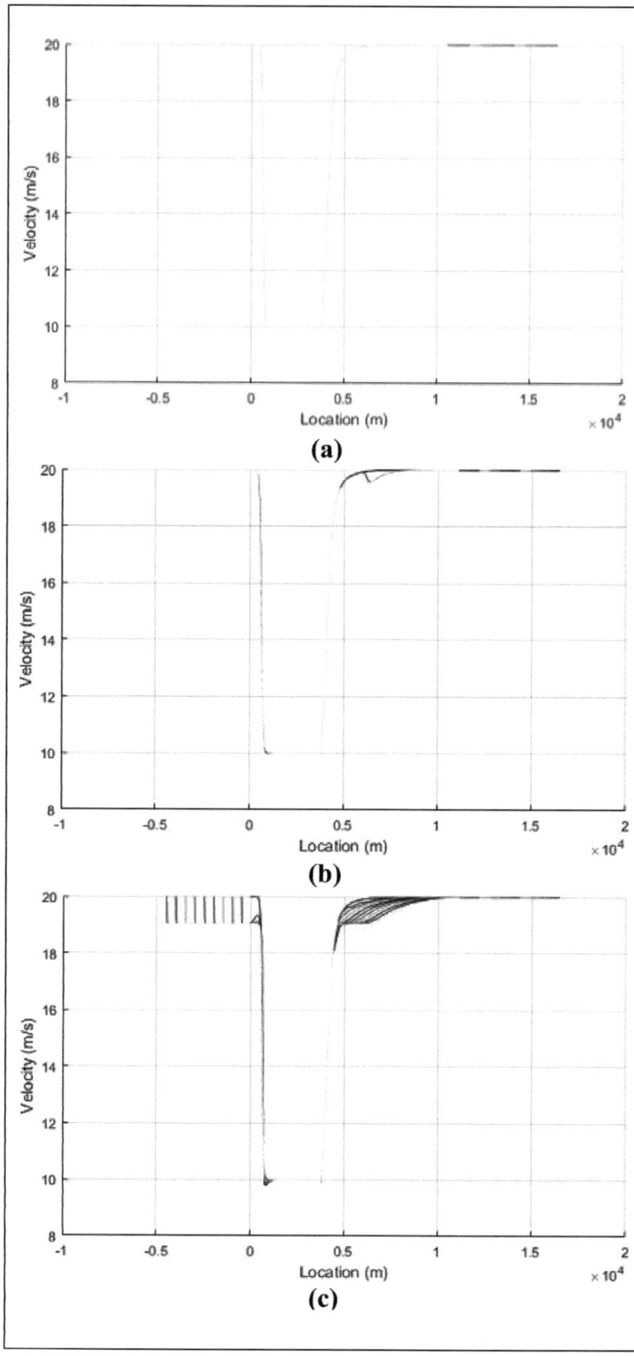

Figure 3. Simulation results for Gipps' model. Velocity profile for vehicles 10, 20, 30, 40, 50, 60, 70, 80, 90, and 100, for T = 3 (or more), T = 2.7, and T = 2.5 are illustrated in (*a*), (*b*), and (*c*), respectively.

Discussion

Figure 3 illustrates that a human driver system may operate well for some initial conditions in which vehicles are not interacting (i.e., vehicles are far away and the driver has enough time to adjust to the change in the speed limit without any influence from the immediate leader). In this situation, vehicles continue to follow a desired velocity profile accurately (i.e., free-flow condition). However, for some initial conditions, in which drivers' reaction times are similar or higher than the time headway, human

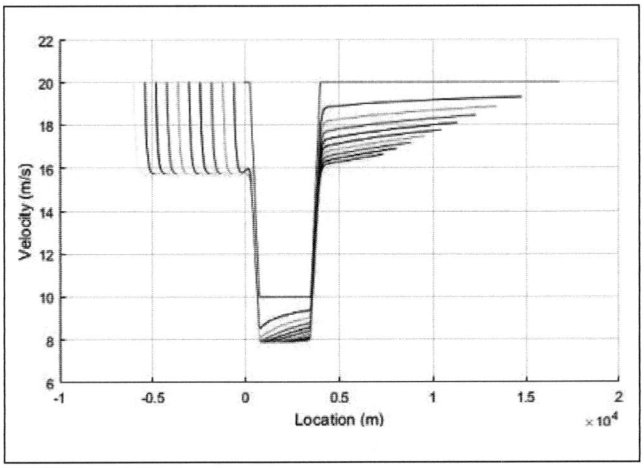

Figure 4. Simulation results for IDM: velocity profile for vehicles 10, 20, 30, 40, 50, 60, 70, 80, 90, and 100, for $T = 2$ s.

behavior could potentially lead to shockwaves propagating upstream and a change in the traffic flow regime. Additionally, simulation results with IDM showed the occurrence of shockwaves in the presence of a speed drop, as shown in Figure 4. However, simulation results also showed that the presented controller and the resulting automated vehicle system could enhance the operation even for small time headway (a 1-s time headway was selected in this study). Figure 5 illustrates that, over time, the velocity profile of vehicles following the control law of the form of Equation 1, approaches the desired velocity profile. This fact guarantees that at the steady state condition, vehicles follow the velocity profile and keep a safe distance of the form $T * \dot{x}_i(t)$ from the leader. Such a system provides an opportunity to improve safety and efficiency throughout the transportation system (e.g., enhancing travel time reliability, improving congestion, and reducing emissions and energy consumption).

Note that the simulation results showed that IDM cannot provide an equilibrium point when vehicles are supposed to follow a fixed velocity profile (i.e., the velocity profile is the same for two successive vehicles). In a special case when the relative speed is zero and all vehicles are traveling at the steady state condition, IDM's acceleration/deceleration command is not zero ($\dot{v}_n = -a$). Thus, it does not yield a desirable velocity profile.

Simulation results presented in Figure 6 show that a mixed string of automated and human-driven vehicles follow the speed profile better than a platoon of regular vehicles. By increasing the penetration rate of automated vehicles, the vehicles' velocity profiles become closer to the desired profile, and travel time reliability is marginally improved. However, by increasing the number of vehicles, the backward shockwave produced in the

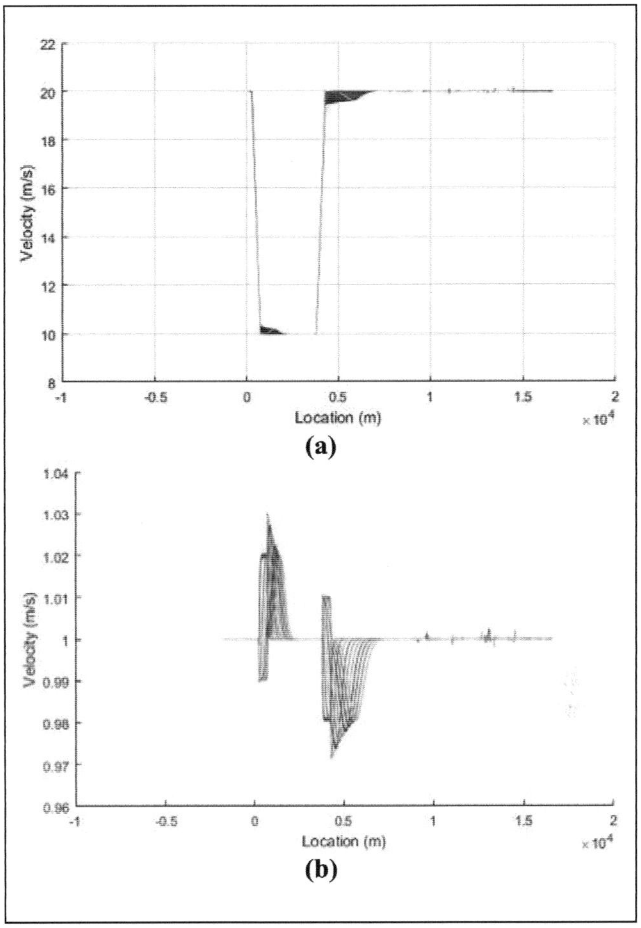

Figure 5. (*a*) Simulation results for autonomous vehicles, for vehicles 10, 20, 30, 40, 50, 60, 70, 80, 90, and 100, and time headway 1 s, and (*b*) time headway evolution over distance for the same vehicles.

velocity profile propagates upstream. These observations lead to the conclusion that such a system is not string stable.

Conclusion

Car-following models based on safe-distance assumptions may not be able to effectively track a velocity profile and keep a safe distance from the leader. Therefore, travel time reliability is reduced and drivers' efforts to follow the velocity profile may result in safety issues and high-risk behaviors. A fully automated system, however, has the potential to enhance safety as well as travel time reliability by accurately tracking the velocity profile. As a result, in the presence of a speed drop (e.g., workzone locations), implementing a fully automated system may result in travel time reduction and enhance the reliability of the transportation system. Accordingly, this study introduced a controller capable of accurately tracking a

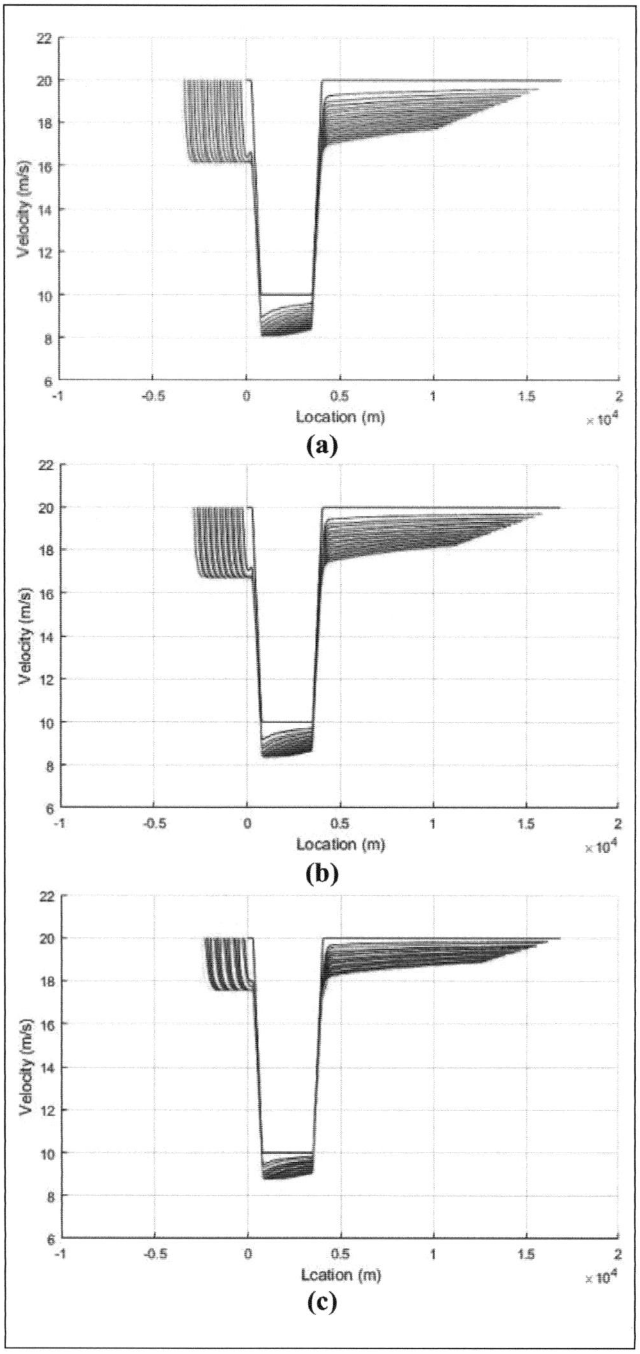

Figure 6. Simulation results for a mixed string of regular and automated vehicles for vehicles 10, 20, 30, 40, 50, 60, 70, 80, 90, and 100. Time headway is set to 1 s between automated vehicles and 2 s between regular vehicles and others. (*a*), (*b*), and (*c*) correspond to a 25%, 50%, and 75% market penetration rate of automated vehicles, respectively.

velocity profile in the presence of speed drops. The designed state feedback controls the maximum value among speed and spacing error terms for every vehicle. The presented decentralized model reduces the need for communication infrastructure. Simulation results showed the benefits of implementing such a controller to regulate travel time and track a velocity profile. Moreover, the results were compared to platoons of regular vehicles modeled with the IDM and Gipps' model. The simulation results showed that the IDM and Gipps' model result in shockwave formation when dealing with a speed drop. Finally, mixed platoons of automated and regular vehicles were simulated and the results showed that such systems can follow the velocity profile better that a platoon of regular vehicles. This result indicates that the proposed controller is implementable at multiple levels of automation from level II to IV. Overall, the simulation results suggest that the proposed platooning strategy is string stable. The analytical investigation of string stability was undertaken by the authors of the paper (*22*).

Author Contributions

The authors confirm contribution to the paper as follows: study conception and design: SA, AT; analysis and interpretation of results: SA, AT; draft manuscript preparation: SA, AT. All authors reviewed the results and approved the final version of the manuscript.

References

1. Mahmassani, H., T. Hou, and M. Saberi. Connecting Networkwide Travel Time Reliability and the Network Fundamental Diagram of Traffic Flow. *Transportation Research Record: Journal of the Transportation Research Board*, 2013. 2391: 80–91.
2. Talebpour, A., H. Mahmassani, and S. Hamdar. Multiregime Sequential Risk-Taking Model of Car-Following Behavior: Specification, Calibration, and Sensitivity Analysis. *Transportation Research Record: Journal of the Transportation Research Board*, 2011. 2260: 60–66.
3. Gazis, D. C., R. Herman, and R. W. Rothery. Nonlinear Follow-the-Leader Models of Traffic Flow. *Operations Research*, Vol. 9, No. 4, 1961, pp. 545–567.
4. Kumamoto, H., K. Nishi, K. Tenmoku, and H. Shimoura. Rule-Based Cognitive Animation Simulator for Current Lane and Lane Change Drivers. *Proc., Second World Congress on ATT*, Yokohama, Japan, 1995, 1746–1752.
5. Helly, W. *Dynamics of Single Lane Vehicular Traffic Flow*. Research Report No. 2. Center for Operations Research, Massachusetts Institute of Technology, 1959.
6. Hoyer, R., and M. Fellendorf. Parametrization of Microscopic Traffic Flow Models through Image Processing. *IFAC Proceedings Volumes*, Vol. 30, No. 8, 1997, Elsevier, pp. 889–894.
7. McDonald, M., J. Wu, and M. Brackstone. Development of a Fuzzy Logic Based Microscopic Motorway Simulation Model. *Proc., IEEE Conference on Intelligent Transportation System, 1997, ITSC'97*, Boston, Mass., IEEE, New York, 1997.

8. Brackstone, M., and M. McDonald. Car-Following: A Historical Review. *Transportation Research Part F: Traffic Psychology and Behaviour*, Vol. 2, No. 4, 1999, pp. 181–196.
9. Rothery, R. W. Car Following Models. *Trac Flow Theory*, 1992. Retrieved from https://ocw.mit.edu/courses/civil-and-environmental-engineering/1-225j-transportation-flow-systems-fall-2002/lecture-notes/carfollowinga.pdf.
10. Swaroop, D., and J. Hedrick. Constant Spacing Strategies for Platooning in Automated Highway Systems. *Journal of Dynamic Systems, Measurement, and Control*, Vol. 121, No. 3, 1999, pp. 462–470.
11. Swaroop, D., and K. Rajagopal. A Review of Constant Time Headway Policy for Automatic Vehicle Following. *Proc., 2001 IEEE, Intelligent Transportation Systems, 2001*, Oakland, Calif., IEEE, New York, 2001.
12. Peters, A. A., R. H. Middleton, and O. Mason, Leader Tracking in Homogeneous Vehicle Platoons with Broadcast Delays. *Automatica*, Vo. 50, No. 1, 2014, pp. 64–74.
13. Öncü, S., J. Ploeg, N. Van de Wouw, and H. Nijmeijer. Cooperative Adaptive Cruise Control: Network-Aware Analysis of String Stability. *IEEE Transactions on Intelligent Transportation Systems*, Vo. 15, No. 4, 2014, pp. 1527–1537.
14. Milanés, V., S. E. Shladover, J. Spring, C. Nowakowski, H. Kawazoe, and M. Nakamura. Cooperative Adaptive Cruise Control in Real Traffic Situations. *IEEE Transactions on Intelligent Transportation Systems*, Vol. 15, No. 1, 2014, pp. 296–305.
15. Ge, J. I., and G. Orosz. Dynamics of Connected Vehicle Systems with Delayed Acceleration Feedback. *Transportation Research Part C: Emerging Technologies*, Vol. 46, 2014, pp. 46–64.
16. Besselink, B., and K. H. Johansson. Control of Platoons of Heavy-Duty Vehicles using a Delay-Based Spacing Policy. *IFAC-PapersOnLine*, Vol. 48, No. 12, 2015, Elsevier, pp. 364–369.
17. Besselink, B., and K. H. Johansson. String Stability and a Delay-Based Spacing Policy for Vehicle Platoons Subject to Disturbances. *IEEE Transactions on Automatic Control*, Vol. 62, No. 9, 2017, pp. 4376–4391.
18. Arefizadeh, S., A. Hasanzadezonuzy, A. Talebpour, and S. Shakkotai. Assessing Strong String Stability of Constant Spacing Policy under Speed Limit Fluctuations. Retrieved from https://arxiv.org/abs/1802.00122.
19. Gipps, P. G. A Behavioural Car-Following Model for Computer Simulation. *Transportation Research Part B: Methodological*, Vol. 15, No. 2, 1981, pp. 105–111.
20. Treiber, M., A. Hennecke, and D. Helbing. Congested Traffic States in Empirical Observations and Microscopic Simulations. *Physical Review E*, Vol. 62, No. 2, 2000, p. 1805.
21. Vasconcelos, L., L. Neto, S. Santos, A. B. Silva, and Á. Seco. Calibration of the Gipps Car-Following Model Using Trajectory Data. *Transportation Research Procedia*, Vol. 3, 2014, pp. 952–961.
22. Arefizadeh, S., A. Talebpour, and I. Zelenko. *Platooning in the Presence of a Speed Drop: A Generalized Control Model. arXiv preprint arXiv:1709.10083*, 2017.
23. Talebpour, A., and H. S. Mahmassani. Influence of Connected and Autonomous Vehicles on Traffic Flow Stability and Throughput. *Transportation Research Part C: Emerging Technologies*, Vol. 71, 2016, pp. 143–163.
24. Talebpour, A., H. S. Mahmassani, and S. H. Hamdar. Effect of Information Availability on Stability of Traffic Flow: Percolation Theory Approach. *Transportation Research Procedia*, Vol. 23, 2017, pp. 81–100.

The Standing Committee on Traffic Flow Theory and Characteristics (AHB45) peer-reviewed this paper (18-06130).

TRR
JOURNAL OF THE TRANSPORTATION RESEARCH BOARD

Article

Transportation Research Record
2018, Vol. 2672(20) 162–177

DOI: 10.1177/0361198118780704
journals.sagepub.com/home/trr
SAGE

A New Control Strategy Integrated into the Desired Safety Margin Car-Following Model Considering the Disturbance Level

Junjie Zhang[1], Yunpeng Wang[1], and Guangquan Lu[1]

Abstract

A vehicular communication system can provide motion information based on various sensors. Thus, adaptive cruise control (ACC) systems based on a reliable communication system can relieve traffic congestion in a platoon. However, acquiring vehicle information entails uncertainties because of the disturbance of traffic environment and vehicular communication systems, thereby influencing vehicle control performance. In this study, the desired safety margin (DSM) model is employed to investigate the influence of uncertainty on car-following performance, such as starting, emergency braking, and car-following processes. Based on the DSM model, the disturbance level on perceived safety margin is introduced to characterize the uncertainty of vehicle information. The stability criterion of the DSM model with the disturbance level is derived via linear stability theory. Analytical results indicate that a negative value of disturbance level will enlarge the stable region. By contrast, a positive value is conducive for maintaining the consensus state and achieving the high acceleration and deceleration of following vehicles in the starting process. Findings show that the disturbance in the vehicular system significantly influences car-following performance. To enhance the smoothness and stability of traffic flow evolution, a new control strategy is proposed in this study. Numerical experiments illustrate the effectiveness of the proposed control strategy in stabilizing traffic flow. This study highlights the need to resolve the stability of sensors and vehicular communication systems, and to develop the vehicular controller to help ACC systems improve vehicle control performance in the car-following process.

Traffic congestion increasingly threatens the environment and economic development of modern society. Driving behavior and traffic environment are two important determinants of traffic flow stability. The influence of driver behavior characteristics and various traffic environment factors on car-following performance is always intensely investigated, especially in terms of the adaptive cruise control (ACC) system or automotive platoon driving. To capture the mechanism of traffic congestion behind vehicular traffic flow from the microscopic viewpoint, various car–following models have been proposed over a period covering 60 years.

The car-following concept was first proposed by Pipes (*1*) in 1953. Chandler et al. (*2*) conceptualized the first linear car-following model using only relative velocity as stimulus–response. Subsequently, the Gazis-Herman-Rothery (GHR) model, suggested by Gazis et al. (*3*), assumes that the following-vehicle motion is determined by the velocity, relative velocity, and headway of two successive vehicles. Certain scholars have conducted extensive research in the calibration of the parameters and validation for the GHR model (*4*). In 1995, the optimized velocity approach proposed by Bando et al. (*5*) postulates that car-following is based on the desire of the driver to maintain optimal velocity (OV) in a traffic stream. Helbing and Tilch (*6*) introduced the generalized force (GF) model to surmount the deficiencies of the OV model. Jiang et al. (*7*) further extended the GF model to overcome its disadvantages, resulting in the full velocity difference (FVD) model. In addition, the intelligent driver model (*8*) is widely used in ACC and the intelligent transportation system as a time-continuous car-following model.

[1]Beijing Advanced Innovation Center for Big Data and Brain Computing, Beijing Key Laboratory for Cooperative Vehicle Infrastructure Systems and Safety Control, Jiangsu Province Collaborative Innovation Center of Modern Urban Traffic Technologies, School of Transportation Science and Engineering, Beihang University, Beijing, China

Corresponding Author:
Address correspondence to Guangquan Lu: lugq@buaa.edu.cn

Subsequent models were recommended to investigate the influence of different stimulus–response factors on traffic flow stability. Among the extended car-following models based on the OV, FVD, and GHR are the models on multiple ahead and velocity difference (*9*); full velocity and acceleration difference (*10*), multiple velocity difference (*11*), and multiple headway, velocity, and acceleration difference (*12*). Orosz et al. (*13, 14*) investigated two types of car-following model as an OV and a nonlinear car-following model incorporating the reaction-time delay of drivers. Zhu and Dai (*15*) developed an extended car-following model by considering the delay of the driver's response in sensing headway, which plays an important role in jamming transition. Low (*16*) examined the degree of chaos in a deterministic car-following model with a non-zero driver-vehicle reaction time and an inner-vehicle separation term based on desired headway time. Tang et al. (*17–24*) presented some new car-following models by considering certain traffic environment and human factors, such as the attribution of drivers, the driving range of electric vehicles, the reliability of inter-vehicle communication, and the driver's forecast effect and driving resistance. Li et al. (*25*) suggested a generalized model to analyze the influence of a two-sided lateral gap under the non-lane-discipline environment. Ge et al. (*26*) proposed a backward-looking anticipation OV model under certain conditions in traffic flow. Yu and Shi (*27*) conceived an FVD considering headway changes with memory. Peng et al. (*28*) also constructed a new FVD model by considering the effect of the OV changes with memory. From a recent comprehensive review of various car-following models, Saifuzzaman and Zheng (*29*) recommended that human factors be considered in car-following modeling for a realistic representation of car-following behavior in complex driving situations to improve traffic safety and better understand traffic flow phenomena. This paper claims that the subjective risk perception of drivers from the view of the human factors should be considered in car-following modeling. Moreover, driving behavior is largely affected by the risk perception of drivers in actual traffic. A desired safety margin (DSM) car-following model proposed by Lu et al. (*30*) not only can simulate the physiological and psychological characteristics of different drivers as the driver's response time, the sensitivity factor for acceleration or deceleration, and the DSM, but also provides a novel method to explain the car-following process. This DSM model was also verified effective for the simulation of real-world car-following. The aforementioned car-following models have confirmed that various traffic information play an important role in determining the qualitative dynamic of vehicles in the car-following process.

The following vehicle usually adjusts its acceleration or deceleration by using the motion information of the preceding vehicle in the platoon. Thus, vehicle information is a vital determinant for car-following performance. However, the vehicle information introduced in the aforementioned car-following models is determinate. In reality, vehicle information is affected by uncertain traffic environment. Consequently, most car-following models cannot reveal the influence of uncertain traffic conditions on traffic flow stability. Nevertheless, connected vehicle (CV) technology has been observed worldwide in the past decade. CV technology affixes diagnostic sensors onto vehicles and infrastructures and enables the wireless transmission of collected data between vehicles and nearby infrastructures. Sensor accuracy and stability (e.g., the accuracy of GPS, communication delay, and lidar) influence the accuracy of vehicle movement data, which include but are not limited to, velocities, positions, acceleration, and deceleration of vehicles. Uncertain traffic environment also affects the accuracy and stability of on-board sensors. Therefore, uncertain traffic environment and perception error of on-board sensors will affect the decision making of an ACC in automotive platoon driving. Furthermore, it is asserted that the decision making of an ACC system is determined by the decision variables and their uncertainties. That is, once sensor error, communication delay, GPS error, or equipment failure occurs in the car-following process, the decision variables with uncertainties influence the car-following performance, thereby affecting the traffic flow stability in the platoon. In this study, those uncertainties are integrated to form the disturbance level to the decision variables of the ACC system. At present, only a few works acknowledge the influence of the disturbance level on traffic flow stability. Therefore, the disturbance level of the decision variables of an ACC system is introduced in this study to characterize the uncertainty of vehicle information. To this end, the DSM car-following model is employed to investigate the influence of disturbance level on car-following performance.

Normative traffic flow control has been proposed in recent years to address traffic congestions and smoothen traffic flow. Several control strategies are usually integrated into a car-following model to stabilize traffic flow. Konishi et al. (*31*) proposed a dynamic version of decentralized delayed feedback control based on an OV model to suppress traffic jams. Hu et al. (*32*) presented a new feedback control car-following model based on the pioneering work of Konishi et al. Zhao et al. (*33*) proposed a velocity difference feedback control to reduce traffic gridlocks based on a coupled map (CM) model. A speed feedback control mechanism was introduced into the system to improve the dynamic performance of traffic flow (*34*). Li et al. (*35*) developed an acceleration feedback control strategy to suppress traffic congestion based on an FVD model under an open boundary. On the basis of an OV model with the reaction time of drivers, Jin and Xu (*36*) introduced a headway and velocity difference feedback control mechanism to stabilize unstable traffic flow. Zheng et al. (*37*) constructed a novel control scheme to reduce the influence range of traffic congestion based on a CM model. Considering the role of control strategy as discussed above, the development of a new efficient controller of vehicular traffic flow is attempted that enhances traffic flow stability during the interruption of vehicle information.

This paper is organized as follows. First, the DSM car-following model considering the disturbance level is specified. The stability and instability regions of the DSM model considering disturbance level (designated as the LDDSM model) are then obtained by using linear stability theory. Next a sliding-mode control strategy is introduced to stabilize unstable traffic flow. The numerical simulations that verify the theoretical results are provided and the comparison and analysis of the effects in the numerical results of the LDDSM model with or without the control strategy are described. Finally, conclusions are drawn.

τ_2 = response time of the brake system;

$v_n(t)$ and $v_{n-1}(t)$ = velocity of the n th and $(n-1)$th vehicle at time t, respectively;

$\Delta x_n(t) = x_{n-1}(t) - x_n(t)$ = headway between the $(n-1)$th and the nth vehicles;

$d_n(t)$ and $d_{n-1}(t)$ = deceleration of the $(n-1)$th and the nth vehicles at time t, respectively; and l_{n-1} = length of the preceding vehicle.

In this study, a platoon of n vehicles running on a straight road is considered. All vehicles are presumed to be equipped, and each vehicle can receive movement information about itself and the preceding vehicle. To simplify the DSM model in ACC, it is assumed that the sensitivity coefficients of acceleration and deceleration are similar with a fixed value. Moreover, it is considered that the vehicle adjusts its acceleration or deceleration by judging whether the risk level (safety margin) is smaller or larger than its acceptable level (DSM). Accordingly, the disturbance level is introduced into the DSM model, and the LDDSM model is proposed as follows:

$$a_n(t+\hat{\tau}) = \alpha\left[(1+\xi)\left(1 - \frac{v_n(t)\cdot\tau_2 + [v_n(t)]^2/2d_n(t)}{x_{n-1}(t)-x_n(t)-l_{n-1}} + \frac{[v_{n-1}(t)]^2/2d_{n-1}(t)}{x_{n-1}(t)-x_n(t)-l_{n-1}}\right) - SM_D\right], \quad (3)$$

LDDSM Model

Safety margin, a new quantitative indicator of the subjective perception risk of car-following, was proposed by Lu et al. (*38*). In 2014, DSM model was used as a car-following strategy for the rear-end avoidance system of CVs in an actual vehicle test platform (*39*). The DSM of car-following was proposed by Lu et al. (*30*) as

$$a_n(t+\tau) = \begin{cases} \alpha_1(SM_n(t) - SM_{nDH}), SM_n(t) > SM_{nDH} \\ \alpha_2(SM_n(t) - SM_{nDL}), SM_n(t) < SM_{nDL} \\ 0, else \end{cases}, \quad (1)$$

$$SM_n(t) = 1 - \frac{v_n(t)\cdot\tau_2 + \left([v_n(t)]^2/2d_n(t)\right)}{x_{n-1}(t)-x_n(t)-l_{n-1}} + \frac{[v_{n-1}(t)]^2/2d_{n-1}(t)}{x_{n-1}(t)-x_n(t)-l_{n-1}}, \quad (2)$$

where

$a_n(t)$ = acceleration of a following vehicle at time t;

SM_{nDH} = upper limit of the DSM in the car-following condition;

SM_{nDL} = lower limit of the DSM in the car-following condition;

α_1 and α_2 = sensitivity factors for acceleration and deceleration, respectively;

τ = response time;

where

α is the sensitivity coefficient for acceleration or deceleration;

SM_D is the DSM, considered to be a fixed value of acceptable risk;

$\hat{\tau}$ is the delay time of the vehicular controller, which is smaller than the response time τ; and

ξ is defined as the disturbance level of vehicular system for perceived safety margin in ACC or automotive platoon driving.

For convenience, the symbol of $\hat{\tau}$ is still marked as τ.

As suggested by Chowdhury et al. (*40*), for convenience of qualitative analysis, the general form for the LDDSM model can be described as

$$\begin{cases} \dfrac{dv_n(t+\tau)}{dt} = f(v_{n-1}, s_n, v_n) \\ \dfrac{ds_n(t)}{dt} = v_{n-1}(t) - v_n(t) \end{cases}, \quad (4)$$

where the function $f(v_{n-1}, s_n, v_n) = \alpha[(1+\xi)SM_n(t) - SM_D]$ indicates the response to the stimulus received through risk perception, and s_n is the gap between vehicles n and $(n-1)$.

Stability Analysis

In this section, the linear stability analysis method is adopted to the proposed LDDSM model (Equation 3).

Specifically, in the steady-state uniform flow solution, the velocity considered is expressed as $v_n = v_{n-1} = \cdots = v^*$ and the same time-independent gap is defined by $s_n = s_{n-1} = \cdots = s^*$ which is calculated by $f(v^*, s^*, v^*) = 0$. Let εv_n and εs_n denote respectively the small perturbations of the velocity and gap around the steady-state uniform flow solution: $v_n = v^* + \varepsilon v_n$ and $s_n = s^* + \varepsilon s_n$.

The first-order Taylor expansion of Equation 4 yields

$$\begin{cases} \dfrac{d\varepsilon v_n(t+\tau)}{dt} = f_n^{v_{n-1}} \varepsilon v_{n-1}(t) + f_n^{s_n} \varepsilon s_n(t) + f_n^{v_n} \varepsilon v_n(t) \\ \dfrac{d\varepsilon s_n(t)}{dt} = \varepsilon v_{n-1}(t) - \varepsilon v_n(t) \end{cases}, \tag{5}$$

where

$$f_n^{v_n} = \left.\frac{\partial f}{\partial v_n}\right|_{(v^*, s^*)}, f_n^{s_n} = \left.\frac{\partial f}{\partial s_n}\right|_{(v^*, s^*)}, f_n^{v_{n-1}} = \left.\frac{\partial f}{\partial v_{n-1}}\right|_{(v^*, s^*)}. \tag{6}$$

That is,

$$\begin{pmatrix} \dfrac{d\varepsilon v_n(t+\tau)}{dt} \\ \dfrac{d\varepsilon s_n(t)}{dt} \end{pmatrix} = \begin{pmatrix} f_n^{v_n} & f_n^{s_n} \\ -1 & 0 \end{pmatrix} \begin{pmatrix} \varepsilon v_n(t) \\ \varepsilon s_n(t) \end{pmatrix} + \begin{pmatrix} f_n^{v_{n-1}} \\ 1 \end{pmatrix} \cdot \varepsilon v_{n-1}(t). \tag{7}$$

After the Laplace transformation,

$$\begin{pmatrix} \lambda e^{\lambda\tau} \\ \lambda \end{pmatrix} \begin{pmatrix} \varepsilon V_n(\lambda) \\ \varepsilon S_n(\lambda) \end{pmatrix} - \begin{pmatrix} \varepsilon V_n(0) \\ \varepsilon S_n(0) \end{pmatrix} = \begin{pmatrix} f_n^{v_n} & f_n^{s_n} \\ -1 & 0 \end{pmatrix} \begin{pmatrix} \varepsilon V_n(\lambda) \\ \varepsilon S_n(\lambda) \end{pmatrix} + \begin{pmatrix} f_n^{v_{n-1}} \\ 1 \end{pmatrix} \cdot \varepsilon V_{n-1}(\lambda), \tag{8}$$

A velocity relationship between vehicles $(n-1)$ and n is obtained and extracted using $\varepsilon V_n(\lambda)$, as shown:

$$T(\lambda) = \frac{\varepsilon V_n(\lambda)}{\varepsilon V_{n-1}(\lambda)}, \tag{9}$$

where $T(\lambda)$ denotes the transfer function. The transfer function $T(\lambda)$ is

$$T(\lambda) = \frac{f_n^{v_{n-1}} \cdot \lambda + f_n^{s_n}}{\lambda^2 e^{\lambda\tau} - f_n^{v_n} \cdot \lambda + f_n^{s_n}}. \tag{10}$$

For the specifications of the LDDSM model above, it can be clearly demonstrated that

$$\begin{aligned} f_n^{v_n} &= \left.\frac{\partial f}{\partial v_n}\right|_{(v^*, s^*)} = -\alpha(1+\xi)\frac{\tau_2 d_n + v^*}{s^* d_n} \\ f_n^{s_n} &= \left.\frac{\partial f}{\partial s_n}\right|_{(v^*, s^*)} = \alpha(1+\xi)\frac{v^*\tau_2 + (v^*)^2/2d_n - (v^*)^2/2d_{n-1}}{(s^*)^2}. \\ f_n^{v_{n-1}} &= \left.\frac{\partial f}{\partial v_{n-1}}\right|_{(v^*, s^*)} = \alpha(1+\xi)\frac{v^*}{s^* d_{n-1}} \end{aligned} \tag{11}$$

The characteristic quasi-polynomial is

$$P(\lambda) = \lambda^2 e^{\lambda\tau} + \alpha(1+\xi)\frac{\tau_2 d_n + v^*}{s^* d_n} \cdot \lambda + \alpha(1+\xi) \frac{v^*\tau_2 + (v^*)^2/2d_n - (v^*)^2/2d_{n-1}}{(s^*)^2}. \tag{12}$$

Therefore, $P(\lambda)$ is approximated by the first-order expansion, that is,

$$P(\lambda) = \tau\lambda^3 + \lambda^2 + \alpha(1+\xi)\frac{\tau_2 d_n + v^*}{s^* d_n} \cdot \lambda + \alpha(1+\xi)\frac{v^*\tau_2 + (v^*)^2/2d_n - (v^*)^2/2d_{n-1}}{(s^*)^2}. \tag{13}$$

As suggested in (*36*, *41*), the stability of system (4) requires that: (1) $P(\lambda)$ is stable and that (2) the infinite norm of transfer function of system (4) is less than 1.

To ensure that $P(\lambda)$ is stable, the following forms should be satisfied by using the Routh–Hurwitz criterion

$$1/\tau > 0, \alpha(1+\xi)\frac{v^*\tau_2}{\tau(s^*)^2} > 0, \alpha(1+\xi)\frac{\tau_2 d + v^*}{\tau^2 s^* d} - \alpha(1+\xi)\frac{v^*\tau_2}{\tau(s^*)^2} > 0, \tag{14}$$

where $d = d_n = d_{n-1}$.

With reference to the test data in (*42*), the decelerations in the calculation of risk, denoted as $d_n(t)$ and $d_{n-1}(t)$, are set as $0.75g$, where g is the acceleration due to gravity. Sokolovskij (*43*) found that "the values of the time of deceleration increase of most of the Japanese and Western vehicles have not exceeded 0.2 s and remained within the limits of 0.1–0.2 s." Under emergency situations, the driver may quickly and strongly apply the brakes. Thus, τ_2 is set as 0.15 s.

On the basis of the above analysis, all conditions of Equation 14 are satisfied. That is, $P(\lambda)$ is stable. Consider $\|T(\lambda)\|_\infty \leq 1$ and set $\lambda = iz$, then

$$|T(iz)| = \sup_{z\in[0,+\infty)} \left| \frac{\alpha(1+\xi)\frac{v^*}{s^*d}\cdot iz + \alpha(1+\xi)\frac{v^*\tau_2}{(s^*)^2}}{-i\tau\cdot z^3 - z^2 + i\alpha(1+\xi)\frac{\tau_2 d + v^*}{s^*d}\cdot z + \alpha(1+\xi)\frac{v^*\tau_2}{(s^*)^2}} \right| \leq 1. \quad (15)$$

Thus, the sufficient condition is

$$4\tau^2\left(\alpha(1+\xi)\frac{v^*}{s^*d}\right)^2 + 8\tau^2\alpha(1+\xi)\frac{v^*\tau_2}{(s^*)^2} - 4\tau\alpha(1+\xi)\frac{\tau_2 d + v^*}{s^*d} + 1 < 0. \quad (16)$$

That is,

$$\alpha < -\frac{2\frac{v^*\tau_2}{(s^*)^2} - \frac{\tau_2 d + v^*}{\tau s^* d}}{2(1+\xi)\left(\frac{v^*}{s^*d}\right)^2} + \frac{1}{1+\xi} \sqrt{\left(\frac{2\frac{v^*\tau_2}{(s^*)^2} - \frac{\tau_2 d + v^*}{\tau s^* d}}{2\left(\frac{v^*}{s^*d}\right)^2}\right)^2 - \frac{1}{4\left(\frac{v^*\tau}{s^*d}\right)^2}}, \quad (17)$$

where $s^* = \frac{(1+\xi)\tau_2 v^*}{1+\xi - \mathrm{SM}_D}$.

In (38), the average DSM of general drivers is approximately 0.9. To satisfy safety performance and comfort, SM_D is set as 0.9. It is assumed that the leading vehicle runs at a fixed speed $v^* = 20\mathrm{m/s}$. To illustrate the effect of the disturbance level ξ on the traffic flow stability, the following two disturbance levels are selected: $\xi = -0.02$ and $\xi = 0.02$. As $\xi = 0$, the stable condition agrees with that of DSM model. Thus, the neutral stability curves can be described in the parameter space $(\tau;\alpha)$ with $\mathrm{SM}_D = 0.9$ (Figure 1). Figure 1 shows that the delay time–sensitivity space is divided into the stable and unstable regions by a neutral curve. The stable region lies at the bottom left of the neutral stability curve, and traffic congestions do not occur within it. By contrast, the unstable region lies at the top right of the neutral stability curve, and backward waves emerge in it as the velocity perturbations of the leading vehicle are inserted.

Results show that the disturbance level of vehicle information plays an important role in traffic flow stability, as shown in Figure 1. Thus, the following concern is worth examining: how does the disturbance level of vehicle information in the car-following process affect dynamic performance? A new control strategy must also be considered in smoothing traffic flow.

New Control Scheme

To reduce the influence range of traffic congestion, a new sliding-mode controller of vehicular traffic flow is introduced in this study. In the car-following process, the vehicle adjusts its velocity to retain a DSM in the ACC system.

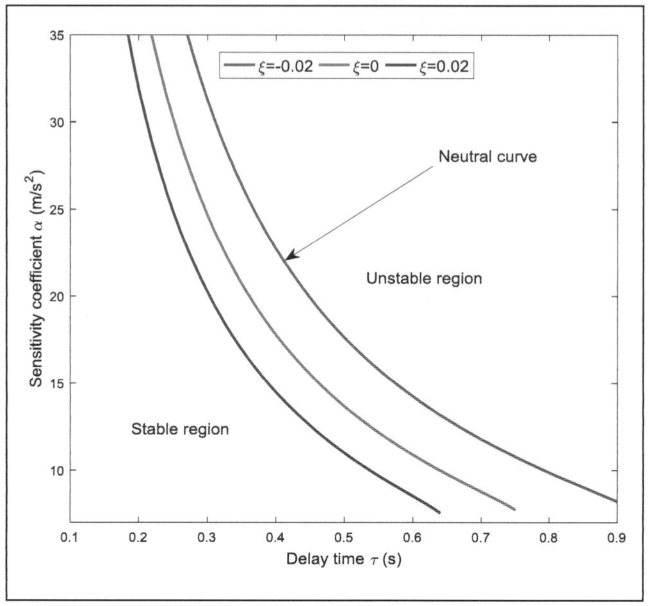

Figure 1. Stability diagrams showing the effect of the disturbance level on traffic dynamics with $SM_D = 0.9$.

That is, all vehicles of the platoon maintain a desired spacing with the same velocity in homogeneous traffic flow. Thus, it is supposed that the error can be defined as

$$\Theta_0 = \varepsilon s_n = s_n - \frac{\tau_2 v^*}{1 - \mathrm{SM}_D}. \quad (18)$$

The goal of the sliding-mode controller is $\lim_{t\to t_0}\Theta_0 = 0$, where $t_0 > 0$ is a finite time constant.

Let the sliding-mode surface be defined (44) as

$$\Theta = \Theta_0 + k_1\dot{\Theta}_0, \quad (19)$$

where $k_1 \in R$ and $k_1 > 0$. Then

$$\begin{aligned}\dot{\Theta} &= \dot{\Theta}_0 + k_1\ddot{\Theta}_0 = \varepsilon v_{n-1}(t) - \varepsilon v_n(t) - k_1\varepsilon\dot{v}_n(t) + k_1\varepsilon\dot{v}_{n-1}(t) \\ &= (1 - k_1\frac{\alpha(1+\xi)v^*}{s^*d})\varepsilon v_{n-1}(t) - k_1\frac{\alpha(1+\xi)v^*\tau_2}{(s^*)^2}\varepsilon s_n(t) \\ &\quad + (k_1\frac{\alpha(1+\xi)(\tau_2 d_n + v^*)}{s^* d_n} - 1)\varepsilon v_n(t) \\ &\quad + k_1\frac{d\varepsilon v_n(t)}{dt} - k_1 u_n(t).\end{aligned} \quad (20)$$

where the dot denotes differentiation with respect to time.

As suggested in (45–47), the candidate Lyapunov function of the controller can be selected as

$$\Sigma = \frac{1}{2}\Theta^2. \quad (21)$$

Hence,

$$\dot{\Sigma} = \Theta\dot{\Theta} = \Theta\left((1 - k_1\frac{\alpha(1+\xi)v^*}{s^*d})\varepsilon v_{n-1}(t) - k_1\frac{\alpha(1+\xi)v^*\tau_2}{(s^*)^2}\varepsilon s_n(t) + (k_1\frac{\alpha(1+\xi)(\tau_2 d_n + v^*)}{s^* d_n} - 1)\varepsilon v_n(t) + k_1\frac{d\varepsilon v_n(t)}{dt} - k_1 u_n(t)\right) \quad (22)$$

Based on Equation 22, the sliding-mode controller for vehicular traffic system can be described as

$$u_n(t) = (\frac{1}{k_1} - \frac{\alpha(1+\xi)v^*}{s^*d})\varepsilon v_{n-1}(t) - \frac{\alpha(1+\xi)v^*\tau_2}{(s^*)^2}\varepsilon s_n(t) + (\frac{\alpha(1+\xi)(\tau_2 d + v^*)}{s^*d} - \frac{1}{k_1})\varepsilon v_n(t) + \frac{d\varepsilon v_{n-1}(t)}{dt} + \frac{1}{k_1}(\rho\Theta + \gamma \mathrm{sgn}(\Theta)). \quad (23)$$

where ρ and γ are positive constants.

Substituting Equation 23 into Equation 22 yields

$$\dot{\Sigma} = \Theta\dot{\Theta} = \Theta(-\rho\Theta - \gamma \mathrm{sgn}(\Theta)) = -\rho\Theta^2 - \gamma|\Theta| \leq 0. \quad (24)$$

Therefore, the proposed sliding-mode controller is stable based on Lyapunov stability theory. This, in turn, implies that Θ and Θ_0 asymptotically converge to zero (*48*).

Numerical Experiments and Discussion

In this section, it is assumed that $d = 0.75g$, $\tau_2 = 0.15$ s, and the length of all vehicles is $l_n = 5$ m, $n = 1, 2, \cdots, N$. Numerical experiments are provided to discuss the effect of the disturbance level on dynamic performance via the LDDSM model under three scenarios, namely, starting, emergency braking, and evolution processes. Furthermore, a new sliding-mode control strategy is integrated into the LDDSM model to analyze its role in car-following performance.

Starting Process

The starting process is established as follows. Initially, the traffic signal is red, and 10 vehicles wait behind the signal with the same gap. Vehicles start to run at time $t = 0$ as the signal is switched to green. The leading vehicle then starts to accelerate until it reaches the OV of $v^* = 20$m/s. As suggested in (*49*), the leading vehicle starts to accelerate with typical values of a comfortable acceleration at 1.5 m/s^2. Other parameters are set as $SM_D = 0.9$, $\tau = 0.3$s, and $\alpha = 10$m/s^2. From Figure 2, all following vehicles can essentially maintain the same movement as the preceding vehicle but with some delay time. The lowest delay time is obtained for the disturbance level of 0.02, followed by the disturbance levels of 0 and −0.02. The delay time is smaller when the disturbance level has a positive value compared to that of a negative ξ value in the starting process.

As shown in Figure 3, the highest magnitude of acceleration for the LDDSM model is obtained with the disturbance level of 0.02, (Figure 3(*a*)), followed by without the disturbance level (Figure 3(*b*)) and with the disturbance level of −0.02 (Figure 3(*c*)). In addition, the vehicles in the LDDSM model with the disturbance level of 0.02 (Figure 3(*a*)) can finish the acceleration phase quickly, and its following vehicle can better follow the consensus state of the leading vehicle compared with those in Figure 3(*b*) and (*c*). The results show that the vehicles with positive ξ value can maintain the consensus state and complete the acceleration phase in a short time.

Emergency Braking Process

The emergency braking process is established as follows. The leading vehicle applies emergency braking if a traffic accident occurs. Initially, 10 vehicles are running on a single lane with the same velocity and gap. The sensitivity, delay time of the vehicular controller, DSM, gap, and velocity are selected as $\alpha = 10$ m/s^2, $\tau = 0.3$ s, $SM_D = 0.9$, $s^* = 35$ m, and $v^* = 20$ m/s, respectively, to ensure that the car-following process is stable, as shown in Figure 1. As suggested in (*50*), the leading vehicle brakes sharply with −8 m/s^2 until it reaches a full stop. The other vehicles follow the deceleration of the leading vehicle until all vehicles come to a full stop. For comparison, the deceleration and velocity profiles with different ξ values of −0.02, 0, and 0.02 are shown in Figures 4 and 5, respectively.

Figure 4 indicates that the lowest magnitude of deceleration is obtained for the LDDSM model with the disturbance level of −0.02, followed by the disturbance level of 0 and 0.02. Figure 5 illustrates that all the following vehicles can basically duplicate the velocities of the leading vehicles but with some delay time. Moreover, the delay time simulated by the LDDSM model with the disturbance level of 0.02 is shorter than that with the disturbance levels of 0 and −0.02. Although the following vehicles decelerate more quickly under the disturbance level of 0.02 than the other two disturbance levels, some

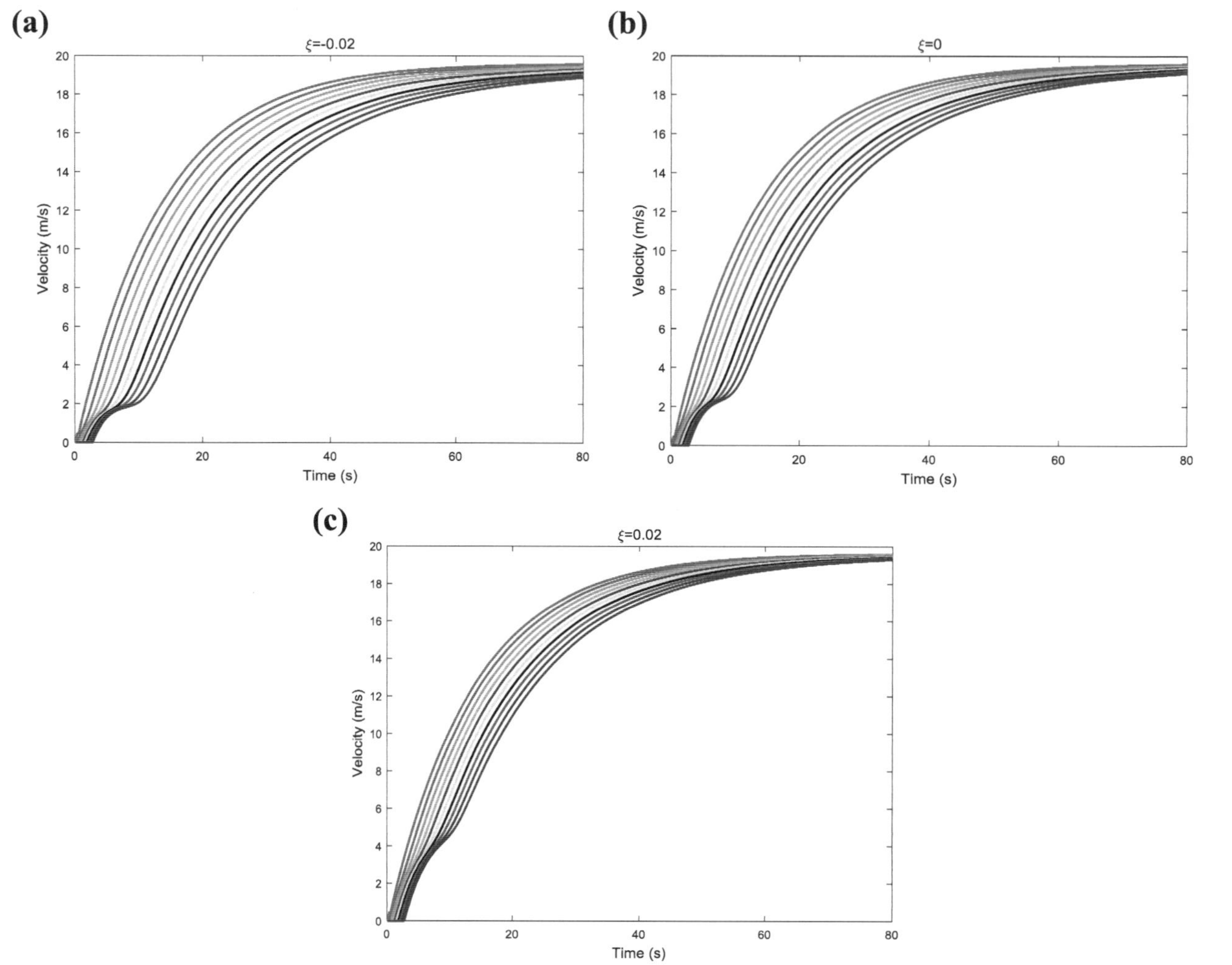

Figure 2. Motions of vehicle nos. 1–10, starting from a green signal for LDDSM model at different ξ values. Each curve shows the velocity of each vehicle.

following vehicles accelerate at the beginning of the emergency braking process. This outcome suggests that the disturbance of vehicle information plays an important role in the dynamic performance of vehicular system.

Conclusions based on the above two scenarios demonstrate that a positive level of disturbance in the vehicular system is conducive to the dynamic performance in the starting process. However, the disturbance of vehicle information is not favorable to the dynamic performance in the emergency braking process. This finding implies that a positive disturbance level in the vehicular information system for ACC or vehicle platoon control can generate the consensus state and cause a high acceleration of following vehicles in the starting process. Moreover, for the speed control system in the platoon, accurate and reliable vehicle information must be provided in the emergency braking process.

Effect on Traffic Flow Evolution

In this subsection, several simulations are conducted under an open boundary condition using the LDDSM model considering the disturbance level to explore how such level affects the traffic flow evolution process with an initial small perturbation. Furthermore, the sliding-mode controller for vehicular traffic system is introduced in the LDDSM to verify its effectiveness on stabilizing traffic flow.

Generally, all the vehicles move on a single lane with the vehicle spacing $L = 40$ and vehicle number $N = 50$.

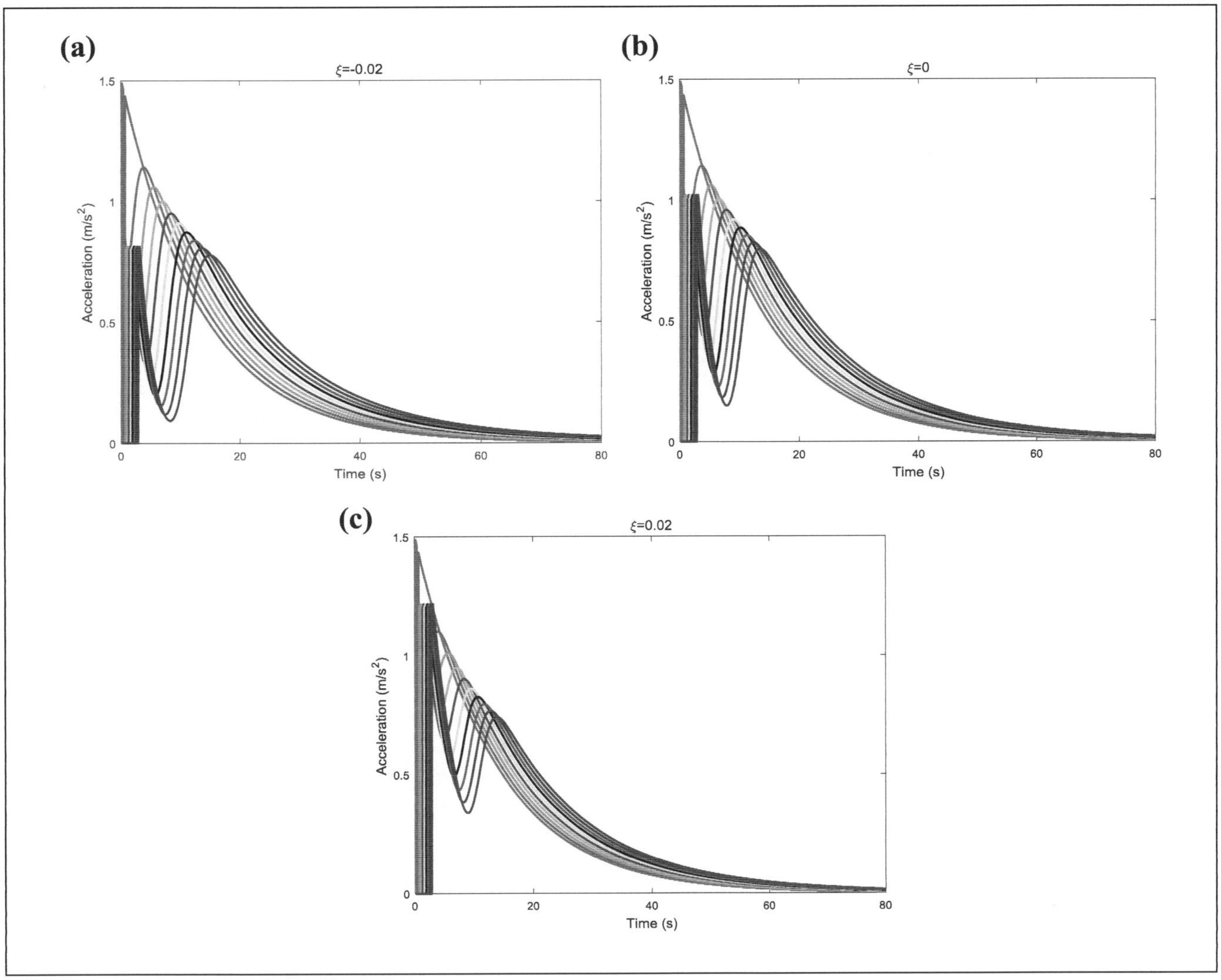

Figure 3. Acceleration profiles of vehicle nos. 1–10, starting from a green signal for LDDSM model at different ξ values. Each curve shows the acceleration of each vehicle.

The time evolution of the velocity is also investigated because a small perturbation is introduced in the leading vehicle by selecting the initial conditions

$$\begin{cases} x_1(0) = L \cdot N, v_1(0) = 20, \dot{v}_1(0) = 0 \\ \dot{v}_1(t) = \varepsilon_1(\hat{t}), t \geq \hat{t} \\ x_n(0) = L \cdot (N - n + 1), v_n(0) = 20, \dot{v}_n(0) = 0, n = 2, \cdots, N \end{cases}, \tag{25}$$

where $x_n(0)$, $v_n(0)$, and $\dot{v}_n(0)$ are the initial position, velocity, and acceleration of the n th vehicle in the system at time 0, respectively; and $\varepsilon_1(\hat{t})$ is a small acceleration fluctuation of the first vehicle in the system after time delay $\hat{t}$, which follows the uniform random distribution $U(-1, 1)$ with amplitude 5×10^{-2}.

Effect of Disturbance Level on Traffic Flow Stability. In this subsection, the influence of the disturbance level on traffic flow stability is investigated. The sensitivity coefficient is set as $\alpha = 25$ m/s^2 and delay time as $\tau = 0.4$ s. Figure 6(a), (c), and (e) illustrate the spatial–temporal pattern of vehicular traffic flow from 0 s to 1000 s under different disturbance levels. Figure 6(b), (d), and (f) display the variations of velocities of the 1st, 25th, and 50th vehicles after 800 s under three disturbance levels of 0, 0.02, and −0.02, respectively. The results suggest that the degree of headway fluctuation is smaller in Figure 6(a) compared with those in Figures 6(c) and 6(e). Figure 6(a)–(d) demonstrate that the traffic flow fluctuation is not ebbing but is growing under the disturbance levels of 0 and 0.02. Moreover, the amplitude of velocity

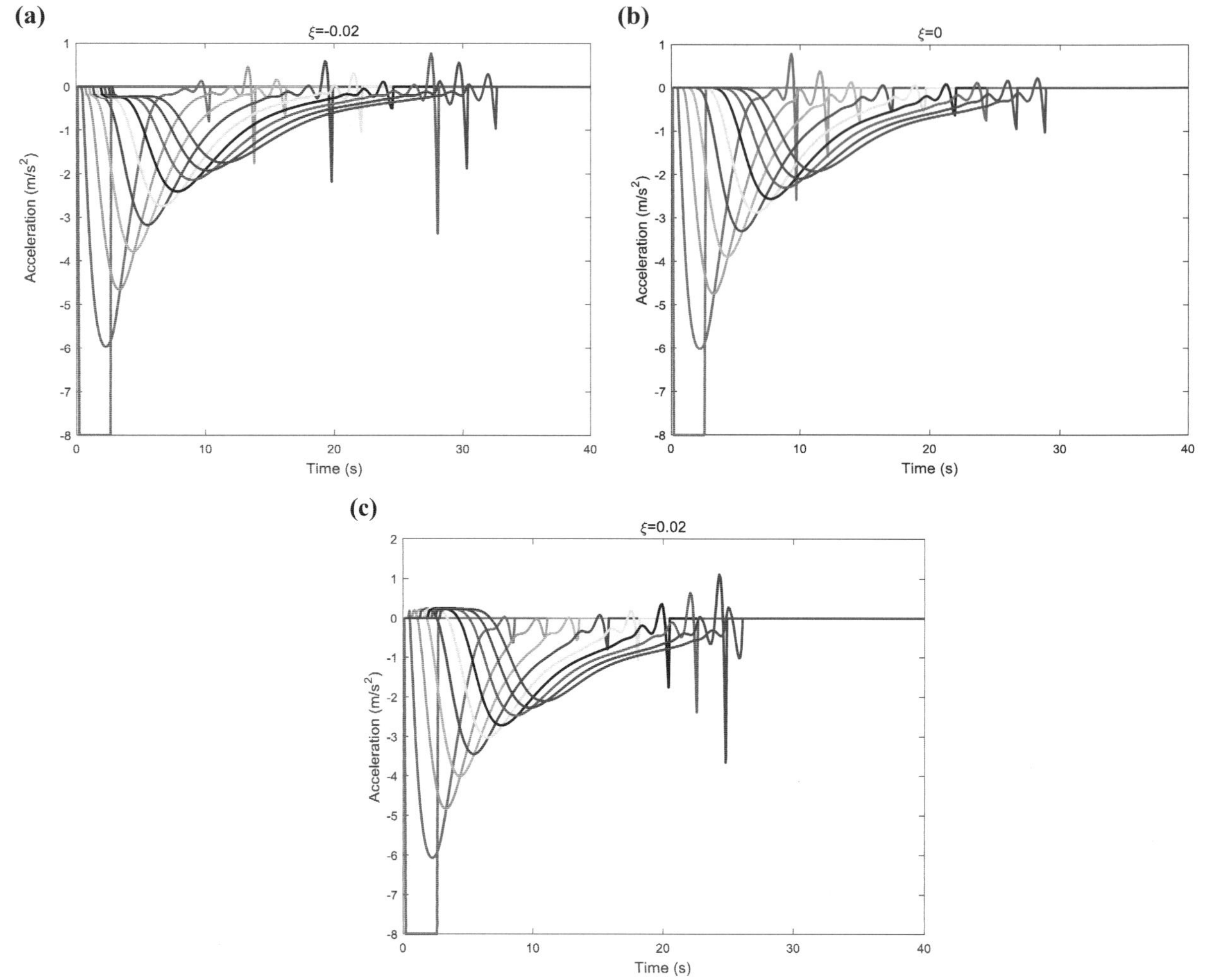

Figure 4. Deceleration profiles of vehicle nos. 1–10 in an emergency for LDDSM model at different ξ values. Each curve shows the acceleration of each vehicle.

fluctuation increases gradually as the vehicle number increases and the stop-and-go wave occurs. Figure 6(*e*) and (*f*) proves that when the stability conditions are satisfied (Figure 1) for the LDDSM model with the disturbance level of −0.02, no oscillating behavior and traffic congestion occur and the vehicles almost run constantly with the initial velocity.

In summary, these simulations show that disturbance level significantly affects traffic flow stability, and a smaller rather than a larger perceived safety margin is more useful in stabilizing traffic flow. Results of the simulation suggest that the disturbance of vehicle information exerts an important influence in traffic flow stability.

Performance of the New Control Strategy in Stabilizing Traffic Flow. In this subsection, the performance of the sliding-mode control strategy in stabilizing traffic flow is investigated. Based on the above analysis, traffic flow is unstable when the disturbance level of the system is 0.02, and the traffic flow remains unstable without disturbance in the vehicular system. The designed sliding-mode controller is introduced in the vehicular system, as shown in Equation 23. The parameters of the sliding-mode controller are $k_1 = 0.01$, $\rho = 0.1, \gamma = 0.2$, and $k = 0.05$, where k is the adjustable gain of sliding-mode controller.

The snapshots of the vehicles' headways profiles under different disturbance levels at $t = 500$ s and $t = 1000$ s are shown in Figures 7–9. These figures demonstrate that the sliding-mode control strategy adequately stabilizes traffic flow. From Figures 7(*a*), 8(*a*), and 9(*a*), the fluctuations of the vehicles' headways increase with the number of vehicles under the disturbance levels of 0 and 0.02. Comparing Figures 7(*b*), 8(*b*), and 9(*b*) reveals

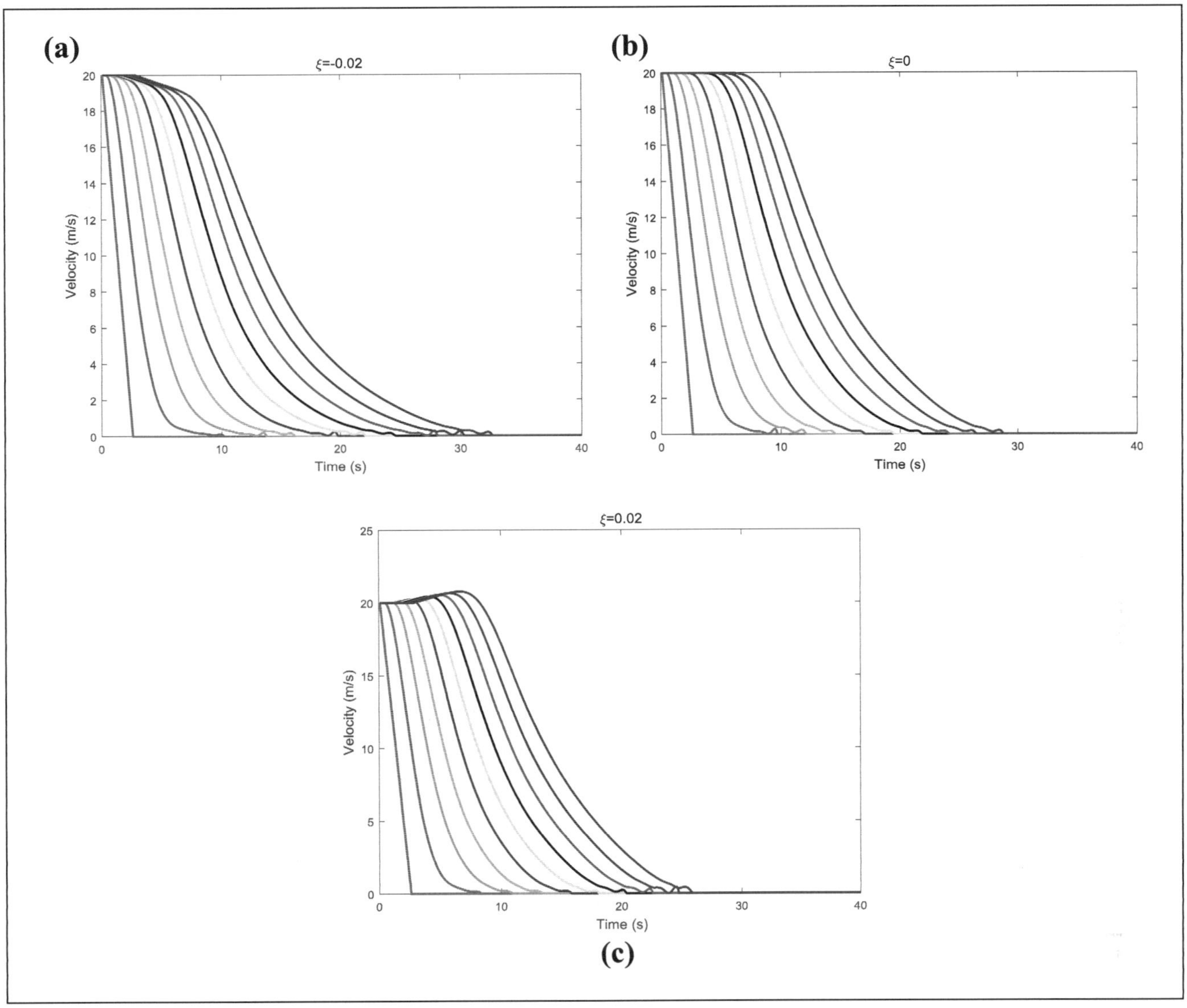

Figure 5. Motions of vehicle nos. 1–10 in an emergency for LDDSM model at different ξ values. Each curve shows the velocity of each vehicle.

that the vehicles' headway profiles fluctuate around the constant headway at 500 s or 1000 s, with a small range between 60.248 m and 61.389 m. This result confirms that the sliding-mode control strategy can improve traffic flow stability and smoothness and increase the spacing between the two vehicles to enhance the safety of car-following in ACC and automotive platoon driving.

The hysteresis loops obtained from the 50th vehicle simulated by the LDDSM model for $\xi = 0$, $\xi = -0.02$, and $\xi = 0.02$ are shown in Figure 10. The hysteresis loops for $\xi = -0.02$ almost degenerates to a point compared to the said loops for $\xi = 0$ and $\xi = 0.02$ in Figure 10(*a*). However, when the sliding-mode control strategy is introduced in the LDDSM model, the hysteresis loops for the three disturbance levels degenerates to a point, as shown in Figure 10(*b*). The hysteresis loops further prove that the traffic flow stability with the sliding-mode control strategy is superior to that obtained without a control strategy. In other words, according to the case studies, unstable traffic flow can be stabilized by the sliding-mode control strategy, whereas stable traffic flow can be further improved. Therefore, simulation results demonstrate that the sliding-mode controller is effective in stabilizing vehicular traffic flow when the disturbance of vehicle information exists in the ACC system.

Conclusions

In a car-following situation, the following vehicle adjusts its speed according to the motion information of the

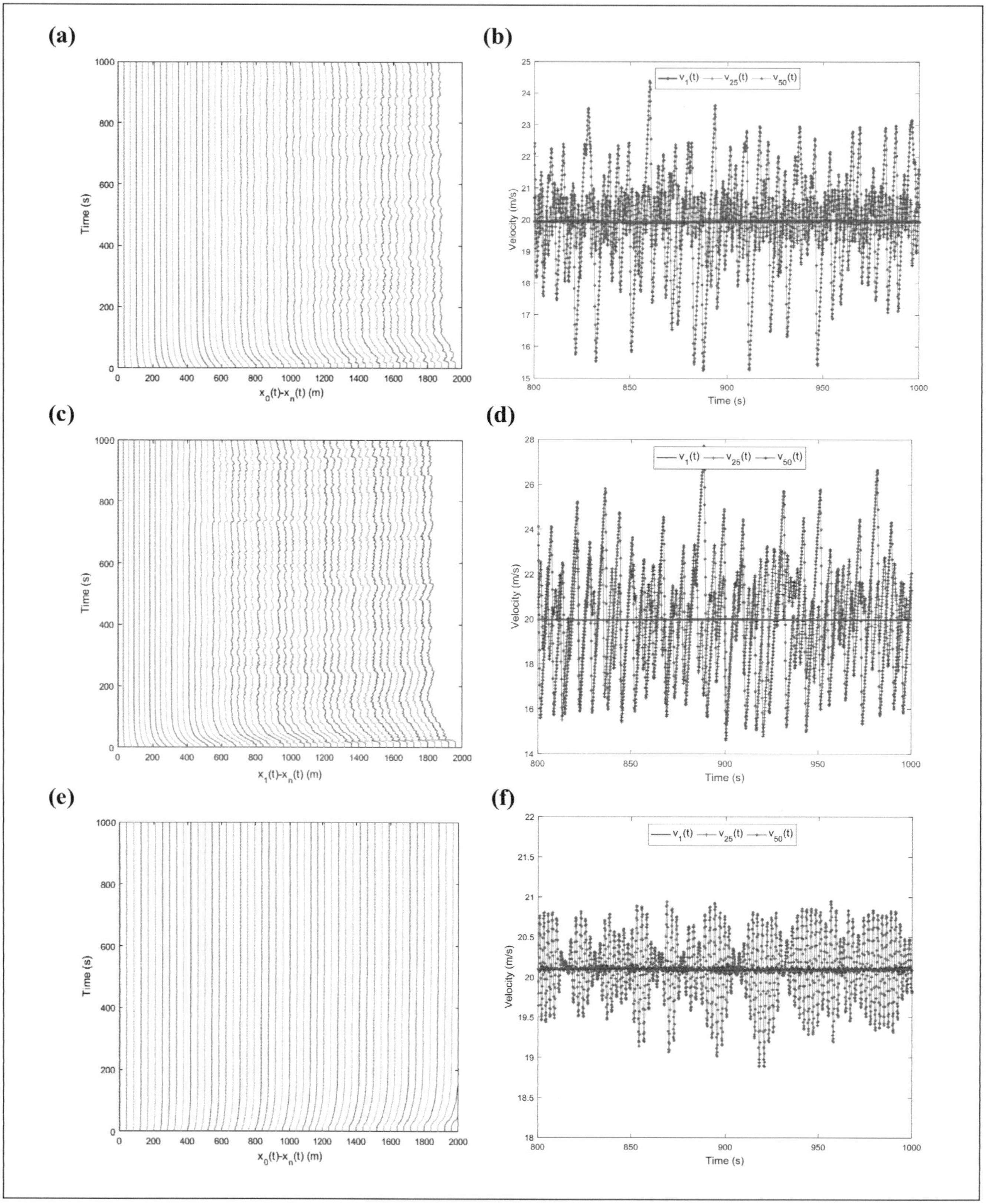

Figure 6. Numerical simulations for the LDDSM model under different the disturbance level. (*a*) Spatial–temporal pattern of vehicular traffic flow without the disturbance level; (*b*) The velocity profiles of the 1st, 25th, and 50th vehicle without the disturbance level; (*c*) Spatial–temporal pattern of vehicular traffic flow with the disturbance level 0.02; (*d*) The velocity profiles of the 1st, 25th, and 50th vehicle with the disturbance level 0.02; (*e*) Spatial–temporal pattern of vehicular traffic flow with the disturbance level −0.02; (*f*) The velocity profiles of the 1st, 25th, and 50th vehicle with the disturbance level −0.02.

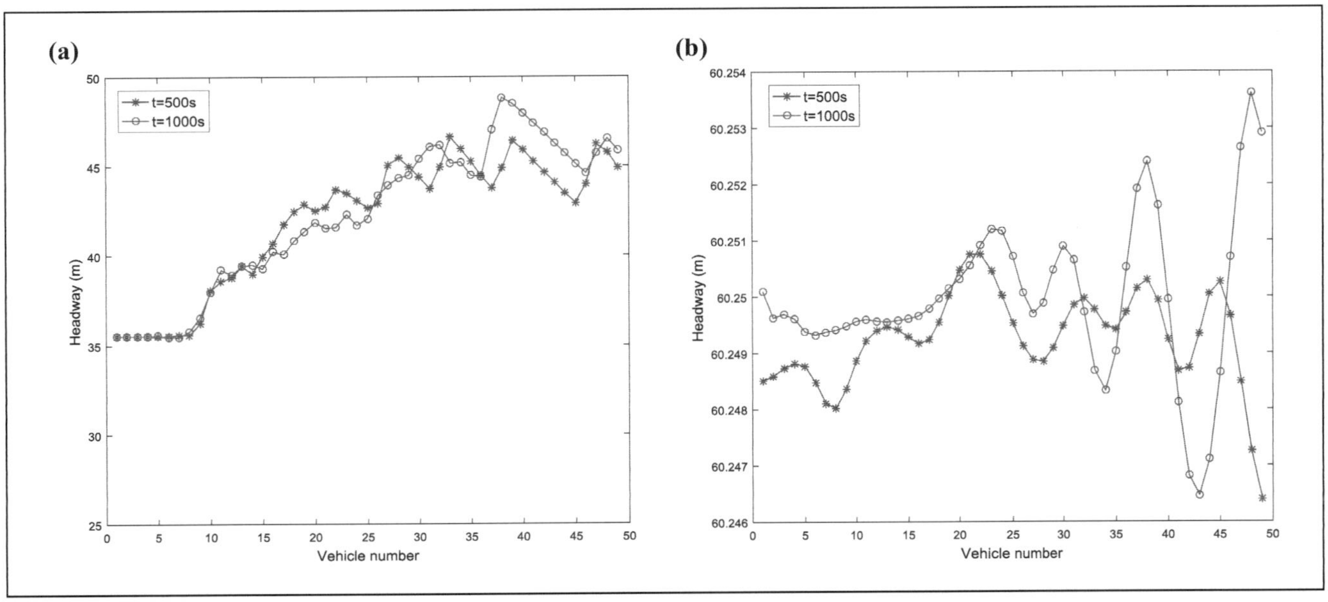

Figure 7. Snapshots of the vehicles' headway profiles under the disturbance level of 0.02 at t = 500 s and t = 1000 s: (a) without the control strategy, and (b) with the sliding mode control strategy.

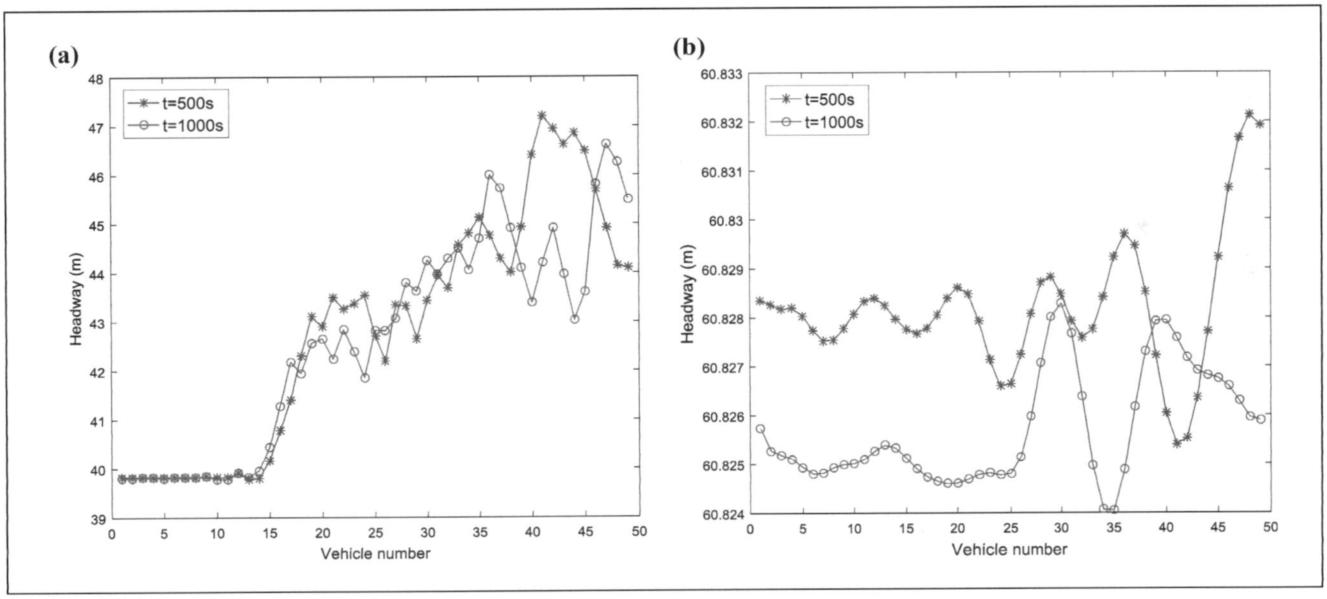

Figure 8. Snapshots of the vehicles' headway profiles without the disturbance level at t = 500 s and t = 1000 s: (a) without the control strategy, and (b) with the sliding mode control strategy.

preceding vehicle (and its own particulars) in terms of position, speed, and acceleration or deceleration in ACC. This information is integrated as a subjective perception of risk, namely, perceived safety margin, in the DSM car-following model based on risk homeostasis theory. Thus, examining the disturbance of vehicle information using the DSM model achieves generality. To explore the influence of the disturbance of vehicle information on car-following performance, the disturbance level is introduced into the perceived safety margin. From an analytical standpoint on linear stability theory, the disturbance level plays an important role in car-following performance. The crucial conclusions of this study can be summarized as follows:

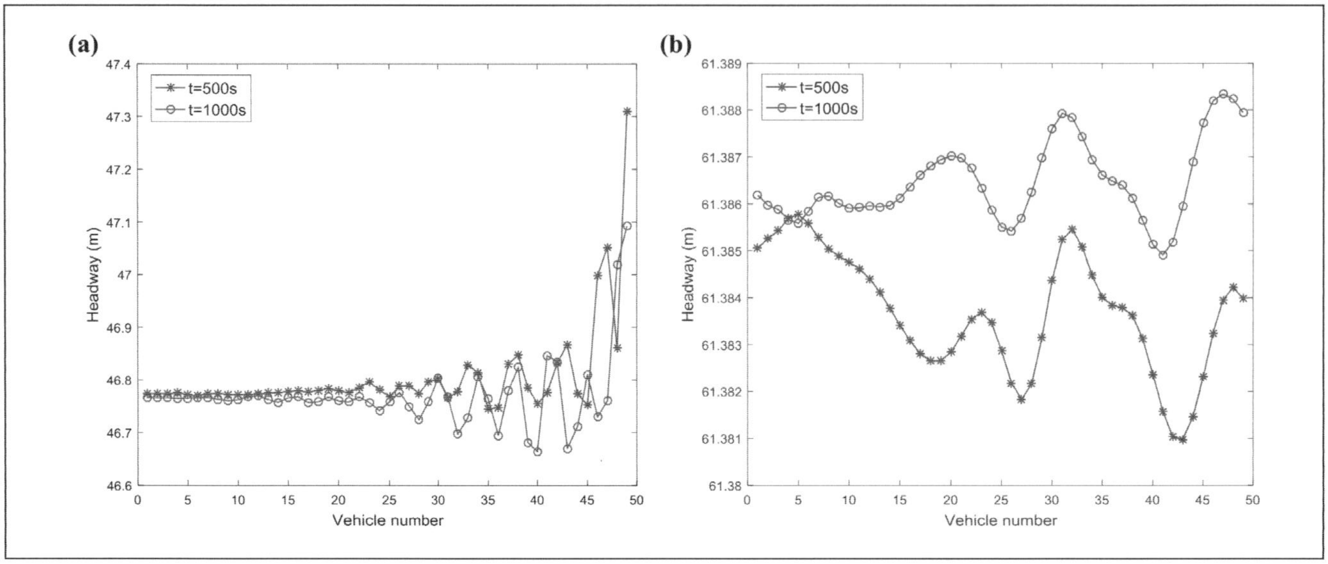

Figure 9. Snapshots of the vehicles' headway profiles under the disturbance level of −0.02 at t = 500 s and t = 1000 s: (*a*) without the control strategy, and (*b*) with the sliding mode control strategy.

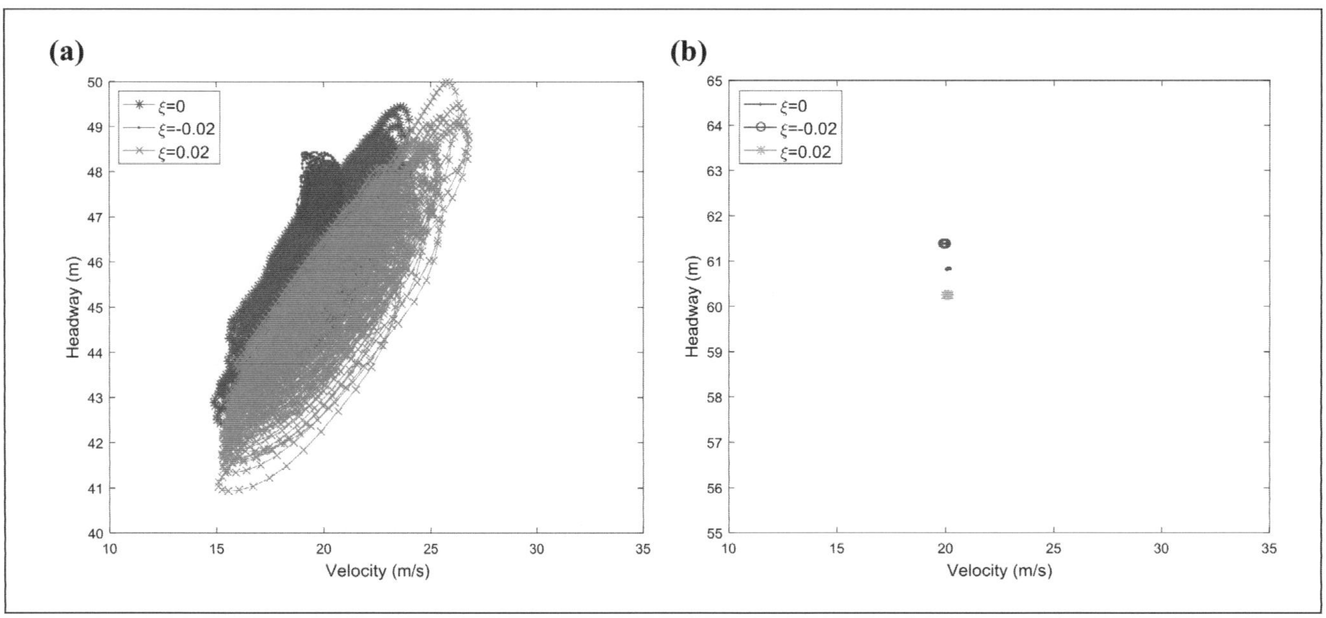

Figure 10. Hysteresis loops obtained from the 50th vehicle simulated by LDDSM model with different disturbance levels: (*a*) without the control strategy, and (*b*) with the sliding mode control strategy.

(1) In the starting process, delay time is minimal under a positive value of disturbance level in the vehicular system. On the contrary, a negative value of the said level leads to considerable delay time. A positive disturbance level in the vehicular information system for ACC may be conducive toward attaining the consensus state and a high acceleration of following vehicles in the starting process. However, the disturbance of vehicle information is not beneficial to the dynamic performance in the emergency braking process regardless of positive or negative value of disturbance level. In addition, the speed control system of the vehicular controller in the platoon requires reliable vehicle information in the emergency braking process.

(2) Given the delay time of the vehicular controller, DSM, and acceleration factor, the traffic flow becomes smoother with the decrease of perceived safety margin. Hence, a negative value of disturbance level can enhance traffic flow stability. By contrast, traffic flow stability weakens with a positive value of disturbance level. Thus, a large value of perceived safety margin is not conducive in stabilizing the traffic flow under the same condition. These results prove that uncertain traffic information plays an important role in traffic flow stability.

Based on the numerical experiments, the influence of the sliding-mode controller on traffic flow stability is further analyzed. The results are as follows:

(1) In comparison with the absence of a control strategy, the use of the sliding-mode controller for vehicular traffic flow can improve traffic flow stability and promote the homogeneity of traffic flow.
(2) Although the disturbance of vehicle information exists in ACC, the control strategy can eliminate the effect of such disturbance on traffic flow stability and can further enhance the said stability.

As for its limitation, this study assigns the same disturbance values for all vehicles. Thus, its outcomes may not apply if the disturbances follow a random distribution. Therefore, this aspect should be considered in future research. Overall, the disturbance of vehicle information influences the vehicle dynamic and traffic flow stability in ACC. Traffic flow stability is improved by constructing the sliding-mode controller. These results are useful in considering the disturbance of the vehicular system for ACC or vehicle platoon control to stabilize traffic flow.

Acknowledgments

This work was supported by the National Natural Science Foundation of China (Grant U1564212, U1664262).

Author Contributions

The authors confirm contribution to the paper as follows: study conception and design: Junjie Zhang, Guangquan Lu, Yunpeng Wang; data collection: Guangquan Lu; analysis and interpretation of results: Guangquan Lu, Junjie Zhang, Yunpeng Wang; draft manuscript preparation: Junjie Zhang. All authors reviewed the results and approved the final version of the manuscript.

References

1. Pipes, L. A. An Operational Analysis of Traffic Dynamics. *Journal of Applied Physics*, Vol. 24, No. 3, 1953, pp. 274–281.
2. Chandler, R. E., R. Herman, and E. W. Montroll. Traffic Dynamics: Studies in Car Following. *Operations Research*, Vol. 6, No. 2, 1958, pp. 165–184.
3. Gazis, D., R. Herman, and R. W. Rothery. Nonlinear Follow-the-Leader Models of Traffic Flow. *Operations Research*, Vol. 9, No. 4, 1900. pp. 545–567.
4. Ranjitkar, P., T. Nakatsuji, and A. Kawamua. Car-Following Models: An Experiment Based Benchmarking. *Journal of the Eastern Asia Society for Transportation Studies*, Vol. 6, 2005, pp. 1582–1596.
5. Bando, M., K. Hasebe, A. Nakayama, A. Shibata, and Y. Sugiyama. Dynamical Model of Traffic Congestion and Numerical Simulation. *Physical Review E Statistical Physics Plasmas Fluids & Related Interdisciplinary Topics*, Vol. 51, No. 2, 1995, p. 1035.
6. Farin, G. E., and D. Hansford. Generalized Force Model of Traffic Dynamics. *Physical Review E*, Vol. 58, No. 1, 1998, pp. 133–138.
7. Jiang, R., Q. Wu, and Z. Zhu. Full Velocity Difference Model for a Car-Following Theory. *Physical Review E*, Vol. 64, No. 1 Pt 2, 2001, p. 017101.
8. Treiber, M., A. Hennecke, and D. Helbing. Congested Traffic States in Empirical Observations and Microscopic Simulations. *Physical Review E Statistical Physics Plasmas Fluids & Related Interdisciplinary Topics*, Vol. 62, No. 2 Pt A, 2000, p. 1805.
9. Li, Y. Car-Following Model Based on the Information of Multiple Ahead & Velocity Difference. *Systems Engineering - Theory & Practice*, Vol. 30, No. 7, 2010, pp. 1326–1332.
10. Zhao, X., and Z. Gao. A New Car-Following Model: Full Velocity and Acceleration Difference Model. *The European Physical Journal B*, Vol. 47, No. 1, 2005, pp. 145–150.
11. Tao, W. Multiple Velocity Difference Model and Its Stability Analysis. *Acta Physica Sinica-Chinese Edition*, Vol. 55, No. 2, 2006, pp. 634–640.
12. Li, Y., D. Sun, W. Liu, M. Zhang, M. Zhao, X. Liao, and L. Tang. Modeling and Simulation for Microscopic Traffic Flow Based on Multiple Headway. *Velocity and Acceleration Difference. Nonlinear Dynamics*, Vol. 66, No. 1–2, 2011, pp. 15–28.
13. Orosz, G., B. Krauskopf, and R. E. Wilson. Bifurcations and Multiple Traffic Jams in a Car-Following Model with Reaction-Time Delay. *Physica D Nonlinear Phenomena*, Vol. 211, No. 3, 2005, pp. 277–293.
14. Orosz, G., and G. Stépán. Subcritical Hopf Bifurcations in a Car-Following Model with Reaction-Time Delay. *Proceedings of the Royal Society A Mathematical Physical & Engineering Sciences*, Vol. 462, No. 2073, 2006, pp. 2643–2670.
15. Zhu, H. B., and S. Q. Dai. Analysis of Car-Following Model Considering Driver's Physical Delay in Sensing Headway. *Physica A Statistical Mechanics & Its Applications*, Vol. 387, No. 13, 2008, pp. 3290–3298.
16. Low, D., and P. Addision. Chaos in a Car-Following Model with a Desired Headway Time. *Proc., 30th International Symposium on Automotive Technology & Automation*, Florence, Italy, ATT/ITS Advances for Enhancing Passenger Freight and Intermodal Transportation Systems, Croydon, England, 1997.

17. Tang, T. Q., C. Y. Li, and H. J. Huang. A New Car-Following Model with the Consideration of the Driver's Forecast Effect. *Physics Letters A*, Vol. 374, No. 38, 2010, pp. 3951–3956.
18. Tang, T., Y. Wang, X. Yang, and Y. Wu. A New Car-Following Model Accounting for Varying Road Condition. *Nonlinear Dynamics*, Vol. 70, No. 2, 2012, pp. 1397–1405.
19. Tang, T. Q., Y. H. Wu, L. Caccetta, and H. J. Huang. A New Car-Following Model with Consideration of Roadside Memorial. *Physics Letters A*, Vol. 375, No. 44, 2011, pp. 3845–3850.
20. Tang, T., W. Shi, H. Shang, and Y. Wang. A New Car-Following Model with Consideration of Inter-Vehicle Communication. *Nonlinear Dynamics*, Vol. 76, No. 4, 2014, pp. 2017–2023.
21. Tang, T. Q., W. F. Shi, H. Y. Shang, and Y. P. Wang. An Extended Car-Following Model with Consideration of the Reliability of Inter-Vehicle Communication. *Measurement*, Vol. 58, No. 11, 2014, pp. 286–293.
22. Chuanyao, L. I., T. Q. Tang, H. J. Huang, and H. Y. Shang. A New Car-Following Model with Consideration of Driving Resistance. *Chinese Physics Letters*, Vol. 28, No. 3, 2011, pp. 38902–38905(38904).
23. Tang, T. Q., J. He, S. C. Yang, and H. Y. Shang. A Car-Following Model Accounting for the Driver's Attribution. *Physica A Statistical Mechanics & Its Applications*, Vol. 413, No. C, 2014, pp. 583–591.
24. Tang, T. Q., L. Chen, S. C. Yang, and H. Y. Shang. An Extended Car-Following Model with Consideration of the Electric Vehicle's Driving Range. *Physica A Statistical Mechanics & Its Applications*, Vol. 430, 2015, pp. 148–155.
25. Li, Y., L. Zhang, S. Peeta, H. Pan, T. Zheng, Y. Li, and X. He. Non-Lane-Discipline-Based Car-Following Model Considering the Effects of Two-Sided Lateral Gaps. *Nonlinear Dynamics*, Vol. 80, No. 1–2, 2015, pp. 227–238.
26. Ge, H. X., H. B. Zhu, and S. Q. Dai. Effect of Looking Backward on Traffic Flow in a Cooperative Driving Car Following Model. *The European Physical Journal B*, Vol. 54, No. 4, 2006, pp. 503–507.
27. Yu, S., and Z. Shi. An Improved Car-Following Model Considering Headway Changes with Memory. *Physica A Statistical Mechanics & Its Applications*, Vol. 421, 2015, pp. 1–14.
28. Peng, G., W. Lu, H. He, and Z. Gu. Nonlinear Analysis of a New Car-Following Model Accounting for the Optimal Velocity Changes with Memory. *Communications in Nonlinear Science & Numerical Simulation*, Vol. 40, 2016, pp. 197–205.
29. Saifuzzaman, M., and Z. Zheng. Incorporating Human-Factors in Car-Following Models: A Review of Recent Developments and Research Needs. *Transportation Research Part C*, Vol. 48, 2014, pp. 379–403.
30. Lu, G., B. Cheng, Y. Wang, and Q. Lin. A Car-Following Model Based on Quantified Homeostatic Risk Perception. *Mathematical Problems in Engineering*, Vol. 2013, No. 6, 2013, pp. 165–185.
31. Konishi, K., H. Kokame, and K. Hirata. Coupled Map Car-Following Model and Its Delayed-Feedback Control. *Physical Review E Statistical Physics Plasmas Fluids & Related Interdisciplinary Topics*, Vol. 60, No. 4 Pt A, 1999, p. 4000.
32. Hu, J. J., Y. Zhang, and T. Zhou. A New Feedback Control Car-Following Model Considering the Steady Desired Velocity Effect. *Icic Express Letters*, Vol. 9, No. 6, 2015, pp. 1603–1608.
33. Zhao, X., and Z. Gao. A Control Method for Congested Traffic Induced by Bottlenecks in the Coupled Map Car-Following Model. *Physica A Statistical Mechanics & Its Applications*, Vol. 366, No. 1, 2006, pp. 513–522.
34. Zhu, W. X., and L. D. Zhang. A Speed Feedback Control Strategy for Car-Following Model. *Physica A Statistical Mechanics & Its Applications*, Vol. 413, No. C, 2014, pp. 343–351.
35. Li, Y., D. Sun, and W. Liu. Feedback Control of Traffic Jam Based on the Full Velocity Difference Car-Following Model. *Journal of Information & Computational Science*, Vol. 9, No. 3, 2012, pp. 719–730.
36. Jin, Y., and M. Xu. Stability Analysis in a Car-Following Model with Reaction-Time Delay and Delayed Feedback Control. *Physica A Statistical Mechanics & Its Applications*, Vol. 459, 2016, pp. 107–116.
37. Zheng, L., T. Zhou, W. Liu, D. Chen, and D. Sun. A New Control Method Integrated into the Coupled Map Car-Following Model for Suppressing Traffic Jams. *Nonlinear Dynamics*, 2016, pp. 1–9.
38. Lu, G., B. Cheng, Q. Lin, and Y. Wang. Quantitative Indicator of Homeostatic Risk Perception in Car Following. *Safety Science*, Vol. 50, No. 9, 2012, pp. 1898–1905.
39. Li, L., G. Lu, Y. Wang, and D. Tian. A Rear-End Collision Avoidance System of Connected Vehicles. *Proc., IEEE 17th International Conference on Intelligent Transportation Systems*, IEEE, Qingdao, China, 2014, pp. 63–68.
40. Chowdhury, D., L. Santen, and A. Schadschneider. Statistical Physics of Vehicular Traffic and Some Related Systems. *Physics Reports*, Vol. 329, No. 4–6, 2000, pp. 199–329.
41. Shi, Y. F., L. C. Yang, S. X. Hao, and Q. Liu. Clustered Car-Following Strategy for Improving Car-Following Stability under Cooperative Vehicles Infrastructure Systems. *IET Intelligent Transport Systems*, Vol. 10, No. 3, 2016, pp. 141–147.
42. Vangi, D., and A. Virga. Evaluation of Emergency Braking Deceleration for Accident Reconstruction. *Vehicle System Dynamics*, Vol. 45, No. 10, 2007, pp. 895–910.
43. Sokolovskij, E. Experimental Investigation of the Braking Process of Automobiles. *Transport*, Vol. 20, No. 3, 2010, pp. 91–95.
44. Li, Y., B. Yang, T. Zheng, Y. Li, M. Cui, and S. Peeta. Extended-State-Observer-Based Double-Loop Integral Sliding-Mode Control of Electronic Throttle Valve. *IEEE Transactions on Intelligent Transportation Systems*, Vol. 16, No. 5, 2015, pp. 2501–2510.
45. Sun, N., Y. Fang, and H. Chen. A New Antiswing Control Method for Underactuated Cranes with Unmodeled Uncertainties: Theoretical Design and Hardware Experiments. *IEEE Transactions on Industrial Electronics*, Vol. 62, No. 1, 2015, pp. 453–465.
46. Lee, J. D., and R. Y. Duan. Cascade Modeling and Intelligent Control Design for an Electromagnetic Guiding

System. *IEEE/ASME Transactions on Mechatronics*, Vol. 16, No. 3, 2011, pp. 470–479.
47. Pisano, A., A. Davila, L. Fridman, and E. Usai. Cascade Control of Pm Dc Drives Via Second-Order Sliding-Mode Technique. *IEEE Transactions on Industrial Electronics*, Vol. 55, No. 11, 2008, pp. 3846–3854.
48. Utkin, V. I., J. Guldner, and J. Shi. *Sliding Mode Control in Electro-Mechanical Systems*, 2nd ed. CRC Press, Boca Raton, 2009.
49. Treiber, M., A. Kesting, and C. Thiemann. *Traffic Flow Dynamics: Data, Models and Simulation*. Springer, Berlin, 2013.
50. *Highway Capacity Manual*. TRB, National Research Council, Washington, D.C., 2000.

The Standing Committee on Traffic Flow Theory and Characteristics (AHB45) peer-reviewed this paper (18-02590).

TRR
JOURNAL OF THE TRANSPORTATION RESEARCH BOARD

Article

Transportation Research Record
2018, Vol. 2672(20) 178–192

DOI: 10.1177/0361198118786823
journals.sagepub.com/home/trr
SAGE

A Macroscopic Multi-Lane Traffic Flow Model for ACC/CACC Traffic Dynamics

Anargiros I. Delis[1], Ioannis K. Nikolos[1], and Markos Papageorgiou[1]

Abstract

An extended second-order macroscopic traffic flow model is presented that describes multi-lane traffic dynamics and also incorporates the effects of adaptive cruise control (ACC) or cooperative ACC (CACC). The extended model equations stem from a recently proposed multi-lane gas-kinetic traffic flow (GKT) model that can simulate lane changes due to vehicle interactions as well as spontaneous ones. The proposed extension that models the effects of ACC/CACC satisfies the time-gap principle of such systems on each lane and allows for consideration of mixed traffic comprising both manual and ACC/CACC vehicles. Numerical simulations are performed for a particular three-lane motorway stretch in the United Kingdom, where recurrent traffic congestion is observed during the morning peak hours, so as to compare the effects of ACC/CACC in the traffic flow conditions with those resulting from manual driving.

Vehicle automation and communication systems (VACS) have the potential to improve traffic operations, safety, driver convenience, and mobility (*1*, *2*). Basic automated vehicle technology has been slowly introduced on high-end production vehicles, for example, driver assist features such as adaptive cruise control (ACC), cooperative ACC (CACC), and lane-changing assist systems, while more advanced automation is primarily being implemented on prototype vehicles that are not yet production-ready. Although some research suggests that autonomous vehicles may enter the market as soon as 2020, with wider market penetration by 2040–2050 (*2*), there still remain many ethical, policy, and logistical hurdles before their implementation becomes feasible (*2*, *3*). Likewise, connected vehicle technology is being explored in parallel and has the advantage of reducing the sensing requirements for each individual vehicle, thus lowering costs and potentially leading to a more rapid deployment (*4*). While these developments are encouraging, there are still questions that remain to be answered, including the extent to which automated and connected vehicles can improve traffic efficiency, safety, and cost (*1*). For both automated and connected vehicle technologies, it is particularly difficult, but at the same time important, to quantify their impact during the phase-in period when traffic flow consists of an evolving mix of existing manual vehicles and vehicles equipped with VACS technology (*5*). Therefore, accurate models that reflect the traffic flow implications of VACS, such as ACC/CACC, are needed to enable realistic predictions of their effects on highway capacity and traffic dynamics.

In an ACC system, a desired speed can be chosen via the driver interface, but when a leading vehicle is detected by its sensors, the vehicle speed is adapted to that of the leading vehicle based on a pre-selected time gap. However, the application of ACC systems, for certain parameter settings, may also induce some negative effects on traffic dynamics and capacity. The next generation of longitudinal control systems is now focused on adding vehicle-to-vehicle (V2V) communications to take advantage of more extensive information from multiple preceding vehicles.

Regarding the modeling of the effects of ACC/CACC systems in traffic flow dynamics, two main approaches can be distinguished in the current literature, namely microscopic (usually car-following) and macroscopic. From the microscopic modeling perspective, research efforts suggest that ACC vehicles may help improve traffic stability and dissipate shock waves, but one cannot draw clear conclusions about their impact on capacity and performance (*6–9*). One of the modeling challenges for ACC systems is that its current implementation is

[1]School of Production Engineering and Management, Technical University of Crete, University Campus, Chania, Crete, Greece

Corresponding Author:
Address correspondence to Ioannis K. Nikolos: jnikolo@dpem.tuc.gr

primarily directed toward driver comfort rather than traffic performance, which creates a discrepancy between the technology's current behavior and its potential capabilities. In terms of microscopic modeling of CACC, recent works suggest that, assuming V2V communication, closer following distances between equipped vehicles are feasible, and thus, significant improvements of traffic flow stability and efficiency at mid-to-high market penetration rates can be realistically expected (*10–12*).

Macroscopic models characterize traffic in terms of aggregate density, space-mean speed, and flow as opposed to microscopic ones that describe interactions between individual vehicles. Essentially, macroscopic models are governed by the continuity equation, which is derived from vehicle conservation and describes the evolution of density based on flow gradients. Recently, second-order models, which include a second partial differential equation to describe how speed changes dynamically as a function of density, local speed, and nonlocal effects, have been introduced to model ACC/CACC systems. Some parameters from this second equation can be used to differentiate the characteristics between manual vehicles and those equipped with ACC and/or CACC systems. Although the literature following this modeling approach is currently rather limited, it has been used to model ACC (*13*, *14*), V2V communication (*15*, *16*), and CACC (*17–19*). Furthermore, it was found (*17–19*) that at 100% penetration, CACC reduced traffic congestion by improving highway capacity and throughput, and attenuated traffic flow instabilities to a greater extent compared with ACC. Although the vast majority of literature following this approach focuses on homogeneous traffic, that is, assuming 100% penetration rate, there exist some very recent attempts that consider varying penetration rates of ACC/CACC systems (*20*, *21*). The results (*20*, *21*) confirm that, if only a fraction of vehicles (around 30–40%) are equipped with CACC, enhanced stabilization of traffic flow occurs, and traffic congestion can be reduced by improving highway capacity and throughput.

In the present work, we extend the ACC/CACC macroscopic modeling approach (*19*, *21*) to a recently developed multi-lane second-order gas-kinetic traffic flow (GKT) model (*22*, *23*). To the best of the authors' knowledge, this is the first time that an ACC/CACC modeling approach has been incorporated in a multi-lane second-order macroscopic model. Thus, the aims of this work are twofold: to develop a numerical simulation tool based on the multi-lane GKT ACC/CACC model and to test its behavior under realistic traffic flow conditions. To this end, a high-resolution relaxation finite-volume scheme is utilized for the numerical approximation, in space and time, of the resulting partial differential equations (*19*, *24*) and a real motorway stretch in the United Kingdom, part of the M56 motorway, is utilized to compare the effect of ACC/CACC on the traffic flow conditions compared with those resulting from manual driving, for various penetration rates.

The Multi-Lane ACC/CACC GKT Model

We present a novel extension of the multi-lane second-order GKT model, recently developed (*22*), so as to incorporate the effects of ACC/CACC. The GKT model (*22*) is able to simulate multi-lane traffic flow dynamics, based on a nonlinear system of conservation laws with additional source/sink terms that take into account lane changes due to vehicle interactions as well as spontaneous ones. Assuming a highway with N lanes, which are numbered by $l = 1, 2, \ldots, N$, (with $l = 1$ denoting the slowest lane) the extended ACC/CACC multi-lane GKT model proposed here is written for each lane, l, with initial conditions as

$$\begin{aligned} &\partial_t\rho_l + \partial_x(\rho_l u_l) = r_{rmp}\delta_{l1} + w_l^1; \\ &\partial_t(\rho_l u_l) + \partial_x\left(\rho_l u_l^2 + \theta_l\rho_l\right) = h_{rmp}\delta_{l1} + w_l^2 \\ &\quad + \left(\frac{\rho_l V_{e,l}^* - \rho_l u_l}{\tau_l}\right)[1 - pF_l(\rho_l)] + p\mathcal{V}_{(c)acc,l}; \\ &\rho_l(x,0) = \rho_{l,0}(x); u_l(x,0) = u_{l,0}(x),\ l = 1, 2, \ldots, N, \end{aligned} \tag{1}$$

where, for each lane l,

ρ_l = traffic density;
u_l = average speed of vehicles;
r_{rmp} = effective source density, from on/off-ramps;
$w_l^{1,2}$ = lane-changing source terms;
θ_l = pressure-like term;
h_{rmp} = effective source term from on/off-ramps;
$V_{e,l}^*$ = dynamic equilibrium speed;
τ_l = relaxation time for manual driving;
p = ACC/CACC penetration rate;
$F_l(\rho_l)$ = Fermi function; and
$\mathcal{V}_{(c)acc,l}$ = ACC/CACC model source terms.

Further, we denote as $q_l = \rho_l u_l$ the traffic flow rate (number of vehicles per unit of time). In the momentum (flow) equation we introduce here the modeling of ACC/CACC vehicles via terms $[1 - pF_l(\rho_l)]$ and $p\mathcal{V}_{(c)acc,l}$, which will be explained later. For $p = 0$ the original multi-lane GKT model for manual cars (*22*) is recovered. In the modeling equations, we have the pressure-like term $\theta_l = A_l(\rho_l)u_l^2$, with $A_l(\rho_l)$ given by the Fermi function as

$$A_l(\rho_l) = A_{0,l} + \delta A_l\left[1 + \tanh\left(\frac{\rho_l - \rho_{cr,l}}{\delta\rho_l}\right)\right], \tag{2}$$

in which $\rho_{cr,l}$ is the critical density in the l-th lane, reflecting the boundary for the transition from free-flow to congested traffic; $A_{0,l}$ and $A_{0,l} + 2\delta A_l$ are the variance pre-factors between the aforementioned two states; and $\delta\rho_l$ is a transition width. Typical ranges of values for the constant parameters $A_{0,l}$, δA_l, and $\delta\rho_l$, along with other typical used model parameters for the GKT model are available (*21–24*).

Terms r_{rmp} and h_{rmp} model the incoming (outgoing) traffic flow from on-ramps (or to off-ramps) on the first (slowest) lane. Specifically, the term r_{rmp} denotes the effective source density that is only active within the merging (diverging) sections of the first lane, with length l_{rmp} and inflow $q_{rmp}>0$ from (or outflow $q_{rmp}<0$ to) the ramps, determined as

$$r_{rmp}(x,t) = \begin{cases} \frac{q_{rmp}(t)}{l_{rmp}} & |\textit{if } x \textit{ inside the merging zone,} \\ 0 & |\textit{ elsewere,} \end{cases} \tag{3}$$

while the term h_{rmp} describes changes of the macroscopic local speed by assuming that on-ramp vehicles merge to the main road at speed $u_{rmp}<$u. The drivers considered to leave the main road reduce their speed to u_{rmp} before they diverge to the off-ramp. Therefore this term is expressed as

$$h_{rmp}(x,t) = \frac{q_1 \cdot r_{rmp}}{\rho_1} + \frac{(u_{rmp} - u_1)|q_{rmp}|}{l_{rmp}}. \tag{4}$$

The model also includes a relaxation term that maintains the concentration of velocity in equilibrium state at each lane, with $V^*_{e,l}(\rho_l, u_l, \rho_{\alpha,l}, u_{\alpha,l})$ being the dynamic equilibrium speed, with relaxation time τ_l, depending not only on the local density ρ_l and mean speed u_l, but also on the non-local traffic state $(\rho_{\alpha,l}, u_{\alpha,l})$. Thus, the non-local and dynamic equilibrium speed, toward which the average speed relaxes, is determined as

$$V^*_{e,l} = u_{max,l}\left[1 - \frac{\theta_l + \theta_{\alpha,l}}{2A_l(\rho_{max,l})}\left(\frac{\rho_{\alpha,l}T_l}{1 - \rho_{\alpha,l}/\rho_{max,l}}\right)^2 B(\delta u_l)\right]. \tag{5}$$

where, for each lane l,

$u_{max,l}$ = maximum desired speed;
T_l = desired time gap for manual driving;
$\rho_{max,l}$ = maximum density;
$\rho_{a,l}$ = density at interaction location $x_{\alpha,l}$; and
$B(\delta u_l)$ = Boltzmann interaction factor.

According to Equation 5, for manual driving, the dynamic equilibrium speed is given by the maximum desired speed $u_{max,l}$, reduced by a braking non-local term in response to necessary deceleration maneuvers in traffic flow at the downstream interaction location $x_{\alpha,l} = x_l + \gamma_l(1/\rho_{max,l} + T_l \cdot u_l)$, with γ_l being a scale factor. Finally, $B(\delta u_l)$ contains the standard normal distribution and the Gaussian error function, given as

$$B(\delta u_l) = 2\left[\delta u_l \frac{e^{-\delta u_l^2/2}}{\sqrt{2\pi}} + (1 + \delta u_l^2)\int_{-\infty}^{\delta u_l} \frac{e^{-y^2/2}}{\sqrt{2\pi}}dy\right]. \tag{6}$$

This term describes the dependence of the braking interaction on the dimensionless velocity difference $\delta u_l = (u_l - u_{a,l})/\sqrt{\theta_l + \theta_{\alpha,l}}$, taking into account the velocity and variance at the actual location x_l and the interaction location $x_{\alpha,l}$, respectively.

Further, according to model Equation 1 for multi-lane traffic, the source terms $w_l^{1,2}$ represent the sources and sinks due to lane-changing, resulting in a weakly coupled system of $2N$ equations for $v_l^1 = \rho_l$ and $v_l^2 = q_l$. These lane-changing terms that represent interactive lane changes ("overtaking maneuvers") are defined as

$$\begin{aligned} w_l^{1,2} = &\left(\frac{1}{T_{l-1}^L}v_{l-1}^{1,2} - \frac{1}{T_l^R}v_l^{1,2}\right)(1 - \delta_{l1}) \\ &+ \left(\frac{1}{T_{l+1}^R}v_{l+1}^{1,2} - \frac{1}{T_l^L}v_l^{1,2}\right)(1 - \delta_{lN}), \end{aligned} \tag{7}$$

with $\frac{1}{T^L}$ and $\frac{1}{T^R}$ being the lane-changing rates from lane l to lane $l+1$ and $l-1$, respectively, and δ_{ij} the Kronecker delta. The lane-changing rates are defined as

$$\begin{aligned} \frac{1}{T_l^L} &= P_L(\rho_{l+1})\sigma(\rho_l) + S_l^L, \\ \frac{1}{T_l^R} &= P_R(\rho_{l-1})(1 - P_L(\rho_{l+1}))\sigma(\rho_l) + S_l^R, \end{aligned} \tag{8}$$

where,

$P_{R,L}(\rho_l)$ = lane-changing probabilities in response to vehicle interactions;
$\sigma(\rho_l)$ = interaction frequencies regarding deceleration and acceleration; and
$S_l^{L,R}$ = spontaneous lane-changing terms.

We set $\sigma(\rho_l) = \sigma_f(1 - \rho_l)\rho_l^2$ and, for simplicity, we assume that $P_R(\rho_l) = P_L(\rho_l)$. The shape of $P_{R,L}(\rho_l)$ depends on a characteristic density value (ρ_P), while that of $\sigma(\rho_l)$ depends on the value of σ_f. More details on these functions are available (*22, 23*).

Furthermore, we assume here that the spontaneous lane changes, not caused by vehicle interactions, which are described by the terms $S_l^{R,L}$, can be formulated as

$$S_l^{L,R} = k_l^{L,R}\left(1 - \frac{\rho_{l\pm1}}{\rho_{max,l\pm1}}\right)^{\beta}, \tag{9}$$

where

$k_l^{R,L}$ = spontaneous lane-changing parameters at the l-th lane; and
β = spontaneous lane-changing parameter.

For the spontaneous lane-changing terms in Equation 9, we adopt the European rule of primarily using the slowest lane at low densities (*22, 23, 25, 26*). Calibration results have shown that spontaneous lane-changing influences mainly low-density regimes. Setting $\beta = 8$ in Equation 9, a smooth correction pre-factor G_{Eu}, with $0 < G_{Eu}(\rho) < 1$, is used to account for the European traffic rule by modifying S_l^L as $S_l^L G_{Eu}$ and S_l^R as S_l^R / G_{Eu} (*22, 25*). It is noted here that in this modeling approach there are, in general, more interactive and spontaneous lane changes at relatively small densities (at about 20–30% of $\rho_{max,l}$), while for lower density values the lane-changing events drop rapidly.

The terms ($[1 - pF_l(\rho_l)]$ and $p\mathcal{V}_{(c)acc,l}$) in the momentum Equation 1 are related to the proposed extension of the multi-lane model, so as to incorporate the effects of ACC/CACC in traffic dynamics. This proposed extension is based on the necessary control objectives that ACC systems should adopt, and applies similarly to CACC systems, as they have similar control objectives, with the important difference being in their desired time gaps. These objectives are, according to Shladover et al (*10*):

- *Speed control mode:* to travel close to the pre-set (by the driver) maximum velocity, in cases where no leading vehicles are detected by the sensor systems, or leading vehicles are detected but their speeds are higher than the driver-defined maximum velocity.
- *Gap control mode:* to adjust the velocity of the equipped vehicle to the velocity of the leading vehicle to maintain the user-defined desired gap, in cases that the leading vehicle is identified by the sensor systems and its velocity is lower than the pre-set by the user maximum speed.

The two aforementioned control modes are required to operate with robust, stable, and smooth transitions between them, to avoid any discomfort to the passengers caused by sudden, abrupt changes in velocity. Moreover, the proposed model is induced by the usual control policy for ACC (and in extension CACC) systems, the "Constant Time Headway policy," where the desired inter-vehicle distance is proportional to the vehicle's velocity (*27*). To satisfy simultaneously the aforementioned objectives, at densities explicitly below a critical threshold $\rho_{acc,l}$ (being lower than or equal to the critical density $\rho_{cr,l}$) the additional terms to the multi-lane GKT model have no influence, since it is assumed that the drivers react by setting their maximum desired speeds as in manual driving. At densities around $\rho_{acc,l}$ a smooth but fast transition between the previous manual case and the ACC/CACC model is established, using the Fermi function:

$$F_l(\rho_l) = 0.5\left[1 + tanh\left(\frac{\rho_l - \rho_{acc,l}}{\Delta\rho_l}\right)\right] \quad (10)$$

where

$\rho_{acc,l}$ = critical density for ACC/CACC activation at the l-th lane; and

$\Delta\rho_l$ = transition width for ACC/CACC driving at the l-th lane.

When the gap control mode is activated, the desired constant time gap T^* is imposed through its corresponding influence on a desired lane density ρ_l^* that is given as:

$$\rho_l^* = \frac{1}{1/\rho_{max,l} + T^* u_l^*} \quad (11)$$

where

$1/\rho_{max,l}$ = vehicle length at the l-th lane.

The denominator reflects the desired lane space headway, and $u_l^* = u_l(x_l^*)$ is the velocity of the preceding vehicle, computed at "interaction" position

$$x_l^* = x_l + \gamma_l^*\left(1/\rho_{max,l} + T^* u_l\right),\ \gamma_l^* \in [1, 2]. \quad (12)$$

where

T^* = desired time gap for ACC/CACC; and

γ_l^* = anticipation factor for ACC/CACC at the l-th lane.

The formulation above reflects that, after a relaxation time τ_l^*, the desired speed relaxes to the speed of the preceding vehicle u_l^*. Accordingly, the corresponding source term that reflects the behavior of ACC systems can be expressed mathematically as

$$V_{acc,l}(\rho_l, u_l, \rho_l^*, u_l^*) = F_l(\rho_l)\left(\frac{\rho_l^* u_l^* - \rho_l u_l}{\tau_l^*}\right) \quad (13)$$

where

τ_l^* = relaxation time for ACC at the l-th lane; and

u_l^* = velocity of the preceding vehicle, computed at "interaction" position x_l^*.

Following ISO 15622, 2010 standards, indicated values for ACC-equipped vehicles should be $T^* \in [0.8, 2.2]$s, while $\tau_l^* \approx 1$s.

Further, to model the impact of CACC vehicles in traffic flow dynamics, a similar approach is utilized, with the only difference being that the vehicle can exchange information with M multiple preceding vehicles, each of which induces a different relaxation time. As a result, the corresponding source term is modified as in Equation 14 by taking into consideration the velocities of the respective preceding vehicles. This feature, for exchanging information from farther downstream, enables a considerable reduction in the size of time gaps, without compromising safety. Thus, the corresponding source term of the momentum equation for lane l takes the form

$$V_{cacc,l}\left(\rho_l, u_l, \rho_l^*, u_{i,l}^*\right) = F_l(\rho_l)\sum_{i=1}^{M}\left(\frac{\rho_l^* u_{i,l}^* - \rho_l u_l}{\tau_{i,l}^*}\right) \quad (14)$$

Table 1. Variables/Terms Definition, and Calibrated Optimal Model Parameters and Cost Function Value for the Manual Driving Case

$A_{0,l}$	Variance pre-factor for free traffic in lane l
$F_l(\rho_l)$	Fermi-like function
h_{rmp}	Effective source term for the flow equation, from on/off-ramps
$k_i^{L,R}$	Spontaneous lane-changing parameters
l_{rmp}	Ramp length
p	ACC/CACC penetration rate
$P_{R,L}(\rho_l)$	Lane-changing probabilities in response to vehicle interactions
q_{rmp}	Inflow/outflow, from/to on/off-ramps
r_{rmp}	Effective source density, from on/off-ramps
$S_l^{L,R}$	Spontaneous lane-changing terms
T_l	Desired time gap in lane l
T^*	Desired gap for ACC/CACC
$1/T^{L,R}$	Lane-changing rates
u_l	Speed in lane l
$u_{\alpha,l}$	Speed at location $x_{\alpha,l}$ in lane l
$u_{max,l}$	Maximum desired speed in lane l
u_{rmp}	Exiting or entering ramp speed
u_l^*	Preceding vehicles speed at position x_l^* in lane l
$\mathcal{V}_{(c)acc,l}$	ACC/CACC source terms in lane l
$w_l^{1,2}$	Lane-changing source terms
$x_{\alpha,l}$	Downstream interaction location in lane l
x_l^*	Downstream interaction location for ACC/CACC in lane l
β	Spontaneous lane-changing parameter
γ_l	Anticipation factor in lane l
γ_l^*	Anticipation factor for ACC/CACC in lane l
δA_l	Pre-factor in lane l
$\delta\rho_l$	Transition width for manual driving in lane l
$\Delta\rho_l$	Transition width for ACC/CACC driving in lane l
ρ_l	Density in lane l
$\rho_{cr,l}$	Critical density in lane l
$\rho_{\alpha,l}$	Density at location $x_{\alpha,l}$ in lane l
$\rho_{acc,l}$	Critical density for ACC/CACC activation in lane l
$\rho_{max,l}$	Maximum density in lane l
ρ_P	Characteristic density value for lane-changing probabilities
ρ_l^*	Desired lane density for ACC/CACC in lane l
$\sigma(\rho_l)$	Interaction frequencies regarding deceleration and acceleration
σ_f	Interaction frequencies parameter
τ_l	Relaxation time for manual driving in lane l
$\tau_l^*, \tau_{i,l}^*$	Relaxation times for ACC and CACC, respectively

	u_{max}	ρ_{max}	ρ_{cr}	T	γ	τ	A_0	δA	$\delta\rho$	σ_f	ρ_P	$\left(\frac{u_{rmp}}{u_1}\right)$	k_1^L	k_2^R	k_2^L	k_3^R
Lanes	$\frac{km}{h}$	$\frac{veh}{km}$		(s)		(s)	–	–	$\frac{veh}{km}$	–	–	–		$\frac{Events}{sec \cdot lane}$		
Lane1	96	140	30	1.4	1.5	28	0.009	0.01	20	0.44	0.02	0.76	0.003	0.007	0.001	0.04
Lane2	118	142	59	2	1.5	23	0.0065	0.032	10							
Lane3	128	140	60	2	1.5	21	0.0064	0.032	14							
Cost Function %: 0.029																

where $u_{i,l}^* = u_l\left(x_{i,l}^*\right)$ with $x_{i,l}^* = x_l + i \cdot \gamma_l^* \ (1/\rho_{max,l} + T^* \cdot u_l)$, $i = 1, \ldots, M$, $\tau_{i,l}^*$ being the relaxation times for CACC, and

$$\rho_l^* = \frac{1}{1/\rho_{max,l} + T^* u_{1,l}^*}. \quad (15)$$

Indicated values for CACC systems, used in this work, are $M = 3$, while $\left[\tau_{1,l}^*, \tau_{2,l}^*, \tau_{3,l}^*\right] = [2, 3, 6]$.

All model terms and parameters are summarized in Table 1. To numerically approximate the ACC/CACC multi-lane GKT model, an explicit conservative higher-order finite-volume relaxation scheme is applied (*19, 24*). This finite volume scheme is based on a fifth-order weighted essential non-oscillatory-type reconstruction for the spatial discretization, while for the time integration a third-order implicit-explicit Runge-Kutta method is utilized. The numerical scheme is stable under the

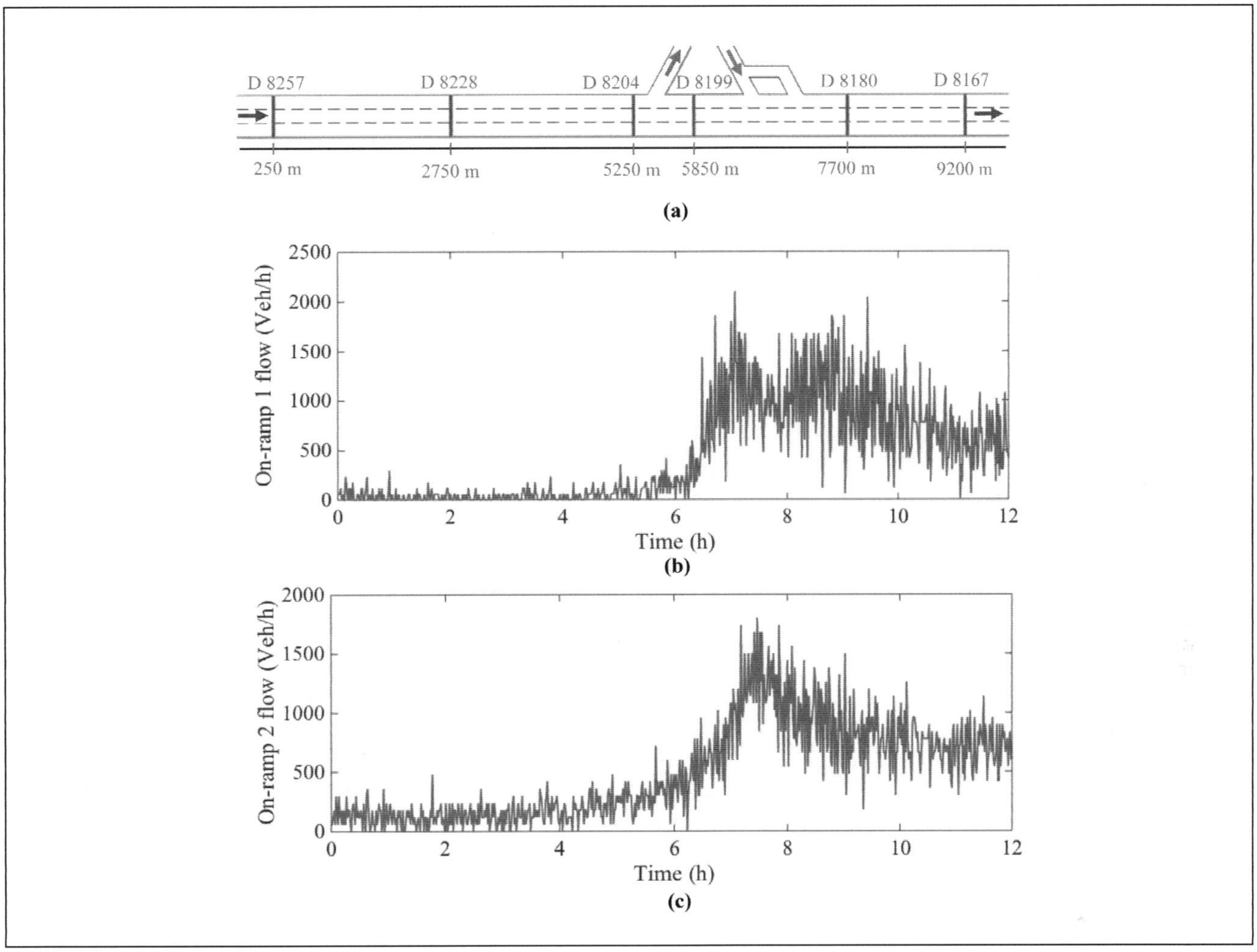

Figure 1. A graphical representation of the UK's M56 motorway stretch considered (*a*) and the measured on-ramp flows (*b, c*).

usual Courant-Friedrichs-Lewy (CFL) stability condition for explicit discretization schemes. The accuracy (compared with low-order schemes), simplicity, and robustness of this higher-order scheme, for numerical approximation of various second-order traffic flow models has been recently demonstrated, and all the details of the numerical scheme and its implementation are available (*19*).

Test Network and Traffic Data

The multi-lane GKT model for manual driving had been previously applied to a particular three-lane motorway stretch in the United Kingdom, and its parameters have been calibrated under recurrent traffic flow conditions (*23*). The multi-lane model requires for different parameter values to be introduced for each lane. Specifically, the model's parameter vector is $\boldsymbol{X} = [u_{max,l}, \rho_{max,l}, \rho_{cr,l}, T_l, \gamma_l, \tau_l, A_{0,l}, \delta A_l, \delta\rho_l, (u_{rmp}/u_1), \nu_f, \rho_P, k_1^L, k_2^R, k_2^L, k_3^R]$, $l = 1,2,3$, (34 parameters in total), which were calibrated to minimize the deviation between simulated and observed traffic data, reflected via an appropriate cost function.

The considered network, presented in Figure 1a, is a 9.45-km-long motorway stretch and is part of the M56 motorway, with direction from Chester to Manchester. The official speed limit is 70 mph (113 km/h). The stretch includes one off-ramp and a two-lane on-ramp, which, before merging into the mainstream, splits into two separate lanes (Figure 1*a*). The measured traffic data at six detectors (D-locations in Figure 1*a*), as provided by MIDAS database (*28*), corresponds to speed and flow measurements per lane with a time resolution of 60 s. Performing a qualitative analysis of real-time traffic data for several days, a recurrent congestion, originated from high on-ramp flows during the morning peak hours, was observed. The real measurements, used for the model calibration for manual driving, were collected on June 3,

2014. Figure 1, *b* and *c*, presents the measured on-ramp flows for the calibration date. Traffic dynamics were simulated for 2.5 morning peak hours (i.e. from 6:30 a.m. to 9 a.m.), using in the numerical scheme a space discretization of $\Delta x = 100$ m and a CFL value equal to 0.4.

A recently developed parallel differential evolution algorithm, combined with two artificial neural networks serving as surrogate models, has been applied to solve the demanding optimization problem and provide the optimal values of the relatively large number of design variables (*23*). The optimal model parameters and the minimum value of the cost function, for the manual driving case, are given in Table 1 (*bottom*).

Figure 2 displays the space-time evolution of the simulated speed, contrasted with the observations for the calibration day. The spatiotemporal contour plots shown in Figure 2 (*top*) display the space-time evolution of the real speed measurements and it can be observed that traffic congestion is formed upstream of the second on-ramp between 7 and 8 a.m.

Next, and by keeping the optimized parameters fixed, a series of simulations is presented for the extended ACC/CACC model, for various penetration rates, to study the effect of the introduction of ACC/CACC cars on the traffic dynamics. The values $\rho_{acc,l} = 0.9\rho_{cr,l}$ and $\Delta\rho_l = 0.006$ were used in the transition function (Equation 10), while $\gamma_l^* = 1.0$ was used by Milanes and Shladover (*12*). For brevity, we present results only with time gap $T^* = 1.2$s for ACC and $T^* = 1.0$s for CACC. These values correspond to more conservative driving behavior, compared with lower values that correspond to a more aggressive one. In general, for V2V CACC vehicles the additional information from more than one preceding vehicle allows for the use of lower values of time gaps, without compromising safety (*10, 11*).

Figure 3 presents the spatiotemporal evolution of speed for different values of penetration rate *p* of ACC vehicles in the system. The time and space extent of the congestion caused by the on-ramps' inflows for manual driving is progressively reduced with increasing ACC-equipped vehicles' penetration rate, though not completely disappearing, even for the case with 100% penetration rate of ACC vehicles.

Similarly, Figure 4 presents the spatiotemporal evolution of speed for different values of penetration rate *p* of CACC vehicles in the system. In this case, the time and space extent of the congestion is more rapidly reduced with increasing penetration of CACC vehicles in the flow. Even for a penetration rate of CACC vehicles equal to 40% the induced congestion begins to disappear, while for higher values of the penetration rate the traffic returns to almost free-flow, avoiding the traffic breakdown.

Since our study concerns a multi-lane model, it is important to investigate the potential effect that the introduction of ACC/CACC vehicles has on the lane changes performed during the simulation. Figure 5*a* presents the evolution of the lane-changing events per hour per km between adjacent lanes for manual driving, while Figure 5*b* contains the corresponding simulation results for ACC vehicles for $p = 60\%$. The on-ramps' presence at a position around 6.5 km has a clear effect on the lane changes from lane 1 to lane 2 and from lane 2 to lane 3, as to distribute the incoming flow to all three lanes. The presence of the congestion has an effect of increasing lane changes at the congestion front, as this propagates upstream. The decrease of the congestion extent with the presence of ACC vehicles is manifested also in the lane changes between adjacent lanes at the congestion front. For the manual case, an increased rate of lane changes is observed downstream of the congestion region. The same is observed for the 60% ACC case, albeit with a smaller extent of the congestion and a lower rate of lane changes, compared with the manual case. The latter observation will be commented on later in this section.

Figure 6*a* presents the evolution of the lane-changing events per hour per km between adjacent lanes for the CACC case with 60% penetration rate, while Figure 6*b* contains the corresponding results for CACC vehicles for $p = 100\%$. The CACC vehicles' presence has the effect of decreasing the lane changes from lane 1 to lane 2 and from lane 2 to lane 3, induced from the on-ramps' inflow. As the CACC vehicles' penetration rate increases, the lane-changing events are decreasing. Although the congestion has almost disappeared for these two cases, an increased number of lane-changing events can be observed downstream of the bottleneck, because of the increased flow rate (capacity), but for a much shorter time period, compared with the manual and ACC cases.

Finally, Figure 7*a* contains the distribution of net lane-changing events for manual driving, while Figure 7, *b* and *c*, presents the distribution of net lane-changing events for 40% CACC and 100% CACC penetration rates. In addition, Figure 7*d* shows the variation of the net lane-changing events (between adjacent lanes) for the complete stretch and time duration of the simulations, from lane 1 to lane 2 (L1 to L2) and from lane 2 to lane 3 (L2 to L3), for the different values of ACC/CACC penetration rates. The increase of the corresponding penetration rates produces a decrease to the lane-changing events, with the effect in the CACC case being more pronounced. For example, for the 100% CACC case, the lane-changing events from lane 2 to lane 3 are reduced by almost 25%, compared with the manual case.

The lane-changing modeling approach consists of two mechanisms, namely the interactive lane-changing (Equations 7 and 8) and the spontaneous lane-changing

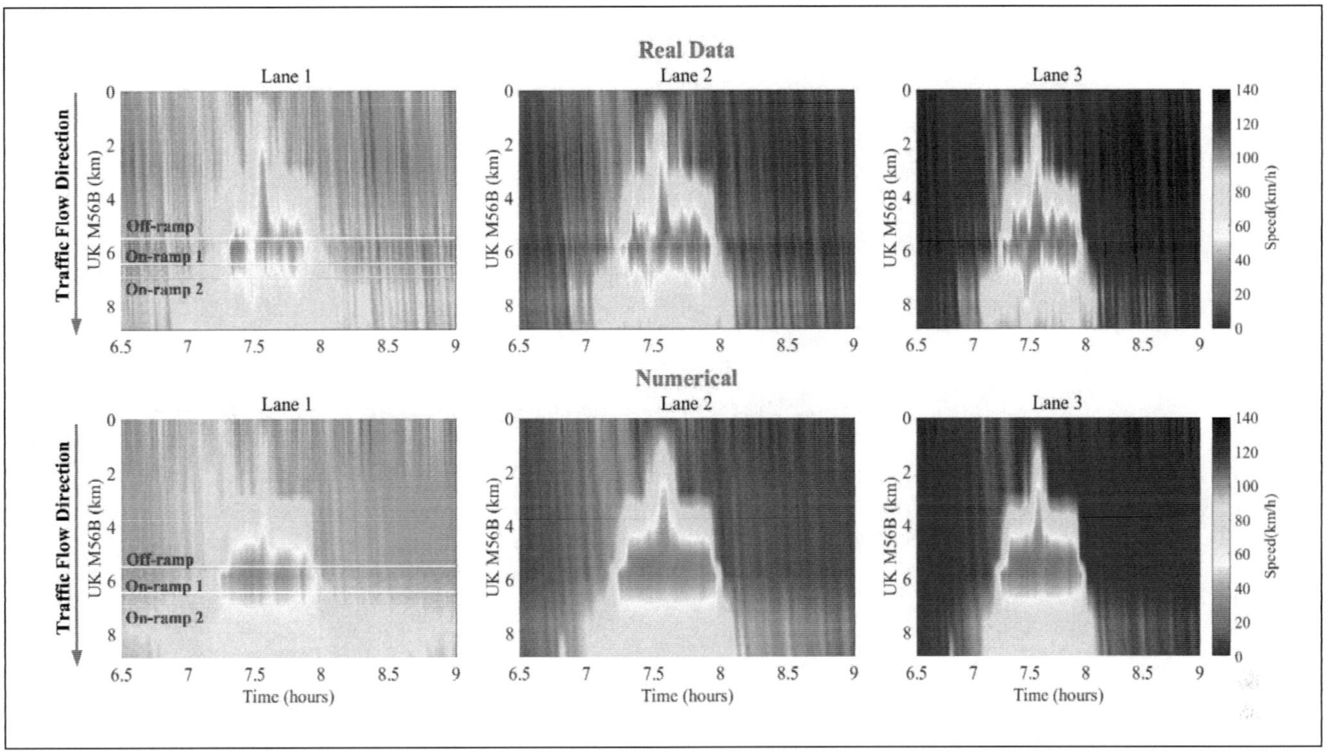

Figure 2. Phase space speed dynamics of manual cars for real measured speed (*top*) and multi-lane model prediction (*bottom*).

(Equation 9), which follows the European driving rule. The interactive lane-changing mechanism tends to send the vehicles toward the fastest lane, while the spontaneous one tends to send the vehicles toward the slowest lane, both at relatively small densities (at about 20–30% of $\rho_{max,l}$). At the congestion front, for manual driving, and referring to Figure 7*a*, high values of lane-changing events are observed, toward the slowest lane, mainly due to the spontaneous lane-changing mechanism (as smaller values of density have been observed in the real measurements in the slowest lane). In the on-ramps region, strong lane-changing maneuvers take place toward the fastest lane, to distribute the incoming flow across all three lanes. The mechanism that implements this effect is the interactive lane-changing, in this case.

When ACC/CACC vehicles are introduced, a reduction of the spontaneous lane-changing events (toward the slowest lane direction) at the (reduced) congestion front takes place (see Figure 7, *b* and *c*, for net lane-changing events and also Figures 5 and 6 for lane changes toward both directions). This is a realistic behavior, resulting from the increased homogenization of the flow for all lanes, especially for CACC vehicles, and the reduction of the congestion extent. Further, this is supported by the reduction of the density upstream of the on-ramps to an average of 10% of $\rho_{max,l}$ (compared with 20–30% in the manual driving case), as a result of which, the lane-changing events drop rapidly upstream of the on-ramps-induced bottleneck. Moreover, when ACC/CACC vehicles are introduced, a reduction in lane-changing events is observed in the on-ramps region, toward the opposite direction (toward the fastest lane) (see Figure 7, *b* and *c*, for net lane-changing events, and also Figures 5 and 6, for lane changes toward both directions). This lane-changing reduction is attributed to the increase of the slowest lane capacity, with increasing ACC/CACC penetration rates, which reduces the need for lane-changing toward the fastest lane.

In Figures 5 and 6 and regarding lane-changing events from lane 2 to lane 1 and from lane 3 to lane 2, some peaks appearing at $x = 0$km are most probably due to the inlet boundary conditions, imposed by using the measured data for manual traffic. During the peak hours, an increase in the inflow is observed, while the oscillations of the measured data (especially their differences between adjacent lanes) may be the reason for high spontaneous lane changes, which cause the resulting peaks. Further downstream on the motorway, the model predictions are smoother than the measured ones, and the differences between adjacent lanes become less pronounced, leading to lower values of spontaneous lane changes. The U-shaped increase just upstream of the bottleneck (Figures 5 and 6) is due to the upstream-moving congestion front. The lane changes from lane 2 to lane 1 (slower lane) and

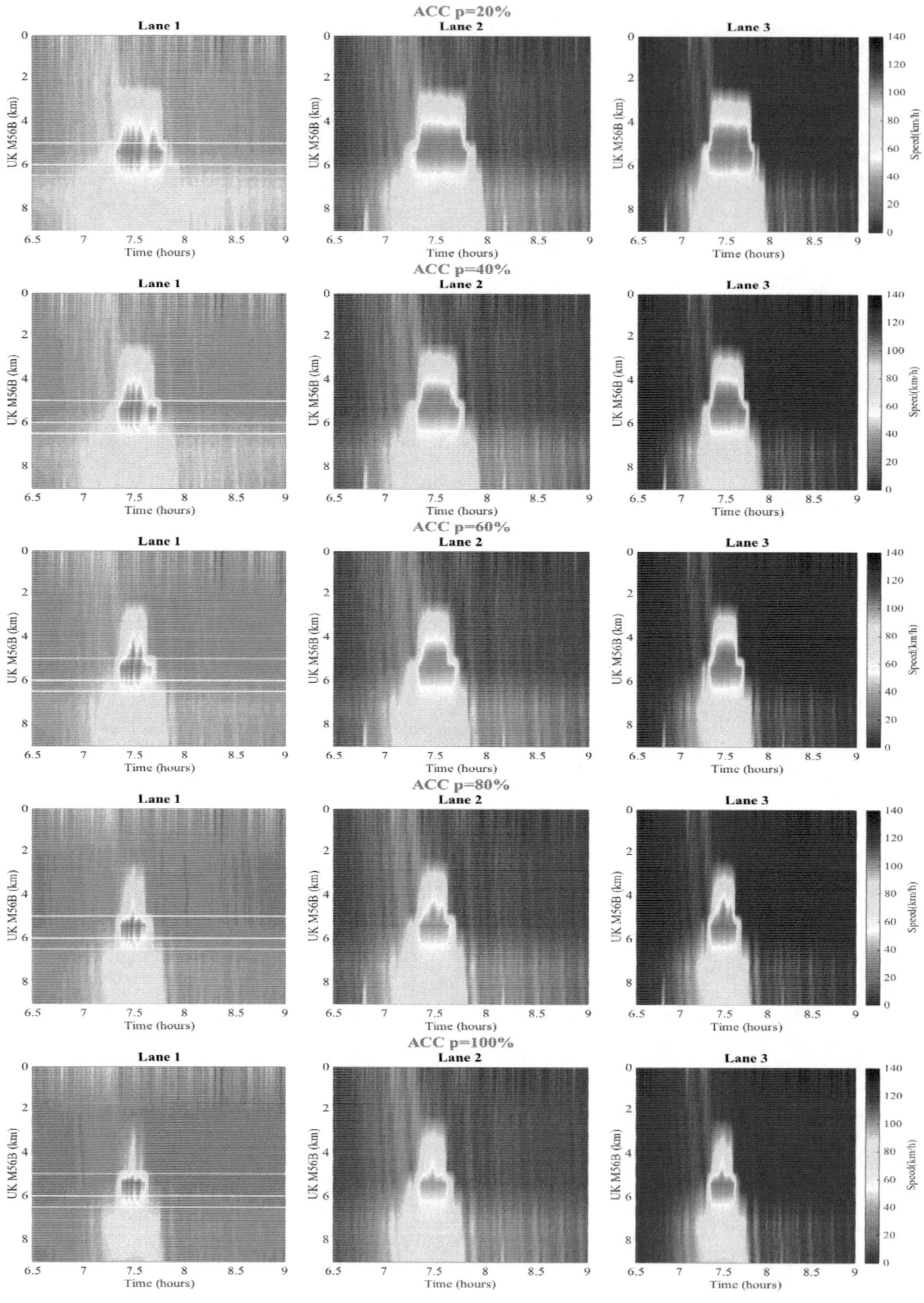

Figure 3. Phase space speed dynamics for multi-lane model prediction for different values of ACC penetration rate.

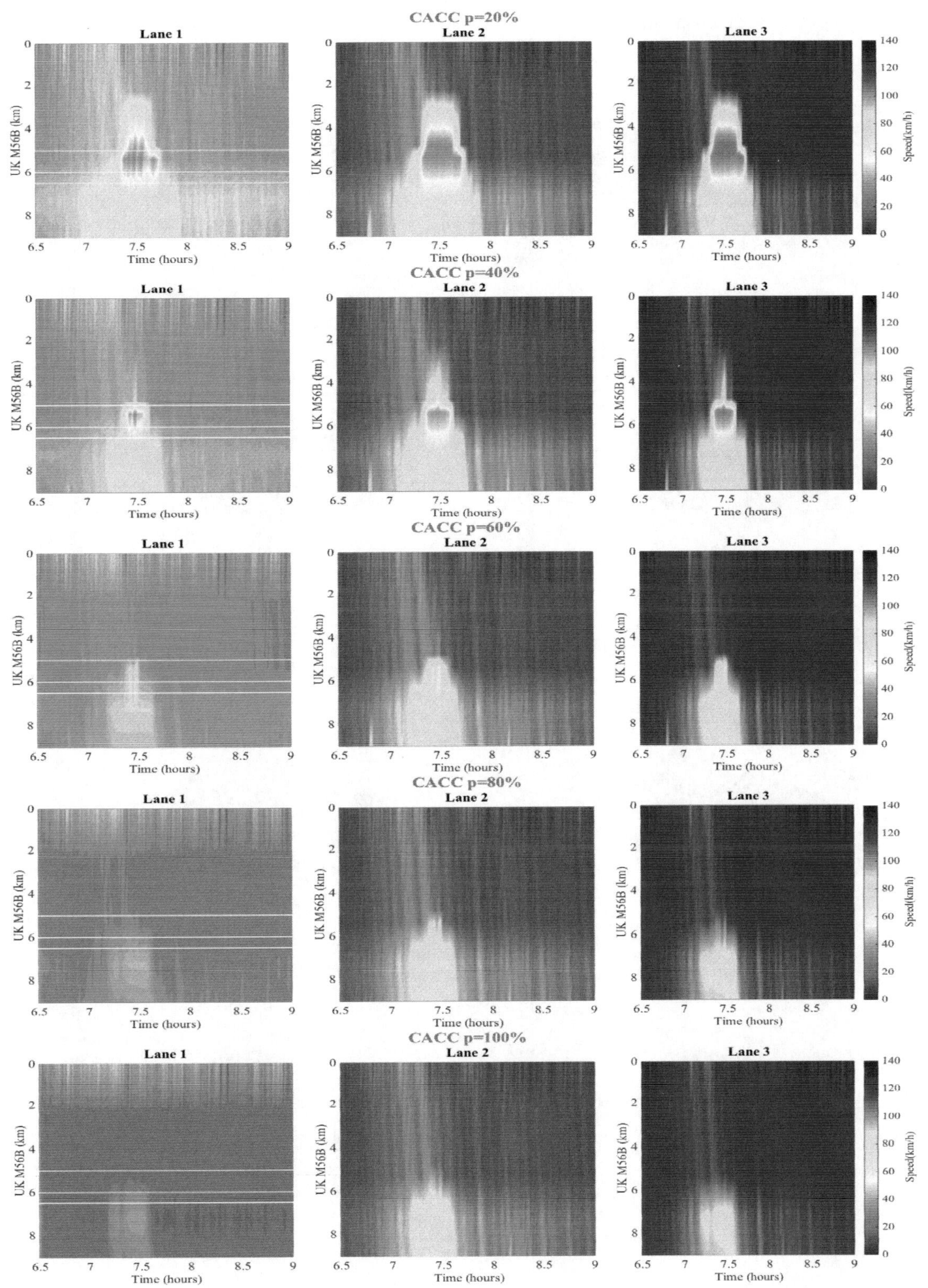

Figure 4. Phase space speed dynamics for multi-lane model prediction for different values of CACC penetration rate.

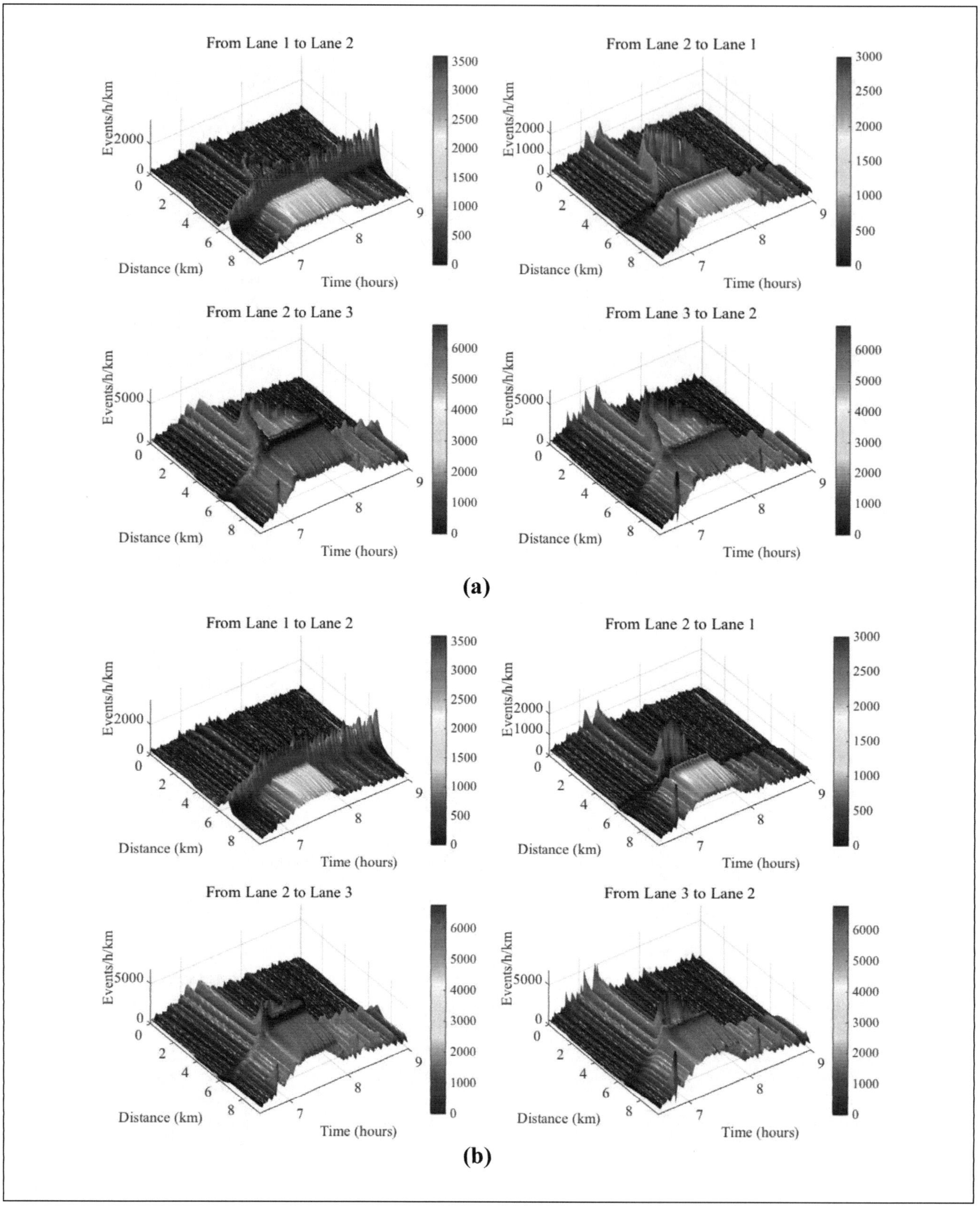

Figure 5. Spatiotemporal distribution of lane-changing events between lanes (*a*) for manual driving and (*b*) for *p* = 60% ACC.

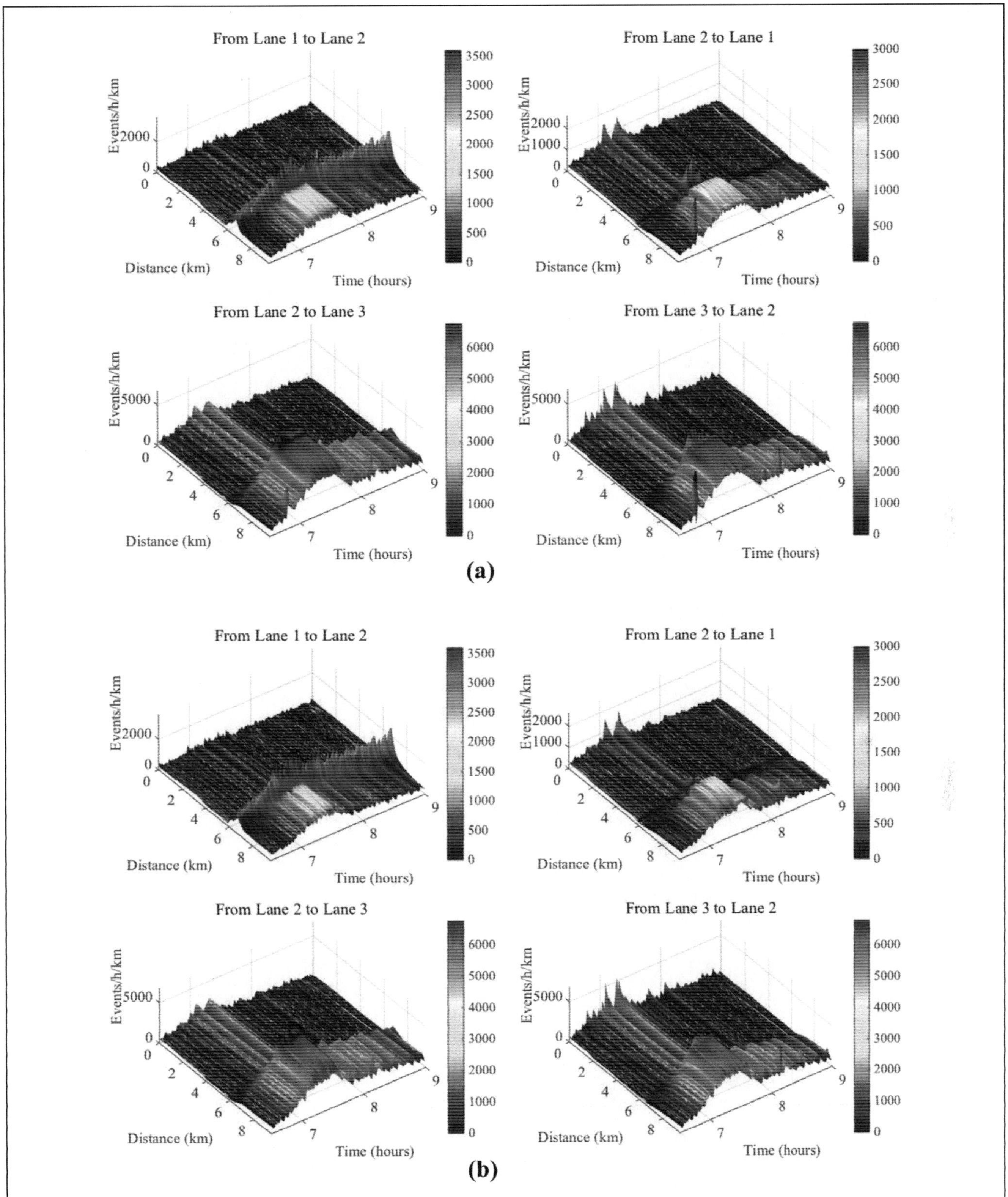

Figure 6. Spatiotemporal distribution of lane-changing events between lanes: (*a*) for *p* = *60%* CACC and (*b*) for *p* = *100%* CACC.

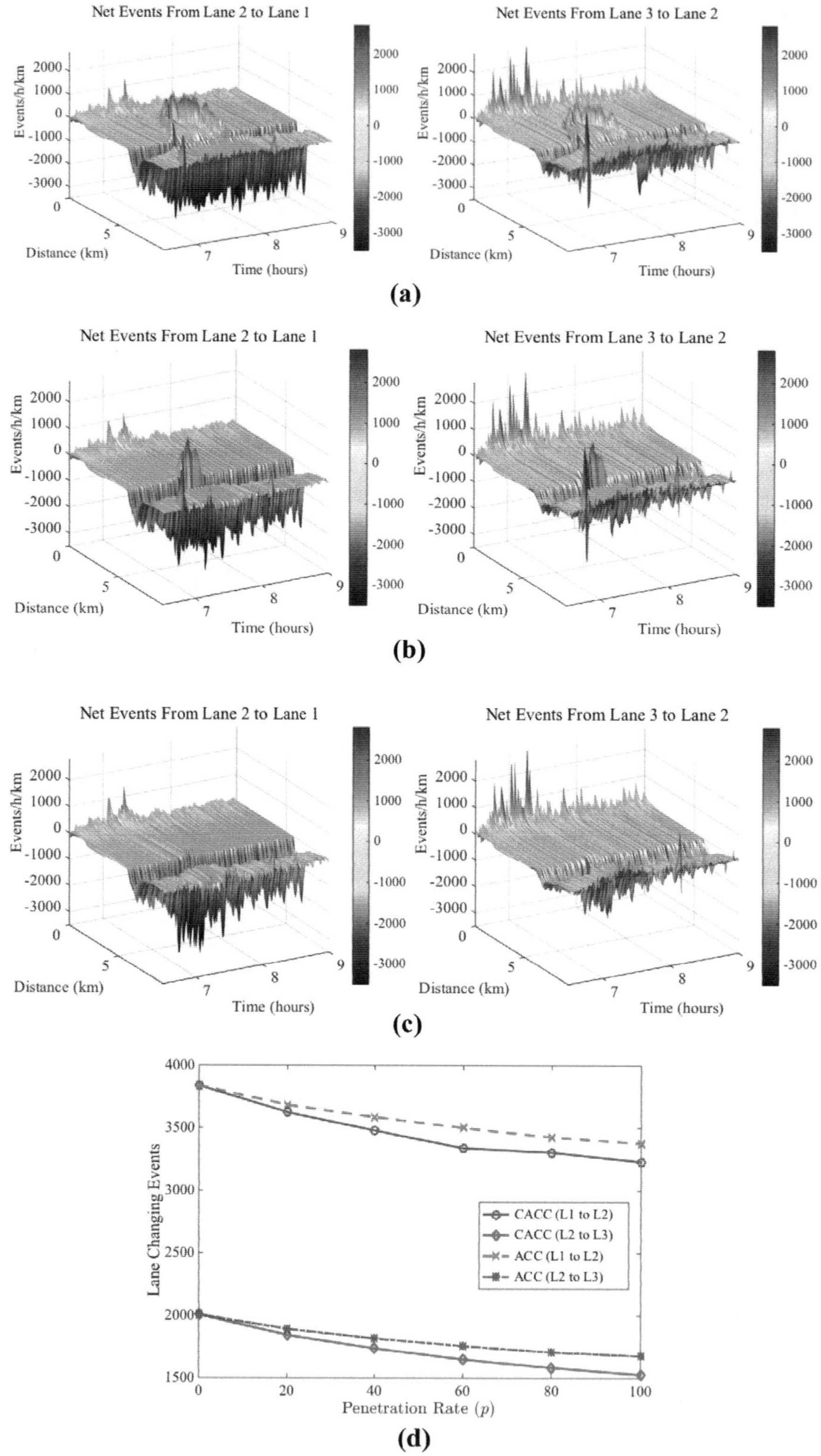

Figure 7. Spatiotemporal distribution of net lane-changing events: (*a*) for manual driving, (*b*) 40% CACC, and (*c*) 100% CACC; (*d*) the effect of ACC/CACC penetration rates (%) on the aggregated net lane-changing events, for the complete stretch and time duration of the simulations.

from lane 3 to lane 2 are generally lower than those in the opposite directions. Such lane changes are due to the combined effect of both interactive and spontaneous lane-changing terms, which produce vehicle movements to both directions, even at congested flow conditions.

Conclusions

A novel extension of a macroscopic multi-lane second-order GKT model has been presented, enabling potential modeling of the effects of ACC/CACC in traffic dynamics at different penetration rates. This extension is based on the introduction of proper relaxation terms in the momentum equations (for each lane) that satisfy the time-gap principle of ACC systems. While in the case of an ACC system the relaxation time is only related to the leading vehicle, it is distributed over multiple vehicles in the CACC system. A numerical investigation has been performed to assess the impact in traffic dynamics by the introduction of such systems. To this end, a real motorway network has been utilized, where a recurrent congestion has been observed and real traffic measurements are available for manual driving. The core parameters of the multi-lane GKT model for manual driving, which were previously calibrated and validated, have been kept fixed, and simulations have been performed for various penetration rates for the ACC/CACC systems under the assumption of a constant time-gap setting, different for each system. The numerical simulations suggest that, even at high penetration rates, ACC systems may not suffice to totally eliminate potential congestions at a bottleneck situation; however, they obviously reduce the extent of the resulting congestion. On the other hand, and due to a dynamical increase of the local capacity near a bottleneck, CACC systems, even at relatively low penetration rates, manage to slow down congestion propagation and duration; while at mid-to-high penetration rates, jam avoidance can be achieved, thus influencing traffic operations. ACC/CACC systems may influence the lane changes performed during an entire simulation; such lane changes can be reduced up to 25%. The presented modeling approach can be further extended to incorporate variable time-gap setting (set by different groups of drivers) as well as other in-homogeneities present in real life.

Acknowledgments

This research was supported by TRAffic MANagement for the 21st century (TRAMAN21) ERC Advanced Investigator Grand under the European Union's Seventh Framework Program (FP/2007-20013). The authors are grateful to Highways England for providing access to the MIDAS data.

Author Contributions

The authors confirm contributing to the paper as follows: study conception and design: A. I. Delis, I. K. Nikolos, M. Papageorgiou; data collection: A. I. Delis, I. K. Nikolos; analysis and interpretation of results: A. I. Delis, I. K. Nikolos, M. Papageorgiou; draft manuscript preparation: A. I. Delis, I. K. Nikolos, M. Papageorgiou. All authors reviewed the results and approved the final version of the manuscript.

References

1. Diakaki, C., M. Papageorgiou, I. Papamichail, and I. K. Nikolos. Overview and Analysis of Vehicle Automation and Communication Systems from a Motorway Management Perspective. *Transportation Research Part A: Policy and Practice*, Vol. 75, 2015, pp. 147–165.
2. Lari, A., F. Douma, and I. Onyiah. Self-Driving Vehicle and Policy Implications: Current Status of Autonomous Vehicle Development and Minnesota Policy Implications. *Minnesota Journal of Law, Science & Technology*, Vol. 16, 2015, pp. 735–769.
3. Fagnant, D. J., and K. Kockelman. Preparing a Nation for Autonomous Vehicles: Opportunities, Barriers and Policy Recommendations. *Transportation Research Part A: Policy and Practice*, Vol. 77, 2015, pp. 167–181.
4. *USDOT Connected Vehicle Research Program: Vehicle-to-Vehicle Safety Application Research Plan*. U.S. Department of Transportation, National Highway Traffic Safety Administration, 2011, Technical report, DOT HS 811 373.
5. Mahmassani, H. S. 50th Anniversary Invited Article - Autonomous Vehicles and Connected Vehicle Systems: Flow and Operations Considerations. *Transportation Science*, Vol. 50, 2016, pp. 1140–1162.
6. Marsden, G., M. McDonald, and M. Brackstone. Towards an Understanding of Adaptive Cruise Control. *Transportation Research Part C: Emerging Technologies*, Vol. 9, 2001, pp. 33–51.
7. Treiber, M., and D. Helbing. Microsimulations of Freeway Traffic Including Control Measures. *Automatisierungstechnik*, Vol. 49, 2001, pp. 478–484.
8. Davies, L. Effect of Adaptive Cruise Control Systems on Traffic Flow. *Physical Review E*, Vol. 69, 2004, pp. 066110-1–066110-8.
9. Kesting, A., M. Treiber, M. Schonhof, and D. Helbing. Adaptive Cruise Control Design for Active Congestion Avoidance. *Transportation Research Part C: Emerging Technologies*, Vol. 16, 2008, pp. 668–683.
10. Shladover, S. E., D. Su, and X.-T. Lu. Impacts of Cooperative Adaptive Cruise Control on Freeway Traffic Flow. *Transportation Research Record: Journal of the Transportation Research Board*, 2012. 2324: 63–70.
11. Shladover, S. E., C. Nowakowski, X.-T. Lu, and R. Ferlis. Cooperative Adaptive Cruise Control: Definitions and Operating Concepts. *Transportation Research Record: Journal of the Transportation Research Board*, 2015, 2489: 145–152.
12. Milanes, V., and S. E. Shladover. Modeling Cooperative and Autonomous Adaptive Cruise Control Dynamic

Responses Using Experimental Data. *Transportation Research Part C: Emerging Technologies*, Vol. 48, 2014, pp. 285–300.

13. Wang, J., and R. Rajamani. Should Adaptive Cruise-Control Systems be Designed to Maintain a Constant Time-Gap Between Vehicles? *IEEE Transactions on Vehicular Technology*, Vol. 53, 2004, pp. 1480–1490.
14. Yi, J., and R. Horowitz. Macroscopic Traffic Flow Propagation Stability for Adaptive Cruise Controlled Vehicles. *Transportation Research Part C: Emerging Technologies*, Vol. 14, 2006, pp. 81–95.
15. Zheng, L., P. J. Jin, and H. Huang. An Anisotropic Continuum Model Considering Bi-directional Information Impact. *Transportation Research Part B: Methodological*, Vol. 75, 2015, pp. 36–57.
16. Ngoduy, D., and D. Jia. Multi Anticipative Bidirectional Macroscopic Traffic Model Considering Cooperative Driving Strategy. *Transpormetrica B: Transport Dynamics*, Vol. 5, 2017, pp. 100–114.
17. Ngoduy, D., S. P. Hoogendoorn, and R. Liu. Continuum Modeling of Cooperative Traffic Flow Dynamics. *Physica A*, Vol. 388, 2009, pp. 2705–2716.
18. Ngoduy, D. Instability of Cooperative Adaptive Cruise Control Traffic Flow: A Macroscopic Approach. *Communications in Nonlinear Science and Numerical Simulation*, Vol. 18, 2013, pp. 2838–2851.
19. Delis, A. I., I. K. Nikolos, and M. Papageorgiou. Adaptive Cruise Control for a Macroscopic Traffic Flow Model: Development and Numerical Solution. *Computers and Mathematics with Applications*, Vol. 70, 2015, pp. 1921–1947.
20. Ngoduy, D. Application of Gas-Kinetic Theory to Modeling Mixed Traffic of Manual and ACC Vehicles. *Transpormetrica*, Vol. 8, 2012, pp. 2705–2716.
21. Delis, A. I., I. K. Nikolos, and M. Papageorgiou. Simulation of the Penetration Rate Effects of ACC and CACC on Macroscopic Traffic Dynamics. IEEE Conference on Intelligent Transportation Systems, Proceedings ITCS, art. No 7795576, 2016, pp. 336–341.
22. Delis, A. I., I. K. Nikolos, and M. Papageorgiou. Macroscopic Modeling and Simulation of Multi-lane Traffic. IEEE Conference on Intelligent Transportation Systems, Proceedings ITSC, art. No. 7313449, 2015, pp. 2213–2218.
23. Porfyri, K. N., A. I. Delis, I. K. Nikolos, and M. Papageorgiou. Calibration and Validation of a Macroscopic Multi-lane Traffic Flow Model using a Differential Evolution Algorithm. Presented at 96th Annual Meeting of the Transportation Research Board, Washington, D.C., 2017.
24. Delis, A. I., I. K. Nikolos, and M. Papageorgiou. High-resolution Numerical Relaxation Approximations to Second-order Macroscopic Traffic Flow Models. *Transportation Research Part C: Emerging Technologies*, Vol. 44, 2014, pp. 318–349.
25. Shvetsov, V., and D. Helbing. Macroscopic Dynamics of Multi-lane Traffic. *Physical Review E - Statistical Physics, Plasmas, Fluids, and Related Interdisciplinary Topics*, Vol. 56, 1999, pp. 6328–6339.
26. Ngoduy, D., S. P. Hoogendoorn, and J. W. C. van Lint. Modeling Traffic Flow Operation in Multilane and Multiclass Urban Networks. *Transportation Research Record: Journal of the Transportation Research Board*, 2005. 1923: 73–81.
27. Zhou, J., and H. Peng. Range Policy of Adaptive Cruise Control Vehicles for Improved Flow Stability and String Stability. *IEEE Transactions on Intelligent Transportation Systems*, Vol. 6, 2005, pp. 229–237.
28. *Motorway Incident Detection and Automatic Signaling (MIDAS) Design Standard.* Highways England, 1st ed. Bristol, U.K., 2007.

The Standing Committee on Traffic Flow Theory and Characteristics (AHB45) peer-reviewed this paper (18-01024).

TRR
JOURNAL OF THE TRANSPORTATION RESEARCH BOARD

Article

Transportation Research Record
2018, Vol. 2672(20) 193–202

DOI: 10.1177/0361198118784169
journals.sagepub.com/home/trr
SAGE

Modeling and Evaluating Traffic Flow at Sag Curves When Imposing Variable Speed Limits on Connected Vehicles

Reza Vatani Nezafat[1], Ehsan Beheshtitabar[1], Mecit Cetin[1], Elizabeth Williams[2], and George F. List[2]

Abstract

Sag curves, road segments where the gradient changes from downwards to upwards, generally reduce the roadway capacity and cause congestion. This results from a change in longitudinal driving behavior when entering a sag curve as drivers tend to reduce speeds or increase headways as vehicles reach the uphill section. In this research, a control strategy is investigated through manipulating the speed of connected vehicles (CVs) in the upstream of the sag curve to avoid the formation of bottlenecks caused by the change in driver behavior. Traffic flow along a sag curve is simulated using the intelligent driver model (IDM), a time-continuous car-following model. A feedback control algorithm is developed for adjusting the approach speeds of CVs so that the throughput of the sag curve is maximized. Depending on the traffic density at the sag curve, adjustments are made for the speeds of the CVs. A simulation-based optimization method using a meta-heuristic algorithm is employed to determine the critical control parameters. Various market penetration rates for CVs are also considered in the simulations. Even at relatively low market penetration rates (e.g., 5–10%), significant improvements in travel times and throughput are observed.

Motivated by their significant safety and mobility impacts, connected vehicles (CVs) equipped with two-way wireless communications capabilities are expected to be widely available in the near future. Given the expected availability of I2V or V2V communications with increasingly more connected vehicles in the traffic stream, it is essential to study how various types of traditional infrastructure-centric traffic flow control methods would operate in a CV environment. For example, I2V communication technologies could be a replacement for the dissemination of information through variable message signs. The primary focus of this paper is to investigate how potential implementation of variable speed limit (VSL) through CVs could impact traffic flow. A segment of the road upstream of the sag curve would be considered as a control section. On entering the control section, each connected vehicle (CV) is instructed about the speed limit only once. Compared to VSL implementations through a variable message sign (VMS), the proposed system is fundamentally the same except information is communicated only to the CVs. The inflow to a sag curve on a freeway would be regulated to avoid capacity drops at the uphill section which might be caused by the inability of drivers to compensate for the gradient. Therefore, for the analyses presented here, CVs do not need autonomous driving functionalities such as adaptive cruise control. However, it should be mentioned that autonomous vehicle (AV) technologies could also be employed to avoid the formation of congestion within a sag curve. Analyzing the impacts of AVs in this context is not within the scope of this paper.

The simulation environment consists of three building blocks: (i) a longitudinal driving behavior model describing car-following dynamics; (ii) a proportional feedback control law for setting VSLs in the upstream; (iii) a framework for optimization of the model parameters. An essential element of the first component pertains to the characterization of driving behavior of vehicles in the sag curve, especially as they climb uphill. For this purpose, a recent car-following model developed in 2012 by

[1]Department of Civil & Environmental Engineering, Old Dominion University, Norfolk, VA
[2]Department of Civil, Construction, and Environmental Engineering, North Carolina State University, Raleigh, NC

Corresponding Author:
Address correspondence to Reza Vatani Nezafat: rvata001@odu.edu

Goñi-Ros et al. which incorporates the impacts of gradient on vehicle acceleration is adapted (*1*). Similar to the strategy implemented by Goñi-Ros et al. in 2014, an ALINEA proportional feedback control law is applied to the second component (*2*, *3*). For the last component of the simulation environment, a meta-heuristic algorithm is developed to find the optimum parameters.

Literature Review and Background

Traffic dynamics are recreated in a simulation environment by categorizing vehicle behavior into longitudinal and lateral components. Longitudinal behavior fundamentally corresponds to the acceleration of a vehicle while lateral behavior corresponds to choosing and changing lanes. In this study, the lateral behavior will not be considered because the investigated network contains a single-lane freeway.

The fundamental macroscopic relationship between traffic flow, density, and speed depends on the characteristics of longitudinal driving behavior. Empirical observations show that vehicles move at the free flow speed when density is low. As density increases, they maintain free flow speed until it reaches a certain critical density point. Beyond this point, further increases in density would result in lower speeds and consequently lower traffic flow. Capacity is then defined as the maximum flow which can be reached. Researchers have noted that capacity of congested traffic conditions is significantly lower than that of free flow traffic conditions, and the magnitude of the drop is dependent on traffic flow and relative speeds (*4*, *5*). This difference is called the capacity drop phenomena. Daganzo explained that bottlenecks arise at sections with abrupt increases of demand such as on ramps or at tunnels and weaving sections where capacity would drop suddenly (*6*). Based on the Transportation Research Board's (TRB) fifth edition of the Highway Capacity Manual, transient events such as weather conditions or accidents can also cause a local drop in the capacity of freeways leading to bottlenecks (*7*). In hilly regions, sag curves are one of the most common causes of bottlenecks. Researchers, such as Okamura in 2000 and Xing in 2010, have shown that the capacity of a sag curve is 10–25% lower than level sections (*8*, *9*). Later in 2014, Xing et al. noted that sag curves are the main reason behind 60% of bottlenecks on Japanese intercity freeways (*10*).

A sag curve is defined as a downhill followed by an uphill or as a section where the gradient of roadway increases gradually in the direction of traffic. Previous studies showed that drivers reduce their desired speeds at sags (*11*, *12*). In 2012, Yoshizawa et al. suggested that the main reason for this reduction is related to poor acceleration behaviors on sags (*13*). However, Laval et al. put forward that when power to mass ratio is considerably large, the reason may be related to insufficient acceleration capability of drivers (*14*).

Many researchers have developed microscopic models to reproduce the characteristics of traffic flow at sags. These car-following models are designed to take into account the influence of gradient on vehicle acceleration (*15*). Yoshizawa et al. noticed that when drivers reach a sag curve, they cannot fully compensate for the increase in the slope (*13*). Other studies have added compensation for the limiting effect which increases in gradient has on vehicle acceleration (*16*, *17*). Empirical observations have shown that models which consider compensation for drivers can reproduce longitudinal driving behavior and traffic dynamics at sag curves more accurately. More recently, in 2012, Goni-Ros et al. introduced another term in the acceleration equation of the Intelligent Driver Model (IDM) to incorporate this effect of compensation for driving behavior (*1*, *18*). The developed model can capture the effect of compensation on a vertical curve. They have assumed drivers would compensate the gradient linearly along the sags in the direction of uphill.

There are several active traffic management (ATM) strategies for reducing congestion at sags, such as queue warning and VSL. Some field tests and simulation experiments have been conducted to evaluate the effectiveness of these approaches. The goals of these strategies are increasing the capacity of sags, preventing the formation of congestion at sags, and increasing the queue discharge capacity of sags. The capacity of the sag could be increased by utilizing adaptive cruise control systems to perform the acceleration task more efficiently than humans (*19*). One of the most popular ATM strategies is VSLs. For instance, Goni-Ros et al. evaluated the possibility of maximizing the throughput of a sag bottleneck by implementing VSL at the upstream of the bottleneck through a VMS (*2*). They have shown that VSL can be integrated with control strategies. It was assumed that all drivers would comply with the VSL, which is not a realistic assumption. In this paper, the VSL is imposed only on the CVs, for which the market penetration is varied from 0% to 100%. It is assumed all CVs would comply with the VSL. Therefore, the results of this paper could also be used to analyze the impacts of compliance rate on the system performance.

The control strategy used in this study is inspired by one of the most popular ramp-metering algorithms called ALINEA (*3*). However, the parameters of this algorithm have to be calibrated to achieve the maximum system efficiency. To do so, one common technique is to employ meta-heuristics, such as a Genetic Algorithm (GA), which are used in various models (*20*). Genetic algorithms are very popular and easy to use, but they have slow convergence and are weak in the local search

(*21*). Another meta-heuristics method, Particle Swarm Optimization (PSO), is very suitable for complex continuous problems because of its simplicity and because it has few parameters. Despite its advantages, PSO usually fails to control its velocity step size for better tuning in search space, resulting in inappropriate results (*22*). Many researchers use a combination of these algorithms to overcome the drawbacks of each. This study uses an algorithm called HPSOGA which is a combination of GA and PSO proposed by Duan et al. in 2013 (*23*). It captures the advantages of both algorithms.

Longitudinal Driving Behavior

In this study, the model Goni-Ros et al. proposed for sag curves is used (*1*).This model accounts for the influence of vertical curves on vehicle acceleration. It calculates acceleration from the summation of two terms as presented in Equation 1. The first component corresponds to car-following behavior and the second one calculates behavior on a uphill:

$$v'(t) = fr(t) + fg(t) \tag{1}$$

The first acceleration term uses speed (v), relative speed (Δv), the desired speed (v_{des}), and spacing to the vehicle ahead (s) to calculate acceleration for the following car. This is a modified version of the IDM:

$$fr(t) = a \times \min\left[1 - \left(\frac{v(t)}{v_{des}}\right)^4, 1 - \left(\frac{s_{des}(v(t), \Delta v(t))}{s(t)}\right)^2\right] \tag{2}$$

In Equation 2, s_{des} is the desired spacing which is computed using Equation 3. The main influencing factor is the safe gap to the lead vehicle:

$$s_{des}(v(t), \Delta v(t)) = s_0 + v(t).\tau(v(t)) + \frac{v(t).\Delta v(t)}{2\sqrt{a.b}} \tag{3}$$

The parameter a is the maximum acceleration, b is the maximum comfortable deceleration, s_0 is the gap at the standstill, and τ is the safe time headway as a function of speed. Based on the traffic state, the safe time headway (τ) changes as shown in Equation 4:

$$\tau(v(t)) = \begin{cases} \tau_f & v(t) \geq v_{crit} \\ \gamma.\tau_f & v(t) < v_{crit} \end{cases} \tag{4}$$

The second term ($fg(t)$) in Equation 1 captures the influence of gradient on vehicle acceleration. This influence is equal to the difference between the gradient at the position of the vehicle ($G(x(t))$) and the compensated gradient by the driver at the time ($G_c(t)$) multiplied by gravity acceleration. This is shown in Equation 5:

$$fg(t) = -g.(G(x(t)) - G_c(t)) \tag{5}$$

Ros et al. assumed that drivers would compensate linearly for any increase in freeway gradient with maximum gradient compensation rate defined by parameter c:

$$G_c(t) = \begin{cases} G(x(t)) & G(x(t)) \leq G(t_c) + c(t - t_c) \\ G(t_c) + c(t - t_c) & \textit{otherwise} \end{cases} \tag{6}$$

Where:

$$t_c = \max[t | G_c(t) = G(x(t))] \tag{7}$$

If the increase in grade over time is lower than c, then $G_c(t)$ is equal to $G(x(t))$ and $fg(t)$ is zero. Therefore, the acceleration of vehicle is not affected and the driver fully compensates for the gradient.

Control Strategy

The objective of the control strategy is to eliminate congestion in sags and improve the performance of highways in hilly regions. For networks not influenced by other control measures, minimizing the total time that vehicles spend in the system is equivalent to maximizing the exit flow (*24*). As mentioned previously, under uncongested traffic conditions, flow or throughput is higher than that under congested conditions. The capacity of the freeway on a sag section (q_{Sag}) is less than other sections ($q_{Capacity}$). Therefore, the network's exit flow is limited by the capacity of the sag bottleneck:

$$q_{Exit} \approx q_{Sag} < q_{Capacity} \tag{8}$$

One way to maximize the exit flow is to prevent traffic from becoming congested at the bottleneck. Keeping traffic state uncongested at the bottleneck is possible if the inflow of the sag is regulated at a controlled section at the upstream. The inflow of sag is approximately equal to the outflow of the control section and based on the fundamental relationship between speed and flow, changing the speed on the control section changes the inflow of the sag. By dynamically modifying the speed at the control section, it is possible to keep the inflow to the bottleneck slightly below its free flow capacity. This will increase the time-weighted sum of exit flow. When the demand in the upstream is large enough, the congestion would not be prevented completely. As a result, the control section and upstream would become congested instead of the sag curve, but the outflow from the controlled part will be higher than the queue discharge capacity of the sag.

The controller which calculates the speed limit for the control section is inspired by the ramp-metering control algorithm called ALINEA and proposed by

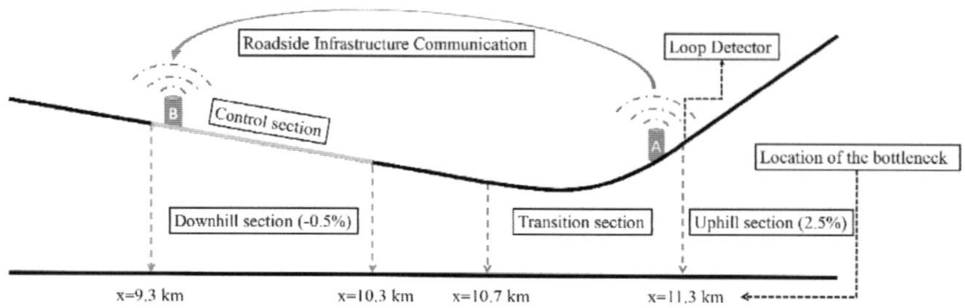

Figure 1. Geometry of the sag curve.

Table 1. Characteristics of the Car-Following Model

$v_{des}(^{km}/_{h})$	$a(^{m}/_{s^2})$	$b(^{m}/_{s^2})$	$\tau(s)$	$s_0(m)$	$v_{crit}(^{km}/_{h})$	$\gamma(-)$	$c(s^{-1})$	$\Delta t(s)$
120	1.45	2.1	1.2	3	65	1.15	0.0001	0.5

Note: $v_{des}(^{km}/_{h})$ = desired speed; a = maximum acceleration; b = maximum comfortable deceleration; s_0 = gap at the standstill; τ = safe time headway in uncongested traffic condition; $v_{crit}(^{km}/_{h})$ = critical speed; $\gamma(-)$ = congestion factor on safe time headway; $c(s^{-1})$ = driver's compensation rate of the gradient; $\Delta t(s)$ = time step of the simulation.

Papageorgiou et al. (*3*). This algorithm is based on the basis of a proportional feedback control law. It calculates the variable speed limit based on Equation 9. The target density (ρ_{Target}) is slightly lower than the critical density of the fundamental diagram, and real-time density (ρ_b) is the estimated density at the sag curve calculated every T_c seconds. The algorithm would change the speed limit as a proportion (κ) of the difference between target and measured density every time that a new density is calculated:

$$v_{Limit}(t) = v_{Target} + \kappa \times (\rho_{Target} - \rho_b(t-1)) \quad (9)$$

As evident from Equation 9, in high demand conditions, the controller would keep the density at the bottleneck close to the target density to prevent breakdown. Whenever demand decreases, the measured density would be significantly less than the target density which leads the controller to impose a higher speed limit and, on the contrary, if demand increases the measured density would be substantially more than the target density which leads the controller to enforce a lower speed limit. The controller always uses the previously estimated density so that drivers would have enough time to cover the distance between the control section and the bottleneck.

Two roadside units, shown in Figure 1, are needed for system operation. As the only input to Equation 9 is the density, the roadside unit A is connected to a typical loop detector that measures occupancy and estimates density at the bottleneck every T_c seconds. Loop detectors are able to sense all vehicles (CVs and non-CVs). Therefore, even at lower penetration rates of CVs the system can estimate the density accurately. This information is then transmitted to the roadside unit B in the upstream. Based on Equation 9, the roadside unit B would update the speed limit every T_c seconds and broadcast it to the connected vehicles (CVs) when they arrive at the control section every second. The geometry of the network and traffic flow parameters are explained below.

Simulation Setup

The simulation model comprises a one-lane freeway with a sag in the middle. The length of the facility is 12 km. The road starts with a constant-gradient downhill section followed by a vertical sag curve, and at the end, a constant-gradient uphill section (see Figure 1). The downhill section has a constant gradient equal to -0.5% and the uphill section has a constant slope equal to 2.5%. At the vertical sag, the slope increases linearly from -0.5% to $+2.5\%$, and the length of the vertical curve is 600 m between x = 10.7 km and x = 11.3 km. The downhill section is long enough to make sure the queue would not reach the entry point of the simulation. The speed limit is 120 km/h. Characteristics of the vehicles and drivers, as defined by the IDM model, are assumed to be homogeneous to prevent the emergence of other types of bottlenecks in the simulation. These model parameters are shown in Table 1.

The distribution of demand over time is illustrated in Figure 2. The first 10 minutes is a transition from 0 to

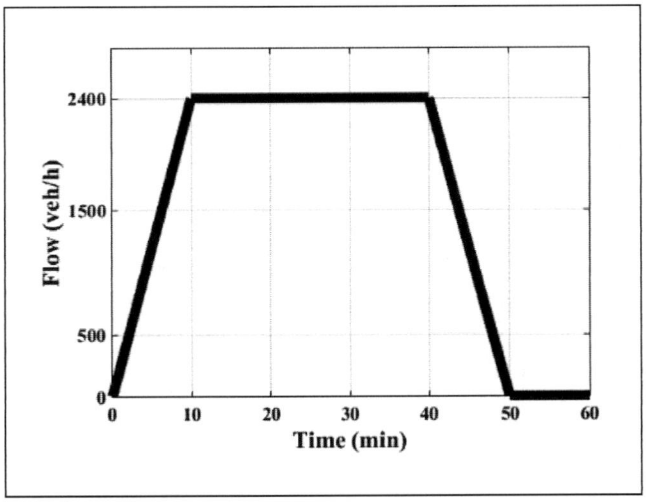

Figure 2. Demand profile over time.

2400 veh/h, a capacity higher than the bottleneck capacity. The demand stays at 2400 veh/h for 30 minutes then transitions back to zero across 10 minutes. Beyond this, the demand remains zero until all vehicles have exited the facility.

The control section in Figure 1 is 1.0 km long. In this section, only connected vehicles would be informed of the calculated speed limit. Notably, it is assumed that all CVs would comply completely. The control section is between x = 9.3 km and x = 10.3 km. The downstream end of the controlled section is 0.4 km away from the beginning of the transition section. As soon as CVs leave the control section, their speed will revert to the default 120 km/h to make sure the vehicles traverse the uphill with the maximum speed possible. In other words, the desired speed of CVs is only varied while they are within the control section.

With the given parameters above, a microsimulation model was created in MATLAB. In Figure 3, a heat map along with sample vehicle trajectories are shown (middle chart) for the base case, that is, when no control strategy is implemented. At the very beginning of the simulation, the effect of the uphill is not significant enough to cause a breakdown at the bottleneck. After a while, a shockwave starts to propagate backwards starting at the bottleneck with constant speed. As the breakdown is caused by the geometry of the road, this shockwave continues to propagate until in-flow decreases. The second shockwave emerges shortly after the first one because of constant over-capacity demand. It shows that the model can reproduce stop-and-go waves at sags which is consistent with the literature (*25*).

Four other charts are included in Figure 3 to provide additional performance measures for analysis. In the top left, the input demand over time (green line), as well as exit flow rates, are depicted for the base case (red line). A hypothetical scenario (black line) in which the uphill has no influence whatsoever on the traffic flow is also shown. This last scenario is included as a reference to show the maximum possible system performance if the effects of sag are eliminated. This phenomenon could perhaps be achieved through automated driving, but this is left for future research. The second chart, density versus time plot in the lower left, shows the measured density by the loop detector at the uphill (see Figure 1). The chart on the top right shows the observed speed (red line) at the control section as well as the imposed VSL (It is assumed that when the control section is empty the average speed is 120 km/h). The last chart at the bottom right reports the total number of vehicles in the system, that is, for the entire corridor, over time for the base case without a VSL system (red line) as well as for the hypothetical scenario (black line). These charts are reproduced to summarize the effects of VSL under various CV market penetrations as presented later in the paper.

Finding Optimal Control Parameter Values

The VSL control strategy described by Equation 9 requires three key parameters to achieve optimal performance. Here, the total delay in the system is used as the objective function. The delay for a vehicle is computed in reference to the hypothetical scenario mentioned above, in which the sag curve is assumed to not influence the traffic flow. For the simulated demand, vehicles travel through the corridor at free-flow speeds. For a given scenario, the total delay is the sum of the individual vehicle delays. For example, when there is no VSL control system, the system performs as shown in Figure 3. This total delay (TD) is then taken as a reference and compared to the total delay for theVSL control strategies. Consequently, the objective function is defined as shown in Equation 10:

$$ObjectiveFunction = \frac{TD_{NoControl} - TD_{Controlled}}{TD_{NoControl}} \quad (10)$$

Optimizing three parameter values is the objective: the gain (κ), the period for sampling occupancy at bottleneck (T_c), and the target speed (v_{Target}). As a mathematical formulation for this optimization problem does not exist, the authors used a meta-heuristic. The hybrid particle swarm optimization genetic algorithm (HPSOGA) proposed by Duan et al. was employed (*23*).

The PSO simulates the behavior of a swarm of particles moving to a potential well with an analogy to the flocking of birds or fish schooling. Random parameters for all particles are generated, and the objective function for each particle is calculated. For this problem, a good position for a particle corresponds to a lower value of the

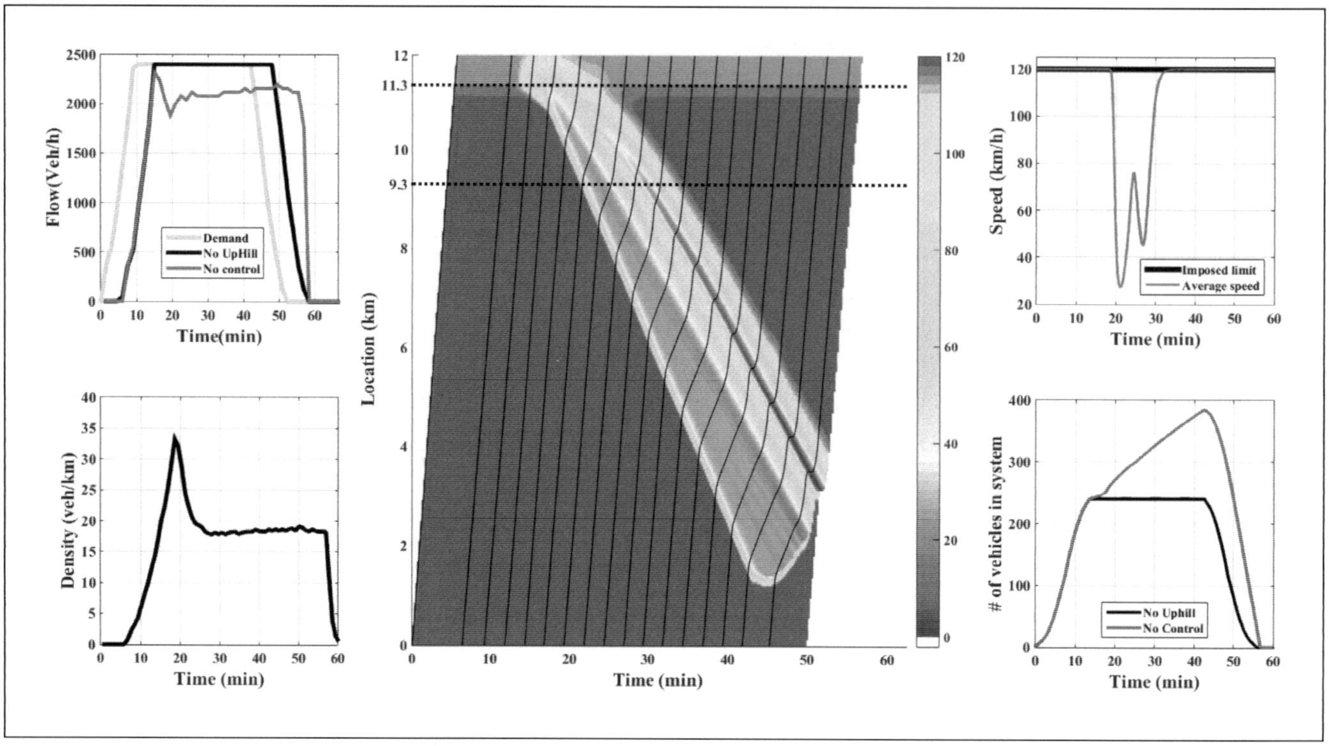

Figure 3. Input demand and exit flows (top left), density measured by loop detector (bottom left), heat map with sample trajectories (middle), average vehicle speeds in the control section (top right), and number of vehicles in the system (bottom right) without a VSL control system.

objective function. The particles move in the search space of the problem to find the lowest value of the objective function. The particles are guided by their own best-known experience in the search-space as well as the entire population's best experience. For each iteration the best experiences are updated both individually and globally.

The solution from the PSO becomes the initial population of the GA. The GA has three stages: Stage 1: Creating an initial population; Stage 2: Evaluating an objective function, and Stage 3: Producing a new population. GA operators manipulate each member. The first operator is a crossover which selects two members of the population as parents and produces two offspring by swapping elements of the parents. Participating in a crossover depends on the value of each member's objective function, which means members with higher values participate in crossovers more often. The second operator is a mutation operator. It is used to increase the space explored. The mutation rate is low. In the end, a new population is selected from the output of these two operators, and the process continues. If identical solutions are obtained, the best solution is assumed to be determined.

The three control parameters were optimized using a HPSOGA algorithm with a population size of 50 and a maximum iteration number of 100. The optimal parameters found by the algorithm are presented below:

- Target speed, $v_{Target} = 95(km/h)$
- Gain, $\kappa = 4.68(km^2/h/veh)$
- Updating period, $T_c = 50(s)$

The resulting system performance is shown in Figure 4. The market penetration rate is 100%. The VSL reduced the total delay by 49% compared to the no control scenario. The shockwave is moved upstream of the control section. The density at the bottleneck stays below 25 veh/km at the onset of congestion, and the system can reach a steady-state density and approximately constant VSLs after 35 minutes. In the top-right chart, it is clear that vehicles adhere to the imposed VSLs as their average speeds follow the VSLs. In the bottom-right graph, the blue-line approximately splits the area between the black and red lines into two equal pieces, visually demonstrating that the total delay is reduced by about 50%. It should also be noted that the queue dissipation time is shorter under the VSL control system.

Sensitivity to the Market Penetration Rate

In this section, the market penetration level of the CVs is varied to understand the impacts on system performance. As arrivals of CVs are assumed to be random, each market penetration level has been simulated 50 times to account for the variability. The results for

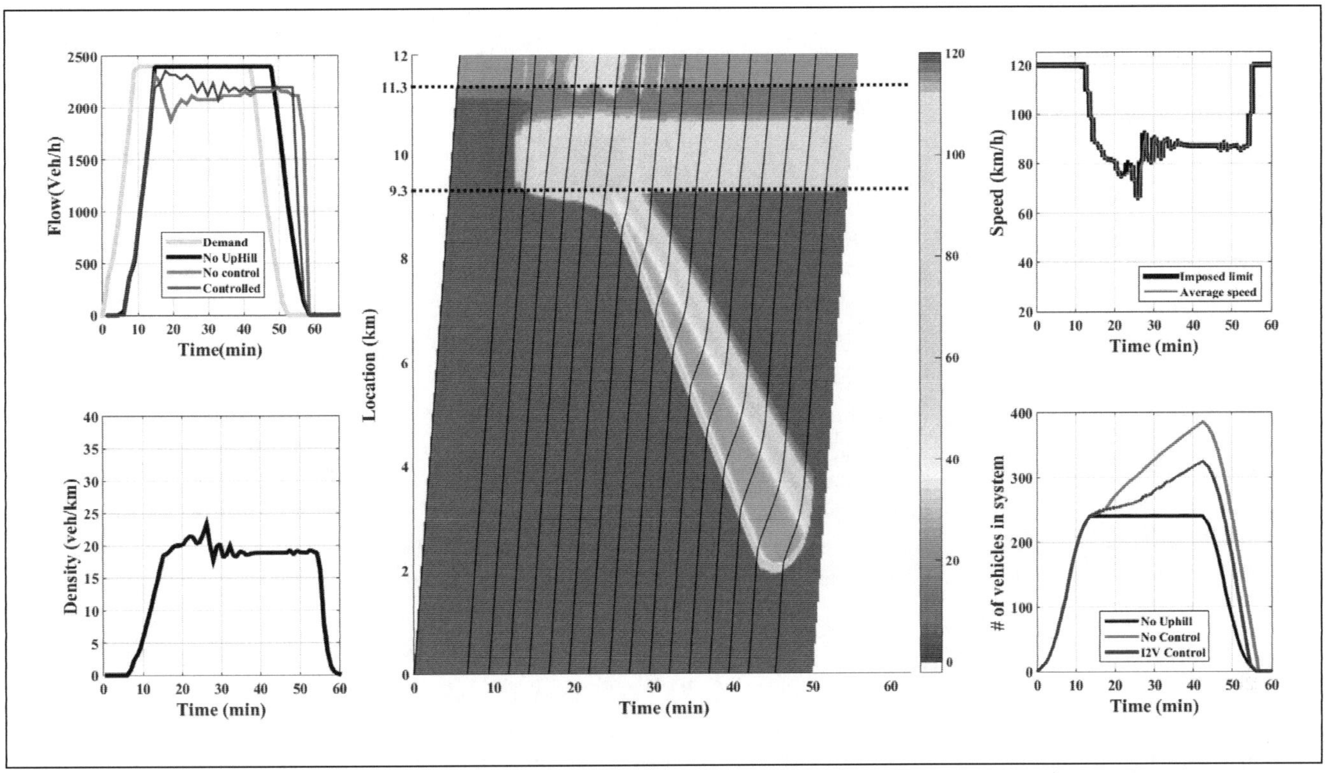

Figure 4. Input demand and exit flows (top left), density measured by loop detector (bottom left), heat map with sample trajectories (middle), average vehicle speeds in the control section (top right), and number of vehicles in the system (bottom right) for a VSL control system at 100% CVs market penetration (delays are 49% lower than no control).

reductions in total delays are presented as notched boxplots in Figure 5. Firstly, as the percentage of CVs increases, the average improvement also increases but at a decreasing rate. Beyond a relatively small value of 15% market penetration, the average improvements are not substantially different from the maximum possible value of 49%. Secondly, there is substantial variation in the low market penetration rates, which rapidly diminishes with increasing CVs in the traffic stream.

Additional analyses are performed to understand the reasons behind the significant variations in the low market penetration levels. As shown in Figure 5, the highest variation is at the 5% penetration level which has improvements as high as 45%, or worse than the no control scenario. These two extreme cases are presented in Figures 6 and 7 respectively. Each vertical red line added to the density diagrams (bottom left) indicates the exact time that a CV enters the control section. It is noticeable that the magnitude of improvement is correlated with the distribution of the arrivals of CVs over time. Whenever there is a large gap between the arrival of CVs, density tends to rise with a delay.

The density diagram in Figure 3 (bottom left) shows how the density at the bottleneck would change if there is no control strategy. The most critical period for the control strategy is between 15 and 25 minutes from the beginning of the simulation which is the transition period from the uncongested to the congested traffic. After this period, congestion starts to propagate and grow

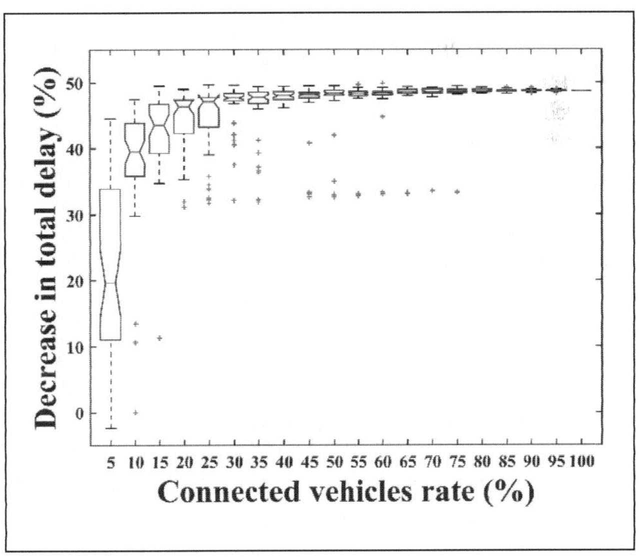

Figure 5. Notched boxplot of sensitivity analysis.

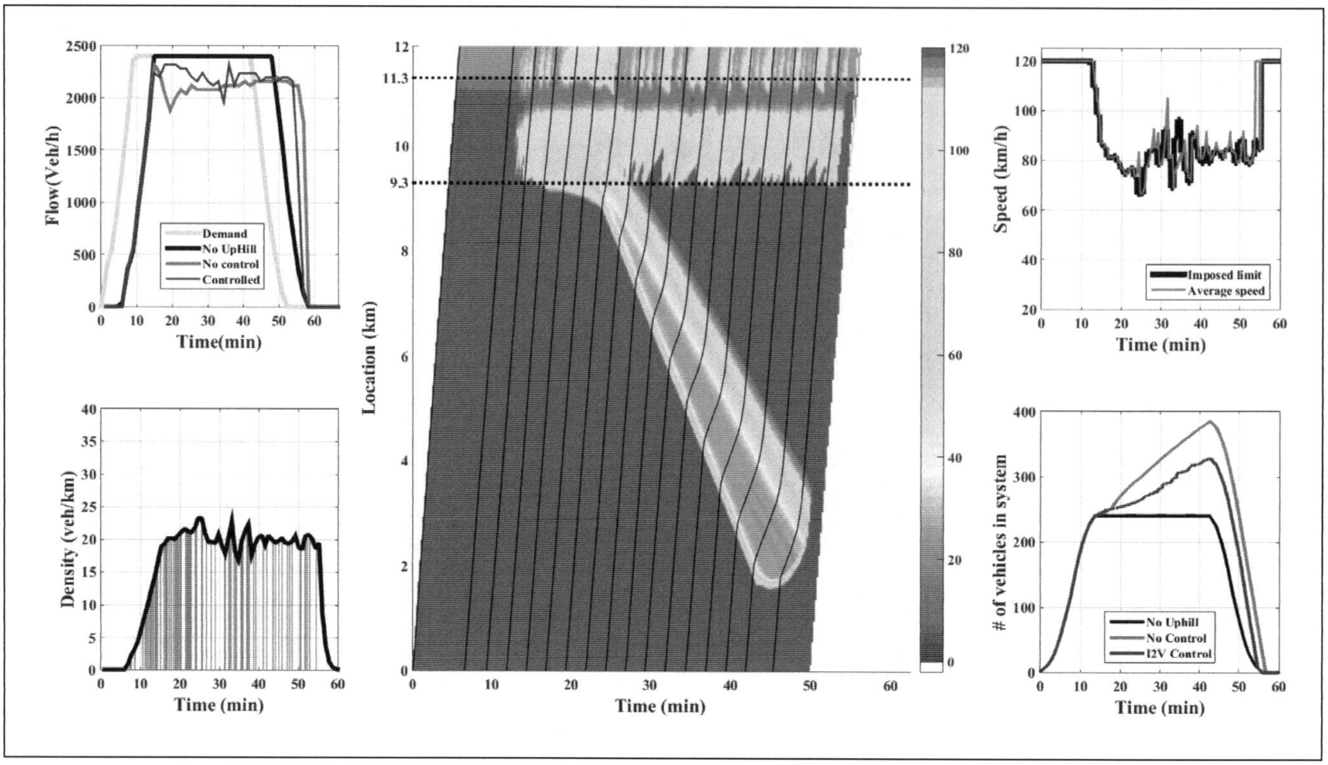

Figure 6. Input demand and exit flows (top left), density measured by loop detector (bottom left), heat map with sample trajectories (middle), average vehicle speeds in the control section (top right), and number of vehicles in the system (bottom right) for a VSL control system with 5% CVs—a random scenario with good performance (delays are 44% better than with no control).

backward. If the density of the CVs over time within this period is not sufficient to mitigate the initiation of the breakdown, the queue will extend from the bottleneck location (i.e., uphill) to the control sections. This will make a recovery to normal operations almost impossible. The density diagram (bottom left) in Figure 6 shows a denser arrival distribution around 15–25 minutes which kept the density at the bottleneck less than the critical point, and that is the main reason this sample has such a good performance. On the contrary, the density diagram (bottom left) in Figure 7 has relatively large gaps in the arrival of CVs around this period which leads to a significant spike in the density of the bottleneck and clearly the controller is not able to stabilize the system operation. This example illustrates the importance of the distribution of CVs within the traffic stream and how uneven or clustered arrivals can negatively impact system performance.

Conclusion

In this paper, a VSL control strategy was developed to regulate the density of the bottleneck on a sag curve. To improve system efficiency, CVs in the upstream control sections were instructed via V2I communications to adjust their speeds based on a VLS algorithm. The VSL prevents traffic from breaking down by using a proportional feedback control law. The optimal parameters are determined by a meta-heuristic algorithm called HPSOGA to obtain the best performance of the strategy. The results show that with optimal parameters for control strategy nearly half of the delay caused by the uphill can be eliminated. A sensitivity analysis shows that even with a low market penetration rate (e.g., 15%) the system can reduce delays significantly. In the simulated network, there is a single lane and no lane changing behavior. Consequently, at low market penetrations, when adjustments are made to a few vehicles these are transferred to the following vehicles more effectively. However, the control strategy is sensitive to the arrival times of CVs. If arrival times of CVs are not dense enough at the beginning of high demand, the performance of the system will drop considerably. It is demonstrated that measures of central tendency are not sufficient to understand the sensitivity of the system to market penetration rates of CVs because of large fluctuations in the system performance, especially at 5–10% levels. For example, at 5% market penetration rate, the median improvement is 20%, but the variance is so high that it includes negative performance as well as a 45% improvement in total delays.

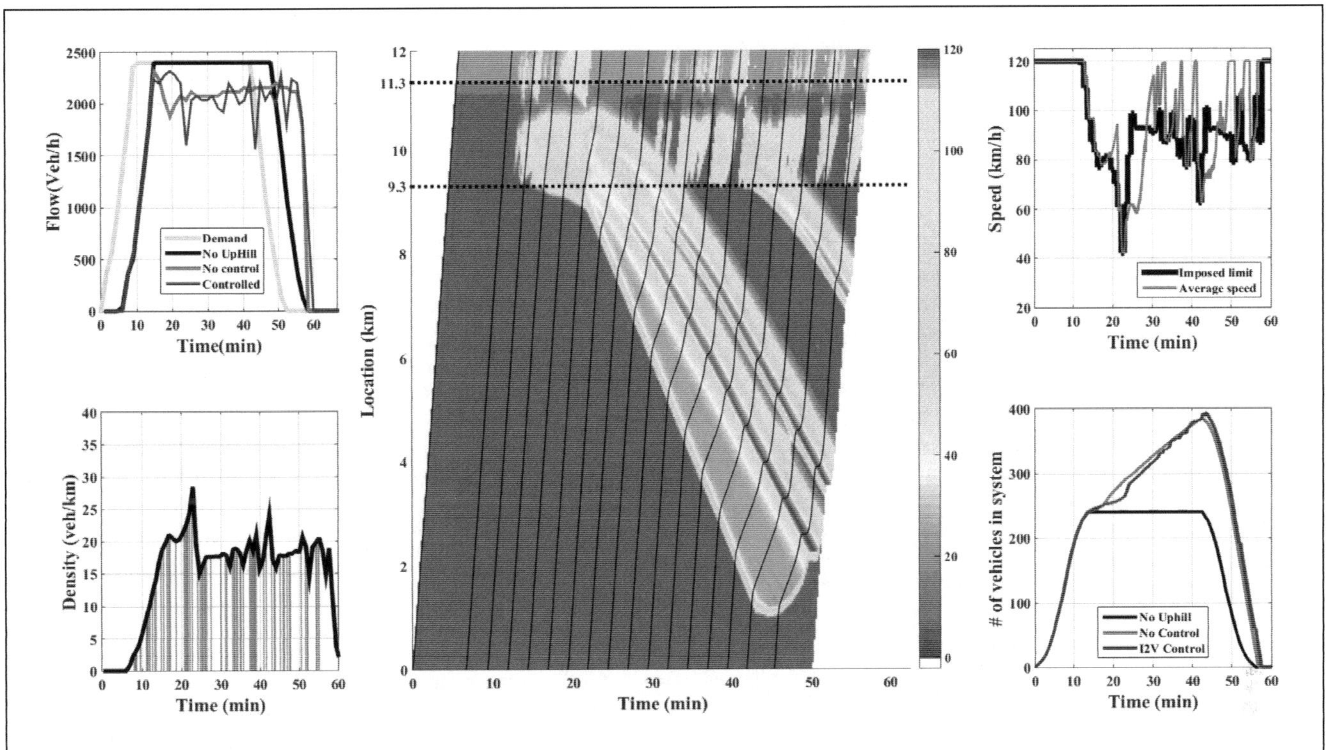

Figure 7. Input demand and exit flows (top left), density measured by loop detector (bottom left), heat-map with sample trajectories (middle), average vehicle speeds in the control section (top right), and number of vehicles in the system (bottom right) for a VSL control system with 5% CVs—a random scenario with bad performance (delays are 2% worse than no control).

A similar control strategy can be applied to other network discontinuities, in which a capacity drop may occur. Further evaluation of the strategy will require investigating heterogeneous traffic and a multilane network as well as lane changing behavior (an important factor affecting capacity at uphill segments). More complex networks such as networks with ramps or other types of bottlenecks should also be considered to explore the applicability of the proposed control strategy in more general cases.

Acknowledgments

The research presented in this paper was funded by a USDOT-UTC grant to The National Center for Strategic Transportation Policies, Investments, and Decisions, a consortium of seven universities which is led by the University of Maryland (UMD).

Author Contributions

The authors confirm contribution to the paper as follows: study conception and design: Vatani, Cetin, List; data collection: Vatani, Beheshtitabar; analysis and interpretation of results: Vatani, Cetin, List; draft manuscript preparation: Vatani, Williams, Beheshtitabar. All authors reviewed the results and approved the final version of the manuscript.

References

1. Goni Ros, B., V. L. Knoop, B. Van Arem, and S. P. Hoogendoorn. Car-Following Behavior at Sags and Its Impacts on Traffic Flow. Presented at 92nd Annual Meeting of the Transportation Research Board, Washington, D.C., 2013.
2. Goni Ros, B., V. Knoop, B. van Arem, and S. Hoogendoorn. Mainstream Traffic Flow Control at Sags. *Transportation Research Record: Journal of the Transportation Research Board*, 2014. 2470: 57–64.
3. Papageorgiou, M., H. Hadj-Salem, and F. Middelham. Alinea Local Ramp Metering: Summary of Field Results. *Transportation Research Record: Journal of the Transportation Research Board*, 1997. 1603: 90–98
4. Hall, F. L., and K. Agyemang-Duah. Freeway Capacity Drop and the Definition of Capacity. *Transportation Research Record: Journal of the Transportation Research Board*, 1991. 1320: 91–98.
5. Sohrabi, S., A. Ermagun, and R. Ovaici. Finding Optimum Capacity of Freeways Considering Stochastic Capacity Concept. Presented at 96th Annual Meeting of the Transportation Research Board, Washington, D.C.
6. Daganzo, C. *Fundamentals of Transportation and Traffic Operations*. Pergamon Press, Oxford, 1997.
7. *Highway Capacity Manual 2010*. Transportation Research Board of the National Academies, Washington, D.C., 2010.

8. Okamura, H., S. Watanabe, and T. Watanabe. An Empirical Study on the Capacity of Bottlenecks on the Basic Suburban Expressway Sections in Japan. *Proc., 4th International Symposium on Highway Capacity*, 2000, pp. 120–129.
9. Xing, J., K. Sagae, and E. Muramatsu. Balance Lane Use of Traffic to Mitigate Motorway Traffic Congestion with Roadside Variable Message Signs. *Proc., ITS World Congress*, Busan, Korea, 2010.
10. Xing, J., E. Muramatsu, and T. Harayama. Balance Lane Use with VMS to Mitigate Motorway Traffic Congestion. *International Journal of Intelligent Transportation Systems Research*, Vol. 12, No. 1, 2014, pp. 26–35.
11. Furuichi, T., S. Yamamoto, M. Kotani, and M. Iwasaki. Characteristics of Spatial Speed Change at Motorway Sag Sections and Capacity Bottlenecks. Presented at 82nd Annual Meeting of the Transportation Research Board, Washington, D.C., 2003.
12. Brilon, W., and A. Bressler. Traffic Flow on Freeway Upgrades. *Transportation Research Record: Journal of the Transportation Research Board*, 2004. 1883: 112–121.
13. Yoshizawa, R., Y. Shiomi, N. Uno, K. Iida, and M. Yamaguchi. Analysis of Car-Following Behavior on Sag and Curve Sections at Intercity Expressways with Driving Simulator. *International Journal of Intelligent Transportation Systems Research*, Vol. 10, No. 2, 2012, pp. 56–65.
14. Laval, J. A. Effects of Geometric Design on Freeway Capacity: Impacts of Truck Lane Restrictions. *Transportation Research Part B: Methodological*, Vol. 43, No. 6, 2009, pp. 720–728.
15. Komada, K., S. Masukura, and T. Nagatani. Effect of Gravitational Force Upon Traffic Flow with Gradients. *Physica A: Statistical Mechanics and Its Applications*, Vol. 388, No. 14, 2009, pp. 2880–2894.
16. Oguchi, T., and R. Konuma. Comparative Study of Car-Following Models for Describing Breakdown Phenomena at Sags through Driving Simulator Experiments. *Proc., 16th ITS World Congress*, Stockholm, Sweden, 2009.
17. Yokota, T. K. A Study of AHS Effects on Traffic Flow at Bottlenecks. Towards the New Horizon Together. *Proc., 5th World Congress on Intelligent Transport Systems*, Seoul, Korea, 12–16 October 1998, Paper No. 3200, 1998.
18. Treiber, M., A. Hennecke, and D. Helbing. Congested Traffic States in Empirical Observations and Microscopic Simulations. *Physical Review E*, Vol. 62, No. 2, 2000, p. 1805.
19. Ozaki, H. Assistance of Drivers to Mitigate Highway Capacity Problem. In *Steps Forward,* Intelligent Transport Systems World Congress, 1995.
20. Cetin, M., P. Foytik, S. Son, A. J. Khattak, R. M. Robinson, and J. Lee. Calibration of Volume-Delay Functions for Traffic Assignment in Travel Demand Models. Presented at 91st Annual Meeting of the Transportation Research Board, Washington, D.C.
21. Li, K., L. Kang, W. Zhang, and B. Li. Comparative Analysis of Genetic Algorithm and Ant Colony Algorithm on Solving Traveling Salesman Problem. *Proc., WSCS'08 IEEE International Workshop on Semantic Computing and Systems*, IEEE, 2008, pp. 72–75.
22. Bai, Q. Analysis of Particle Swarm Optimization Algorithm. *Computer and Information Science*, Vol. 3, No. 1, 2010, p. 180.
23. Duan, H., Q. Luo, Y. Shi, and G. Ma. Hybrid Particle Swarm Optimization and Genetic Algorithm for Multi-Uav Formation Reconfiguration. *IEEE Computational Intelligence Magazine*, Vol. 8, No. 3, 2013, pp. 16–27.
24. Papageorgiou, M., C. Diakaki, V. Dinopoulou, A. Kotsialos, and Y. Wang. Review of Road Traffic Control Strategies. *Proceedings of the IEEE*, Vol. 91, No. 12, 2003, pp. 2043–2067.
25. Ros, B. G., V. L. Knoop, B. Van Arem, and S. P. Hoogendoorn. Empirical Analysis of the Causes of Stop-and-Go Waves at Sags. *IET Intelligent Transport Systems*, Vol. 8, No. 5, 2014, pp. 499–506.

The Standing Committee on Traffic Flow Theory and Characteristics (AHB45) peer-reviewed this paper (18-05319).

Article

TRR
JOURNAL OF THE TRANSPORTATION RESEARCH BOARD

Transportation Research Record
2018, Vol. 2672(20) 203–214

DOI: 10.1177/0361198118794717
journals.sagepub.com/home/trr
SAGE

Traffic Shockwave Detection in a Connected Environment using the Speed Distribution of Individual Vehicles

Amr Elfar[1], Connie Xavier[2], Alireza Talebpour[2], and Hani S. Mahmassani[1]

Abstract

Traffic shockwaves reflect a transition from the free-flow traffic state to the congested state. They create potentially unsafe situations for drivers, increase travel time, and significantly reduce freeway capacity. Several shockwave detection methods based on loop detector data and other traditional databases have been around for years. However, these methods face certain accuracy and reliability issues, many of which are due to the nature and accuracy of available data. Connected-vehicles technology is expected to provide reliable and accurate data about individual vehicles that can be potentially used for shockwave detection. Accordingly, this paper presents a novel method to identify shockwave formation and track its propagation based on the speed distribution of individual vehicles available through connected-vehicles technology. In addition, this paper analyzes the impact of partial connectivity on shockwave identification and compares the accuracy of the proposed method to a wavelet transformation-based method. Vehicle trajectories from the Next Generation Simulation (NGSIM) US-101 dataset were analyzed. The analysis shows a consistent pattern in which shockwave formation, indicated by a drop in speed propagating over space and time, is associated with a sharp increase in the value of speed standard deviation (SSD). Furthermore, the analysis shows that shockwaves can be accurately identified using vehicle trajectory data from connected vehicles at minimum 30% market penetration rates. Finally, the results show that the SSD of individual vehicles is more responsive to shockwave formation than the mean speed wavelet transformation, which can lead to improved shockwave detection accuracy.

Traffic shockwaves reflect a transition from the free-flow traffic state to the congested state (*1*, *2*) and entail several negative impacts on safety, performance, and emissions (*3*). They could create unsafe conditions as drivers might have to decelerate sharply when approaching a stream of slow-moving vehicles. Additionally, the frequent acceleration/deceleration behavior of vehicles in a propagating shockwave (traffic jam) increases fuel consumption and emissions. Shockwaves cause a significantly lower throughput than the nominal capacity of the freeway (*3*, *4*).

The traditional approach to detecting shockwaves on a freeway is to track changes in speed and density over space and time based on fundamental diagrams (*1*, *3*, *5*, *6*) and in accordance with the Lighthill–Whitham–Richards first-order continuum traffic theory (*7*, *8*). The main limitation of this approach is that density along a freeway is difficult to measure in practice using traditional data collection methods such as loop detectors. It is rather estimated from other measures such as occupancy. Furthermore, locating the starting point of shockwaves may not be accurate as tracking changes in density over freeway segments depends on the number of installed loop detectors and the spacing between them (i.e., data resolution).

Wavelet transformation is a different approach proposed by Zheng et al. (*9*) to detect freeway bottlenecks and oscillations. The technique is a time–frequency decomposition tool that extracts useful information from stationary time-series data, and it has been used in different intelligent transportation system applications (*9*). In their study, the authors used wavelet transformation to

[1]Civil and Environmental Engineering, Northwestern University, Transportation Center, Evanston, IL
[2]Zachry Department of Civil Engineering, Texas A&M University, College Station, TX

Corresponding Author:
Address correspondence to Hani S. Mahmassani: masmah@northwestern.edu

identify the time and location of a bottleneck activation by tracking the changes in average speeds recorded by loop detectors. They also proposed another application to wavelet transformation in which they identified traffic oscillations (shockwaves) by tracking the acceleration–deceleration cycles of individual vehicle trajectories using NGSIM data. Assuming that such data will be available through connected-vehicles technology, Talebpour et al. (*10*, *11*) adopted a similar wavelet-based approach to identify shockwaves and to activate a speed harmonization system.

Connected-vehicles technology provides various opportunities for enhancing the efficiency, productivity, sustainability, safety, and reliability of transportation systems (*10*, *12*). One of the potential applications of the new technology is generating detailed vehicle trajectories through vehicle-to-vehicle (V2V) and vehicle-to-infrastructure (V2I) communications. Connected vehicles can send out basic safety messages that include real-time updates of a vehicle's status, including its location, speed, and acceleration to other vehicles or to the infrastructure and receive the same types of information in return. The detailed vehicle movement can be used to improve current traffic control algorithms (*13*, *14*), such as speed harmonization (*11*, *15*) and queue warning. An example of such improvement is the possibility to accurately detect shockwaves at the onset of formation, providing more flexibility in mitigating congestion.

Perturbations in traffic flow, a main factor causing traffic congestion (*2*, *16*), can be identified using V2V/V2I communication technology. Unlike aggregate datasets generated by loop detectors, detailed vehicle trajectories can be used to measure disturbances caused by individual vehicles at a much higher resolution. One way to estimate traffic disturbances is through estimating the speed variation among individual vehicles in a road segment (*14*). Previous studies suggested, on the basis of both measurements and theory using stochastic macroscopic models (*17*, *18*) and microscopic models (*19*), that the widening of the speed distribution can serve as an early indicator of traffic breakdown. Kühne suggested using a radar to measure the speed variation of vehicles over space as an early warning of traffic breakdown (*17*, *18*). The limitation in this method, however, is that it can only detect breakdown at the radar location and cannot track shockwave propagation across a segment.

Accordingly, this paper presents a novel method to identify shockwave formation and track its propagation based on the speed distribution of individual vehicles available through connected-vehicles technology. In addition, this paper analyzes the impact of partial connectivity on shockwave identification and compares the accuracy of the proposed method to a wavelet transformation-based method (*9*).

This paper is organized as follows: the next section provides a brief description of the vehicle trajectory data used for this study. After that, the Methodology section describes how the traffic properties were defined and estimated. Next, the Results and Analysis section discusses the SSD patterns found in the data and their relationship to shockwave formation for both a fully and partially connected transportation system. In addition, the section provides a comparison between the accuracy of using SSD waves and speed wavelet transformation for shockwave detection and tracking. Finally, the Conclusion section summarizes the study's major findings, applications, and future work.

Data Description

The Next Generation Simulation (NGSIM) dataset was utilized in this study (*20*). The vehicle trajectories were collected on the southbound direction of US-101/Hollywood Freeway in Los Angeles, CA on June 15, 2005, from 7:50 to 8:35 a.m. The study segment contained five main lanes, one on-ramp, one off-ramp, and was approximately 2,100 ft long. The final dataset includes individual vehicles' geographical coordinates, speeds, lane positions, and accelerations at every one-tenth of a second.

Preprocessing

Montanino and Punzo's (*21*) multistep corrections to NGSIM's vehicle trajectories were applied as the first step for data preprocessing; however, no outliers were identified using the acceleration threshold suggested by the authors and therefore interpolation was not necessary. The second step was to define spatial and temporal boundaries to remove errors on the aggregated estimates of traffic properties. The identified spatial boundary was the first 100 ft of the 2,100 ft segment since mounted cameras might not have recorded all vehicles entering the segment.

Methodology

The 2,000 ft segment (after identifying the spatial boundary) was divided into 200 ft sections to easily track traffic shockwaves (in this context, the propagation of speed drops over space and time). Different section lengths can be used for this analysis. While increasing the section length may reduce noise in aggregated traffic properties, it would reduce the accuracy of tracking shockwaves as the change in speed would be tracked over longer sections. Lu et al. suggested using 500 ft (150 m) sections for aggregating NGSIM trajectories to estimate fundamental diagrams (*22*); however, that would be too long for the purpose of tracking shockwaves as the whole segment length is only 2,000 ft. To easily track shockwaves over time, traffic properties were aggregated over 10 s time steps, an aggregation duration that was suggested in the Lu et al. study as well (*22*).

Flow, mean speed, and density for each section at each time step were calculated using Edie's generalized definitions for individual facilities (*23*, *24*):

$$q(A) = \frac{d(A)}{|A|} \quad (1)$$

$$k(A) = \frac{t(A)}{|A|} \quad (2)$$

$$v(A) = \frac{d(A)}{t(A)} \quad (3)$$

where $q(A)$, $k(A)$, $v(A)$ denote flow, density, and mean speed for observed vehicles in section A, respectively, $d(A)$ and $t(A)$ represent the total distance traveled and total time spent by all vehicles in section A, respectively. Finally, $|A|$ is the area covered by section A (200 ft section multiplied by a 10 s interval).

The SSD of individual vehicles in the main five lanes was calculated for each section at each time step. The following two steps were followed for this analysis: (1) calculating the average speed of each vehicle over a 10 s period in a specific section; and (2) calculating the standard deviation of the average speeds of all individual vehicles in that section. This definition estimates spatial deviation in speeds among the vehicles in the five main lanes. The definition is slightly different from previous studies (*17*, *18*), which used radar to measure individual vehicle speeds crossing a specific point over time. After estimating the traffic properties, the first and last time steps (10 s) for each section were discarded as those time steps did not have complete trajectory information for the full 10 s duration and flow/density values were underestimated.

The above calculations were performed for both full and partial connectivity cases. For the first case, all vehicles were assumed to be able to transmit their trajectories (location, speed, acceleration, and so forth) at all times. However, full connectivity at the early stage of technology deployment is highly unlikely. Therefore, in the second case, the mean speed and the speed standard deviation were calculated assuming partial connectivity for different market penetrations. In the partial connectivity case, a percentage of vehicles were assumed to be connected and able to transmit their trajectories. Therefore, the calculations were performed for a sample of randomly selected vehicles. For example, at 30% market penetration, 30% of vehicles in the dataset were randomly selected using the vehicle ID and the mean speed/standard deviation were calculated for this sample only for all sections at all time steps.

Note that the mean speed and the speed standard deviation estimates calculated for a sample of vehicle trajectories (partial connectivity) using the above-mentioned method can be very close to the true estimate (full connectivity), as will be discussed in the following section. However, estimating density and flow for the partially connected stream will give values that are less than the true estimates. The reason is that only connected vehicles will be counted in this case instead of counting the actual number of vehicles, which will result in a lower density/flow estimate (unless model-based imputation techniques are invoked). The focus of this analysis, however, was to find out whether the SSD calculated at partial connectivity can still be used to detect traffic shockwaves. This will be discussed further in the following section.

Results and Discussion

SSD as an Indicator of Shockwave Formation

The first step in the analysis is to explore the existence of patterns between the SSD and shockwave formation in the dataset. To visualize those patterns, a time–space diagram was created for the mean speed and the SSD, as shown in Figure 1. The direction of travel is from section 1 to section 10. The figure indicates that shockwave formation is associated with a jump in the SSD. Figure 1*a*, for example, shows six backward-propagating shockwaves that start in section 6 in the period 7:50–8:05 a.m. and a seventh shockwave that starts outside the segment of study. Figure 1*b* shows that each shockwave that started in section 6 is associated with a jump in the SSD that is also backward-propagating. The other periods show the same pattern even though it is much noisier than the first period as the sections become more congested.

Another way to look at the relationship between speed and SSD is through a time-series graph. Figure 2 shows the mean speed changes over time compared to SSD changes during the time periods (a) 7:50–8:05 a.m., and (*b*) 8:05–8:20 a.m. The blue line refers to the mean speed on the left *y*-axis, while the orange line refers to the SSD on the right *y*-axis. This figure shows that speed drops are associated with the peaks in SSD values that usually occur before the speed drops start. This correlation is clearer in the first time period (Figure 2*a*) than in the second one (Figure 2*b*), where most shockwaves in the second period start outside the study segment, as seen in Figure 1*c*. Therefore, Figures 1 and 2 show that peaks in the SSD are better indicators of shockwaves that are starting to form than those which have already formed and are propagating.

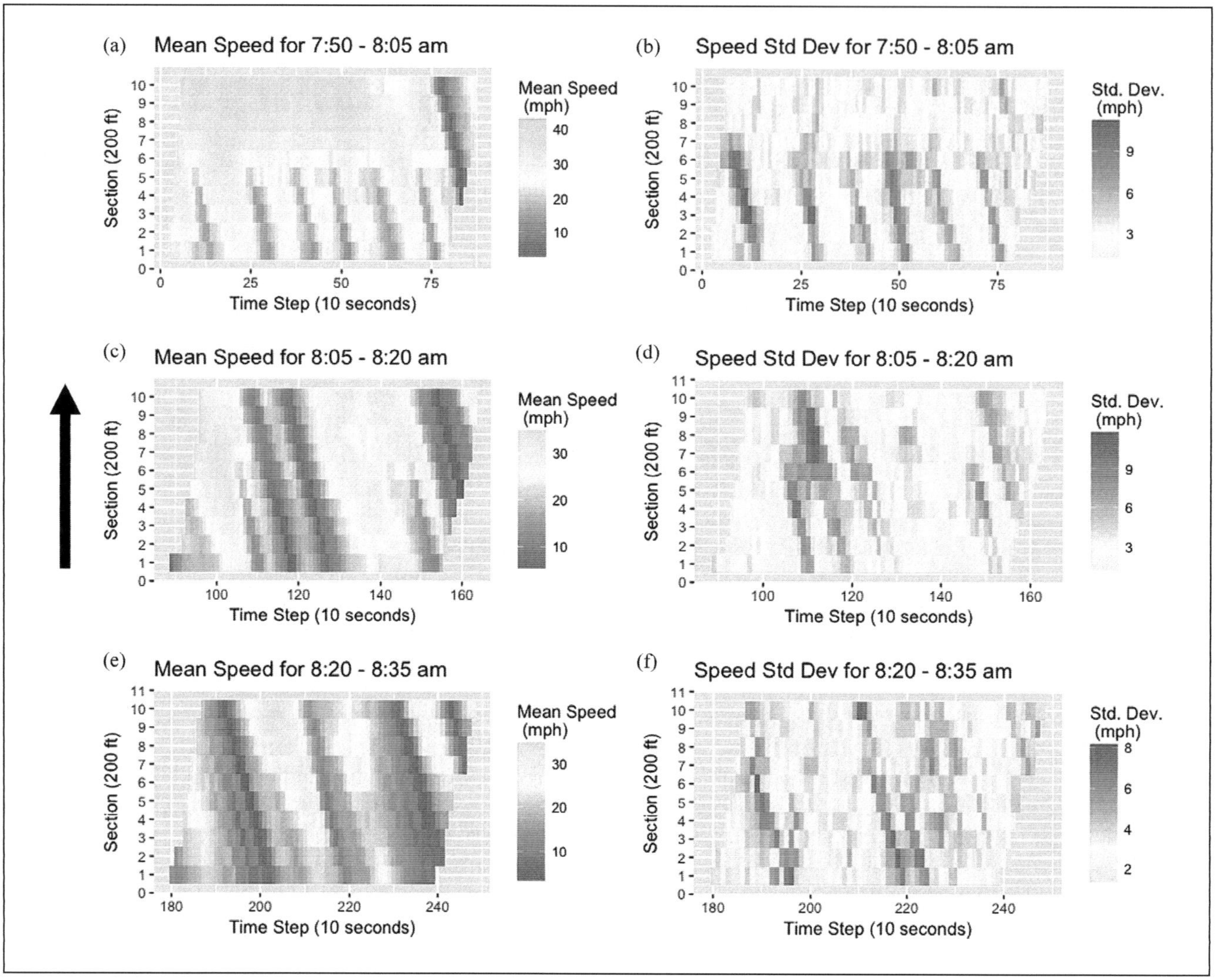

Figure 1. Time–space diagrams for (*a*) mean speed 7:50–8:05 a.m., (*b*) SSD 7:50–8:05 a.m., (*c*) mean speed 8:05–8:20 a.m., (*d*) SSD 8:05–8:20 a.m., (*e*) mean speed 8:20–8:35 a.m., and (*f*) SSD 8:20–8:35 a.m.

Propagation of SSD Waves versus Propagation of Shockwaves

While previous figures showed a pattern between SSD and mean speed, the figures do not show in which section SSD waves start and how they propagate backward compared to shockwaves. Figure 3 shows the SSD waves (orange line) and speed shockwaves (blue line) in sections 3 through 7. In section 7 (where speed shockwaves have not started yet), the SSD waves are starting to form, signaling a reduction in speed in upstream sections. In section 6, the SSD waves show a clear signal while speed shockwaves are starting to form. In sections 5 through 3, the speed shockwaves and the SSD waves propagate backward simultaneously at the same speed. Two conclusions can be drawn from this figure: (1) SSD waves start to form before speed shockwaves, signaling a reduction in speed in upstream sections; and (2) the SSD waves propagate in conjunction with speed shockwaves at almost the same speed. The propagation speed of SSD waves was estimated to be 10.2 mph, which is the slope of the lines passing through the wave peaks. This is close to the 11.4 mph shockwave propagation speed that was estimated by Lu and Skabardonis in their analysis of NGSIM trajectories (*25*).

Flow–Density–SSD Relationship

Figure 4*a* shows a flow–density diagram represented by the blue dots in addition to the SSD plotted against

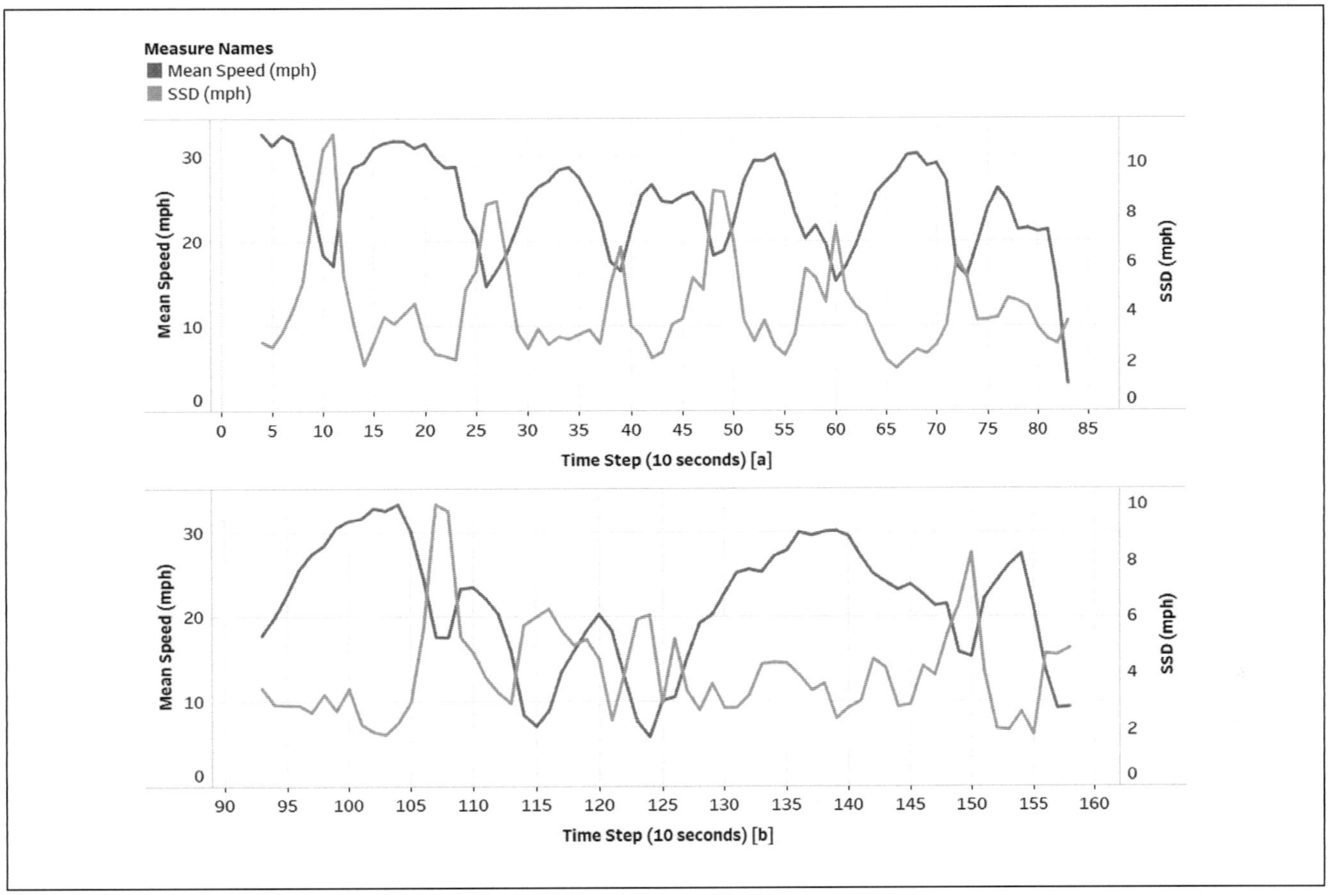

Figure 2. Time-series graph for mean speed and SSD in section 4 during (*a*) period 7:50–8:05 a.m. and (*b*) period 8:05–8:20 a.m.

density on the right *y*-axis, represented by the orange dots. Each dot represents a value calculated for a section and a time step. The flow–density diagram (left *y*-axis in Figure 4*a*) shows that the traffic is operating close to capacity before the breakdown, as expected in the morning peak hour. The SSD–density diagram (right *y*-axis in Figure 4*a*) shows that SSD increases around 80–90 vpmpl (vehicles per mile per lane) density value, which is a typical range for critical density before traffic breakdown (*22*). The increase in SSD and the scatter in the diagram reflects the instability of traffic flow during the transitioning phase from the uncongested to the congested state (*14*), as evident from the oscillations in Figure 1*a*. After breakdown, SSD values drop, reflecting less variation in speed of vehicles moving in the congested traffic regime. Note that this graph confirms Kühne's findings in his study on freeway speed distribution (*18*). Figure 4*b* shows the empirical relationship between traffic flow and SSD. The trend line in the graph suggests that flow generally decreases at higher SSD values, where traffic is usually unstable. However, it is important to note that the graph is highly scattered and deriving a definitive relationship requires further research.

Speed–Density–SSD Relationship

Figure 4*c* shows the speed–density graph, represented by red dots, in addition to the SSD plotted against density on the right *y*-axis, represented by the orange dots. As in the case of the above flow–density diagram, each dot represents a value calculated for a section and a time step. The speed–density diagram shows typical traffic behavior where the mean speed drops as the traffic density increases. Adding SSD to the graph shows around which speed the SSD starts to increase. In this case, SSD increases sharply right before a speed of 15–20 mph and drops after breakdown at around 10 mph. As in the previous case, the graph is highly scattered and the above-mentioned values are only approximations. Figure 4*d* shows the relationship between mean speed and SSD. Similar to the trend in Figure 4*b*, the traffic mean speed decreases at higher SSD values. Both of the aforementioned trends suggest that SSD can be used as an indicator of traffic disturbances. As the graphs show high scatter, however, deriving representative mathematical relationships requires further analysis and more data.

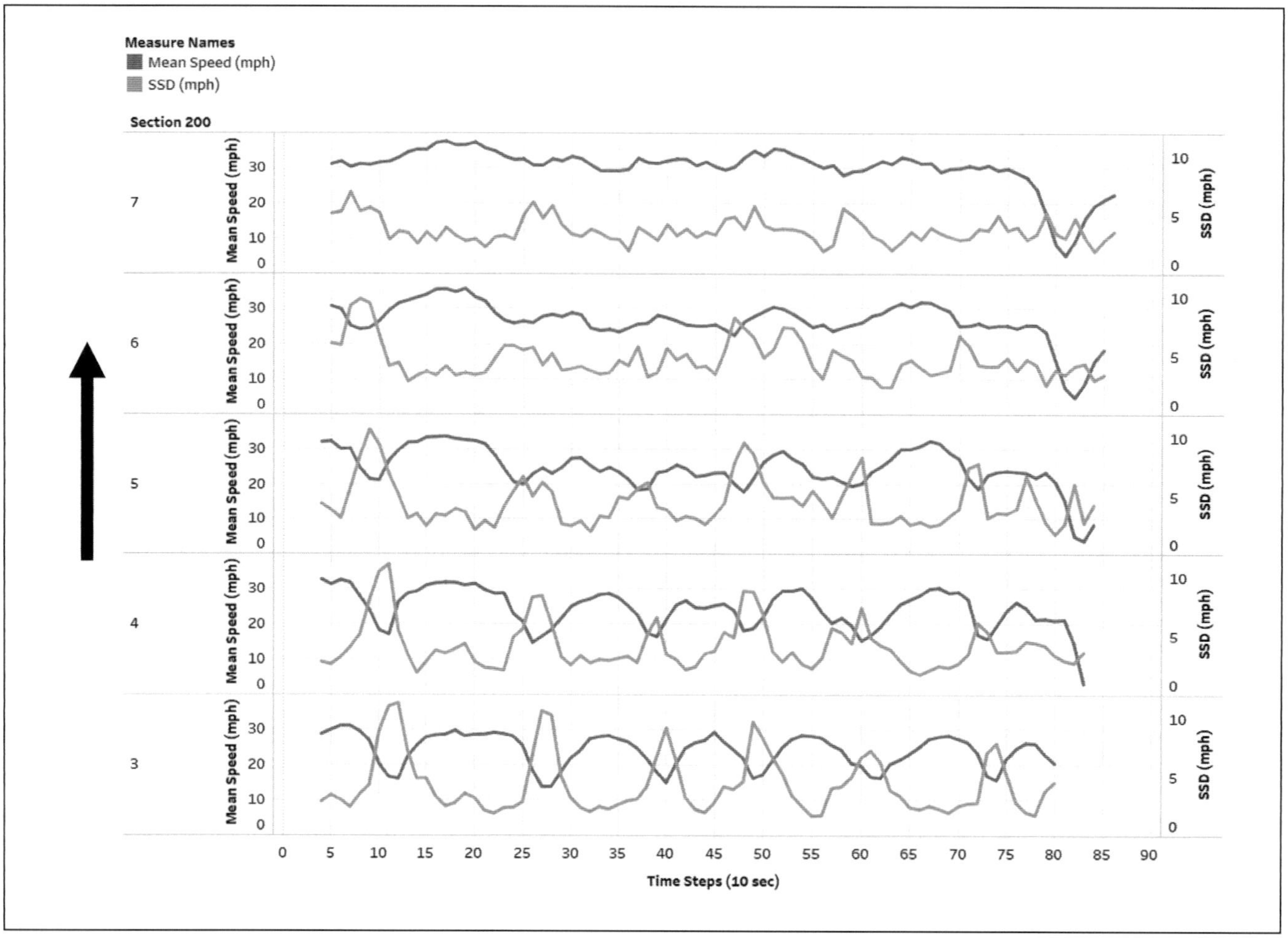

Figure 3. Propagation of SSD spikes compared to the propagation of shockwaves during the period 7:50–8:05 a.m. Direction of travel is from section 3 to section 7.

SSD Waves with Partial Connectivity

The above-mentioned graphs and discussions assume full connectivity of vehicles, in which case all vehicles are able to transmit their trajectories (speed, location, lane, and so forth). However, it is unlikely that traffic streams would be fully connected, at least in the early stages of technology deployment. Therefore, the remaining sections will examine the SSD waves calculated using a sample of the provided trajectories, simulating scenarios in which only part of the vehicles in the traffic stream is connected and able to transmit their trajectories. For example, in the 30% connectivity scenario, the SSD will be calculated using trajectories of 30% of the vehicles in the dataset. Those vehicles are selected randomly using the vehicle ID. More importantly, the analysis will show whether the SSD waves detected for partially connected traffic can still be an indicator of shockwave formation.

Figure 5 shows SSD waves (colored lines) calculated for 10%, 20%, 30%, 70%, and 100% market penetrations. The mean speed in the graph (blue lines) is calculated at 100% connectivity to compare SSD waves to the actual speed shockwaves. The figure shows that at low market penetrations, specifically 10% and 20%, SSD could not be estimated for some time steps because no connected vehicles were detected (see disconnected lines in Figure 5). For market penetration of greater than 30%, SSD could be estimated for all time steps. Furthermore, the SSD–speed shockwave patterns are clear in the figure for medium to high market penetrations, specifically above 30%. This suggests that SSD waves can still be an indicator of shockwave formation at partial market penetrations above 30%. Note that choosing different seeds to randomly select the sample of vehicles may show slightly different results. Moreover, it can be observed from the graph that at some time steps at very low market penetrations, SSD is either overestimated or underestimated, which means that low market penetrations can affect the quality of the SSD estimate.

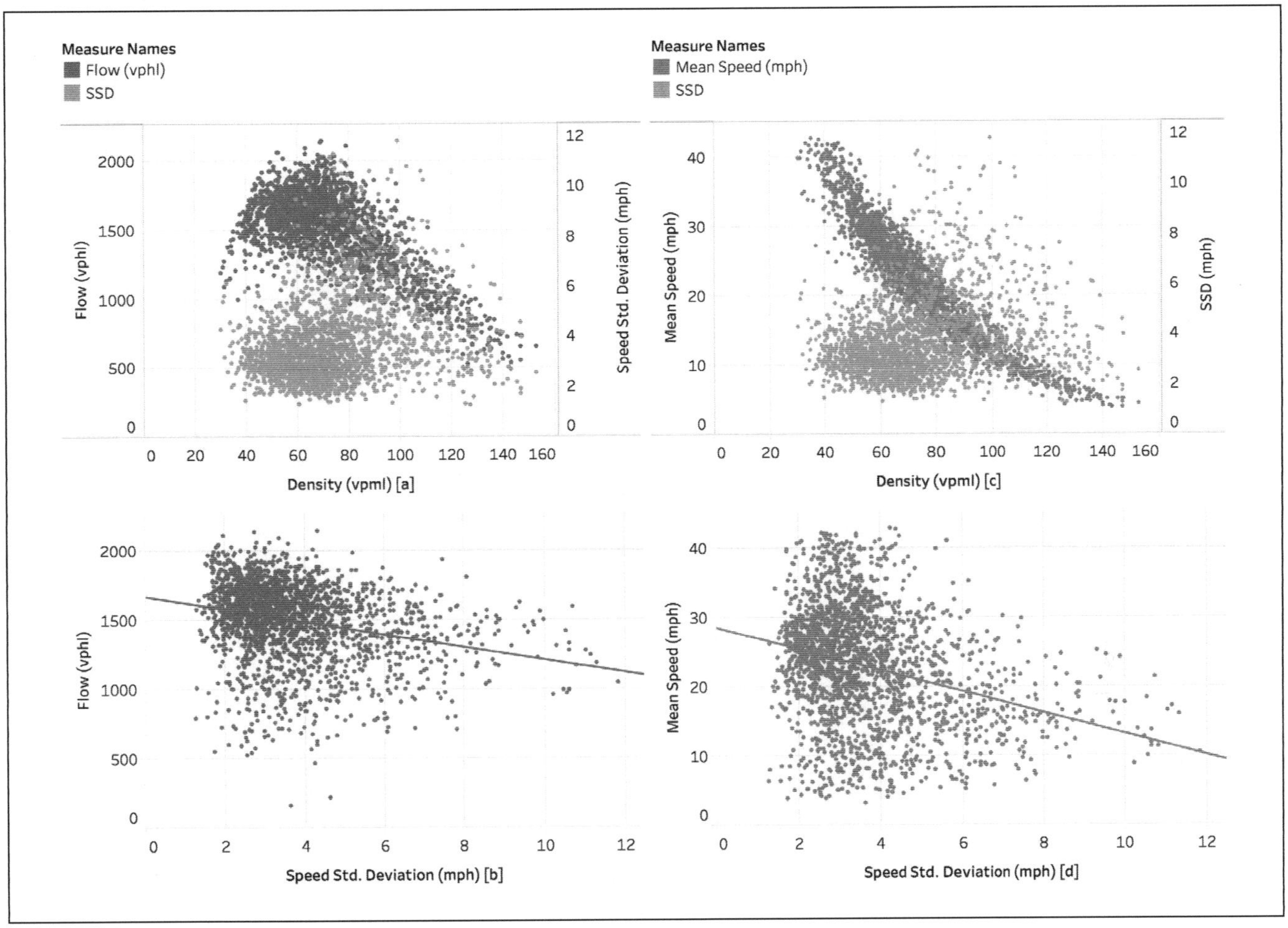

Figure 4. (*a*) Flow–density–SSD; (*b*) flow–SSD; (*c*) mean speed–density–SSD; (*d*) mean speed–SSD for all periods (7:50–8:35 a.m.).

As an indicator for the quality of SSD estimates, Figure 6 shows the number of vehicles used to estimate SSD at different market penetrations. Generally, a higher number of observations (the number of vehicles in this case) used to estimate a statistic leads to a higher quality estimate. At 10% and 30% market penetrations, the figure shows that the number of vehicles used to estimate SSD is below 15 (red lines), which suggests a low-quality estimate. At 50% market penetration, the quality increases, with 15–25 observations used to estimate SSD for most time steps. Above 70% market penetration, the quality of the estimate is high, with more than 25 observations used to estimate SSD for most time steps. These results support the pattern found in Figure 5.

Propagation of SSD Waves at Different Market Penetrations

Figure 7 shows the SSD waves estimated at 30% (green lines) and 50% (purple lines) market penetrations for sections 2 through 5 during the period 7:50–8:05 a.m. The figure confirms that even for partially connected traffic streams, SSD waves can still serve as an indicator of shockwave formation because (1) the propagation of waves can be tracked over space and time, as seen in this figure, and (2) the pattern where a significant reduction in speed is associated with a peak in SSD is still evident.

Comparison between SSD Waves and Speed Wavelet Transformation for Detecting Shockwaves

In a study by Zheng et al. (*9*), the authors proposed a wavelet transformation (WT) technique to identify bottleneck locations and activation times on freeways. WT is a time–frequency decomposition tool that extracts useful information from non-stationary time-series (*9*). In this method, the location of shockwaves and their formation time was extracted from transformed speed time-series. This is done by capturing local changes over time—a speed drop, for example—through moving the wavelet's location while squeezing and dilating the wavelet's window. In their study, the authors used WT to

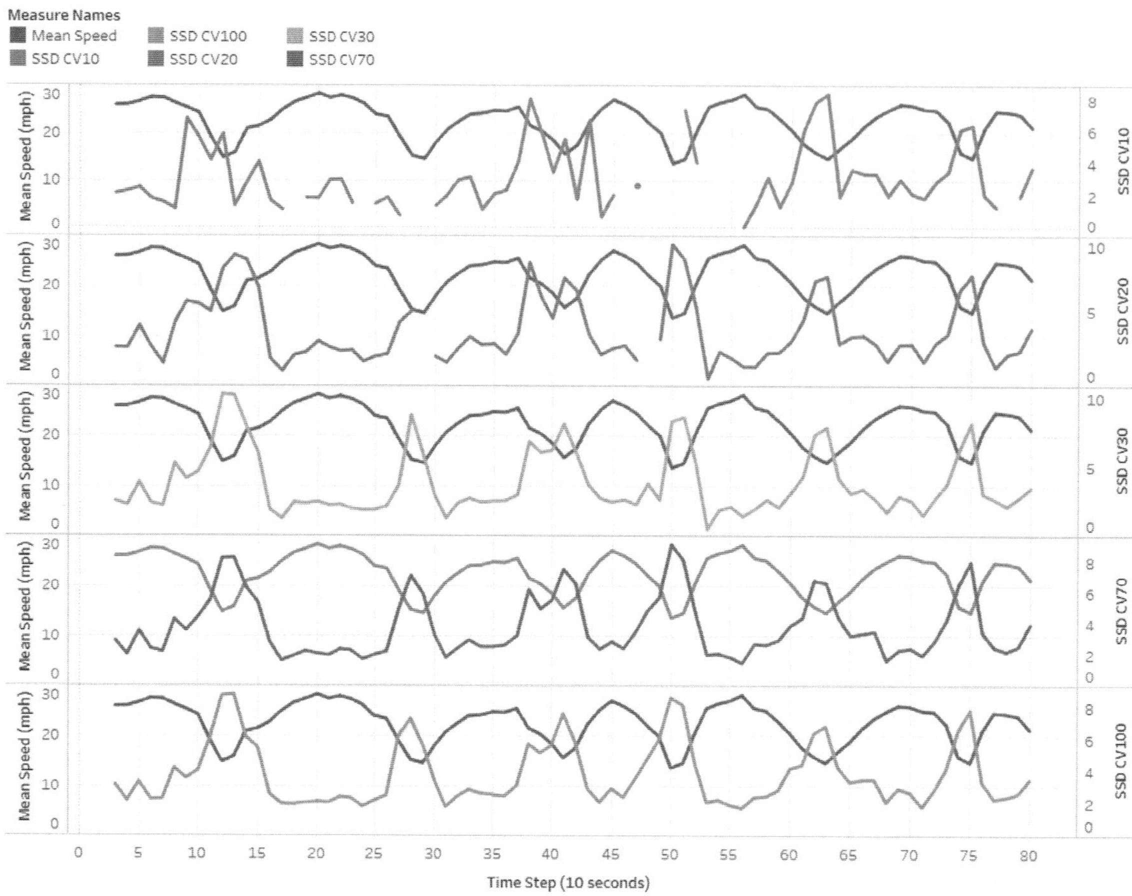

Figure 5. Speed standard deviation waves versus speed shockwaves at different market penetrations (10%, 20%, 30%, 70%, and 100%) in section 2 during the period 7:50–8:05 a.m.

detect shockwave (bottleneck) formation on US-101 by transforming speed time-series of 10 loop detectors. They identified the starting point of the shockwave by tracking the wavelet energy peaks that result from significant reductions in speed.

Figure 8 compares speed WT (green line) to the SSD wave (orange line) estimated for section 3 during the period 7:50–8:05 a.m. To estimate the wavelet energy distribution, the Mexican hat wavelet was used to transform the speed time-series and the distribution was estimated by averaging the wavelet energy over scales 1 through 64, as suggested by Zheng et al. (*9*) and Talebpour and Mahmassani (*10*). As seen in the figure, the wavelet energy is sensitive to every significant change in speed, whether it is a decrease or an increase. To identify shockwaves in this case, cycles of energy peaks need to be tracked. For example, the first peak identifies a shockwave (reduction of speed) and the second peak identifies the increase in speed that follows. Then, a third peak identifies a new shockwave. In the case of SSD waves, however, the peaks are only sensitive to reductions in speed. Therefore, identifying shockwaves can be done more efficiently by tracking SSD peaks only.

Another feature to compare between speed WT and SSD is the responsiveness to speed reductions. Figure 9 shows the propagation of wavelet energy and SSD estimated for sections 3 through 6 during the period 7:50–8:05 a.m. In section 6, where the shockwaves start to form, the wavelet energy (green line) does not show any sign of disruption (except for the seventh shockwave, which started outside the segment as seen in Figure 1*a*). The SSD values, on the other hand, are increasing, which indicates a reduction in speed that is about to occur in the following upstream section. In section 5, the speed reductions have started and the WT energy is responding to them. However, the full energy peak cycles (refers to an energy peak responding to a speed reduction and another peak responding to the recovery of speed after a

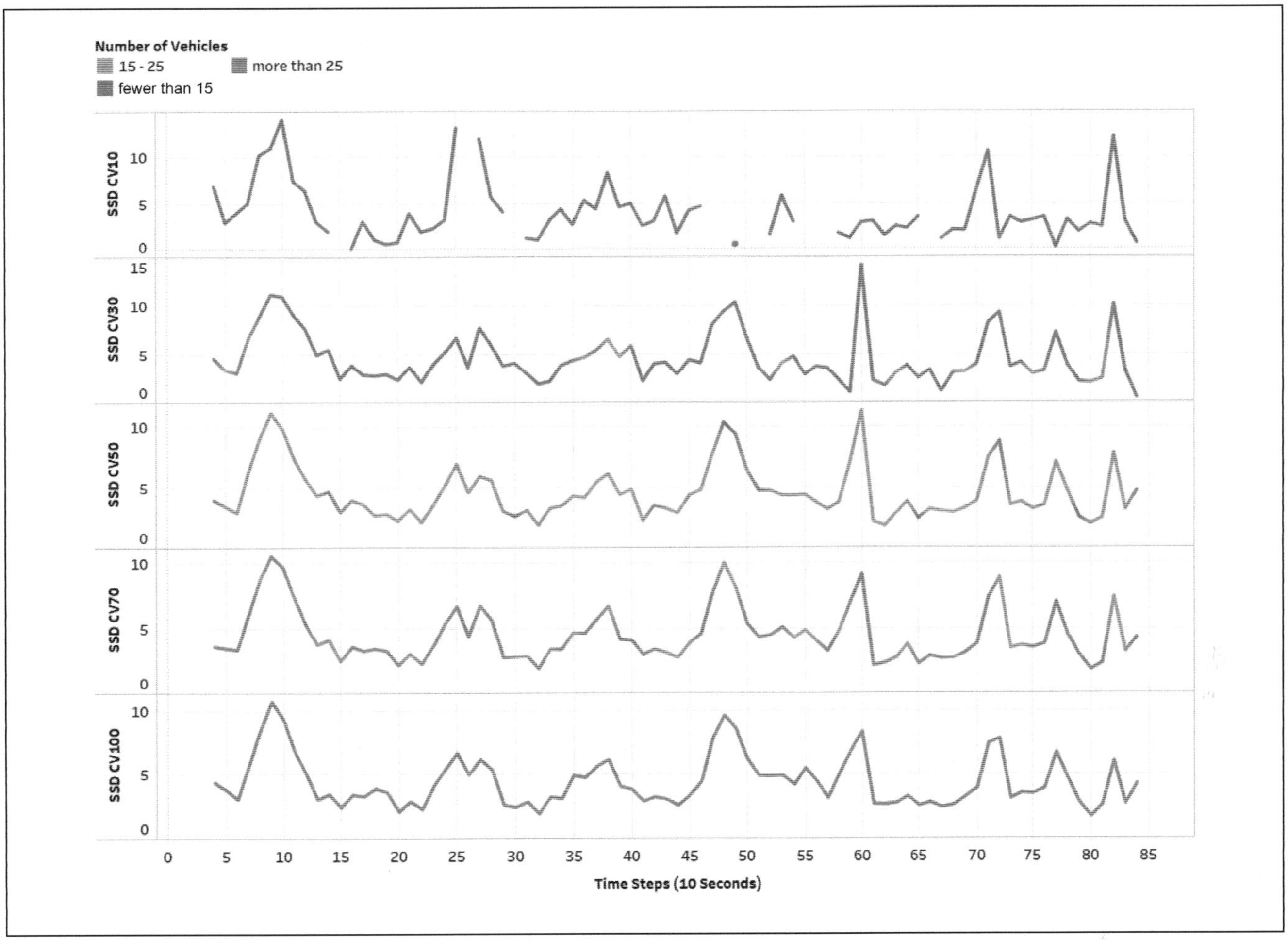

Figure 6. Number of vehicles used to estimate the SSD at different market penetrations for section 2 in the period 7:50–8:05 a.m.

shockwave is resolved—see Figure 8) have not developed yet. The SSD values in section 5 indicate the occurrence of shockwaves clearly. In section 4, the reductions in speed are more significant. Both speed WT and SSD reflect that. Again, however, the full peak cycles responding to speed reduction/increase have still not developed in section 4. Finally, in section 6, speed WT and SSD waves clearly reflect the formation of shockwaves. This graph suggests that while both techniques can be used to detect shockwaves, SSD waves are more responsive to the reductions in speed than speed WT, which can result in a higher detection accuracy.

Conclusions and Future Work

Shockwave detection at the onset of congestion formation can contribute to more effective active traffic management techniques. Shockwave detection is traditionally done by tracking changes in speed and density of road segments using loop detectors. The accuracy of this method is relatively low as it would depend on the number of loop detectors installed in a road segment and the spacing between them. Speed WT is another method to track shockwaves by decomposing speed time-series. The technique can be used on speeds estimated by loop detectors, in which case it suffers from the same drawbacks as the previous method, or it can be used on speeds of individual vehicles. In the latter case, however, a high percentage of vehicles need to be able to transmit their trajectories to track the vehicle that starts a shockwave accurately. This may be difficult in practice as the traffic flow is unlikely to be highly connected, at least not in the near term.

To address the above-mentioned limitations, this paper explored the effectiveness of a method to detect shockwaves using the speed distribution of individual vehicles in a connected environment. The method is based on previous theoretical and empirical findings connecting the speed distribution to the onset of flow breakdown, though not specifically related to shockwave formation (*17*, *18*). The main advantages of this method

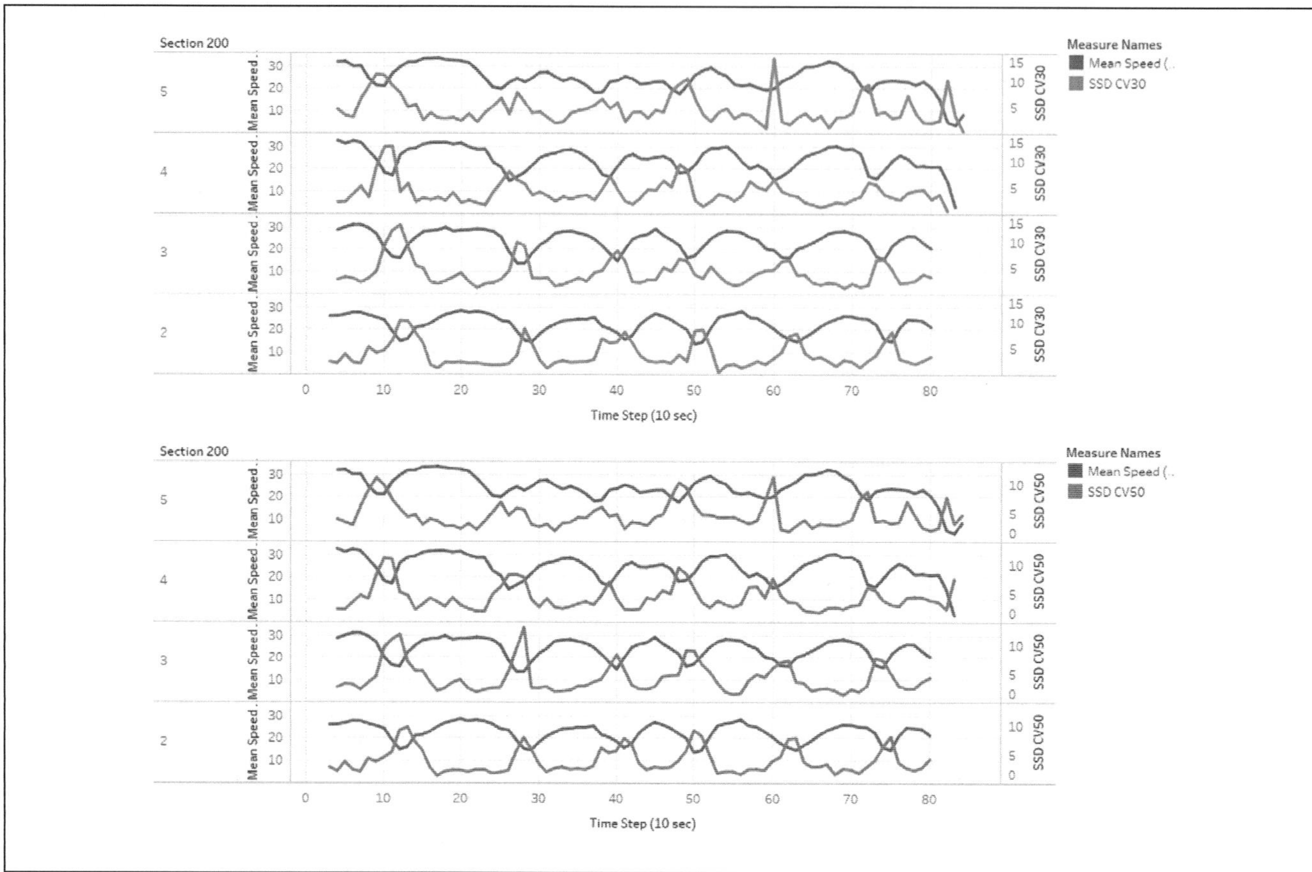

Figure 7. Propagation of SSD waves at partial market penetrations.

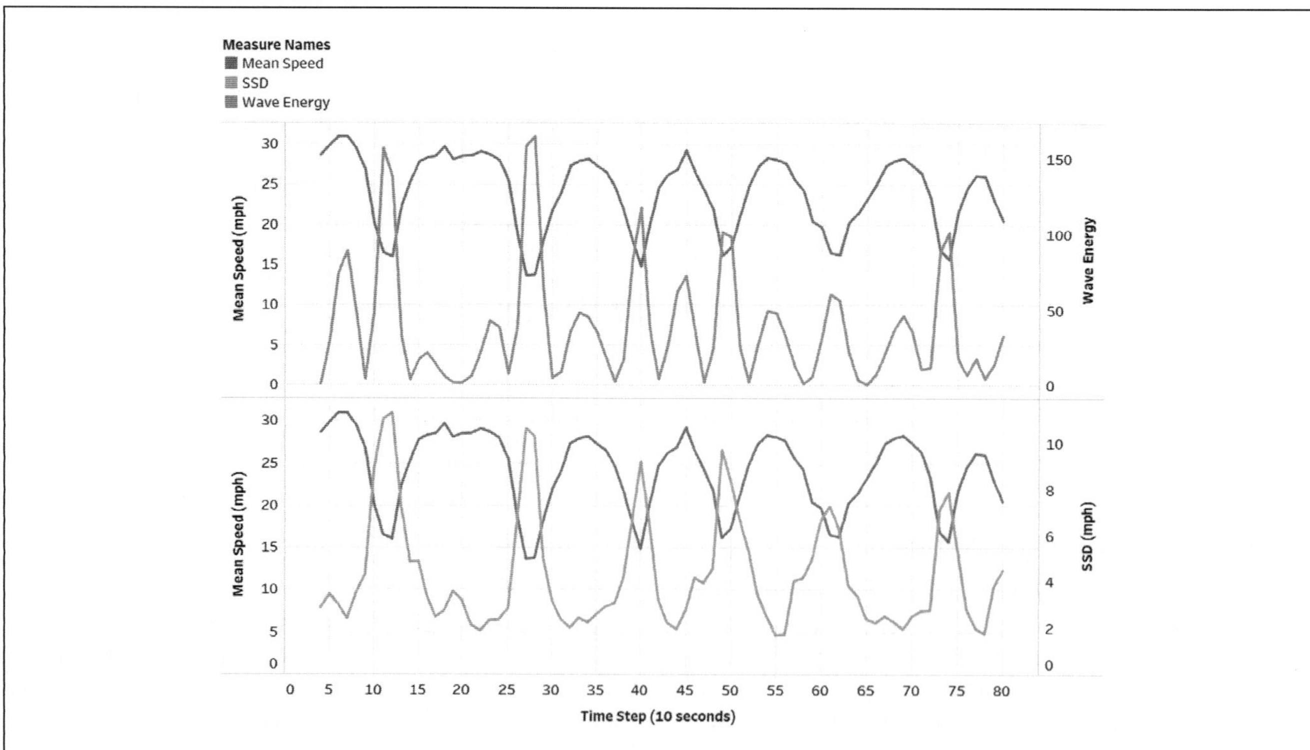

Figure 8. WT of mean speed versus SSD waves estimated for section 3 during the period 7:50–8:05 a.m.

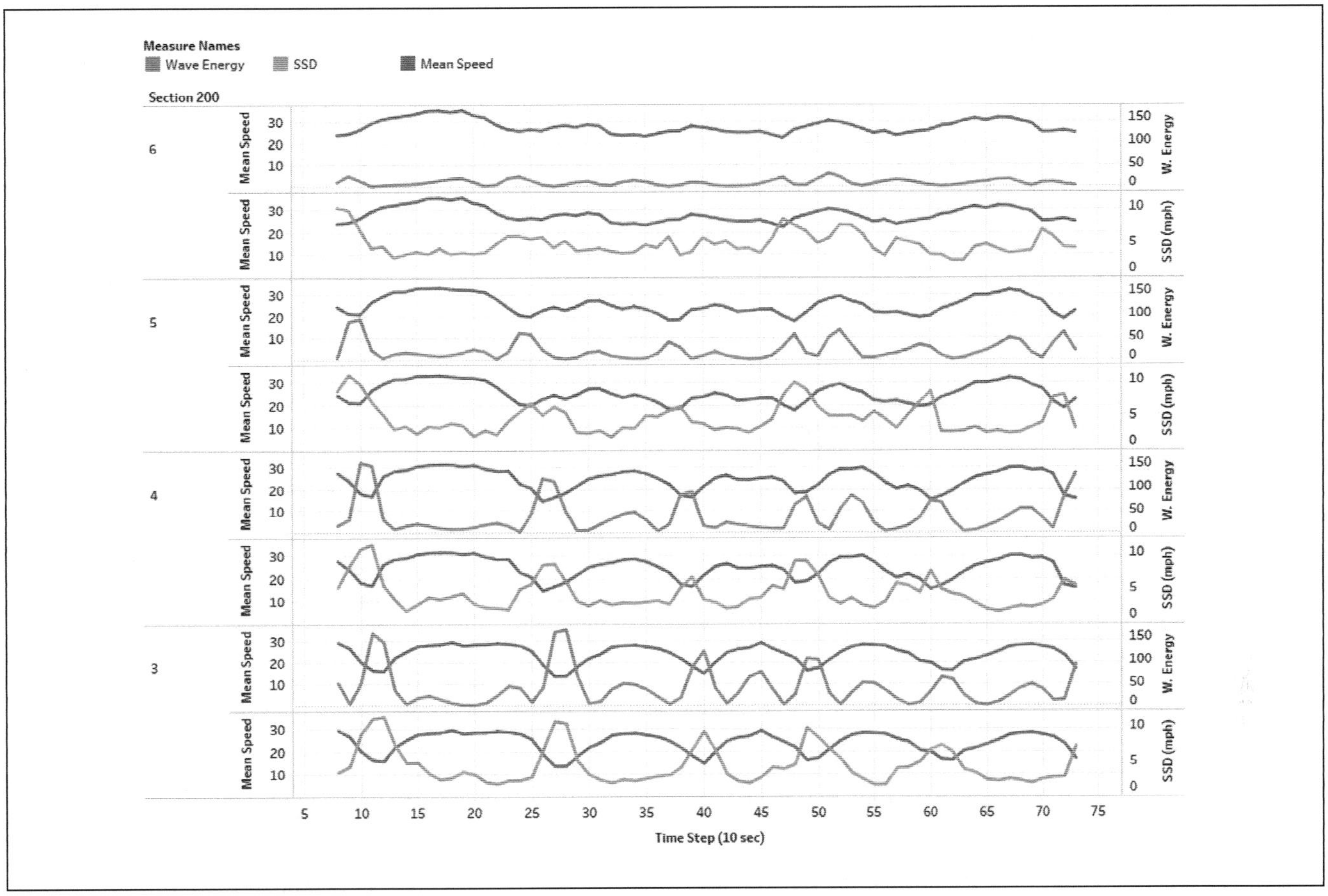

Figure 9. Propagation of wavelet energy and SSD over sections 3–6 during the period 7:50–8:05 a.m.

are: (1) shockwaves can be tracked accurately over small sections of a freeway (e.g., 200 ft segments were used in this analysis); and (2) shockwaves can be detected clearly for partially connected traffic streams. To test this method, vehicle trajectories from the NGSIM program were analyzed using the US-101 dataset for a 2,100 ft study segment in Los Angeles, CA. Two main cases were evaluated in this study: full connectivity and partial connectivity. In the case of full connectivity, it was assumed that all vehicles are able to transmit their detailed trajectories. In the partial connectivity case, only a percentage of vehicles were able to transmit their detailed trajectories and shockwaves were tracked using those trajectories only.

The results showed a consistent pattern in which shockwave formation (significant reduction in speed propagating over space and time) is associated with a sharp increase in the speed standard deviation value (SSD) that usually occurs before the start of the shockwave development. Results also showed that SSD waves propagate in conjunction with speed shockwaves at almost the same speed, confirming that SSD waves can serve as an indicator for shockwave formation and propagation. These patterns were evident in both the full and partial connectivity cases. For quality of the SSD estimate, the analysis showed that the quality improves at higher market penetration as the number of vehicles (observations) used to calculate the SSD is higher. Compared to the WT method, the analysis showed that SSD waves are more responsive to reductions in speed than speed WT, which can result in higher shockwave detection accuracy. The higher accuracy in detecting shockwaves in time and space can help improve the effectiveness of active traffic management techniques. It is worth mentioning, however, that this method was tested for a non-equilibrium traffic state in which traffic was transitioning from an uncongested to a congested state. In equilibrium (steady-state) traffic, a shockwave can occur with zero variance in speed and therefore may not be detected using this method.

Author Contributions

The authors confirm contribution to the paper as follows: study conception and design: AW, CX, AT, HSM; data collection: AW, CX, AT, HSM; analysis and interpretation of results:

AW, CX, AT, HSM; draft manuscript preparation: AW, CX, AT, HSM. All authors reviewed the results and approved the final version of the manuscript.

References

1. Treiber, M., and A. Kesting. *Traffic Flow Dynamics: Data, Models and Simulation*. Springer-Verlag, Berlin, Germany, 2013.
2. Treiber, M., A. Hennecke, and D. Helbing. Congested Traffic States in Empirical Observations and Microscopic Simulations. *Physical Review E*, Vol. 62, No. 2, 2000, p. 1805.
3. Hegyi, A., S. Hoogendoorn, M. Schreuder, H. Stoelhorst, and F. Viti. SPECIALIST: A Dynamic Speed Limit Control Algorithm Based on Shock Wave Theory. *Proc., 2008 11th International IEEE Conference on Intelligent Transportation Systems*, Beijing, China, IEEE, New York, NY, 2008, pp. 827–832.
4. Kerner, B. S., and H. Rehborn. Experimental Features and Characteristics of Traffic Jams. *Physical Review E*, Vol. 53, No. 2, 1996, p. R1297.
5. Hegyi, A., and S. Hoogendoorn. Dynamic Speed Limit Control to Resolve Shock Waves on Freeways: Field Test Results of the SPECIALIST Algorithm. *Proc., 13th International IEEE Conference on Intelligent Transportation Systems*, Funchal, Portugal, IEEE, New York, NY, 2010, pp. 519–524.
6. Jabari, S. E., and L. Wynter. Sensor Placement with Time-to-Detection Guarantees. *EURO Journal on Transportation and Logistics*, Vol. 5, No. 4, 2016, pp. 415–433.
7. Lighthill, M. J., and G. B. Whitham. On Kinematic Waves: II. A Theory of Traffic Flow on Long Crowded Roads. *Proc., Royal Society of London A: Mathematical, Physical and Engineering Sciences*, No. 229, The Royal Society, 1955, pp. 317–345.
8. Richards, P. I. Shock Waves on the Highway. *Operations Research*, Vol. 4, No. 1, 1956, pp. 42–51.
9. Zheng, Z., S. Ahn, D. Chen, and J. Laval. Applications of Wavelet Transform for Analysis of Freeway Traffic: Bottlenecks, Transient Traffic, and Traffic Oscillations. *Transportation Research Part B: Methodological*, Vol. 45, No. 2, 2011, pp. 372–384.
10. Talebpour, A., and H. S. Mahmassani. Modeling Acceleration Behavior in a Connected Environment. Presented at Midyear Meetings and Symposium Celebrating 50 Years of Traffic Flow Theory, Portland, OR, 2014.
11. Talebpour, A., H. Mahmassani, and S. Hamdar. Speed Harmonization: Evaluation of Effectiveness Under Congested Conditions. *Transportation Research Record: Journal of the Transportation Research Board*, 2013. 2391: 69–79.
12. Mahmassani, H. S. 50th Anniversary Invited Article: Autonomous Vehicles and Connected Vehicle Systems—Flow and Operations Considerations. *Transportation Science*, Vol. 50, No. 4, 2016, pp. 1140–1162.
13. Mahmassani, H., H. Rakha, E. Hubbard, and D. Lukasik. *Concept Development and Needs Identification for Intelligent Network Flow Optimization (INFLO): Assessment of Relevant Prior and Ongoing Research*. U.S. Department of Transportation ITS Joint Program Office Research and Innovative Technology Administration (RITA), 2012.
14. Treiber, M., A. Kesting, and D. Helbing. Understanding Widely Scattered Traffic Flows, the Capacity Drop, and Platoons as Effects of Variance-Driven Time Gaps. *Physical Review E*, Vol. 74, No. 1, 2006, p. 016123.
15. Ma, J., X. Li, S. Shladover, H. A. Rakha, X.-Y. Lu, R. Jagannathan, and D. J. Dailey. Freeway Speed Harmonization. *IEEE Transactions on Intelligent Vehicles*, Vol. 1, No. 1, 2016, pp. 78–89.
16. Helbing, D., M. Treiber, A. Kesting, and M. Schönhof. Theoretical vs. Empirical Classification and Prediction of Congested Traffic States. *The European Physical Journal B: Condensed Matter and Complex Systems*, Vol. 69, No. 4, 2009, pp. 583–598.
17. Kühne, R. Macroscopic Freeway Model for Dense Traffic: Stop–Start Waves and Incident Detection. *Proc., 9th International Symposium on Transportation and Traffic Theory (ISTTT9)*, Delft, The Netherlands. CRC Press, Boca Raton, FL, 1984, pp. 21–42.
18. Kühne, R. Freeway Speed Distribution and Acceleration Noise: Calculations from a Stochastic Continuum Theory and Comparison with Measurements. *Proc., 10th International Symposium on Transportation and Traffic Theory*, Cambridge, MA. Elsevier, Amsterdam, The Netherlands, 1987, pp. 119–137.
19. Krauß, S., P. Wagner, and C. Gawron. Metastable States in a Microscopic Model of Traffic Flow. *Physical Review E*, Vol. 55, No. 5, 1997, p. 5597.
20. Alexiadis, V., J. Colyar, J. Halkias, R. Hranac, and G. McHale. The Next Generation Simulation Program. *Institute of Transportation Engineers Journal*, Vol. 74, No. 8, 2004, p. 22.
21. Montanino, M., and V. Punzo. Making NGSIM Data Usable for Studies on Traffic Flow Theory: Multistep Method for Vehicle Trajectory Reconstruction. *Transportation Research Record: Journal of the Transportation Research Board*, 2013. 2390: 99–111.
22. Lu, X.-Y., P. Varaiya, and R. Horowitz. Fundamental Diagram Modelling and Analysis Based NGSIM Data. *IFAC Proceedings Volumes*, Vol. 42, No. 15, 2009, pp. 367–374.
23. Edie, L. C. Car-Following and Steady-State Theory for Noncongested Traffic. *Operations Research*, Vol. 9, No. 1, 1961, pp. 66–76.
24. Edie, L. Discussion of Traffic Stream Measurement and Description. *Proc., 2nd International Symposium on the Theory of Road Traffic Flow* (Almond, J. ed.), OECD, Paris, France, 1965.
25. Lu, X.-Y., and A. Skabardonis. Freeway Traffic Shockwave Analysis: Exploring the NGSIM Trajectory Data. Presented at 86th Annual Meeting of the Transportation Research Board, Washington, D.C., 2007.

The Standing Committee on Traffic Flow Theory and Characteristics (AHB45) peer-reviewed this paper (18-03622).

TRR
JOURNAL OF THE TRANSPORTATION RESEARCH BOARD

Article

Transportation Research Record
2018, Vol. 2672(20) 215–227

DOI: 10.1177/0361198118790324
journals.sagepub.com/home/trr
SAGE

Optimizing Transit Signal Priority Implementation along an Arterial

Kan Wu[1] and S. Ilgin Guler[1]

Abstract

Transit signal priority (TSP) is a common method of providing priority to buses at signalized intersections. The implementation of TSP can affect travel time of cars traveling in the same, opposite, and cross directions. The bus delay savings and car travel-time impacts are not expected to increase linearly when considering multiple intersections along an arterial. This paper quantifies the influence of TSP on arterials with dedicated bus lanes considering an arterial-wide approach utilizing variational theory. Existing tools were modified to quantify the change in capacity along an arterial where TSP was implemented and it was shown that this effect was negligible. In addition, the bus delay savings and cross-street capacity losses were determined. Case studies provided insights into the influence of TSP among different network homogeneities and bus frequencies. Using these tools, an optimization framework was developed to determine where to implement TSP along an arterial to maximize the marginal benefits, or minimize marginal costs. In addition, a comparison of evaluating an arterial as a sum of isolated intersections as opposed to evaluating an arterial as a whole is presented. This analysis indicates the necessity of the arterial-based method in considering TSP impacts along corridors.

Public transportation systems can utilize available road space more efficiently than cars by carrying more people in the same amount of space. Therefore, properly implemented public transportation systems are effective tools to combat congestion. By providing priority to buses, their attractiveness to travelers can be increased. This in turn can provide an incentive for car users to switch modes to public transportation, reducing the total number of vehicles on roadways. Transit signal priority (TSP) is one of the most commonly used strategies to provide priority to buses traveling through signalized intersections. The goal of TSP is to allow buses to cross signalized intersections without stopping.

TSP strategies can be passive, active, or adaptive. Passive strategies change the signal timing based on historic data and are not real-time strategies (*1*). Active and adaptive strategies work in real-time and change the signal timing once a bus is detected in the proximity of the traffic signal. Adaptive or active strategies' goals are to minimize bus delay, or the total delay of both bus and car travelers, respectively. Typically used signal-timing modifications are green extension, red truncation, phase insertion, or phase rotation. Various studies have analyzed the impacts of active or adaptive TSP on the performance of buses and cars at isolated intersections (*1*, *2*). These studies generally confirm that car operations are negatively affected by TSP implementation. However, when considering the system as a whole, TSP implementation can improve overall passenger delays (*1*). Some researchers have considered implementing TSP along an arterial. Often this work is limited in scope as either the focus is on signal-timing optimization (*3*, *4*) or the work is based on simulations of case studies (*5*, *6*). In addition, to the authors' knowledge, there has been one study considering TSP on a network (*7*). This work considered only green extension as a priority strategy and analyzed a single route with a specific geometric configuration and demand pattern. Hence, the results of this work are not generalizable.

To theoretically study capacity along an arterial or network, the macroscopic fundamental diagram (MFD) can be used. MFD identifies the relationship between average capacity and average density across an arterial or a network. It has been shown to represent real networks based on empirical data (*8*), and has been used to analytically study general network properties (*9–17*), along with impacts of adaptive traffic signals (*18*, *19*), signal offsets (*20*), congestion pricing (*21–23*), perimeter

[1]Department of Civil and Environmental Engineering, Pennsylvania State University, University Park, PA

Corresponding Author:
Address correspondence to S. Ilgin Guler: sig123@engr.psu.edu

control (*24–30*), and so forth. Furthermore, this tool has been used to study multimodal networks considering both buses and cars, and space allocation to buses (*31–33*). Therefore, the MFD tool is well suited to evaluate the impacts of TSP on multimodal at an arterial level.

The goal of this paper is to analytically study the impacts of TSP for buses traveling along an arterial in terms of arterial capacity, cross-street capacity, and bus delay savings. The study first explores the influence of bus frequency and signal timings on the capacity of an arterial with TSP implementation. To continue, a framework to determine the optimal location to invest in TSP infrastructure along an arterial is developed. This framework can be used to identify locations along an arterial where the implementation of TSP would achieve the maximum bus delay savings, or minimum car capacity losses. In addition, considering an arterial as a combination of isolated intersections compared with the method proposed in this paper validates the requirements of the methodology presented in this paper.

The remainder of the paper is organized as follows. First, the considered scenario is described and background information on theoretical MFD estimation methods is provided. Next, the methodology is described and a numerical example is provided. Determining optimal locations for TSP infrastructure implementation is described next. The MFD-based method is then compared with an analysis of an arterial as a combination of isolated intersections. Finally, some concluding remarks are presented.

Background

This section first describes the scenario considered in this paper, and then summarizes the basic concepts of using variational theory (VT) to estimate arterial MFD.

Scenario Description

Considered here is an arterial street with k signalized intersections, for example, five in Figure 1*a*. Note that in Figure 1*a*, the signal settings of the first and last intersections (both are labeled as *1*) are exactly the same, that is, the first and the last intersections are the same ones. This allows analysis of an extended version of the finite arterial as an infinite corridor by placing end-to-end an infinite number of copies of the representative finite street (*34*). Traffic on the homogenous road segments upstream and downstream of any intersection is assumed to obey kinematic wave theory with a triangular fundamental diagram (FD) (*35*). Let v_f denote the free-flow speed, $-w$ the backward wave speed, and q_m the capacity of the roadway for both bus travel and the opposite directions. The capacity of the roadway in the cross direction is assumed to be q_c for all intersections. Buses travel within a dedicated lane with speed v_f, and stop at intersections

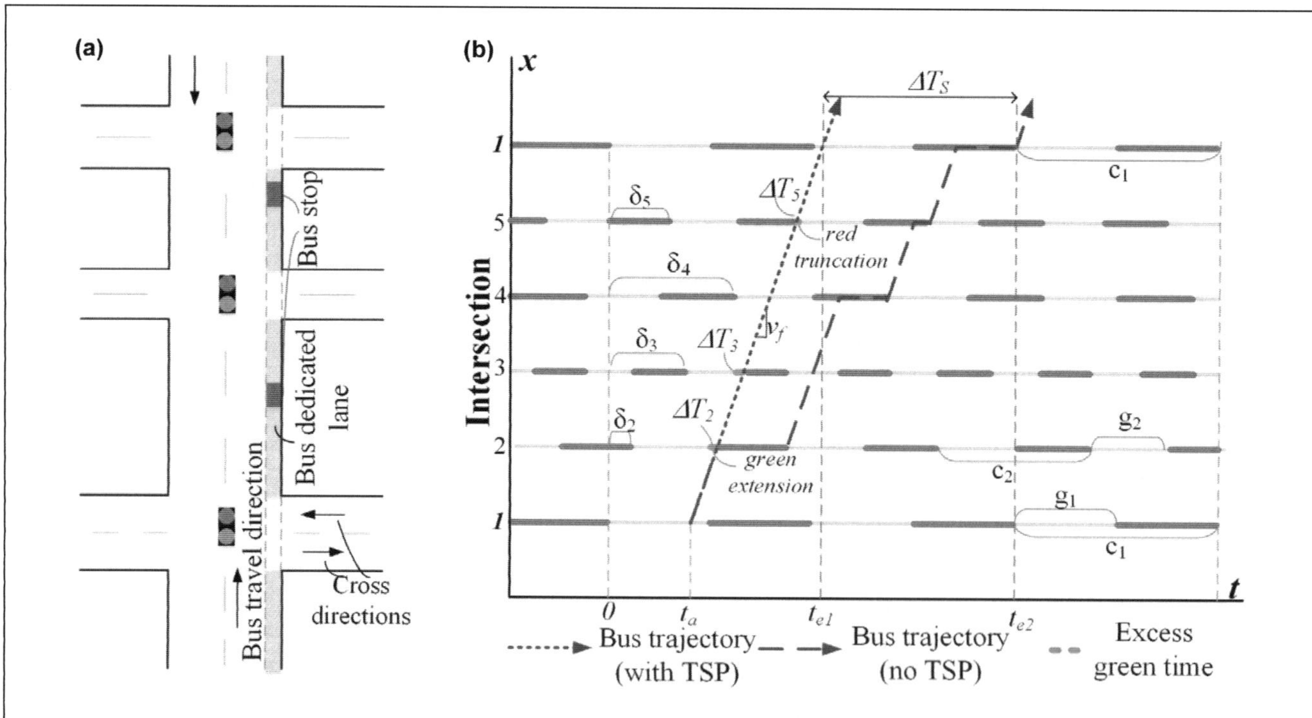

Figure 1. (*a*) Illustration of arterial street network; (*b*) signal-timing diagram with bus trajectories and TSP treatment.

during red times and at bus stops. All buses and cars enter and exit this arterial at Intersection 1 and n, respectively.

The signal timings are described by cycle lengths: $\boldsymbol{C} = [c_1, c_2, \cdots, c_k]$, green times: $\boldsymbol{G} = [g_1, g_2, \cdots, g_k]$, and offset: $\boldsymbol{O} = [0, \delta_2, \delta_3, \cdots \delta_k]$ for intersections $1, 2, \cdots k$, see Figure 1*b*. Note that for simplicity, the bus dwell time at the bus stops is not shown in Figure 1*b*, however, this is considered in the analysis. A bus enters the arterial at Intersection 1 at time t_a, where t_a is uniformly distributed over a cycle (in green phases), that is, $t_a \in [mc_1, (m + g_1)c_1], m = 0, 1, 2, \cdots$. The bus would have departed from the arterial at time t_{e2} if TSP were not implemented. Two candidate TSP strategies are implemented to provide bus priority: green extension and red truncation, which allows the bus to exit the arterial at time t_{e1}. Thus, the bus delay savings can be represented as $\Delta T_S = t_{e2} - t_{e1}$. In this example, the activation of TSP implies that the red phases at intersections 2, 3 and 5 are shortened by $\Delta T_2, \Delta T_3, \Delta T_5$, respectively (see Figure 1*b*). Correspondingly, the green phases for cross direction traffic are also shortened, hence, car throughput on cross streets is reduced by $\Delta N = 2q_c(\Delta T_2 + \Delta T_3 + \Delta T_5)$.

The quantification of bus delay saving ΔT_S and cross-street car throughput, ΔN, are quite straightforward. However, estimating the impacts on the arterial capacity is not easy. An efficient estimation method is to consider the MFDs on arterials with TSP implementation. This method applies VT to an arterial with and without TSP (*36*). The next two subsections and following section will cover this VT-based method at length.

Estimation of the MFD

The VT method is commonly used to estimate theoretical street MFDs (*34*). A summary of the methodology is shown in Figure 2*c*. This method transforms a kinematic wave model into a least-cost path (shortest-path) problem. A valid path starts from the boundary and moves in the (t, x) plane with a mean speed V, where $V \in [-w, v_f]$. Along the path, the cost function $r(u)$ defines the maximum passing rate at which vehicles can overtake an observer moving with speed u. The cost function is linear and decreases from $r(-w) = q_m(1 + w/v_f)$ to $r(v_f) = 0$ when the FD is triangular.

An arterial street usually includes stationary bottlenecks such as signalized intersections or bus stops. Hence, an observer could travel downstream (with speed v_f) and upstream (with speed $-w$), and stop at bottlenecks (with speed 0) to explore all motions of traffic flow presented in the arterial. That is, a moving observer with speeds $u \in \{-w, 0, v_f\}$ can estimate street MFDs sufficiently.

To illustrate, Figure 2*a* exhibits a signal-timing diagram with observers' moving path for a four-intersection arterial. Within this diagram, a path consists of a set of vertices, which are the end of the red times at intersections, and links connecting them to each other. All paths start and end with vertices at the boundary. A link can be backward, stationary, or forward. The associated costs for backward and forward branches of links are $r(-w)$ and zero, respectively, whereas the costs assigned to the stationary branch varies between 0 and q_m in red and green phases. Thus, the total cost, $\Delta(P)$, of a path, P, is the summation of costs of all links involved (see Equation 1a). The least cost, $R(V)$, and mean speed, V, of all possible paths is then represented by equations 1b and 1c, respectively. Calculating the least-cost path with cost $R(V)$ is difficult since it requires exploration of all paths (with mean speed V). Hence a shortest-path algorithm (e.g., Dijsktra's algorithm) is used to determine the least-cost path (*37*).

$$\Delta(P) = \sum_j r(u_j)t_j, j = 1, \cdots J \tag{1a}$$

$$R(V) = \text{inf}_P(\Delta(P)/T_p) \tag{1b}$$

$$V = L/T_p \tag{1c}$$

where

$u_j, t_j, r(u_j)$ represents the speed, travel time and cost within link j, respectively;

L denotes the length of the street; and

$T_p = \sum_j t_j$ is the total travel time on all links, J, within path P.

As per Daganzo and Geroliminis, an MFD relating flow with density can be mathematically represented in Equation 2 (*34*). An MFD is the lower bound of a family of lines with a vertical intercept $R(V)$ and a slope V. These lines are referred to as "cuts" because they individually constrain the maximum flow, Q, at a given density, K, of a street network, see Figure 2*b*.

$$Q = \text{inf}_V(KV + R(V)) \tag{2}$$

To include all possible $R(V)$s, the analysis time window T_w should be large enough. Hence, we usually designate a sufficiently large $T_w \times L$ rectangle in the (t, x) plane (see Figure 2*a*). This rectangle is called a variational graph. All paths within the sufficient variational graph are used to estimate $R(V)$.

A systematic method to estimate the MFD using VT with tight cuts was described in (*34, 37*) and is summarized here. The goal is to calculate the cost of traveling between all possible vertex combinations. To do so, a matrix $\boldsymbol{R}^{n \times n}$ can be derived, where each element r_{ij} represents the cost of the link connecting vertices i and j, and n refers to the total number of vertices. A vertex is

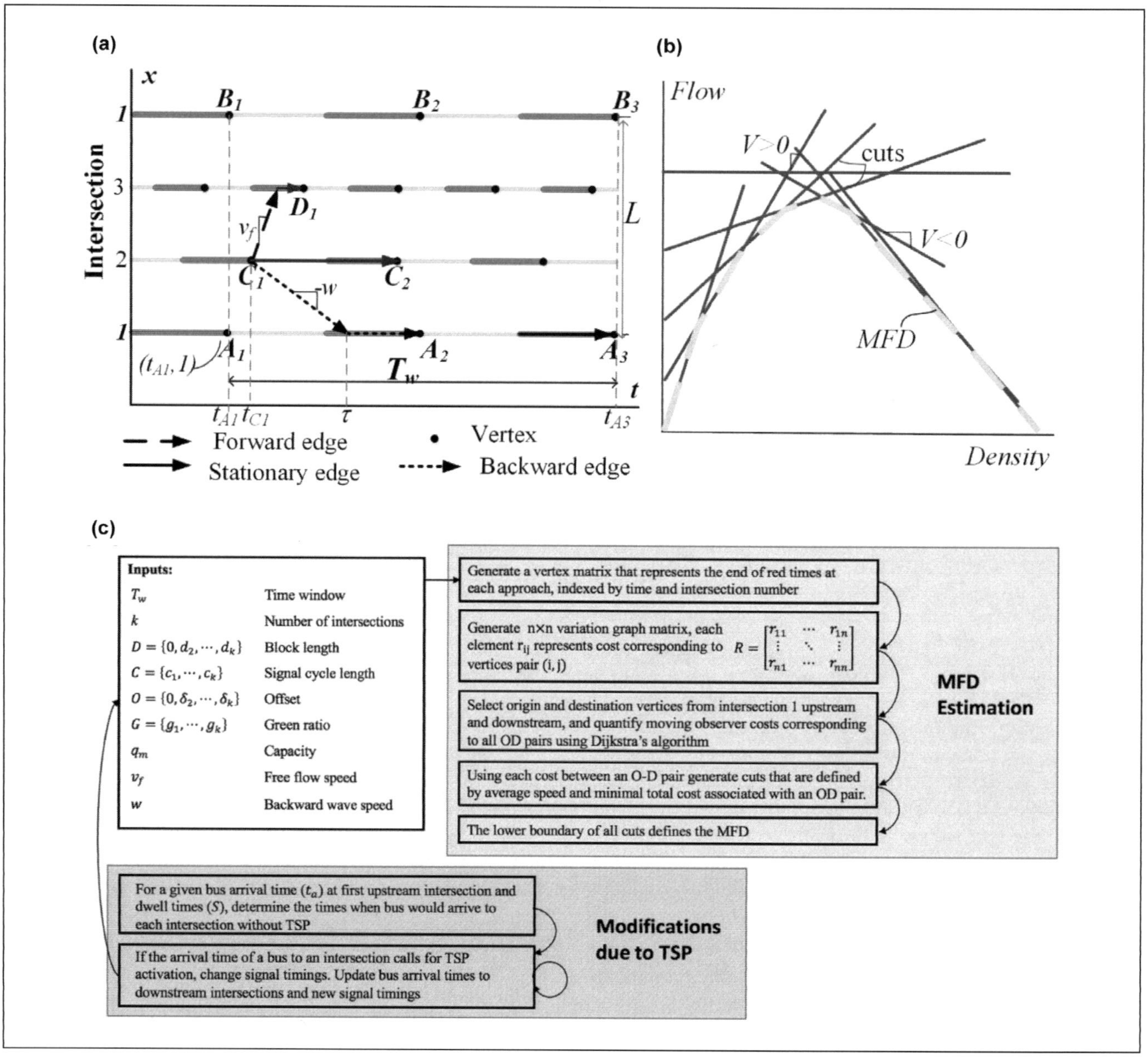

Figure 2. (*a*) Signal-timing diagram with vertices, paths, and relevant costs; (*b*) MFD defined by a family of cuts; (*c*) summary of methodology for MFD estimation (with consideration of TSP).

defined as a point on the variational graph that represents the end of a red period (see black dots in Figure 2*a*). Note that $(r_{11}, r_{22}, \cdots, r_{nn}) = 0$, and $r_{ij} = \infty$ if i is earlier in time than j. Using a given vertex as a starting point, all adjacent vertices that an observer can access will have a finite path cost. In Figure 2*a*, using C_1as an example the adjacent vertices are:

1) C_2, which is the end vertex when the moving observer stays at the intersection for one cycle. The cost of link C_1C_2 is $q_m g_2$;
2) D_1, which is the end vertex when the observer moves forward. The cost of link C_1D_1 is $0 + q_m \cdot 0$, where the first 0 refers to the cost of the free-flow branch of link C_1D_1, and the second 0 represents the duration of green time within the stationary branch of link C_1D_1;
3) A_2, which is the end vertex when the observer moves backward. The cost of link C_1A_2 is $r(-w) \cdot (\tau - t_{C1}) + q_m \cdot 0$, where $(\tau - t_{C1})$ and 0 represent the duration of backward branch and green time within stationary branch of link C_1A_2, respectively.

Once the $\boldsymbol{R}$ matrix is specified, feasible pairs of vertices on Intersection 1at the upstream and downstream ends as origins and destinations of a path are chosen. The shortest-path algorithm then determines a path with least cost $R(V)$ corresponding to each feasible pair of origin and destination vertices.

Note that not all vertices mentioned above are affected by TSP. If TSP were activated, some red phases would end earlier or some green phases would end later, or both. As a result, the graph matrix $\boldsymbol{R}$ needs to be updated. The following section describes the recipe to update $\boldsymbol{R}$ for TSP.

MFD Estimation with TSP

The previous method is adapted to consider the effects of TSP along an arterial. A dedicated bus lane is implemented to prioritize bus movement, which ensures that there is no interaction between buses and other vehicles, allowing buses to travel at a speed v_f when not stopped at bus stops or red signals. The trajectory of a bus is simulated to determine whether a bus would benefit from TSP at any given intersection. The bus trajectory can be simulated independent of traffic queues as it is assumed that buses travel on dedicated lanes. Hence, the arrival time of each bus to each signal can be determined a priori. Two TSP strategies, green extension and red truncation, are considered. It is assumed that TSP is only activated if buses would benefit from it, for example, the green extension is activated for buses that miss the green signal by a little bit, and red truncation is activated for the remaining buses that arrive at the signal during the red time. The maximum change to the signal phase is limited to $\Delta T \leq \Delta T_{max}$ and a constant cycle length is kept with TSP implementation. With all these known inputs, this section provides a description of the analytical methodology and numerical examples.

Extension of the MFD Method to Include TSP Impacts

For illustrative purposes, we consider a case where a bus travels along the arterial without dwelling at bus stops: see the (t, x) plane in Figure 3*a*. The arterial street has $k$signalized intersections, where Intersection 1 and k are identical in signal timings. The analysis time window is set to be $T_w = mc_1$ where m is any integer value (typically greater than 5). Traffic on all approaches obeys kinematic wave theory with the same triangular FD.

To determine the impacts of TSP, a bus's trajectory is simulated and the signal-timing change at any intersection is updated. Then, the previous MFD method is repeated to determine $R(V)$ and V, which are then used to derive the new MFD.

To illustrate the methodology, consider the bus trajectory shown in Figure 3*a* with the dashed line. A bus enters the arterial during the green phase of Intersection 1 at time t_a. The bus would clear the arterial at time t_{B_4} without TSP. However, if TSP is activated the green phase at Intersection 2 is extended by ΔT allowing the bus to clear the intersection at time t_{B_3}. Hence, the bus delay savings can be expressed as $\Delta T_S = t_{B_4} - t_{B_3}$. Correspondingly, the cost of the shortest path between A_2 and B_5 increases by $q_m \Delta t$ due to the change in signal timing at Intersection 2. The resulting cuts with no TSP and TSP are shown in Figure 3, *b*1 and *b*2, respectively. Analytically, cut k moves upward as a result of TSP as $R'(V_k) = R(V_k) + q_m \Delta T$. This results in an MFD with a higher uncongested flow for a range of densities.

Note that TSP can change the shape of MFD in four different ways:

1) TSP does not change the cost of any cuts since it does not alter any of the shortest paths;
2) TSP increases costs associated with noncritical cuts (e.g., Cut 2), hence, the final MFD is not affected;
3) TSP increases costs associated with critical cuts, hence, the uncongested branch, the congested branch, or both branches of the MFD move upward. However, the capacity (maximum flow) remains constrained by another cut (e.g., a stationary cut);
4) TSP increases the capacity of the arterial (but the impact is usually very small).

Through preliminary tests of the algorithm it was found that (1) and (2) are common cases for arterials with homogeneous conditions, that is, intersections with same cycle length and green ratio, and cases (3) and (4) only occur when the arterial network is sufficiently heterogeneous.

In order to account for the randomness in bus arrival times, multiple MFDs are estimated assuming a uniformly distributed bus arrival time to the upstream intersection. Then, the average MFD is simply the arithmetic mean of all MFDs. This measures the average performance in terms of arterial capacity. Note that this is a commonly used method for estimating an average MFD (e.g., [*20*]).

Numerical Examples

This subsection aims to show the application of the previous recipe to an example arterial street consisting of four repeating intersections. Three different arterial settings are considered. Arterial 1 is homogeneous in cycle length and green ratio on all intersections, and the offset between consecutive intersections is d/v_f, where d denotes the length of a block. Such offsets provide a

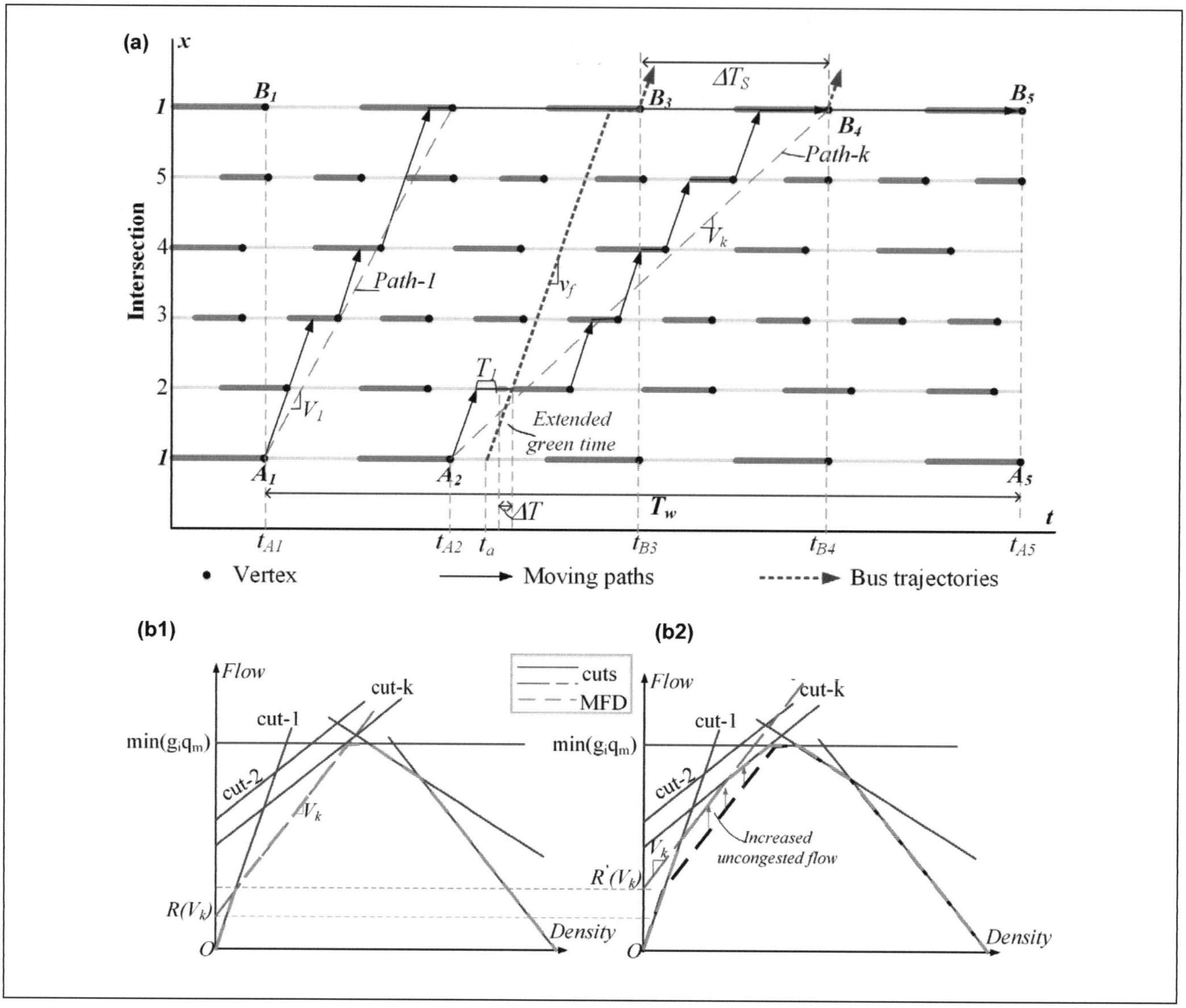

Figure 3. (*a*) Time–space diagram of an arterial with bus trajectories with and without TSP; (*b1*) relevant cuts and MFD without the activation of TSP; (*b2*) relevant cuts and MFD affected by TSP.

maximum green band in the bus travel direction if all vehicles travel at free-flow speed. Arterial 2 has similar signal settings, however the offsets are equal to 0 (i.e., all red phases start at the same time). Arterial 3 consists of intersections with different cycle lengths. In detail, all signal-timing information on the three arterial streets is listed below,

- *Arterial 1*: $c_i = 90, \boldsymbol{O} = \left[0, \frac{d}{v_f}, \frac{2d}{v_f}, \frac{3d}{v_f}\right], g_i = 0.5;$
- *Arterial 2*: $c_i = 90, \delta_i = 0, g_i = 0.5;$
- *Arterial 3*: $\boldsymbol{C} = [90, 60, 75, 120], \delta_i = 0, g_i = 0.5.$

Each arterial is equipped with TSP ($\Delta T_{max} = 0.1\boldsymbol{C}$) at intersections 2, 3, and 4. Note that at the first (and last) intersection, TSP is not implemented. All arterials have dedicated bus lanes and bus stops in between intersections. The lengths of all three blocks are the same, $d = 200$ m. Within each block, buses travel at free-flow speed $v_f = 16\text{m/s}$, and stop at bus stops with dwell time S, which is normally distributed, $S \sim N(30, 10)$. Other parameters are set as follows: $q_m = 0.5\text{veh/s}$, $-\text{w} = -5\text{m/s}$, and $q_c = 0.55\text{veh/s}$.

For each arterial, seven different bus headways $h = c_1, 2c_1, \cdots, 6c_1, >6c_1$, where $c_1 = 90\text{s}$, is tested. The case where TSP is not provided is also included as a baseline and is shown as the case in which the number of buses is zero. Among all cases, the analysis time window is set to be $T = 720\,\text{s}$. Hence, the total number of buses within the arterial for the different headways is $7, 4, 3, 2, 2, 2, 1$, respectively. This analysis assumes a uniformly distributed bus arrival time over the green time at Intersection 1. The results are shown as an average across all these possible bus arrival times.

The numerical results include MFDs, total bus delay savings ΔT_S, and change in total car throughput of cross directions ΔN. All the results are estimated as average values per bus.

The overall MFD remains unchanged for Arterial 1, and hence the results are not shown here. For this arterial, since there is perfect coordination there are multiple moving observer paths that overlap in a cut. Hence, even if TSP changes one of the paths for the moving observer, the remaining cuts still determine the final MFD. When TSP is implemented on Arterial 2, only the congested branch is affected whereas the capacity remains unchanged. The maximum change observed in the congested branch is an increase in flow of 3.21% for the case with bus headway $h = 90s$ compared with the baseline. Hence, these results are also omitted from the paper.

Arterial 3 is affected the most by TSP implementation. Figure 4 exhibits a family of MFDs for Arterial 3 with the seven different headways and the baseline case. TSP increases arterial capacity for the cases when $h < 540s$. As expected, the impact of TSP decreases as bus headway increases. Among all cases, the maximum change in capacity Q_{max} is 1.75% compared with the baseline. The speed v_c varies between 4.56 m/s and 4.60 m/s if vehicles discharge at Q_{max}. An average bus saves between 18.39 s and 25.23 s. Correspondingly, cross-street car throughput loss varies between 3.01 and 5.74 vehicles.

Optimal TSP on Homogeneous Street

The goal is to find the optimal number of intersections that should be equipped with TSP (i.e., have TSP capability), and determine whether implementation on different intersection combinations would lead to different performances. Given this, the problem can be formulated as an integer program in which the decision variable is an array *nt* with binary variables 1 or 0 to indicate TSP being implemented or not at an intersection. The results of the previous section show that the capacity of an arterial is minimally affected by providing TSP, especially on homogeneous arterials. Therefore, when considering the optimal location of TSP implementation, only delay savings and cross-street capacity loss are considered in this paper. However, the methodology can easily account for

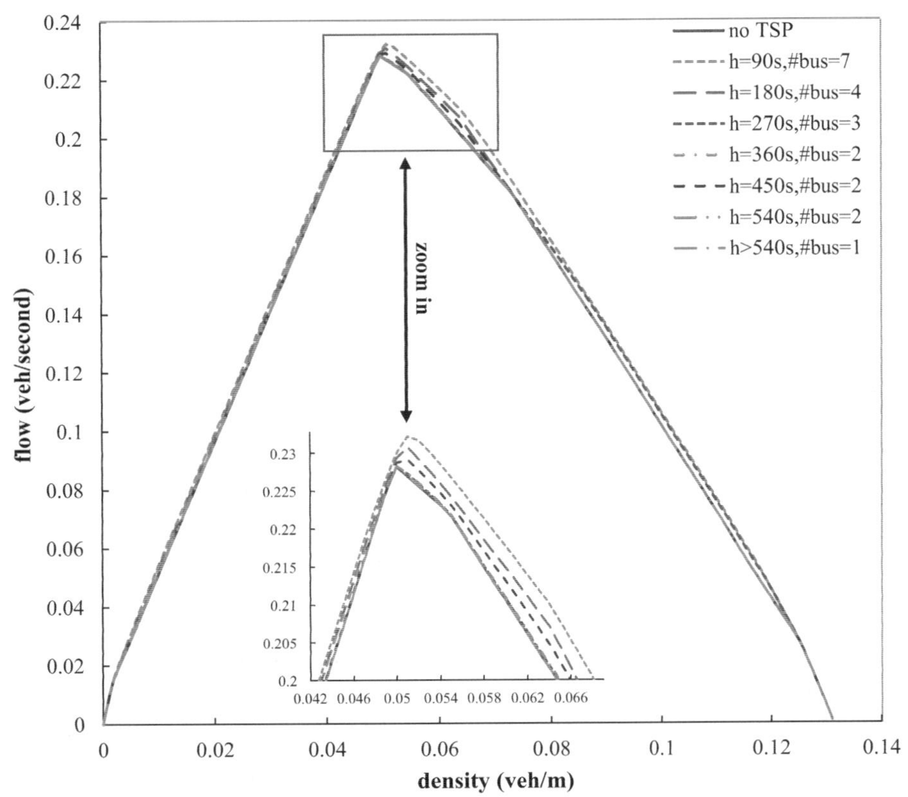

Figure 4. Arterial 3 MFDs for seven different bus headways when TSP is implemented.

changes in the arterial capacity. As a result, different objectives can be determined such as:

1) minimizing the impacts on cross streets (the change in car throughput ΔN);
2) maximizing bus delay savings ΔT_S; and
3) a combination of 1 and 2.

Note that the first and second objectives are normally in conflict with each other. In general, with more intersections that have TSP implemented, the cross direction will have more capacity loss whereas bus delay savings will increase. Therefore, a myopic study would conclude that the best solution is to implement TSP at every intersection. However, the marginal benefit or cost of each implementation may not be the same. Given how expensive TSP implementation can be, the tradeoff between cost of implementation and impact should be carefully considered. Given a limited budget, implementing TSP on one intersection as compared to another one can lead to very different impacts. Therefore, in this paper, the results are discussed in terms of Pareto frontiers.

To solve this optimization problem, all possible TSP strategies in a set **NT** are enumerated. For more complicated scenarios different optimization heuristics such as branch-and-bound can also be used. An example seven-intersection homogeneous arterial with $c_i = 90, g_i = 0.5, \delta_i = 0$ is considered. Note that there is no TSP at the first and last intersections, hence, $nt(1) = 0$, and $\sum nt(i) \leq 6$. An example solution would be $nt = [0, 1, 0, 0, 0, 0, 0] \in \mathbf{NT}$ indicating TSP is implemented only at Intersection 2. For this arterial, a total of 63 candidate solutions, nt, are considered (which excludes the no TSP solution). By applying the previous recipe, all ΔN and ΔT_S corresponding to **NT** are quantified.

Only one bus is included in this analysis. Bus arrival times are uniformly distributed across a cycle at Intersection 1. The results are shown as an average of all possible bus arrival times. All parameters are the same as the previous section: $v_f = 16m/s$, $S \sim N(30, 10)$, $-w = -5m/s$, $q_m = 0.5veh/s$, $q_c = 0.55veh/s$.

Figure 5*a* shows the total change in car throughput of cross streets, ΔN, versus the average total bus delay savings, ΔT_S, across all intersections. In this figure, the data points corresponding to different number of intersections where TSP is implemented are shown with different color and shape markers. As expected, in general, total ΔT_S and ΔN increase simultaneously as the intersection number increases. However, there are certain combinations of intersections on which TSP can be implemented that result in smaller changes to cross-street capacity and larger bus delay savings compared with others. To this end, a Pareto frontier can be identified as shown in Figure 5*a*. The data points on the Pareto frontier are labeled with the corresponding intersection numbers. In addition, note that for certain combinations of intersections, implementing TSP on fewer intersections can result in larger total bus delay savings than implementing it on more intersections. This is especially evident when looking at larger numbers of intersections equipped with TSP. For example, compare implementing TSP on three intersections vs. four intersections. It can be seen that although in most cases implementing TSP on three intersections results in fewer bus delay savings, one specific combination of intersections outperforms implementing TSP on four intersections for 10 different cases. This Pareto graph helps decision makers to choose best-fit local TSP locations under different circumstances. This figure clearly shows that TSP implementation at different intersections does not always result in the same benefits and costs. Note that this discrepancy occurs despite consideration of a homogeneous arterial.

The implementation of TSP for a specific number of intersections (e.g., only four intersections at once) is considered to explore the marginal benefits of extending TSP. For a given number of intersections equipped with TSP, the optimal distribution that provides the maximum bus delay savings, ΔT_S, is determined using an iterative method. This can be thought of as implementing TSP using a limited budget (i.e., only on a limited number of intersections) on intersections where it would have the largest benefits. For these combinations of intersections, the bus delay savings per intersection and the corresponding cross-street capacity change per intersection are plotted against each other, as shown in Figure 5*b*. As can be seen, both the maximum ΔT_S per intersection and ΔN per intersection decrease simultaneously as the number of intersections on which TSP is implemented increase. Hence, each additional implementation of TSP provides buses with less benefit, and also leads to less of a reduction in the cross-street throughput. This means that the marginal impact of TSP decreases as more intersections are equipped with TSP.

Necessity of an Arterial Level Analysis

Given the existing theoretical tools for analyzing the impacts of TSP implementation, the typical method for evaluating an arterial would be to consider it as a summation of independent intersections rather than evaluating the arterial as a whole. This section considers how evaluating an arterial as a whole could change the predicted impacts of TSP implementation as compared to evaluating the arterial as a sum of independent intersections. To do so, two separate methods are utilized. To evaluate the arterial as a sum of independent intersections, the method developed by the authors previously

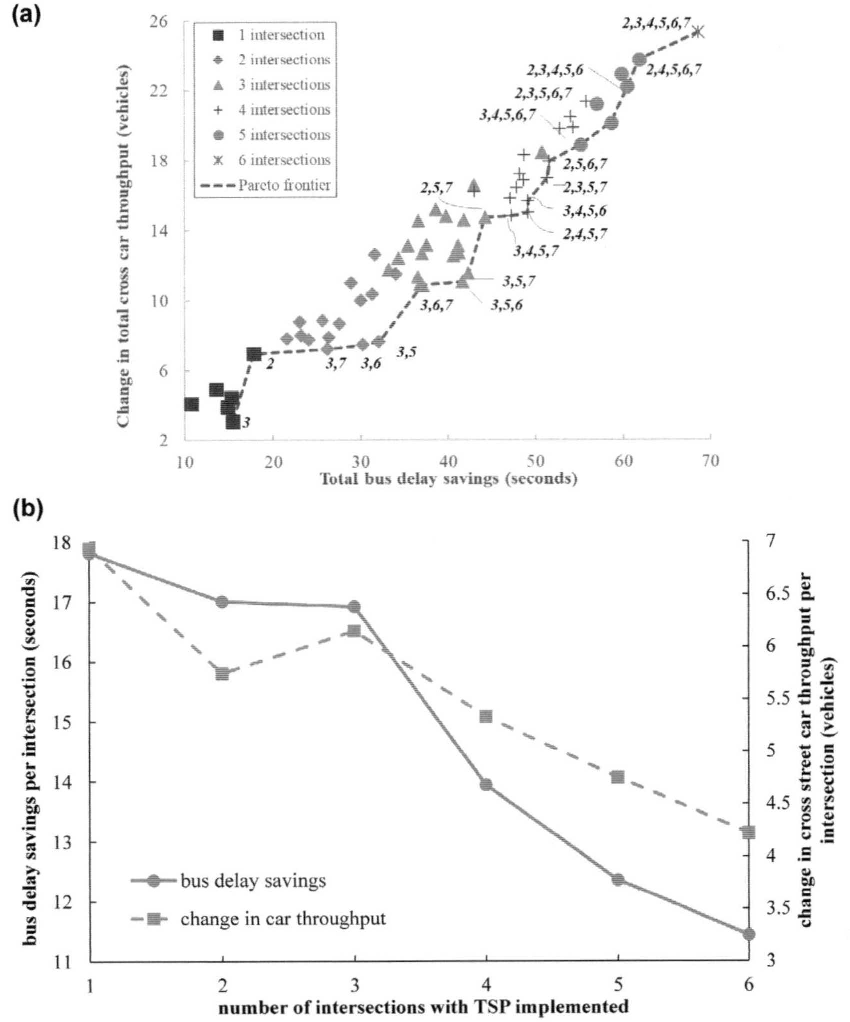

Figure 5. (*a*) Total change in car throughput versus total bus delay savings; (*b*) marginal effect of TSP implementation on multiple intersections.

(*1*) is used to evaluate the car capacity and bus delays resulting in TSP implementation. This methodology assumes that bus arrivals at the intersection have a uniform distribution, and TSP is activated only if buses would benefit from it. VT is used to estimate the average car capacity for all possible bus arrival times. Intersections along an arterial with eight intersections are individually evaluated for the expected impacts of TSP (i.e., each intersection has the same expected car capacity and bus delay as they are assumed to operate independently). To evaluate the arterial as a whole, the MFD recipe presented in this paper is used for an arterial with $c_1 = c_2 \cdots = c_8$, and $g_1 = g_2 \cdots = g_8$. It is assumed that a bus enters an arterial with times uniformly distributed over a cycle. For simplicity the bus is assumed to have zero dwell time at bus stops. All parameters are kept the same as the previous sections.

For the two methods to be comparable, both the change in total car throughput, ΔN, and bus delay savings, ΔT_S, are given as an average value per intersection per cycle. When considering individual intersections, these values are just an output of the methodology and it is assumed that every intersection experiences the same change in capacity and bus delay savings. However, when considering the arterial as a whole, the change in capacity along an arterial is zero, as shown previously. Hence ΔN represents only the average change in cross-street capacity over all the intersections along the arterial. Additionally, ΔT_S represents the bus delay savings per intersection, that is, the total bus delay savings divided by eight.

Two candidate offsets are considered, zero offset $\boldsymbol{O} = 0$, and random offset $\boldsymbol{O} = [0, 5, 20, 10, 20, 10, 15, 5]$. Different scenarios with cycle length, c_i, varying between 60 s and 120 s, and green ratio, g_i, varying between 0.3 and 0.7 are tested. Figure 6 shows the change in car throughput per intersection and bus delay savings per intersection using the arterial methodology with zero offset (blue circles), or a random offset (orange squares), or using the isolated intersection methodology (black diamonds). Figure 6, *a* and *b* show the change in ΔN and ΔT_S, respectively, for varying cycle lengths, and Figure 6, *c* and *d* show the change in ΔN and ΔT_S, respectively, for varying green ratios.

According to Figure 6, *a* and *b*, ΔN and ΔT_S have an increasing trend as the cycle length increases, specifically when considering only an isolated intersection. Intersections with longer cycle lengths provide more opportunities to provide TSP, and hence both the change in car throughput and bus delay savings increase. However, note that considering only an isolated intersection exaggerates the benefits and costs of implementing TSP. The bus delay savings per intersection vary around

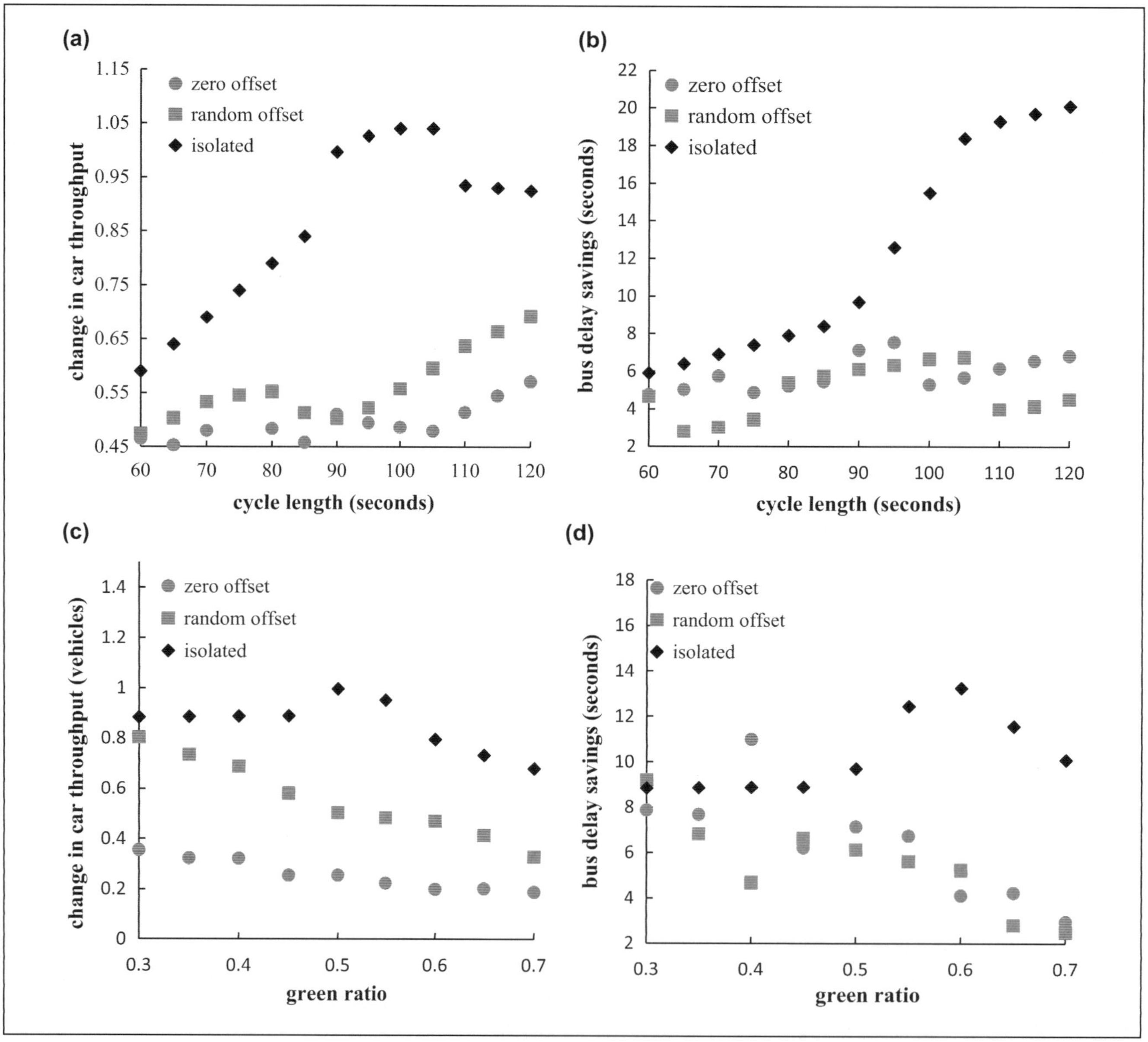

Figure 6. Comparative graphs showing for different cycle lengths (with $\mathbf{g}_i = 0.5$): (*a*) change in car throughput (per intersection); (*b*) change in bus travel time (per intersection); and for different green ratios ($\mathbf{c}_i = 90$); (*c*) change in car throughput (per intersection); (*d*) change in bus travel times (per intersection).

5 s per intersection for both considered offsets when evaluating the arterial as a whole, and are much smaller than predicted with the isolated intersection methodology. When considering the arterial as a whole, even though all possible bus arrival times to the upstream intersection are considered, this results in a limited combination of bus arrival times to other intersections. Hence, at some intersections it is never feasible to implement TSP. On the other hand, when considering an isolated intersection, all possible bus arrival times are considered, hence, TSP can be activated at every intersection. As for the green ratio, the change in car throughput decreases as the green ratio increases as more buses arrive during the green time and do not require preemption. Hence, accordingly the bus delay savings also decrease as the green ratio increases, especially when considering the arterial as a whole. However, the trends are not as evident when looking at an isolated intersection as the results depend on the average times of bus arrivals over a cycle. Again, both ΔN and ΔT_S are over-estimated when only an isolated intersection is considered.

Note that these results are estimated using the parameters as mentioned above. However, considering only an isolated intersection may also underestimate the values under some parameters. This indicates that 1) the two methods are not comparable, and 2) the methodology proposed in this paper has its unique strengths in evaluating the impacts of TSP at an arterial level.

Conclusions

This paper presents a tool to analyze the impacts of TSP implementation along a corridor using an analytical MFD and VT. Implementation of TSP leads to changes in signal timings in terms of increased green time at some intersections, imposing changes to the shape of MFD. Case studies indicate that impact of TSP on arterial capacity varies among different signal settings. The impacts are typically dominated by bus frequency. In general, as bus headway decreases a larger increase in arterial capacity is observed. However, these impacts are usually minimal, especially for a homogeneous street. Therefore, it is sufficient to only consider the impacts on cross-street car throughputs and bus delay savings for a homogeneous roadway.

Using these findings a scheme for determining optimal locations of TSP implementation was proposed. The proposed problem used a binary decision variable indicating whether or not TSP was equipped at an intersection, and multiple objective functions were considered including cross-street car throughput loss and bus delay savings. The results were presented as the change in cross-street throughput as compared to the bus delay savings for TSP implemented at different numbers of intersections. The final Pareto graph (see Figure 5a) displays the optimal TSP locations, considering both bus and cars. This figure shows that investing in more intersections to have TSP capabilities will not always result in the largest bus delay savings. Another observation was that the marginal effect of TSP decreases as more intersections are equipped with TSP capabilities. This can be verified considering the maximum bus delay savings and corresponding change in car throughput per intersection. Finally, this paper showed that considering an intersection in an isolated fashion can result in misleading conclusions regarding the benefits of TSP implementation.

Ongoing research considers further extensions of this work for more complicated network structures and different bus priority strategies. Estimating MFD for an urban network is challenging. A network consists of horizontal and vertical streets, and hence different bus routes will compete for priority. It is important to model this competition accurately to obtain an understanding of the effects of TSP. As for alternative bus priority strategies, this recipe can be adapted to consider more scenarios such as pre-signals, or intermittent bus lanes. If buses are required to mix with cars, the travel speed of buses needs to be modeled.

Author Contributions

The authors confirm contribution to the paper as follows: study conception and design: all authors; data collection: KW; analysis and interpretation of results: all authors; draft manuscript preparation: all authors. All authors reviewed the results and approved the final version of the manuscript.

References

1. Wu, K., S. I. Guler, and V. V. Gayah. Estimating the Impacts of Bus Stops and Transit Signal Priority on Intersection Operations: A Queueing and Variational Theory Approach. *Transportation Research Record: Journal of the Transportation Research Board*, 2017. 2622: 70–83.
2. Gu, W., M. J. Cassidy, V. V. Gayah, and Y. Ouyang. Mitigating Negative Impacts of near-Sde Bus Stops on Cars. *Transportation Research Part B: Methodological*, Vol. 47, 2013, pp. 42–56.
3. Stevanovic, J., A. Stevanovic, P. T. Martin, and T. Bauer. Stochastic Optimization of Traffic Control and Transit Priority Settings in VISSIM. *Transportation Research Part C: Emerging Technologies*, Vol. 16, No. 3, 2008, pp. 332–349.
4. Christofa, E., K. Ampountolas, and A. Skabardonis. Arterial Traffic Signal Optimization: A Person-Based Approach. *Transportation Research Part C: Emerging Technologies*, Vol. 66, 2016, pp. 27–47.
5. Wadjas, Y., and P. Furth. Transit Signal Priority along Arterials Using Advanced Detection. *Transportation Research Record: Journal of the Transportation Research Board*, 2003. 1856: 220–230.

6. Dion, F., H. Rakha, and Y. Zhang. Evaluation of Potential Transit Signal Priority Benefits along a Fixed-Time Signalized Arterial. *Journal of Transportation Engineering*, Vol. 130, No. 3, 2004, pp. 294–303.
7. Ahn, K., and H. Rakha. System-Wide Impacts of Green Extension Transit Signal Priority. Presented at the Intelligent Transportation Systems Conference, 2006.
8. Geroliminis, N., and C. F. Daganzo. Existence of Urban-Scale Macroscopic Fundamental Diagrams: Some Experimental Findings. *Transportation Research Part B: Methodological*, Vol. 42, No. 9, 2008, pp. 759–770.
9. Gayah, V. V., and C. F. Daganzo. Effects of Turning Maneuvers and Route Choice on a Simple Network. *Transportation Research Record: Journal of the Transportation Research Board*, 2011. 2249: 15–19.
10. Daganzo, C. F., V. V. Gayah, and E. J. Gonzales. Macroscopic Relations of Urban Traffic Variables: Bifurcations, Multivaluedness and Instability. *Transportation Research Part B: Methodological*, Vol. 45, No. 1, 2011, pp. 278–288.
11. Gayah, V. V., and C. F. Daganzo. Clockwise Hysteresis Loops in the Macroscopic Fundamental Diagram: An Effect of Network Instability. *Transportation Research Part B: Methodological*, Vol. 45, No. 4, 2011, pp. 643–655.
12. Mahmassani, H. S., M. Saberi, and A. Zockaie. Urban Network Gridlock: Theory, Characteristics, and Dynamics. *Transportation Research Part C: Emerging Technologies*, Vol. 36, 2013, pp. 480–497.
13. Saberi, M., A. Zockaie, and H. S. Mahmassani. Network Capacity, Traffic Instability, and Adaptive Driving: Findings from Simulated Network Experiments. Presented at 1st European Symposium on Quantitative Methods in Transportation Systems, Lausanne, Switzerland, 2012.
14. Saberi, M., and H. S. Mahmassani. Empirical Characterization and Interpretation of Hysteresis and Capacity Drop Phenomena in Freeway Networks. *Transportation Research Record: Journal of the Transportation Research Board*, 2013. 2391: 44–55.
15. Ortigosa, J., M. Menendez, and V. V. Gayah. Analysis of Network Exit Functions for Different Urban Grid Network Configurations. *Transportation Research Record: Journal of the Transportation Research Board*, 2015. 2491: 12–24.
16. Muhlich, N., V. V. Gayah, and M. Menendez. An Examination of MFD Hysteresis Patterns for Hierarchical Urban Street Networks Using Micro-Simulation. *Transportation Research Record: Journal of the Transportation Research Board*, 2015. 2491: 117–126.
17. Knoop, V. L., D. D. Jong, and S. Hoogendoom. Influence of Road Layout on Network Fundamental Diagram. *Transportation Research Record: Journal of the Transportation Research Board*, 2014. 2421: 22–30.
18. Zhang, L., T. M. Garoni, and J. de Gier. A Comparative Study of Macroscopic Fundamental Diagrams of Arterial Road Networks Governed by Adaptive Traffic Signal Systems. *Transportation Research Part B: Methodological*, Vol. 49, 2013, pp. 1–23.
19. Gayah, V. V., X. Gao, and A. S. Nagle. On the Impacts of Locally Adaptive Signal Control on Urban Network Stability and the Macroscopic Fundamental Diagram. *Transportation Research Part B: Methodological*, Vol. 70, No. 1, 2014, pp. 255–268.
20. Girault, J. T., V. V. Gayah, S. I. Guler, and M. Menendez. Exploratory Analysis of Signal Coordination Impacts on Macroscopic Fundamental Diagram. *Transportation Research Record: Journal of the Transportation Research Board*, 2016. 2560: 36–46.
21. Geroliminis, N., and D. M. Levinson. Cordon Pricing Consistent with the Physics of Overcrowding. Presented at the Transportation and Traffic Theory 2009: Golden Jubilee, 2009.
22. Simoni, M., A. Pel, R. A. Waraich, and S. Hoogendoom. Marginal Cost Congestion Pricing Based on the Network Fundamental Diagram. *Transportation Research Part C: Emerging Technologies*, Vol. 56, 2015, pp. 221–238.
23. Zheng, N., R. A. Waraich, K. W. Axhausen, and N. Geroliminis. A Dynamic Cordon Pricing Scheme Combining the Macroscopic Fundamental Diagram and an Agent-Based Traffic Model. *Transportation Research Part A: Policy and Practice*, Vol. 46, No. 8, 2012, pp. 1291–1303.
24. Geroliminis, N. *Increasing Mobility in Cities by Controlling Overcrowding*. Doctor of Philosophy. University of California, Berkeley, Berkeley, California, 2007.
25. Keyvan-Ekbatani, M., A. Kouvelas, I. Papamichail, and M. Papageorgiou. Exploiting the Fundamental Diagram of Urban Networks for Feedback-Based Gating. *Transportation Research Part B: Methodological*, Vol. 46, No. 10, 2012, pp. 1393–1403.
26. Keyvan-Ekbatani, M., M. Papageorgiou, and I. Papamichail. Urban Congestion Gating Control Based on Reduced Operational Network Fundamental Diagrams. *Transportation Research Part C: Emerging Technologies*, Vol. 33, 2013, pp. 74–87.
27. Haddad, J., and N. Geroliminis. On the Stability of Traffic Perimeter Control in Two-Region Urban Cities. *Transportation Research Part B: Methodological*, Vol. 46, No. 9, 2012, pp. 1156–1176.
28. Haddad, J., M. Ramezani, and N. Geroliminis. Optimal Perimeter Control for Two Urban Regions with Macroscopic Fundamental Diagrams: A Model Predictive Control Approach. *IEEE Transactions on Intelligent Transportation Systems*, Vol. 14, No. 1, 2012, pp. 348–359.
29. Haddad, J., M. Ramezani, and N. Geroliminis. Cooperative Traffic Control of a Mixed Network with Two Urban Regions and a Freeway. *Transportation Research Part B: Methodological*, Vol. 54, 2013, pp. 17–36.
30. Ortigosa, J., M. Menendez, and H. Tapia. Study on the Number and Location of Measurement Points for an MFD Perimeter Control Scheme: A Case Study of Zurich. *EURO Journal on Transportation and Logistics*, Vol. 3, No. 3–4, 2014, pp. 245–266.
31. Geroliminis, N., N. Zheng, and K. Ampountolas. A Three-Dimensional Macroscopic Fundamental Diagram for Mixed Bi-Modal Urban Networks. *Transportation Research Part C: Emerging Technologies*, Vol. 42, 2014, pp. 168–181.
32. Ortigosa, J., N. Zheng, M. Menendez, and N. Geroliminis. *Traffic Performance and Road Space Allocation in*

Multimodal Urban Networks with an MFD Representation. No. 17–03540, 2017.
33. Chiabaut, N. Evaluation of a Multimodal Urban Arterial: The Passenger Macroscopic Fundamental Diagram. *Transportation Research Part B: Methodological*, Vol. 81, No. 2, 2015, pp. 410–420. https://doi.org/https://doi.org/10.1016/j.trb.2015.02.005.
34. Daganzo, C. F., and N. Geroliminis. An Analytical Approximation for the Macroscopic Fundamental Diagram of Urban Traffic. *Transportation Research Part B: Methodological*, Vol. 42, No. 9, 2008, pp. 771–781.
35. Lighthill, M. J., and G. B. Whitham. On Kinematic Waves. I. Flood Movements in Long Rivers. II. A Theory of Traffic Flow on Long Crowded Roads. *Proceedings of the Royal Society*, No. A 229, 1955, pp. 281–345.
36. Daganzo, C. F. A Variational Formulation of Kinematic Waves: Basic Theory and Complex Boundary Conditions. *Transportation Research Part B: Methodological*, Vol. 39, No. 2, pp. 187–196.
37. Leclercq, L., and N. Geroliminis. Estimating MFDs in Simple Networks with Route Choice. *Transportation Research Part B: Methodological*, Vol. 57, 2013, pp. 468–484.

The Standing Committee on Traffic Flow Theory and Characteristics (AHB45) peer-reviewed this paper (18-01408).

TRR
JOURNAL OF THE TRANSPORTATION RESEARCH BOARD

Article

Transportation Research Record
2018, Vol. 2672(20) 228–238

DOI: 10.1177/0361198118786641
journals.sagepub.com/home/trr
SAGE

Some Indications on How to Calibrate the Social Force Model of Pedestrian Dynamics

Tobias Kretz[1], **Jochen Lohmiller**[1], **and Peter Sukennik**[1]

Abstract

The social force model of pedestrian dynamics is formulated in such a way that (a) most of its parameters do not have an immediate interpretation in the sense that they cannot be measured directly, (b) often one single parameter has an impact on many aspects of walking behavior, and (c) a certain aspect of walking behavior results from the values of more than one parameter. This makes calibration difficult. The aim of this paper is to give practitioners an indication of how to proceed in the calibration process. For this purpose, by analytical transformations the parameters of the social force model are related to real properties that have a clear and immediate meaning and are also highly relevant result properties of a simulation: extent and clearance time of a queue, respectively maximum density, and capacity flow. The theory for this is presented, and thus this study offers a deeper understanding of the model including its theoretical aspects.

The social force model (SFM) (*1–4*) is one of the most prominent models of pedestrian dynamics. It is formulated in such a way that (a) most of its parameters do not have an immediate interpretation in the sense that they cannot be measured directly, (b) often one single parameter has an impact on many aspects of walking behavior, and (c) a certain aspect of walking behavior results from the values of more than one parameter. In this regard the SFM is very different from a number of car-following models like the Wiedemann model (*5*) where nearly each parameter has an immediate interpretation (can be measured directly) and has only a local effect—"local" in the sense that within the model it affects only certain observable properties of driving behavior, that is, the results.

This structure of the SFM makes calibration challenging since, when the meaning of a parameter is abstract instead of immediate, for someone confronted with the model for the first time it is often unclear which are the most relevant parameters to adjust and even in which direction a certain modification of the value of a certain parameter changes the simulation results. Among models of pedestrian dynamics the SFM is not unique in having parameters with a rather abstract meaning. The cellular automata floor field models, for example, share that property to a similar degree (*6–8*).

Compared with action-point car-following models the SFM has an advantage in that it can be analyzed, transformed, and simplified in a rigid mathematical way. This allows a deeper understanding of the a priori abstract parameters, which leads to the aim of this paper: to give practitioners an indication of how to proceed in the calibration process. For this purpose, by analytical transformations the parameters of the SFM are related to real properties that have a clear and immediate meaning: number of pedestrians, stand still density, and capacity flow. The latter two are highly relevant since they frame the fundamental diagram and thus are important properties in the calibration process.

At the same time these are basic properties, such that a model which reproduces these well does not necessarily work well in more complex situations, like bi-directional flow situations with emergent lane formation. Thus the work done in this contribution is necessary for calibration, but not complete. However, with regard to the numerous variants of the SFM (*3*, *9–16*) and to general force-based models (*17*, *18*) the basic character of the proposed method is beneficial, since it means that it applies to most of these variants.

The outline of this paper is as follows: in the next section we will derive the results step by step and also

[1]PTV Group, Karlsruhe, Germany

Corresponding Author:
Address correspondence to Tobias Kretz: tobias.kretz@ptvgroup.com

document the model simplifications that lead to the results. A reader not interested in the derivation may skip that section and continue on to the "Summary of Results and Discussion" section, after which follows a section with examples.

Derivation of Results

Our starting point is the circular specification of the SFM. This is the simplest version since inter-pedestrian forces depend only on distance between pedestrians. Since in all states of the system that we use in our reasoning the relative velocities between pedestrians are zero, the arguments also hold for the elliptical specification II which, in addition to distance, also considers relative velocity as a determining property for the forces. Elliptical specification I, on the contrary, which takes into account distance and the velocity of the pedestrian who exerts the force, is fundamentally different (3).

The circular specification of the SFM—neglecting forces from walls—is defined as:

$$\ddot{\vec{x}}_i(t) = \frac{\vec{v}_{0,i} - \dot{\vec{x}}_i(t)}{\tau_i} + \tilde{A}_i \sum_j w(\vec{x}_i(t), \vec{x}_j(t), \dot{\vec{x}}_i(t), \lambda_i) e^{-\frac{|\vec{x}_j(t) - \vec{x}_i(t)| - R_i - R_j}{B_i}} \hat{e}_{ij} \quad (1)$$

with

$$w(\vec{x}_i(t), \vec{x}_j(t), \dot{\vec{x}}_i(t), \lambda_i) = \lambda_i + (1 - \lambda_i) \frac{1 + \cos(\theta_{ij}(\vec{x}_i(t), \vec{x}_j(t), \dot{\vec{x}}_i(t)))}{2} \quad (2)$$

where $v_{0,i}$ is the desired speed of pedestrian i. $\tilde{A}_i > 0$, $B_i > 0$, $0 \leq \lambda_i \leq 1$, and $\tau_i > 0$ are parameters of the model. R_i denotes the body radius of a pedestrian. $\hat{e}_{ij}$ has the direction from pedestrian j to pedestrian i. $\vec{x}_i$ is the position of a pedestrian and dots mark time derivatives (i.e. speed and acceleration). The sum runs over all—potentially infinitely many—pedestrians in a simulation scenario. Function $w()$ is there to suppress forces acting from behind. Within it θ_{ij} is the angle between pedestrian i's velocity vector and the line connecting pedestrians i and j.

Abstract parameters in the sense described above are particularly A_i, B_i, and λ_i and—to a lesser degree—τ_i. The desired speed $v_{0,i}$, on the contrary, in this contribution and in accordance with most publications on the SFM, is viewed as a parameter that has an immediate interpretation and can be measured directly by measuring the pedestrian's free, unobstructed walking speed, although it is an unproven hypothesis that for pedestrians there is a situation-independent desired speed that matches the free walking speed. Therefore in the framework of this contribution we assume that there is no freedom in choosing the value of $v_{0,i}$ and count it as one of the empirical properties although at the same time it is a model parameter.

From here on we assume that parameters $\tilde{A}$, B, λ, τ, R, and v_0 have identical values for all pedestrians, so we omit the indices. This allows us to combine $\tilde{A}$ and R into a new parameter:

$$A = \tilde{A} e^{\frac{2R}{B}} \quad (3)$$

It is essential to keep this parameter transformation in mind, since some implementations of the SFM require $\tilde{A}$ and some A as input parameter. Put differently: if there is no $2R$ in the exponent and if "distance between pedestrians" refers to surface to surface one needs to calculate with parameter $\tilde{A}$, while if it is center to center parameter A is required. In this contribution "distance" always means center to center. We use the formulation with A for theoretical investigation, but for all simulations A means $\tilde{A}$, that is, the $2R$ are "added" in the simulation.

Since it is obvious which properties are time dependent, we also omit the "(t)". Equation 1 can thus be written for the one-dimensional case:

$$\ddot{x}_i = \frac{v_0 - \dot{x}_i}{\tau} + A \sum_j w(x_i, x_j, \lambda) e^{-\frac{d_{ij}}{B}} \quad (4)$$

$$d_{ij} = |x_j - x_i| \quad (5)$$

$$w(x_i, x_j, \lambda) = \lambda \text{ if } x_j - x_i < 0 \quad (6)$$

$$w(x_i, x_j, \lambda) = -1 \text{ if } x_j - x_i > 0 \quad (7)$$

with the additional assumption that for all pedestrians and times $v_0, \dot{x} > 0$. This can be done without limiting generality since we are not interested in phenomena far from equilibrium, but only investigate a stationary state of the system (strong equilibrium).

Considering only nearest neighbor interactions (one neighbor $i+1$ ahead and one neighbor $i-1$ behind) and neglecting all other forces, the sum in Equation 4 is reduced:

$$\ddot{x}_i = \frac{v_0 - \dot{x}_i}{\tau} - A\left(e^{-\frac{d_{i,i+1}}{B}} - \lambda e^{-\frac{d_{i,i-1}}{B}}\right) \quad (8)$$

for all pedestrians who have a leading and a following pedestrian. For the first and the last pedestrians in the line one of the two terms in brackets is missing.

Assume now the first pedestrian in line is waiting at a red traffic signal. Let us define its position as being $x_0 = 0$. For a finite system (i.e. there is a last pedestrian) and if $\lambda > 0$, the distances between pedestrians grow toward the end of the queue since the last pedestrian is not experiencing a pushing force from behind, and is therefore exerting a smaller force to his leader who is therefore further away from his leader than others further along the queue. This effect is most pronounced toward the end of the queue, while distances vary only slightly toward the front, particularly if λ has a value close to zero, much smaller than 1. We assume now that there is a long queue of Z pedestrians and we are interested in the distance L from the first to the Nth pedestrian where $N << Z$ when all pedestrians are standing still and for each pedestrians there is zero net force, that is, the left hand side of Equation 8 is zero. Because $N << Z$ for the pedestrians concerned we will not err greatly by assuming—as if the queue was infinitely long—that all distances between pedestrians are identical setting $d = d_{i,i+1} = d_{i,i-1}$:

$$0 = \frac{v_0 - \dot{x}_i}{\tau} - (1-\lambda)Ae^{-\frac{d}{B}} \quad (9)$$

where the left side (acceleration) is zero, since we are at equilibrium. This can be solved easily for d; and, as stated, we are interested in the smallest possible distance d_{min}, that is, when speed is zero:

$$d_{min} = B \ln(\alpha) \quad (10)$$

$$\alpha = \frac{(1-\lambda)A\tau}{v_0} \quad (11)$$

with parameter combination α defined for convenience since this parameter combination will occur often in the remainder.

The inverse of d_{min} is the maximum density ρ_{max}, which we identify with the stand still density, that is, the upper end point in the fundamental diagram. For this and for the abovementioned L results, it is simple to formulate:

$$\rho_{max} = \frac{1}{B \ln(\alpha)} \quad (12)$$

$$L = (N-1)B \ln(\alpha) \quad (13)$$

Equation 13 connects the two properties L and N which have an immediate non-abstract meaning with parameter B and parameter combination α of the SFM and give a first indication as to which parameter modification has which impact on the simulation results.

The second step is to imagine that the traffic light at which the pedestrians were waiting turns green and that the queue of pedestrians discharges. At first pedestrians which cross $x = 0$, that is, the stop line of the signal, are still accelerating, yet after some time the fourth, or fifth, or tenth pedestrian has already accelerated to the speed which is possible at that density. The further discharge occurs with capacity flow j_c at capacity density ρ_c.

For this situation we also assume that distances between pedestrians are constant. We resolve Equation 9 for $\dot{x}_i$ and use it for v in the fundamental equation:

$$j(\rho) = \rho v(\rho) \quad (14)$$

$$= \rho v_0\left(1 - \alpha e^{-\frac{1}{B\rho}}\right) \quad (15)$$

At capacity (ρ_c, j_c) the first derivative of j by ρ vanishes:

$$\frac{\delta j}{\delta \rho} = v_0\left(1 - \alpha\left(1 + \frac{1}{B\rho}\right)e^{-\frac{1}{B\rho}}\right) = 0, \text{for } \rho = \rho_c \quad (16)$$

Next we multiply and at the same time divide the last term by e:

$$0 = v_0\left(1 - \alpha e\left(1 + \frac{1}{B\rho_c}\right)e^{-\left(1 + \frac{1}{B\rho_c}\right)}\right) \quad (17)$$

which prepares the equation to be resolved for ρ_c:

$$\rho_c = -\frac{1}{B\left(1 + W\left(-\frac{1}{\alpha e}\right)\right)} \quad (18)$$

where $W()$ is the Lambert W function (*19–23*) which is defined as the inverse of xe^x, that is:

$$x = W(x)e^{W(x)} \quad (19)$$

To be precise: in this work $W()$ always references the lower real valued branch of the Lambert W function which is denoted as $W_{-1}()$ when there is a need to distinguish different branches.

The lower branch of $W()$ is defined for all $-1/e \le x < 0$. This is the case for the argument $-1/e/\alpha$, since Equation 12 requires $\alpha > 1$. All of its values are $W() < -1$ which is what is also required in Equation 18.

Using this in Equation 15, the capacity flow is:

$$j_c = -\frac{v_0}{B\left(1 + W\left(-\frac{1}{\alpha e}\right)\right)}\left(1 - \alpha e^{1 + W\left(-\frac{1}{\alpha e}\right)}\right) \quad (20)$$

$$= -\frac{v_0}{B\left(1 + W\left(-\frac{1}{\alpha e}\right)\right)}\left(1 - (\alpha e)\frac{W\left(-\frac{1}{\alpha e}\right)}{W\left(-\frac{1}{\alpha e}\right)}e^{W\left(-\frac{1}{\alpha e}\right)}\right) \quad (21)$$

$$= -\frac{v_0}{B\left(1 + W\left(-\frac{1}{\alpha e}\right)\right)}\left(1 - (\alpha e)\frac{-\frac{1}{\alpha e}}{W\left(-\frac{1}{\alpha e}\right)}\right) \quad (22)$$

$$= -\frac{v_0}{B\left(1 + W\left(-\frac{1}{\alpha e}\right)\right)}\left(1 + \frac{1}{W\left(-\frac{1}{\alpha e}\right)}\right) \quad (23)$$

$$= -\frac{v_0}{B}\frac{1}{W\left(-\frac{1}{\alpha e}\right)} \quad (24)$$

where for the second transformation we use the defining Equation 19 of the Lambert W function.

To put it in terms of directly observable properties: the time T for N consecutive pedestrians to cross the stop line with Equation 24 results in the simple formlation:

$$T = \frac{(N-1)}{j_c} \tag{25}$$

$$= -(N-1)\frac{B}{v_0}W\left(-\frac{1}{\alpha e}\right) \tag{26}$$

Equations 12 and 24 or their repackings in terms of L and T can be combined to obtain an equation without parameter B or without parameter α. For this we first define a combination of observable properties:

$$q = \frac{j_c}{v_0 \rho_{max}} = \frac{L}{v_0 T} \tag{27}$$

Then q can be written depending only on α, but not B:

$$q = -\frac{\ln(\alpha)}{W\left(-\frac{1}{\alpha e}\right)} \tag{28}$$

or on B, but not α:

$$\frac{1}{q\rho_{max}} = \frac{Tv_0}{N-1} = -BW\left(-e^{-\left(1+\frac{1}{B\rho_{max}}\right)}\right) \tag{29}$$

The right hand side of Equation 28 has a lower limit of 0 and an upper limit of 1 since $W()$ can be approximated (*24*) by:

$$W(x) \approx \ln(-x) - \ln(-\ln(-x)) \text{ for } x \to 0^- \tag{30}$$

and since:

$$\lim_{\alpha\to\infty}\left[-\frac{\ln(\alpha)}{\ln\left(\frac{1}{\alpha e}\right) - \ln\left(-\ln\left(\frac{1}{\alpha e}\right)\right)}\right] = 1 \tag{31}$$

The lower limit of 0 does not have any relevant implication, yet the upper limit of 1 implies that under the assumptions and simplifications made in this contribution the SFM can only reproduce pedestrian system dynamics where:

$$q < 1 \tag{32}$$

It appears that this is not a relevant limitation since in Seyfried et al. (*25*) roughly the values $v_0 \approx 1.25$ m/s, $j_c \approx 0.8$ s^{-1}, and $\rho_{max} \approx 2.0$ m^{-1} were found such that $q \approx 0.32$.

Equations 28 and 29 can be solved for α and B respectively. Both transformations include steps which are not obviously goal directed. Therefore we will do this step by step, beginning with Equation 28:

$$q = -\frac{\ln(\alpha)}{W\left(-\frac{1}{\alpha e}\right)} \tag{33}$$

$$W\left(-\frac{1}{\alpha e}\right) = -\frac{\ln(\alpha)}{q} \tag{34}$$

$$-\frac{1}{\alpha e} = -\frac{\ln(\alpha)}{q}e^{-\frac{\ln(\alpha)}{q}} \tag{35}$$

$$\frac{1}{e} = \frac{\ln(\alpha)}{q}e^{-\frac{\ln(\alpha)}{q}+\ln(\alpha)} \tag{36}$$

$$\frac{q-1}{e} = \frac{q-1}{q}\ln(\alpha)e^{\frac{q-1}{q}\ln(\alpha)} \tag{37}$$

$$\frac{q-1}{e}e^{-\frac{q-1}{q}\ln(\alpha)} = \frac{q-1}{q}\ln(\alpha) \tag{38}$$

$$e^{\frac{q-1}{e}e^{-\frac{q-1}{q}\ln(\alpha)}} = e^{\frac{q-1}{q}\ln(\alpha)} \tag{39}$$

$$\frac{q-1}{e}e^{-\frac{q-1}{q}\ln(\alpha)}e^{\frac{q-1}{e}e^{-\frac{q-1}{q}\ln(\alpha)}} = \frac{q-1}{e} \tag{40}$$

$$\frac{q-1}{e}e^{-\frac{q-1}{q}\ln(\alpha)} = W\left(\frac{q-1}{e}\right) \tag{41}$$

$$\frac{q-1}{e}\alpha^{-\frac{q-1}{q}} = W\left(\frac{q-1}{e}\right) \tag{42}$$

$$\alpha = \left[-W\left(-\frac{1-q}{e}\right)\frac{e}{1-q}\right]^{\frac{q}{1-q}} \tag{43}$$

Now for Equation 29. It can be written as:

$$\frac{1}{qB\rho_{max}} = -W\left(-e^{-\left(1+\frac{1}{B\rho_{max}}\right)}\right) \tag{44}$$

$$-\frac{1}{qB\rho_{max}}e^{-\frac{1}{qB\rho_{max}}} = -e^{-\left(1+\frac{1}{B\rho_{max}}\right)} \tag{45}$$

$$\frac{1}{B}e^{\frac{1}{B\rho_{max}}-\frac{1}{qB\rho_{max}}} = \frac{q\rho_{max}}{e} \tag{46}$$

$$\frac{1}{B}\left(\frac{1}{\rho_{max}} - \frac{1}{q\rho_{max}}\right)e^{\frac{1}{B\rho_{max}}-\frac{1}{qB\rho_{max}}} = \frac{q\rho_{max}}{e}\left(\frac{1}{\rho_{max}} - \frac{1}{q\rho_{max}}\right) \tag{47}$$

$$\left(\frac{1}{\rho_{max}} - \frac{1}{q\rho_{max}}\right)\frac{1}{B} = W\left(\frac{q\rho_{max}}{e}\left(\frac{1}{\rho_{max}} - \frac{1}{q\rho_{max}}\right)\right) \tag{48}$$

$$B = -\frac{1-q}{q\rho_{max}W\left(-\frac{1-q}{e}\right)} \tag{49}$$

To finish this section we summarize the model simplifications which we had to assume to obtain the results:

1. Homogeneous population: all pedestrians have identical parameters. Microsimulations, by their construction, make it possible to account for a heterogeneous population by assigning individual parameters to each unit. To allow analytical

investigation we had to refrain from doing that. Obviously the values in our analysis would be the average or median value of any distribution. To test the theory, simulations can be run with identical parameters for all pedestrians.

2. One-dimensional movement: reducing the dimensionality from one spatial dimension to two introduces a fixed order between pedestrians which is required for the next simplification step of considering only nearest neighbors.
3. Only nearest neighbor interaction: the SFM originally assumed that each pedestrian in a simulation exerts a force on any other. The reduction to nearest neighbor interactions made the analytical investigation much easier, but obviously this step implies that results cannot be exact. It may be interesting in this regard that it was found that the approach of considering all other pedestrians as having an impact on a certain pedestrian does not lead to the most realistic speed–density relation, but that cutting forces at a certain number or at least suppressing them with more remote neighborhood degree is beneficial (*26*, *27*). Thus the approximation of nearest neighbor interaction can also be interpreted as an improvement of the model. In computer implementations of simulation models the number of forces on a particular pedestrian is usually limited anyway to achieve reasonable computation times, which is another reason why this assumption is not only a simplification, but matches the pragmatic ways in which the SFM is actually applied.
4. Stationary state: it was assumed that movement is in a stationary state, that is, macroscopic properties are stable over time. This is required by the aim of the work to relate *simple* macroscopic properties to model parameters. Without the system being stationary there are no simple properties. This assumption implies that one cannot expect results to describe situations exactly which includes, for example, an acceleration phase.
5. Equal spacing between pedestrians instead of increasing distances toward the rear end of the queue: this is an approximation to which one can get arbitrarily close in a simulation by enlarging the size of the system.
6. By assuming that the queue is a *long* queue we implicitly assumed that the time delay from the first pedestrians in line having to accelerate from zero speed can be ignored. Thus our results cannot be transferred to queues consisting of only a small number of pedestrians.

Summary of Results and Discussion

For N pedestrians who are stationary at the front of a long single-file queue, for example, at a traffic signal or service desk, we find in the SFM with only nearest neighbor interaction for the length L from the first to the Nth pedestrian and the time T it takes for discharge between the first and the Nth pedestrian crossing a given line:

$$L = (N-1)B\ln(\alpha) \tag{50}$$

$$T = -(N-1)\frac{B}{v_0}W\left(-\frac{1}{\alpha e}\right) \tag{51}$$

with parameter combination α:

$$\alpha = \frac{(1-\lambda)A\tau}{v_0} \tag{52}$$

where it should not be forgotten that parameter A is defined differently among various studies in the literature—compare Equation 3.

In terms of maximum (stand still) density ρ_{max} and capacity flow j_c, Equations 50 and 51 can be rewritten:

$$\rho_{max} = \frac{1}{B\ln(\alpha)} \tag{53}$$

$$j_c = -\frac{v_0}{B}\frac{1}{W(-\frac{1}{\alpha e})} \tag{54}$$

Defining a combination of observable parameters:

$$q := \frac{j_c}{v_0\rho_{max}} = \frac{L}{v_0 T} \tag{55}$$

it is possible to rewrite Equations 53 and 54 omitting either α or B:

$$\frac{1}{q} = -\rho_{max}BW\left(-e^{-\left(1+\frac{1}{B\rho_{max}}\right)}\right) \tag{56}$$

$$q = -\frac{\ln(\alpha)}{W(-\frac{1}{\alpha e})} \tag{57}$$

These can be resolved for α and B, giving:

$$\alpha = \left[-W\left(-\frac{1-q}{e}\right)\frac{e}{1-q}\right]^{\frac{q}{1-q}} \tag{58}$$

$$B = -\frac{1}{W(-\frac{1-q}{e})}\frac{1}{\rho_{max}}\frac{1-q}{q} \tag{59}$$

Note that for q it does not make a difference if j_c refers to the total flow and ρ_{max} to the line density or if j_c refers to the specific flow and ρ_{max} to the two-dimensional density, since Equation 55 requires only an additional factor

"1 = width/width" to get from the former to the latter. Thus, in a certain sense parameter q is independent of the dimension.

The first hope would now be that one could measure v_0, ρ_{max}, and j_c in reality, from that compute a q, and with all these use Equations 58 and 59 to compute parameters α and B. Then simulating with α and B the same values as measured would result for ρ_{max} and j_c in the simulation. However, with all the approximations we have made, the chances of success are doubtful. We want to benefit from the results obtained from considering a one-dimensional system for two-dimensional situations as well; and we do not seriously want to consider only the two nearest neighbors in a two-dimensional simulation, but to have at least four, six, or eight pedestrians emitting a force on a certain pedestrian. Doing so would definitely change the results presented above for one-dimensional as well as two-dimensional situations, since the net value of the sum of forces will generally be larger. Thus, we cannot directly apply these equations. However, we can hope that the tendencies which they give are preserved even if we change fundamental properties of the system. In other words, in a calibration process the results derived above might allow an educated guess and indicate in which direction parameters need to be changed so that the results change in a certain direction.

Assume you have already done a simulation and obtained a certain ρ_{max} and a certain j_c. You find that compared with your empirical data both ρ_{max} and j_c are too large. Equations 53 and 54 provide a simple solution: increasing the value of B will lower the value of both result attributes, and the same is true for parameter α: larger α will lead to smaller ρ_{max} and smaller j_c, as is shown in the Appendix.

The case when both ρ_{max} and j_c should result in larger values is just as simple: lower the value(s) of B or α or both.

However, how should the parameters be changed if it is required to increase the value of ρ_{max} but reduce j_c? Should the value of α be raised and that of B be lowered or the other way round? The answer is that to achieve a higher value for ρ_{max} and a lower value for j_c the value of α needs to be reduced and the value of B raised (for the math, see Appendix).

In summary, this suggests the following procedure for calibration of the SFM:

1. Measure or define reasonable values for desired speed v_0, capacity flow j_c (or discharge rate T for a given number N of pedestrians), and maximum (stand still) density ρ_{max} (or queue length L for the same given number N of pedestrians). There is no universally correct value for L and T. Younger people may have different values than older people. In emergency egress situations one may expect to find smaller values for L as well as T than if people are queuing on their way to carry out some rather unpleasant duty. Make sure that $0 < q < 1$ in Equation 55.
2. With the help of Equation 59 compute a value for parameter B. The Lambert W function is implemented in numerical computing and computer algebra software like Matlab ("lambertw(-1, x)") (*28*), Mathematica ("ProductLog[-1, x]") (*29*) or Maple ("lambertw(-1, x)") (*30*). One can also find plugins for Excel to download from the Internet. The easiest way to compute a few values is by using the (online) computational knowledge engine Alpha (*31*). In any case make sure that the value from the lower real valued branch ($W_{-1}()$) is computed.
3. With the help of Equation 58 compute a value for parameter combination α. Make sure that $\alpha > 1$.
4. Choose values for parameters λ, τ, and A, such that Equation 52 holds. Consider what is written above and in Equation 3 about A and $\tilde{A}$. Reasonable boundaries are $0 \leq \lambda < 0.4$ (in Equations 3 and 4 values between 0.02 and 0.2 are given as resulting from empirical studies) and $0.05 \leq \tau \approx < 2.0$ s. Also consider that $4v_0\tau/B \leq 1$ should hold or that at least the left side is only slightly larger than 1 since otherwise a pedestrian may visibly oscillate back and forth when approaching another (*32*).
5. Perform simulations. In the simulation, evaluate ρ_{max} and j_c. And, to repeat:
   - if both ρ_{max} and j_c need to be larger, reduce the value of B and/or α
   - if both ρ_{max} and j_c need to be smaller, increase the value of B and/or α
   - if ρ_{max} needs to be smaller and j_c larger, reduce the value of B and increase the value of α
   - if ρ_{max} needs to be larger and j_c smaller, increase the value of B and reduce the value of α
   - simulate again and evaluate ρ_{max} and j_c
6. Set the values of all other parameters in case you are working with a variant or an extension of the SFM. For this it is typically necessary to include more complex scenarios in the calibration procedure. For example, the width of the distribution of desired speeds as well as the parameter that sets the impact strength of the relative velocity in elliptical specification II (Equation 3) (called *VD* or Δt): both are most relevant in bi- or multi-directional flow situations, so include scenarios with bi-directional and crossing flows. Remember that you are free to change the values of τ, λ, and A without spoiling the calibration, as long as the value of α in Equation 52 is conserved.

Verification and Example

In this section we first check if the math presented above can be correct by testing it with a concrete example and simulating with the same model simplifications that were assumed. In a second step we release some of the simplifications, most importantly applying the calibration procedure in a two-dimensional setting.

We configure a simulation of pedestrian dynamics (*33*) as close as possible to the model simplifications (a pedestrian "sees" forces only from two other pedestrians, very narrow distribution of desired speeds, circular specification of the SFM) and simulate a de facto one-dimensional situation (50-cm wide corridor) where 1,000 pedestrians first walk toward a red traffic signal, stop there, and start moving when it turns green later. We measure the linear density of pedestrians as they wait in front of the red signal (number of pedestrians in a 100 m long section) and the discharge flow after the signal turns green (number of pedestrians passing within 100 seconds, measurement beginning 100 seconds after the signal turns green). We pretend we know nothing about the parameters of the SFM and use the empirical results of (*25*) ($v_0 \approx 1.25$m/s, $j_c \approx 0.8$ s^{-1}, and $\rho_{max} \approx 2.0$ m^{-2}) which give $q = 0.32$ and from that with Equations 58 and 59: $\alpha = 2.7532$ and $B = 0.4937$ m (in the software for this purpose it is $B_{soc,iso} = 0.497$ m whereas $B_{soc,mean}$ is irrelevant since to disable the elliptical specification II it is required to set $A_{soc,mean} = 0$).

The first check is on the freedom to set the parameters τ, A, and λ that make up α. The results, as displayed in Table 1, show that in fact the flow (capacity) is constant and that the maximum density varies only slightly for identical values of α, such that the simulation in this respect confirms the theory presented above.

Looking at the particular values for flow and maximum density we see that the flow matches expectations exactly, which further increases confidence in the above results, but that the maximum density is slightly smaller than expected. This can be due to (at least) two causes: first, it can be a consequence of a violation of the (non-) oscillation condition $4v_0\tau \leq B$ (see [*32*]), that is, pedestrians are not yet at rest at the time of measurement, but—for some of the parameter sets clearly, for others barely visible—they still move back and forth at the time of measurement, hence requiring more space. Among other things, this is one of the reasons not to utilize the circular specification of the SFM alone, but to combine it with, for example, elliptic specification II which by construction suppresses oscillations. Second, our simulation system is not exactly one-dimensional, but it is a two-dimensional simulation system artificially confined to something like a one-dimensional system. This may have certain (small) effects which would be beyond the scope of this discussion. In total we see the results to be in accordance with the theoretical considerations presented above.

The second test is in a two-dimensional situation: 8,000 pedestrians walk down a 4 m wide strip. We aim at reproducing Weidmann's (*34*) properties of pedestrian dynamics ($v_0 = 1.34$ m/s, $j_c = 1.25$ (ms)$^{-1}$, and $\rho_{max} = 5.4$ m^{-2}; implying $q = 0.172$ and with Equation 58 $\alpha = 1.44$). Compared with data from other literature, this capacity flow is small and the maximum density is moderate. We assign pedestrians a radius of 0.213 m ± 10%, implying that 5.4 such (circle-shaped) pedestrians cover an area of 0.77 m^2. This is near the maximum size (radius of 0.231 m) which allows the mentioned maximum density without overlapping (densest packing of spheres). This raises the expectation that it might be difficult to reproduce the desired maximum density of 5.4 m^{-2} with any set of parameters.

Going from one dimension to two dimensions first bears the difficulty of how to transfer Equation 59, since maximum density in two dimensions has a dimension of m$^-$2 while B needs the dimension m. We assumed—which we believe to be the simplest assumption—that the maximum density needs to be divided by a lane width w_l and the explicit ρ_{max} in Equation 59 needs to be replaced by ρ_{max}/w_l. As an initial guess we assume a lane width of $w_l = 0.5$ m. This leads to $B = 1.02$ m.

Going from one dimension to two dimensions it is furthermore anything but certain that the properties and relations derived above still hold. We want to test (a) if, as in one dimension, in two dimensions also capacity flow and maximum density only depend on α, but not on how A, τ, λ, and v_0 contribute to the value of α, (b) if Equations 53 and 54 have a meaning in two dimensions,

Table 1. Parameter Choices and Simulation Results in a One-Dimensional Setting

τ [s]	$A_{Soc,Iso}$ [m/s^2]	λ	ρ_{max} [m^{-1}]	j_c [s^{-1}]
0.4	3.390	0.1	1.71	0.8
0.2	6.780	0.1	1.91	0.8
0.15	9.041	0.1	1.92	0.8
0.4	4.359	0.3	1.82	0.8

Note: The pedestrians all had the same length of 0.456 m and the desired speed was sharply 1.25 m/s, and in all cases $B = 0.4937$ m.

Table 2. Parameter Choices and Simulation Results in a Two-Dimensional Setting

τ [s]	$A_{Soc,Iso}$ [m/s^2]	λ	ρ_{max} [m^{-2}]	j_c [(ms)$^{-1}$]
0.2	6.477	0.1	1.40	0.43
0.4	3.239	0.1	1.40	0.43
0.2	5.830	0.0	1.41	0.43

and (c) if the calibration procedure as defined above can in fact be carried out in a two-dimensional setting.

Another question in a two-dimensional environment is how many pedestrians should be taken into account as "nearest neighbors" for the SFM. Obviously two would be too few. Assuming a near to hexagonal (or triangular) walking formation, we choose six as the number of influencing pedestrians. It is reported by Porzycki et al. (*35*) that the sixth closest neighbor has the highest probability of walking directly in front of a person. This can be seen as an indication in favor of this idea.

The results as given in Table 2 support two conclusions. First, in two dimensions as in one dimension, ρ_{max} and j_c depend only on α and not on its factors. It was not at all clear that this fact would "survive" the expansion from one to two dimensions. Therefore this result is particularly pleasing. Second, the results of both ρ_{max} and j_c are much smaller than expected from the one dimension theory. Obviously concerning this aspect the transformation from one dimension to two dimensions brings a major change and with it a challenge for future work.

The simplest way to bring the simulation results closer to the expected values is to assume larger values for w_l: $w_l = 1.0$ m leads to $\rho_{max} = 4.10$ m^{-2} and $j_c = 1.56$ (ms)$^{-1}$ such that one of the result parameters (ρ_{max}) is smaller than desired and the other is larger. This implies that the best parameters A and B may neither be both smaller nor both larger than the ones applied at this point ($A = 3.9$ m/s^2 and $B = 0.51$ m). Furthermore we apply the process and change parameters A and B following the results of previous simulations as described above, keeping the values of τ and λ constant. Table 3 gives an overview of those simulations where the resulting capacity flow is close to the desired value. It can easily be seen that indeed the value of parameter A needs to be reduced and the value of parameter B increased to get closer to the desired result. However, even for $B \rightarrow \infty$ the desired maximum density cannot be achieved for the given flow capacity.

It may appear disappointing at first that the model parameters cannot be set such that the desired values of result parameters are met. However, this can also be interpreted as a positive property of the model: since the pedestrians in the simulation had a size (radius, to be precise) which implied a maximum density (densest

Table 3. More Parameter Choices and Simulation Results in a Two-Dimensional Setting

$A_{Soc,Iso}$ [m/s^2]	$B_{Soc,Iso}$ [m]	ρ_{max} [m^{-2}]	j_c [(ms)$^{-1}$]
30.0	0.1	1.99	1.19
10.0	0.3	2.82	1.22
5.0	0.5	3.49	1.20
3.0	1.0	4.02	1.28
2.5	2.0	4.04	1.25
2.0	10.0	4.43	1.30
1.9	30.0	4.51	1.33
1.9	50.0	4.49	1.30
1.9	100.0	4.48	1.29
1.9	10,000.0	4.46	1.26

Note: For all simulations, $\tau = 0.2$ s and $\lambda = 0.1$ s.

packing of spheres) near the desired result for ρ_{max} one may simply require from a model that it cannot be tweaked by parameter modifications to produce any result one could think of.

Finally we checked once more that this depends only on α and not the parameters it is made of and varied the values of λ and τ as in Table 2, adjusting $A_{soc,iso}$ such that $\alpha = 0.2566$ is preserved. Again j_c and ρ_{max} remained nearly unchanged.

Summary and Outlook

In this contribution we were able to derive two formulas that assign values to a parameter (B) and a parameter combination (λ, τ, A) for a simplified version of the SFM using as input only observable properties (stand still density, capacity flow, and free speed). Based on this we proposed a calibration procedure that can be applied, which does not require knowledge of either the derivation of the results as given in this work or details of the SFM. We demonstrated the calibration procedure with a concrete example. As a result of the insight into the fundamentals of the SFM presented in this paper, this calibration procedure is an *informed* procedure that exploits the model structure for calibration as opposed to optimization schemes which work irrespective of this. Nevertheless it may well be that a particular general optimization scheme can be combined with the particular information of this article to enhance the performance of an optimization (calibration) process.

Future work may concern mainly the transformation from one to two dimensions as well as the further calibration process. The approach presented in this work leaves freedom to choose some of the parameters. For extended variants of the SFM with additional parameters there is even more such freedom. To calibrate the values of these parameters by empirical data from further walking situations (e.g. bi-directional, crossing) is required.

Appendix

The Function $-1/W(-1/(e\alpha))$

The easiest way to gain some insight into the properties of a function is to plot it, see Figure 1. There the visual impression already suggests strongly that the first derivative by α is always negative. This can be proven easily:

$$f(y(\alpha)) = -\frac{1}{W(y(\alpha))} \tag{60}$$

$$y(\alpha) = -\frac{1}{e\alpha} \tag{61}$$

$$\frac{df(y(\alpha))}{d\alpha} = \frac{1}{W^2(y(\alpha))}\frac{dW(y)}{dy}\frac{dy(\alpha)}{d\alpha} \tag{62}$$

Using:

$$\frac{dW(y)}{dy} = \frac{W(y)}{y(1+W(y))} \tag{63}$$

see for example (*23*), we get:

$$\frac{df(y(\alpha))}{d\alpha} = \frac{1}{W^2(y(\alpha))}\frac{W(y(\alpha))}{y(\alpha)(1+W(y(\alpha)))}\frac{1}{e\alpha^2} \tag{64}$$

$$= \frac{1}{W(y(\alpha))}(-e\alpha)\frac{1}{(1+W(y(\alpha)))}\frac{1}{e\alpha^2} \tag{65}$$

$$= \left(-\frac{1}{\alpha}\right)\left(\frac{1}{W(y(\alpha))}\right)\left(\frac{1}{1+W(y(\alpha))}\right) \tag{66}$$

Since all three top level brackets enclose negative values for all $\alpha>1$ the derivative is negative for all $\alpha>1$.

To Lower the Value of ρ_{max} and Raise the Value of j_c

To answer the question how α and B need to be changed to achieve a smaller ρ_{max} and a larger j_c, we imagine that we already have the right value for ρ_{max} and want to raise the value of j_c without changing ρ_{max}. So we have "old" parameters α and B and "new" parameters α' and B'. Since ρ_{max} should remain constant it is required that:

$$\frac{1}{B'\ln(\alpha')} = \frac{1}{B\ln(\alpha)} \tag{67}$$

$$\alpha' = \alpha^{\frac{B}{B'}} \tag{68}$$

Note that if $B'>B$ then $\alpha'<\alpha$ and vice versa. This allows one to eliminate α' in the new $j_{c'}$:

$$j_{c'} = -\frac{v_0}{B}\frac{1}{W\left(-\frac{1}{e\alpha^{\frac{B}{B'}}}\right)} \tag{69}$$

$$f(\varepsilon) := \frac{j_{c'}}{j_c} = \frac{1}{1+\varepsilon}\frac{W\left(-\frac{1}{e\alpha}\right)}{W\left(-\frac{1}{e\alpha^{\frac{1}{1+\varepsilon}}}\right)} \text{ where} \tag{70}$$

$$1+\varepsilon := \frac{B'}{B} \tag{71}$$

Parameter ε quantifies the difference between old and new B; if $\varepsilon<0$ then $B'<B$, otherwise $B' \geq B$. The hands-on method to see that $f(\varepsilon)>1$ if $\varepsilon<0$ and $f(\varepsilon)<1$ if $\varepsilon>0$ is to plot $f(\varepsilon)$ for various α and thereby see that it appears to decrease strictly monotonically with growing ε and for all α. A more mathematical approach is to compute the linear Taylor approximation of $f(\varepsilon)$:

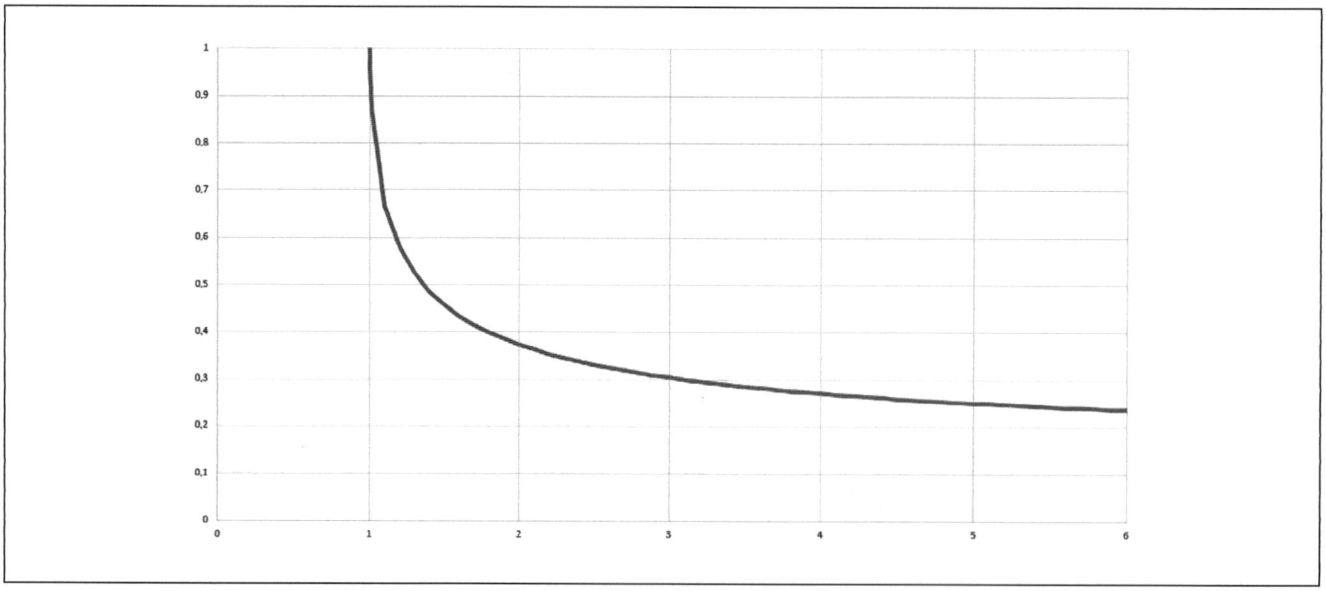

Figure 1. $f(x) = -\frac{1}{W\left(-\frac{1}{ex}\right)}$ with x from 1 to 6.

$$\frac{df(\varepsilon)}{d\varepsilon} = -\left(\frac{1}{1+\varepsilon}\right)^2 \frac{W\left(-\frac{1}{e\alpha}\right)}{W\left(-\frac{1}{e\alpha^{\frac{1}{1+\varepsilon}}}\right)} \left(1 + \frac{1}{1+\varepsilon} \frac{\ln(\alpha)}{1 + W\left(-\frac{1}{e\alpha^{\frac{1}{1+\varepsilon}}}\right)}\right) \tag{72}$$

$$\frac{df}{d\varepsilon}(0) = -\left(1 + \frac{\ln(\alpha)}{1 + W\left(-\frac{1}{e\alpha}\right)}\right) \tag{73}$$

So we have:

$$f(\varepsilon) \approx 1 - \left(1 + \frac{\ln(\alpha)}{1 + W\left(-\frac{1}{e\alpha}\right)}\right)\varepsilon \tag{74}$$

for small values of ε, where the value inside the top level brackets is always positive for all $\alpha > 1$, otherwise there would be some $a > 1$ for which holds:

$$-a = \frac{\ln(\alpha)}{1 + W\left(-\frac{1}{e\alpha}\right)} \tag{75}$$

$$W\left(-\frac{1}{e\alpha}\right) = -\frac{1}{a}\ln(\alpha) - 1 \tag{76}$$

$$-\frac{1}{e\alpha} = \left(-\frac{1}{a}\ln(\alpha) - 1\right)e^{-\frac{1}{a}\ln(\alpha)-1} \tag{77}$$

$$\alpha^{\frac{1}{a}-1} = \frac{1}{a}\ln(\alpha) + 1 \tag{78}$$

which is not possible since in the last equation the left side is < 1 and the right side > 1 since $\alpha > 1$.

Thus we could show that $j_{c'} < j_c$ if $B' > B$ (with a certain $\alpha' < \alpha$). Coming back to the original idea of an unmodified ρ_{max} where if $B' > B$ it needs to be $\alpha' < \alpha$: choose some $\hat{B}$ with $B < \hat{B} < B'$ and it will obviously be:

$$\rho_{max}(\hat{B}, \alpha') > \rho_{max}(B, \alpha) \tag{79}$$

$$j_c(\hat{B}, \alpha') < j_c(B, \alpha) \tag{80}$$

which shows that in order to raise the value of ρ_{max} and lower it for j_c one needs to raise the value of B and lower it for α.

Acknowledgments

For the preparation of this contribution the authors used David Pritchard's Transportation Research Board template for LaTeX: http://davidpritchard.org/archives/116, 2009.

Author Contributions

The authors confirm contribution to the paper as follows: study conception and design: Kretz; data collection: Kretz; analysis and interpretation of results: Kretz, Lohmiller, Sukennik; draft manuscript preparation: Kretz. All authors reviewed the results and approved the final version of the manuscript.

References

1. Helbing, D., and P. Molnar. Social Force Model for Pedestrian Dynamics. *Physical Review E*, Vol. 51, No. 5, 1995, p. 4282.
2. Helbing, D., I. Farkas, and T. Vicsek. Simulating Dynamical Features of Escape Panic. *Nature*, Vol. 407, 2000, pp. 487–490.
3. Johansson, A., D. Helbing, and P. Shukla. Specification of the Social Force Pedestrian Model by Evolutionary Adjustment to Video Tracking Data. *Advances in Complex Systems*, Vol. 10, 2007, pp. 271–288.
4. Helbing, D., and A. Johansson. Pedestrian, Crowd and Evacuation Dynamics. In *Encyclopedia of Complexity and Systems Science* (R. Meyers, ed.), Springer, New York, NY, 2011, pp. 697–716.
5. Wiedemann, R. *Simulation des Strassenverkehrsflusses.* Schriftenreihe des IfV, Karlsruhe, 1974.
6. Burstedde, C., K. Klauck, A. Schadschneider, and J. Zittartz. Simulation of Pedestrian Dynamics Using a Two-Dimensional Cellular Automaton. *Physica A: Statistical Mechanics and its Applications*, Vol. 295, No. 3, 2001, pp. 507–525.
7. Nishinari, K., A. Kirchner, A. Namazi, and A. Schadschneider. Extended Floor Field CA Model for Evacuation Dynamics. *IEICE Transactions on Information and Systems*, Vol. 87, No. 3, 2004, pp. 726–732.
8. Kretz, T., M. Kaufman, and M. Schreckenberg. Counterflow Extension for the FAST-model. *Cellular Automata*, 2008, pp. 555–558.
9. Yu, W., R. Chen, L. Dong, and S. Dai. Centrifugal Force Model for Pedestrian Dynamics. *Physical Review E*, Vol. 72, No. 2, 2005, p. 026112.
10. Lakoba, T., D. Kaup, and N. Finkelstein. Modifications of the Helbing-Molnar-Farkas-Vicsek Social Force Model for Pedestrian Evolution. *Simulation*, Vol. 81, No. 5, 2005, pp. 339–352.
11. Yu, W., and A. Johansson. Modeling Crowd Turbulence by Many-Particle Simulations. *Physical Review E*, Vol. 76, No. 4, 2007, p. 046105.
12. Parisi, D., M. Gilman, and H. Moldovan. A Modification of the Social Force Model Can Reproduce Experimental Data of Pedestrian Flows in Normal Conditions. *Physica A: Statistical Mechanics and its Applications*, Vol. 388, No. 17, 2009, pp. 3600–3608.
13. Steffen, B. A Modification of the Social Force Model by Foresight. *Pedestrian and Evacuation Dynamics 2008*, 2010, pp. 677–682.
14. Zanlungo, F., T. Ikeda, and T. Kanda. Social Force Model with Explicit Collision Prediction. *EPL (Europhysics Letters)*, Vol. 93, No. 6, 2011, p. 68005.
15. Ratsamee, P., Y. Mae, K. Ohara, T. Takubo, and T. Arai. Modified Social Force Model with Face Pose for Human Collision Avoidance. *Proc., 2012 7th ACM/IEEE International Conference on Human-Robot Interaction (HRI)*, IEEE, 2012, pp. 215–216.
16. Kretz, T., J. Lohmiller, and J. Schlaich. The Social Force Model and its Relation to the Kladek Formula. Presented at 95th Annual Meeting of the Transportation Research Board, Washington, D.C., 2016.

17. Chraibi, M., A. Seyfried, and A. Schadschneider. Generalized Centrifugal-force Model for Pedestrian Dynamics. *Physical Review E*, Vol. 82, No. 4, 2010, p. 046111.
18. Campanella, M. Improving the Nomad Microscopic Walker Model. In *12th IFAC Symposium on Transportation Systems*, September 2009, IFAC, Redondo Beach, CA, 2010.
19. Corless, R., D. Jeffrey, and D. Knuth. A Sequence of Series for the Lambert W Function. In *Proc., 1997 International Symposium on Symbolic and Algebraic Computation*, ACM, 1997, pp. 197–204.
20. Barry, D., J.-Y. Parlange, L. Li, H. Prommer, C. Cunningham, and F. Stagnitti. Analytical Approximations for Real Values of the Lambert W-Function. *Mathematics and Computers in Simulation*, Vol. 53, No. 1, 2000, pp. 95–103.
21. Chapeau-Blondeau, F., and A. Monir. Numerical Evaluation of the Lambert W Function and Application to Generation of Generalized Gaussian Noise with Exponent 1/2. *IEEE Transactions on Signal Processing*, Vol. 50, No. 9, 2002, pp. 2160–2165.
22. Scott, T., G. Fee, and J. Grotendorst. Asymptotic Series of Generalized Lambert W Function. *ACM Communications in Computer Algebra*, Vol. 47, No. 3/4, 2014, pp. 75–83.
23. Weisstein, E. *Lambert W-Function*, Online, from MathWorld–A Wolfram Web Resource. http://mathworld.wolfram.com/LambertW-Function.html
24. Veberic, D. Having Fun with Lambert W (x) Function. *arXiv preprint arXiv:1003.1628*, 2010.
25. Seyfried, A., B. Steffen, W. Klingsch, and M. Boltes. The Fundamental Diagram of Pedestrian Movement Revisited. *Journal of Statistical Mechanics: Theory and Experiment*, Vol. 2005, No. 10, 2005, p. P10002.
26. Seyfried, A., B. Steffen, and T. Lippert. Basics of Modelling the Pedestrian Flow. *Physica A: Statistical Mechanics and its Applications*, Vol. 368, No. 1, 2006, pp. 232–238.
27. Kretz, T., J. Lohmiller, and J. Schlaich, The Inflection Point of the Speed-Density Relation and the Social Force Model. In *Traffic and Granular Flow 2015* (V. Knoop, W. Daamen, and S. Hoogendoorn, eds.), 2016-2017, p. 0, accepted for publication; preprint available at http://arxiv.org/abs/1507.04935.
28. Mathworks. *MATLAB Documentation*, 2014, http://de.mathworks.com/help/matlab/.
29. Wolfram, S. *The Mathematica Book*. Wolfram Media, Incorporated, 1996.
30. Maplesoft. *Maple User Manual - Maplesoft*, 2014, www.maplesoft.com/view.aspx?sl = 5883
31. *Wolfram Alpha*, 2016, http://www.wolframalpha.com
32. Kretz, T. On Oscillations in the Social Force Model. *Physica A: Statistical Mechanics and its Applications*, Vol. 438, 2015, pp. 272–285.
33. PTV Group, *PTV Vissim 9 Manual*, 2016, Chapter 9, Viswalk.
34. Weidmann, U. Transporttechnik der Fussgänger. *IVT Schriftenreihe*, Vol. 90, 1992.
35. Porzycki, J., M. Mycek, R. Lubaś, and J. Wąs, Pedestrian Spatial Self-Organization According to its Nearest Neighbor Position. *Transportation Research Procedia*, Vol. 2, 2014, pp. 201–206.

The Standing Committee on Traffic Flow Theory and Characteristics (AHB45) peer-reviewed this paper (18-00416).

Article

TRR
JOURNAL OF THE TRANSPORTATION RESEARCH BOARD

Transportation Research Record
2018, Vol. 2672(20) 239–248

DOI: 10.1177/0361198118788181
journals.sagepub.com/home/trr
SAGE

Introducing a Re-Sampling Methodology for the Estimation of Empirical Macroscopic Fundamental Diagrams

Lukas Ambühl[1], Allister Loder[1], Michiel C. J. Bliemer[2], Monica Menendez[1,3,4], and Kay W. Axhausen[1]

Abstract

The uncertainty in the estimation of the macroscopic fundamental diagram (MFD) under real-world traffic conditions and urban dynamics might result in an inaccurate estimation of the MFD parameters—especially if congestion is rarely observed network-wide. For example, as data normally come from punctual observations out of the whole network, it is unclear how representative these observations might be (i.e., how much is the observed capacity affected by the network's inhomogeneity). Similarly, if the observed data do not exhibit a distinct congested branch, it is hard to determine the network capacity and critical density. This, in turn, also leads to uncertainties and errors in the parametrization of the MFD for applications, for example traffic control. This paper introduces a novel methodology to estimate (i) the level of inhomogeneity in the network, and (ii) the critical density of the MFD, even when no congested branch is observed. The methodology is based on the idea of re-sampling the empirical data set. Using an extensive data set from Lucerne, Switzerland, and London, UK, insights are provided on the performance and the application of the proposed methodology. The proposed methodology is used to illustrate how the level of inhomogeneity is lower in Lucerne than in the three areas of the network of London that are investigated. The proposed measure of the level of inhomogeneity gives city planners the possibility to analyze and investigate how efficiently their road network is utilized. In addition, the analysis shows that, for the network of Lucerne, the proposed methodology allows accurate estimation of the critical density up to 16 times more often than would be possible otherwise. This simple and robust estimation of the critical density is crucial for the application of many traffic control algorithms.

The macroscopic fundamental diagram (MFD) is the upper bound in the macroscopic flow–density (q-k) relationship for vehicular traffic in an urban road network. Similar to a fundamental diagram of a single road, MFDs are characterized by an uncongested branch and a congested branch. In the uncongested branch, flow increases with density, whereas in the congested branch, flow decreases with increasing density until gridlock is reached, and speeds drop to zero. Thus, MFDs can be described by a combination of some of the following parameters: free flow speed, backward wave speed, capacity, jam density, critical density, and a smoothing parameter that describes the change in the slope of the average speed. Notice that for control purposes the critical density is an important parameter, as it marks the difference between the congested and the uncongested branch. The shape of the MFD is determined by the urban road structure, traffic control (*1*) and the level of inhomogeneity in the distribution of traffic (*2–4*), but more or less independent of the demand (assuming trip length remains more or less constant). New applications have been developed exploiting the elegance of the MFD to model traffic dynamics, predict travel behavior, and control traffic in urban networks (*5–16*), all relying on an accurate estimation of the MFD.

The MFD can be either analytically derived or approximated from empirical data (*17*). Daganzo and Geroliminis (*1*) proposed a semi-analytical formulation for urban networks. However, their method relies on technical information (green times, number of links that can be passed by a fast vehicle without stopping, etc.),

[1]IVT, ETH Zürich, Zürich, Switzerland
[2]Institute of Transport and Logistics Studies, University of Sydney, NSW, Australia
[3]Division of Engineering, New York University Abu Dhabi, United Arab Emirates
[4]Tandon School of Engineering, New York University, USA

Corresponding Author:
Address correspondence to Allister Loder: allister.loder@ivt.baug.ethz.ch

which might be highly variant or not available at all. The MFD can also be directly estimated from empirical data, leading to a concave and to some extent well-defined and reproducible relationship. However, homogeneity is rarely found in real road networks, and network-wide congestion is not always apparent. In the case where only a few roads show a different behavior than all the other roads, the observed flow–density relationship lies slightly below the upper bound (*2*). In general, empirical MFDs exhibit scatter that leads to a range of observed flows for any given accumulation, which leads to uncertainty in the estimation of the descriptive parameters (*18–21*). Seminal studies have identified the inhomogeneous spatial distribution of vehicle densities as the reason for this scatter (*2, 4, 22–24*). A partial solution, in case of contiguous and homogeneous sub-regions, is partitioning (*25–27*).

This uncertainty in the estimation of the MFD under real-world traffic conditions and urban dynamics might result in an inaccurate estimation of the MFD parameters—especially if congestion is rarely observed network-wide (*28*). For example, if the observed data do not exhibit a distinct congested branch, it is hard to determine the network capacity and critical density. Similarly, as the data normally come from punctual observations across the whole network, it is unclear how representative these observations might be (i.e., how much is the observed capacity affected by the network's inhomogeneity). This, in turn, also leads to uncertainties and errors in the parametrization of the MFD for applications (e.g., traffic control), reducing the effectiveness of such applications. This leads to the question addressed in this paper: Given day-to-day fluctuations from limited and probably biased empirical data, how can the shape of the MFD, and especially the critical density for traffic control purposes, as well as the level of inhomogeneity, be accurately estimated?

This paper proposes a novel re-sampling methodology to estimate the level of inhomogeneity and the critical density in urban networks under uncertain traffic conditions, even when only limited empirical data are available. Empirical evidence is provided on the performance of the proposed methodology using two large data sets for the cities of Lucerne, Switzerland (1 year) and London, UK (3 weeks).

The results show that the level of inhomogeneity is lower in Lucerne than in the three areas of the network in London that are investigated. Moreover, it is shown that in the case of London, the capacity of the system could be increased by around 20%, under the assumption that all links behave in a similar manner as the 50% best links. In addition, the paper illustrates how the proposed methodology estimates the critical density accurately for Lucerne, even when no congested branch of the MFD is available. All things considered, this new methodology promotes the concept and the use of the MFD in real-world applications.

The remainder of this paper is organized as follows. The next section introduces the re-sampling approach, then the two available data sets are presented and discussed in detail. Thereafter the empirical results are shown. Conclusions are given at the end.

Methodology

Re-Sampling the MFD

This section introduces the re-sampling method, which later will be used to estimate the critical density and the level of inhomogeneity in an urban network. The idea of the re-sampling method is to identify the most homogeneous sub-samples of all roads by first creating many random sub-samples of the network, estimating for each an MFD, and extracting the smooth upper bound from the superposition of all MFDs. When the re-sampling parameters are chosen appropriately, all points on the upper bound represent the most homogeneous traffic states.

As an illustration, consider an urban road network in and around a central business district (CBD) with many roads and vehicles. During the morning commute, most roads are congested, but not all of them; for example, arterial roads leading into the CBD are more congested than those leading out. The resulting averages provide an estimate on the actual mean performance of the road network, but not the potential performance if all links were similarly congested. Choosing a certain sample size and repeating the sampling many times enables a homogeneous set of roads to be found without the need to filter the most homogeneous links manually or apply a more complex partitioning (which usually assumes spatial contiguity of the partitioned regions).

The urban road network of a city is given by N directed links, where N refers to all the monitored links (i.e., N might be any number covering between 0 and 100% of the links in the network). The length of each link i is known as l_i. Following Geroliminis and Daganzo (*18*), the MFD is given by the length-weighted means of flow q, and density k, $q = \sum_i q_i l_i / \sum_i l_i$ and $k = \sum_i k_i l_i / \sum_i l_i$, respectively. As previously discussed, under real conditions a network might not be homogeneously congested, and some links might be more congested than others. If these form a connected subgraph in the road network, the network could be partitioned as, for example, in Ji and Geroliminis (*23*). However, if these links are randomly distributed across the network and the sample varies with time, the MFD can be estimated as follows in order to reduce the influence of inhomogeneity in the measurements: randomly sample Ω-times without replacement $N_s \subset N$, where the ratio N_s/N

denotes the sample size as a proportion of all observations. The maximum number of combinations is given by the binomial coefficient $C = \binom{N}{N_s}$. Notice that sometimes, the number of possible combinations that can be considered might be limited by the computational resources, leading to $\Omega \ll C$.

The intuition behind this statistical sampling is simple. In the case of a network with homogeneous roads (i.e., all exhibiting identical fundamental diagrams) and a homogeneous distribution of congestion, the proof of the equivalence between the upper bound of the full sample and the re-sampled upper bound is trivial. For a more realistic and therefore inhomogeneous network, a representative sub-sample with $N_s \gg 0$ links will never exceed the theoretical upper bound by Daganzo and Geroliminis (*1*); the sub-sample will be equal to the upper bound in case of perfect homogeneous congestion, and below the upper bound in all other cases. In other words, estimating average flows and densities for all combinations C (or a subset of Ω combinations) of a representative subsets of links increases the chance of obtaining for some combinations and time intervals a homogeneous distribution of congestion, leading to an MFD estimate that is less susceptible to inhomogeneities.

Estimating the Level of Inhomogeneity

Following the re-sampling methodology, a measure is proposed to estimate inhomogeneity in a network based on the observed capacities in the re-sampled MFD without explicitly considering spatial and temporal effects. This approach is different from the approaches by Geroliminis and Sun and Ji and Geroliminis (*22*, *23*), which explicitly capture temporal and spatial effects. The idea is to obtain the highest capacity from all investigated combinations in each sample size, and calculate the relative difference in reference to the observed capacity of the full sample. This relative difference is defined here as the additional capacity that the network could handle if all links were to behave similarly to the best sub-sample of the respective sample size.

As an illustration, imagine a perfectly homogeneous road network, where all roads carry the same level of traffic. It is clear that for such a case the additional capacity will be zero for all sample sizes. In all other cases, where roads and traffic states are inhomogeneous, non-zero additional capacities will be observed, as it is expected that some combinations from lower sample sizes will exclude the constraining roads from the sample. In other words, the additional capacity is estimated as a function of the sample size, and the area below this function is then calculated—the larger this area is, the less homogeneous the urban network is based on its currently observed state. Notice that the level of inhomogeneity, as defined here, is a measure which is *relative* to the reference case.

Identifying the Critical Density and Capacity for Empirical MFDs

Since the proposed sampling method is repeated Ω times (i.e., Ω sub-samples of the same sample size are taken), it significantly increases the scatter and range of flow–density relationships, but exhibits a smooth upper bound. To identify such a stable upper bound for the re-sampled MFD, the median of the top M flow values per density bin is used, where M depends on the number of observations in each bin. The median was chosen to avoid any bias caused by outliers. After several empirical trials, most cases were defined as $M = 50$, but this can decrease down to 5 when the number of observations is small. From this upper bound the capacity is defined as the 97.5th percentile of flow to avoid also the influence of outliers. The critical density is then the mean density corresponding to this capacity.

Data

This section presents the available data sets for the cities of Lucerne, Switzerland and London, UK. The cities differ significantly in size, population, and network topology. Lucerne has around 80,000 inhabitants, whereas London has around 100 times as many, with a population density which is around twice as large. The traffic authorities of Lucerne and London operate an extensive network of loop detectors in their cities. Such loop detectors are mainly installed for traffic control and congestion identification purposes. They measure traffic flow (number of vehicles passing the detector) and occupancy (fraction of time the detector is occupied by a vehicle) during a certain observation interval T. As a result of the system design, the loop detectors in London are located more frequently at the upstream end of the link, while in Lucerne the detectors are located further downstream. For London, this study uses three different regions, one around Whitechapel (253 loop detectors), one around Fulham (93 loop detectors), and one around Chelsea (102 loop detectors). For Lucerne, the entire downtown area is considered (158 loop detectors). Figure 1 provides an overview of the two cities, and Table 1 provides some relevant statistics. In the following, how the data sets were prepared is discussed in more detail.

To locate each detector, all spatial information of the loop detector positions is geo-referenced in reference to the whole road network. Connecting the network and the loops has several advantages. First, MFDs for arbitrary shaped perimeters can be estimated, and potential

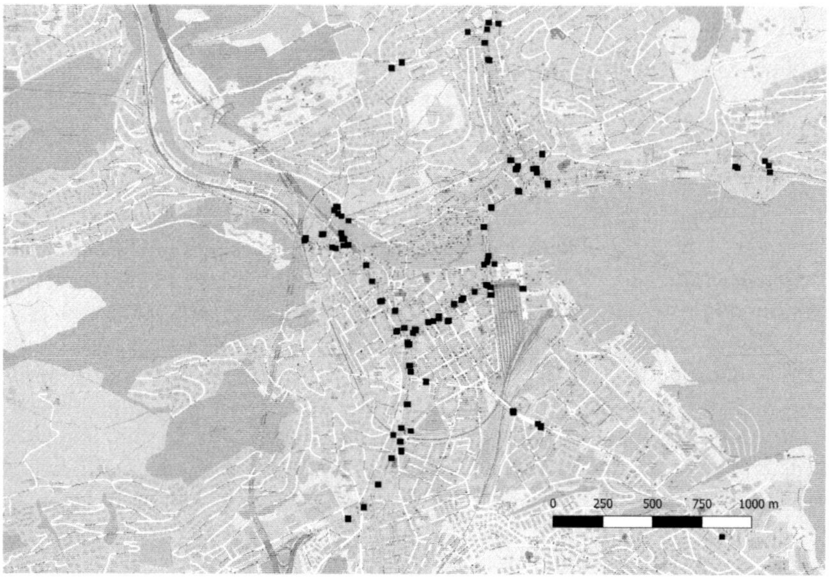

(a) Lucerne.

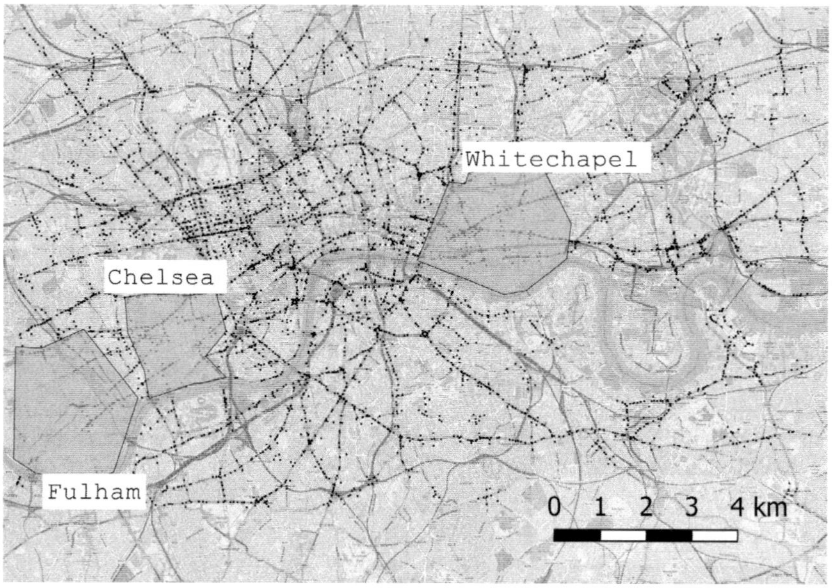

(b) London.

Figure 1. Regions in Lucerne and London which are analyzed. Loop detectors are represented as black squares. Both maps are oriented toward north.

Table 1. Overview of the Two Cities and the Respective Data Set

City	Lucerne (CH)	London (UK)
Total population	81,000	8,500,000
Total number of detectors	158	5,719
Total lane-km covered [km]	26	1,298
Aggregation interval [min]	3	5
Number of working days	258	15

partitioning can be carried out. Second, attributes of the road network, for example road type and speed limit, can be linked to the loop detectors, allowing further filtering, for example removing residential roads (*20*) that generally have no connecting character. Third, the data were enriched with information on the driving direction of each lane covered by a loop detector. This allowed the identification of multiple detectors per lane, and the removal of some to avoid any duplicity. For this analysis, the road network from OpenStreetMap was queried. At the end, for this study only loops located on trunk, primary, secondary, and tertiary roads were selected.

It is clear that empirical datasets are error prone. Both Lucerne and London have an internal system that detects malfunctioning loops. Nevertheless, a set of rules was defined whose objective is to identify potentially malfunctioning loop detectors, as in Bickel et al. (*29*). Measurements were automatically identified as false when no variation of the values were registered during a full day, 80% of the values were zero, either flow or occupancy were zero while the other was not (as long as the occupancy was smaller than 0.95), and obviously when the internal system itself reported an error. In addition, each loop detector's scatter plot was inspected and the results of the filtering verified. The error rate is 30% (23% internal system, 7% verification) for Lucerne and 31% (30% internal system, 1% verification) for London. Finally, the data were further scrutinized by identifying outliers (*30*) and by reducing noise with a moving average technique (*31*), and removing time slots where less than 85% of all loop detectors provide valid measurements.

Occupancy–Density Conversion

Loop detectors measure flow and occupancy. Traffic density can be approximated from occupancy using a scalar conversion (*29*, *32*, *33*). The scalar corresponds to the mean vehicle length as seen by the detector, that is, the sum of vehicle length and detector length (*34*). Hall and Persaud (*33*) and Coifman (*34*) showed that this scalar conversion is a good approximation with small errors in uncongested traffic. Even though previous research predominantly focused on highways, based on recent findings it is assumed that this scalar conversion also holds for urban networks (*20*). In the context of MFD research, values around 0.005 km have been identified for the mean vehicle length: 0.0053 km for Yokohama (*18*) and 0.0063 km for Zurich (*20*). The mean effective vehicle length in Lucerne was estimate to be 0.0063 km, and for London to be 0.006 km.

Results

In this section, the results of the re-sampling methodology using the data described above are presented.

Empirical MFDs

First, see the full-sample MFD for all regions in London and Lucerne. Figure 2 displays the MFD for 258 working days (i.e., one full year) for the city of Lucerne, and 15 working days for Whitechapel, Chelsea, and Fulham (London) all in the respective aggregation interval (see Table 1). Every point represents a macroscopic traffic state in terms of vehicle density and vehicle flow for every aggregation interval during the observation period, indicated in Table 1. All regions show an uncongested branch and the beginning of the congested branch. It is not surprising that Lucerne shows a higher level of scatter, as there are significantly more days included in the dataset. In addition, in a more detailed analysis, some of the working days (less than 6%) showed a slight hysteresis. The reasons for the hysteresis loops have not yet been investigated, especially as the authors do not have access to signal timing, nor signal plans. The stable upper bound was added, as defined previously, in white squares, as well as a line indicating the critical density. The highest capacity is found in Fulham, whereas the lowest is in Whitechapel. The free flow speeds of Lucerne and Fulham are similar and higher than the speeds observed in the other two regions. This is attributed to the difference in the network topology (road hierarchy) and traffic control. More details can be found in Loder at al. (*35*).

Remarkably, all MFDs exhibit a smooth upper bound, which supports the general theory of the MFD, defined as a tight upper bound relatively independent of demand. Notice also that, to the best of the authors' knowledge, this is the first time that a full year MFD is estimated. It shows relatively little variation, which indicates that empirical MFDs can indeed be used for control schemes and other applications demanding a long-term invariance.

Estimating the Level of Inhomogeneity

The first application of the re-sampling method is now discussed. The approach outlined in the previous section was followed. For the re-sampled MFDs a total of 10 sample sizes was chosen (10–100% of the full sample), and the number of draws without replacement Ω was set to 500. As an example, 20% of the loop detectors of a region were randomly selected, the MFDs thereof created, and repeated 500 times, re-selecting another 20% of loop detectors each time. The stable upper bound from the joint set of the 500 samples was found and its capacity derived.

Figure 3 shows the results of measuring the level of inhomogeneity in a city. The relative capacity increases are shown as a function of the sample size for Lucerne and the three different regions in London introduced in Figure 1. As expected, the lower the sample size, the higher is the relative difference. This is attributed to the fact that with smaller sample sizes, it is easier to identify combinations of roads that exhibit a behavior exceeding

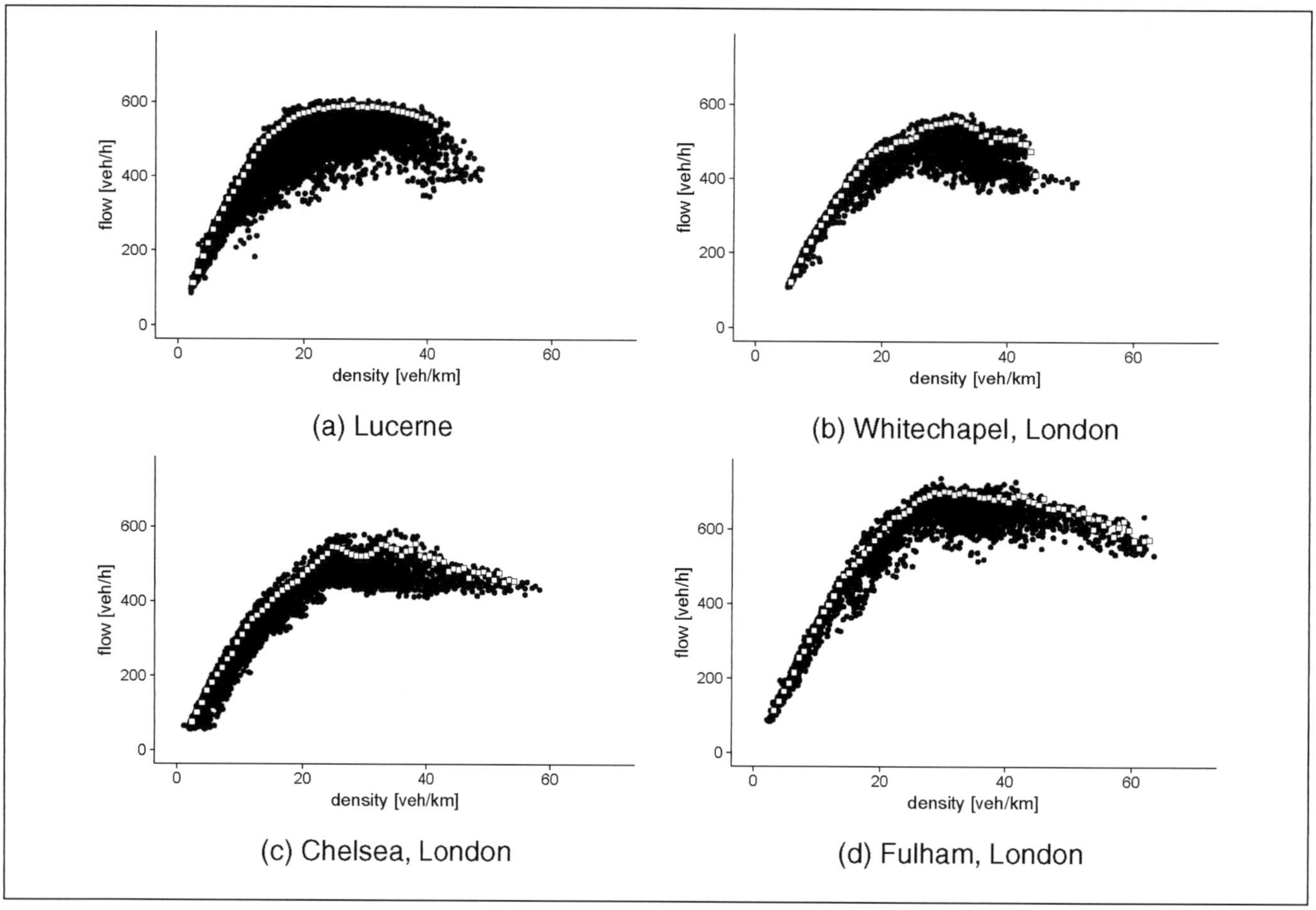

Figure 2. MFDs from all days in which data is available and the upper bound. Estimated critical density is indicated by the vertical line.

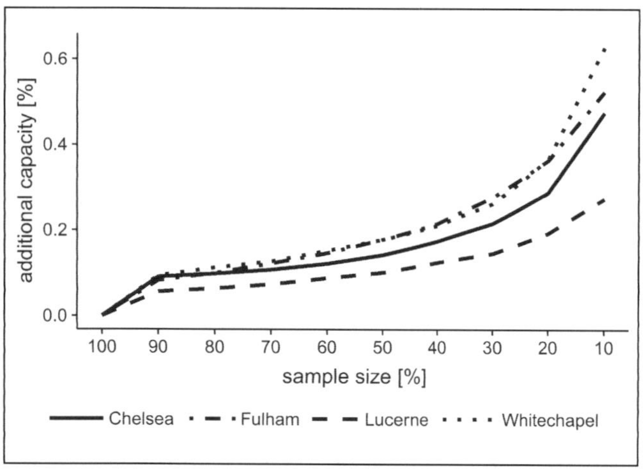

Figure 3. Results of the analysis on the level of inhomogeneity. All results cover a period of 15 consecutive weekdays.

the average in terms of vehicle flow, for example, pockets of congestion. Two issues, however, deserve some consideration: First, it is clear that a certain number of roads need to be included in order to have a representative and meaningful sub-sample for the investigated network. The minimal sample size was set to 10%. Second, the increases in capacity are of theoretical nature, because they would only be observed if all roads were to behave like the best roads in the sub-sample. For example, the region around Whitechapel in London would have a 19% higher capacity if all roads were to behave like the 50% best roads.

Regarding the differences between the different regions in Figure 3, Lucerne has the lowest curve and therefore the smallest area below the curve (i.e., integral of the curve) compared with the three regions in London. Thus, it was concluded that Lucerne exhibits the least amount of inhomogeneity from the investigated sample. This was attributed to (i) the size of the network, which offers only a few higher level roads, and (ii) to the limited feasible route alternatives that are possible in Lucerne in combination with the very limited number of origin–destination (OD) pairs within this small city. Within the investigated perimeter, the south and the north of the city are connected by only two bridges (see

Figure 1*a*); all other bridges are pedestrian bridges. In other words, for the few OD combinations that exist, there are only one or two routes available, constraining route choice. Conversely, for the regions of the city of London, the map in Figure 1*b* shows that the network is connected with more higher level roads, and thus the number of feasible routes is larger. In other words, many route options combined with many OD pairs increases the likelihood of an inhomogeneous distribution of traffic. When the area below the curves is calculated as an indicator for the level of inhomogeneity, the areas in London are around 1.5 to 2 times larger than the one in Lucerne. Interestingly, the three curves for the London area follow the same trend, including a relatively sharp increase at the full sample size (100% sample size). This indicates that a few roads exist in the sample that have a significant impact on the average capacity of the network. In the best case, removing the 10% worst roads (in terms of capacity) from the sample theoretically increases the network capacity by around 15%.

Notice that this approach explicitly quantifies the level of inhomogeneity from the shape of the MFD, without considering spatial or temporal variations in traffic volumes and densities. However, the method does have some clear advantages: it is easy to use, requires very few inputs, and is robust to the placement bias of loop detectors (*19*, *20*). Link-based partitioning methods might include such a bias if they are based on traffic densities. For the re-sampling method, on the other hand, the spatial mean of the densities is considered, as multiple detectors are included in every sample. Therefore, this method represents the first of its kind for assessing and identifying the level of inhomogeneity in an urban road network.

Estimating the Critical Density

The few existing MFD empirical studies show that in reality, data availability is very often limited and only few days can be used for analysis. Empirical studies rely on observation periods that may cover only several days including weekends, which show little or no congestion, making an estimation of the critical density hard or even impossible (e.g., *18–20*, *36*). In the following, how the proposed re-sampling method performs in the estimation of the critical density when only limited data are available is investigated; for example, if only data from 1 day, 3 consecutive days, or 6 consecutive days are available, and not all exhibit a distinct congested branch.

Here, the focus is on only the city of Lucerne and refrains from showing the same results for London, as the MFD for the three regions in London exhibits a congested branch every day. Thus, there is no need to apply the method for identifying the critical density. For this analysis, the MFD in Figure 2*a* serves as the reference MFD. The experimental set-up is similar to the previous section, but now the focus is on the critical density instead of the capacity.

Four sample sizes were chosen (20%, 40%, 60%, and 80% of the full sample), and the number of draws without replacement Ω was set to 500. For a sensitivity analysis vary the observation period for each MFD estimation was also varied, using either 1, 3, or 6 consecutive day(s). These observation periods do not include weekends. In order to apply the methodology to days where congestion is highly unlikely, weekend days (Saturday and Sunday) were further included as an observation period. The critical density for each of the estimated MFDs was then determined. As an example, 20% of Lucerne's loop detectors were randomly selected, the MFDs thereof created, and repeated 500 times, re-selecting another 20% of loop detectors each time for every set of 1, 3 and 6 consecutive day(s) and the weekends. Notice that not every MFD shows a decreasing branch, thus it is not possible to properly determine a critical density for every estimated MFD. It is assumed that the determined value of the critical density is valid only when the MFD shows a decrease in flow of at least 30 veh/h (around 5% of the capacity in the observed cases) for densities higher than the determined value of the critical density. Figure 4 shows the results of the critical density estimation for the MFD of Lucerne. Figure 4*a* shows the percentage of successful critical density estimations for each combination of sample size and observation period. In the case of an observation period of 1 day, the maximum number of estimations of critical densities is 258, in the case of 3 days that number is 86, in the case of 6 days the number is 43, and in the case of weekends 40. From the figure it can first be seen that as the sample size is reduced, the percentage of days where the critical density can be properly determined increases significantly. As a matter of fact, the critical density can be effectively determined over 75% of the time when the sample size is 20% or lower and the number of available days is 1, 3 or 6 workdays. In addition, and not surprisingly, the more days included in the estimation of the MFD, the more likely it is to observe a valid critical density. For example, if there is only 1 day of data, then an estimation of the critical density would only be possible during 5% of all days (13 days) using the full sample (100% sample size), whereas if there are 3 days of available data, the critical density could be estimated in 17% of all cases. For the weekend days as observation period, it is possible to increase the fraction of valid estimations from 2% up to around 45% when considering a sample size of 20%. Some of the weekend days do not show any signs of congestion, even at a link level. Thus, it is not possible to estimate the critical density for such days.

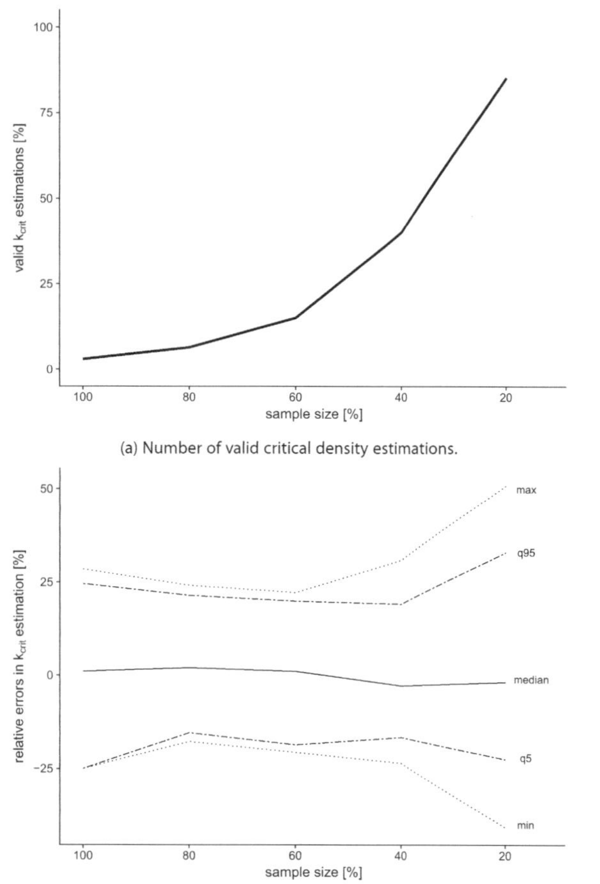

(a) Number of valid critical density estimations.

(b) Errors in the critical density estimation using the re-sampling methodology for an observation period of 1 day.

Figure 4. Results for the critical density estimation.

These results show how the re-sampling methodology can effectively increase the ability to determine the critical density, even when only limited data are available, for example, when no clearly congested branch is apparent in the MFD. The analysis of the weekend days, however, shows that the network must be at least loaded with traffic to some extent, in order for this methodology to work.

Given that it is possible to determine a critical density for a given day, Figure 4*b* shows the range of the relative errors of such densities compared with the reference critical density, identified from the 1-year MFD. For brevity, only the results for the 1-day estimates are shown; the results for 3 and 6 days look similar. In the figure, an overestimation of the critical density has a positive sign. Interestingly, the accuracy remains within a similar range for all sample sizes (the different sample sizes include the number of observations given in Figure 4*a*). In other words, even as the sample size is decreased to increase the number of days where the critical density can be determined, the error in the estimation remains approximately the same. As a matter of fact, for the lowest two sample sizes (20% and 40%) the value of the critical density is only underestimated by less than 2% in the median. To conclude, this clearly indicates that the re-sampling method not only increases the ability to estimate the critical density, but also maintains the level of accuracy of such estimation.

As a further validation, the error distributions were compared from the full sample and the re-sampling method using a Kolmogorow–Smirnov test. A high *p*-value (0.4) was found for all sample sizes, indicating that they all come from the same distribution of errors.

Discussion and Conclusions

In this paper a novel methodology was introduced that allows estimation of the critical density of the MFD, even when no congested branch is observed, and estimation of the level of inhomogeneity within the network. The proposed methodology is based on the idea of re-sampling the empirical data set. Using an extensive data set, which included the cities of Lucerne, Switzerland, (1 year) and London, UK, (3 weeks) insights were provided on the performance and the application of the re-sampling method. First, using the re-sampling method, the level of inhomogeneity in urban networks was quantified. For London, the capacity of the system could be increased by around 20%, under the assumption that all links behave in a similar manner as the 50% best links. For Lucerne this value is around 10%. The level of inhomogeneity was estimated as the area below the curve relating the sample size and the capacity increase. Second, the critical density of the MFD was estimated using limited data; it was found that the number of days on which an estimation is possible can be drastically increased, even if the network does not show network-wide congestion every day. This is important, as very often access is only given to data of a very limited time period. For the network of Lucerne, it was shown that the fraction of days that allow an estimation of the critical density can be increased, based on data of 1 day, from 5% to 80% while keeping a similar level of accuracy. Such robust estimation of the empirical critical density is crucial for many control algorithms. In summary, the re-sampling method presented in this paper is very promising for different application purposes of the MFD. It is easy to use, requires only very few inputs, and is robust against a potential placement bias in loop detector data.

The proposed measure of the level of inhomogeneity gives city planners the possibility to analyze and quantify how efficiently their road network is utilized. This line of questions was started over 50 years ago by Smeed (*37*), but can now be evaluated in more detail with the availability of big data. There have been numerous studies investigating the performance of the network when implementing a perimeter control scheme based on the MFD. However, the estimation of the required critical density has not yet been discussed in an empirical context, and only a few of the cities investigated to date exhibit a strong congested branch. The re-sampling methodology proposed in this paper provides a promising approach to estimate the critical density empirically at large urban scale.

Future research will concentrate on how the re-sampling method and the theoretical upper bound from analytical approaches relate to each other. In other words, it should address the question of whether the theoretical upper bound can be approached with this re-sampling methodology based on the findings of Knoop et al. and Geroliminis and Sun (*2*, *22*). A more detailed investigation on the spatial and temporal distribution of congestion could generate further interesting findings. Furthermore, additional insights could be gained from an application of the methodology to floating car data, or the fusion of multiple data sources (*38*). Ongoing studies are also evaluating the possibility to estimate an infrastructure potential using the MFD, and thereby deepen the understanding of the effects of inhomogeneous traffic in urban networks.

Acknowledgments

This work was supported by ETH Research Grants ETH-04 15-1 and ETH-27 16-1. We especially thank Thomas Karrer and Milena Scherer from the city of Lucerne and Ashley Turner and Andy Emmonds from Transport for London (TfL) for their collaboration.

Author Contributions

The authors confirm contribution to the paper as follows: study conception and design: Lukas Ambühl, Allister Loder, Michiel Bliemer, Monica Menendez; data collection: Lukas Ambühl, Allister Loder; analysis and interpretation of results: Lukas Ambühl, Allister Loder, Michiel Bliemer, Monica Menendez; draft manuscript preparation: Allister Loder, Lukas Ambühl, Monica Menendez, Michiel Bliemer K.W. Axhausen. All authors reviewed the results and approved the final version of the manuscript.

References

1. Daganzo, C. F., and N. Geroliminis. An Analytical Approximation for the Macroscopic Fundamental Diagram of Urban Traffic. *Transportation Research Part B: Methodological*, Vol. 42, No. 9, 2008, pp. 771–781.
2. Knoop, V. L., H. van Lint, and S. P. Hoogendoorn. Traffic Dynamics: Its Impact on the Macroscopic Fundamental Diagram. *Physica A: Statistical Mechanics and its Applications*, Vol. 438, 2015, pp. 236–250.
3. Daganzo, C. F., V. V. Gayah, and E. J. Gonzales. Macroscopic Relations of Urban Traffic Variables: Bifurcations, Multivaluedness and Instability. *Transportation Research Part B: Methodological*, Vol. 45, No. 1, 2011, pp. 278–288.
4. Gayah, V. V., and C. F. Daganzo. Clockwise Hysteresis Loops in the Macroscopic Fundamental Diagram: An Effect of Network Instability. *Transportation Research Part B: Methodological*, Vol. 45, No. 4, 2011, pp. 643–655.
5. Aboudolas, K., and N. Geroliminis. Perimeter and Boundary Flow Control in Multi-reservoir Heterogeneous Networks. *Transportation Research Part B: Methodological*, Vol. 55, 2013, pp. 265–281.
6. Haddad, J., and N. Geroliminis. On the Stability of Traffic Perimeter Control in Two-region Urban Cities. *Transportation Research Part B: Methodological*, Vol. 46, No. 9, 2012, pp. 1159–1176.
7. Geroliminis, N., and D. M. Levinson. Cordon Pricing Consistent with the Physics of Overcrowding. *Transportation and Traffic Theory 2009: Golden Jubilee*, Springer, Boston, MA, 2009, pp. 219–240.
8. Zheng, N., R. A. Waraich, N. Geroliminis, and K. W. Axhausen. A Dynamic Cordon Pricing Scheme Combining a Macroscopic and An Agent-based Traffic Model. *Transportation Research A*, Vol. 46, No. 8, 2012, pp. 1291–1303.
9. Zheng, N., G. Rérat, and N. Geroliminis. Time-dependent Area-based Pricing for Multimodal Systems with Heterogeneous Users in An Agent-based Environment, *Transportation Research Part C: Emerging Technologies*, Vol. 62, 2016, pp. 133–148.
10. Boyaci, B., and N. Geroliminis. Estimation of the Network Capacity for Multimodal Urban Systems. *Procedia - Social and Behavioral Sciences*, Vol. 16, 2011, pp. 803–813.
11. Zheng, N., and N. Geroliminis. On the Distribution of Urban Road Space for Multimodal Congested Networks. *Transportation Research Part B: Methodological*, Vol. 57, 2013, pp. 326–341.
12. Schreiber, A., A. Loder, and K. W. Axhausen. Urban Mode and Subscription Choice - An Application of the Three-dimensional MFD. *Proc., 16th Swiss Transport Research Conference*, Ascona, May 2016.
13. Ampountolas, K., N. Zheng, and N. Geroliminis. Macroscopic Modelling and Robust Control of Bi-modal Multi-region Urban Road Networks, *Transportation Research Part B: Methodological*, Vol. 104, 2017, pp. 616–637.
14. Haddad, J., and B. Mirkin. Coordinated Distributed Adaptive Perimeter Control for Large-scale Urban Road Networks. *Transportation Research Part C: Emerging Technologies*, Vol. 77, 2017, pp. 495–515.
15. Amirgholy, M., and H. O. Gao. Modeling the Dynamics of Congestion in Large Urban Networks Using the Macroscopic Fundamental Diagram: User Equilibrium, System Optimum, and Pricing Strategies. *Transportation Research Part B: Methodological*, Vol. 104, 2017, pp. 215–237.
16. Yang, K., N. Zheng, and M. Menendez. Multi-scale Perimeter Control Approach in a Connected-vehicle Environment. *Transportation Research Part C: Emerging Technologies*, 2017. DOI: 10.1016/j.trc.2017.08.014. (In Press).

17. Leclercq, L., N. Chiabaut, and B. B. Trinquier. Macroscopic Fundamental Diagrams: A Cross-comparison of Estimation Methods. *Transportation Research Part B: Methodological*, Vol. 62, 2014, pp. 1–12.
18. Geroliminis, N., and C. F. Daganzo. Existence of Urban-scale Macroscopic Fundamental Diagrams: Some Experimental Findings. *Transportation Research Part B: Methodological*, Vol. 42, No. 9, 2008, pp. 759–770.
19. Buisson, C., and C. Ladier. Exploring the Impact of Homogeneity of Traffic Measurements on the Existence of Macroscopic Fundamental Diagrams. *Transportation Research Record: Journal of the Transportation Research Board*, 2009. 2124: 127–136.
20. Ambühl, L., A. Loder, M. Menendez, and K. W. Axhausen. Empirical Macroscopic Fundamental Diagrams: New Insights from Loop Detector and Floating Car Data. Presented at 96th Annual Meeting of the Transportation Research Board, Washington, D.C., 2017.
21. Muhlich, N., V. V. Gayah, and M. Menendez. An Examination of MFD Hysteresis Patterns for Hierarchical Urban Street Networks Using Micro-Simulation, *Transportation Research Record: Journal of the Transportation Research Board*, 2015. 2491: 117–126.
22. Geroliminis, N., and J. Sun. Properties of a Well-defined Macroscopic Fundamental Diagram for Urban Traffic. *Transportation Research Part B: Methodological*, Vol. 45, No. 3, 2011, pp. 605–617.
23. Ji, Y., and N. Geroliminis. On the Spatial Partitioning of Urban Transportation Networks. *Transportation Research Part B: Methodological*, Vol. 46, No. 10, 2012, pp. 1639–1656.
24. Daganzo, C. F., V. V. Gayah, and E. J. Gonzales. Macroscopic Relations of Urban Traffic Variables: Bifurcations, Multivaluedness and Instability. *Transportation Research Part B: Methodological*, Vol. 45, No. 1, 2011, pp. 278–288.
25. Saeedmanesh, M., and N. Geroliminis. Clustering of Heterogeneous Networks with Directional Flows Based on "Snake" Similarities. *Transportation Research Part B: Methodological*, Vol. 91, 2016, pp. 250–269.
26. Saeedmanesh, M., and N. Geroliminis. Dynamic Clustering and Propagation of Congestion in Heterogeneously Congested Urban Traffic Networks. *Transportation Research Part B: Methodological*, Vol. 105, 2017, pp. 193–211.
27. Lopez, C., P. Krishnakumari, L. Leclercq, N. Chiabaut, and H. van Lint. Spatiotemporal Partitioning of Transportation Network Using Travel Time Data. *Transportation Research Record: Journal of the Transportation Research Board*, 2017. 2623: 98–107.
28. Gao, X., and V. V. Gayah. An Analytical Framework to Model Uncertainty in Urban Network Dynamics Using Macroscopic Fundamental Diagrams. *Transportation Research Procedia*, Vol. 23, 2017, pp. 497–516.
29. Bickel, P., C. Chen, J. Kwon, J. Rice, E. v. Zwet, and P. Varaiya. Measuring Traffic. *Statistical Science*, Vol. 22, No. 4, 2007, pp. 581–597.
30. Chen, C., and L.-M. Liu. Joint Estimation of Model Parameters and Outlier Effects in Time Series. *Journal of the American Statistical Association*, Vol. 88, No. 421, 1993, pp. 284–297.
31. Gómez, V., and A. Maravall. *Programs Tramo and Seats. Instructions for the User* (Beta Version), 1996.
32. Athol, P. Interdependence of Certain Operational Characteristics within a Moving Traffic Stream. *Highway Research Record*, Vol. 72, 1965, pp. 58–87.
33. Hall, F. L., and B. N. Persaud. Evaluation of Speed Estimates Made with Single-detector Data from Freeway Traffic Management Systems. *Transportation Research Record: Journal of the Transportation Research Board*, 1989. 1232: 9–15.
34. Coifman, B. Improved Velocity Estimation Using Single Loop Detectors. *Transportation Research Part A: Policy and Practice*, Vol. 35, 2001, pp. 863–880.
35. Loder, A., L. Ambühl, M. Menendez, and K. W. Axhausen. Empirics of Multi-modal Traffic Networks – Using the 3D Macroscopic Fundamental Diagram. *Transportation Research Part C: Emerging Technologies*, Vol. 82, 2017, pp. 88–101.
36. Jiyang, B., H. van Zuylen, and L. Shoufeng. Determining the Macroscopic Fundamental Diagram on the Basis of Mixed and Incomplete Traffic Data. Presented at the 95th Annual Meeting of the Transportation Research Board, Washington, D.C., 2016, pp. 1–13.
37. Smeed, R. J. *The Traffic Problem in Towns*. Statistical Society, Manchester, 1961.
38. Ambühl, L., and M. Menendez. Data Fusion Algorithm for Macroscopic Fundamental Diagram Estimation. *Transportation Research Part C: Emerging Technologies*, Vol. 71, 2016, pp. 184–197.

The Standing Committee on Traffic Flow Theory and Characteristics (AHB45) peer-reviewed this paper (18-01047).

TRR
JOURNAL OF THE TRANSPORTATION RESEARCH BOARD

Article

Transportation Research Record
2018, Vol. 2672(20) 249–261

DOI: 10.1177/0361198118791626
journals.sagepub.com/home/trr
SAGE

A Simple Example of a Two-Dimensional Model for Traffic: Discussion about Assumptions and Numerical Methods

Stephane Mollier[1], Maria Laura Delle Monache[1], and Carlos Canudas-de-Wit[1]

Abstract

This article presents a two-dimensional (2D) macroscopic model for traffic flow on a network. We considered a 2D conservation law in which we suggested a new method to estimate the direction of the velocity using only information and properties of the network. We present an algorithm for the computation of the 2D velocity on the network and present numerical simulations. Moreover, the numerical approximation of the 2D equation was carried out using an operator and splitting method and we present the results of some simulations on realistic networks. Lastly, we propose a comparison with the well-known cell transmission model method.

This paper deals with the problem of designing two-dimensional (2D) macroscopic traffic flow models for large-scale traffic networks. Traffic modeling literature is rather vast (see van Wageningen-Kessels et al. [*1*] for a complete description of traffic models evolution). A key observation in traffic flow modeling is that the speed of vehicles can be expressed as a function of vehicle density, see Greenshields et al. (*2*). The flux-density relation is well known as fundamental diagram (FD), and used as a key element in macroscopic traffic model design. Macroscopic models were introduced by Lighthill and Whitham, and Richards (LWR) (*3*, *4*). The LWR is described by a partial differential equation that is able to reproduce several features that are observed in traffic data on highways such as shock and rarefaction waves. The discretized version of the LWR is known as the cell transmission model (CTM), see Daganzo (*5*). Models of junction (*6*) together with some split ratio assumptions allow extension of the CTM and other similar models to urban networks. The junction models ensure a correct propagation of flux locally through an intersection and, together with the CTM, can be used to describe one-dimensional (1D) flow propagation along the network. Modeling each specific road in a large and complex network may be tedious, difficult to tune, and in some cases computationally expensive. These difficulties have motivated the research for aggregated traffic models describing traffic evolution by geographical zones rather than modeling each individual road and intersection in the city network. A first natural approach has been to extend the 1D FD to an urban area. This is called a macroscopic fundamental diagram (MFD), which captures the aggregated number of vehicles in a given area (density/occupancy), and maps it into their respective averaged velocity (flows). The existence of the MFD was first studied analytically by Daganzo and Geroliminis, then numerically by Geroliminis and Sun, and lastly using real field data by Geroliminis and Daganzo (*7–9*). The MFD is then used in the cumulative mass conservation equation:

$$\frac{dn}{dt}(t) = \Phi_{in}(t) - \Phi_{out}(n(t)), \qquad \forall t \in \mathbb{R}^+ \qquad (1)$$

where n represents the cumulative number of car in the considered area, Φ_{in} and Φ_{out} are the area inflow and outflow, respectively. In this set-up, the inflow Φ_{in} is assumed to be known (or measured), whereas the outflow Φ_{out} is computed using the MFD and the assumed known user trip length. These types of model focus on the modeling of the number of vehicles in an area with a single variable. They preserve less information about the spatial distribution of the vehicle than models that consider density. However, Hajiahmadi et al. and Leclercq

[1]Université Grenoble Alpes, CNRS, Inria, Grenoble INP, GIPSA-Lab, Grenoble, France

Corresponding Author:
Address correspondence to Maria Laura Delle Monache:
ml.dellemonache@inria.fr

et al. investigated the extension to multizone models to consider different MFDs for each zone for which the road pattern may differs (*10*, *11*). It is worth noting that these models are set once the modeling zone has been defined, and they are therefore not able to "expand/contract" themselves, meaning the modeling zone cannot be adapted to changes in traffic conditions.

Another approach to modeling traffic in large networks is the two-dimensional model (2D model) in which the density depends on 2D spatial and temporal coordinate systems: a point in the plane (x, y), and time t. The first research on 2D traffic modeling were carried out on a static case, that is, without time dependency (*12*). Contrary to dynamical models, which aim to predict the evolution of the system, static models are interested in specific traffic feature, for example, optimal assignment of traffic. An overview of existing static 2D models is given in Ho andWong (*13*).

In this paper, we will focus on 2D dynamical models. The aim of such models is to simplify the number of individual variables, avoid specific models for junctions, and simplify the tuning of the model parameters. There is also some motivation for simplifying both the control design and the use of the model for predicting the density evolution. Such models can take the following general form:

$$\begin{cases} \frac{\partial \rho(t,x,y)}{\partial t} + \nabla \cdot \vec{\Phi}(\rho(t,x,y), \theta(x,y)) = 0 \\ \rho(0,x,y) = \rho_0(x,y) \end{cases}, \forall t \in \mathbb{R}^+, \forall (x,y) \in \Omega \quad (2)$$

where

ρ is the aggregated density,

Φ the flow defined as the product of the density ρ, and velocity v and θ the velocity direction.

This type of model based on 2D conservation laws is also used in pedestrian modeling, which started with the work of Helbing, and Hughes (*14*, *15*). Several extensions for these pedestrian models have been considered including a second-order model proposed by Jiang et al., and a non-local flux by Colombo et al. (*16*, *17*). A numerical comparison of pedestrian models can be seen in Twarogowska et al., and a review of some of these models is given in Kormanová et al. (*18*, *19*).

The literature concerning 2D traffic models is not broad, only a few works have focused on 2D traffic flow models, see Della Rossa et al., Jiang et al., Du et al., Jiang et al., Romero Perez and Benitez, Saumtally, Sossoe and Lebacque, Chetverushkin et al., and Sukhinova et al. (*20–28*). We assumed for 2D models that vehicles run on a dense network instead of a single road. The key point of the 2D models is to understand how the FD can be translated for 2D traffic. First of all, the FD in 1D holds the assumption that the velocity is a function of only the density allowing therefore the construction of a unique curve that relates traffic flux and traffic density. In 2D the problem is more complex and in the literature several approaches have been proposed. In Romero, Perez, and Benitez, an advection–diffusion equation is introduced and the authors assume that the traffic velocity does not depend on the density but it is instead a function of the space, that is, the network (*24*). In this way the velocity is predetermined and the advection–diffusion equation becomes linear. Other authors, like Della Rossa et al. consider the velocity as a scalar function of the density, its direction predetermined by estimating the shape of the network (*20*). In turn, they then solve a diffusion equation in which the direction of traffic flow is indicated by the direction of the speed but also by the diffusion operator. In research by Jiang et al., and Du et al., the authors consider a model inspired from pedestrian models in which the flux takes the direction of the lowest cost that they determine with the help of an Eikonal equation (*21*, *22*). Lastly in Jiang et al., this model has been extended to a second-order dynamic to ensure a more accurate acceleration model for pollutant estimation (*23*). In work by Saumtally, and Sossoe and Lebacque, they investigate the possibility of constructing a 2D model that takes into account conflicts that occur on intersections (*25*, *26*). In this spirit, they take inspiration from models of junctions in 1D, as in Lebacque and Khoshyaran (*29*). However, the problem is formulated in a discrete setting using the concepts of demand and supply. In Chetverushkin et al. and in Sukhinova et al., the authors consider 2D models for traffic for application in multilane roads instead of road networks (*27*, *28*).

The contribution of this article is to present a new traffic flow model that extends the LWR model to 2D. We suggest a way to estimate the direction of the velocity using information and properties from the network. We also describe in its entirety the numerical process based on operator and dimensional splitting, and show some simulations.

The paper is organized as follows. In the first section, we introduce the model and the assumptions taken to construct this model. We then suggest a methodology to estimate the velocity fields for the model. Finally, we describe the numerical process, show the results of the simulations and present a comparison between our 2D model and the CTM.

From Traffic Flow Model in 1D–2D

Recall of 1D LWR Model

In the 50s, Lighthill, and Whitham and, independently, Richards (*3*, *4*) introduced a hydrodynamic model (LWR) of traffic flow. This model was inspired by fluid dynamics. The main assumption of this model is the

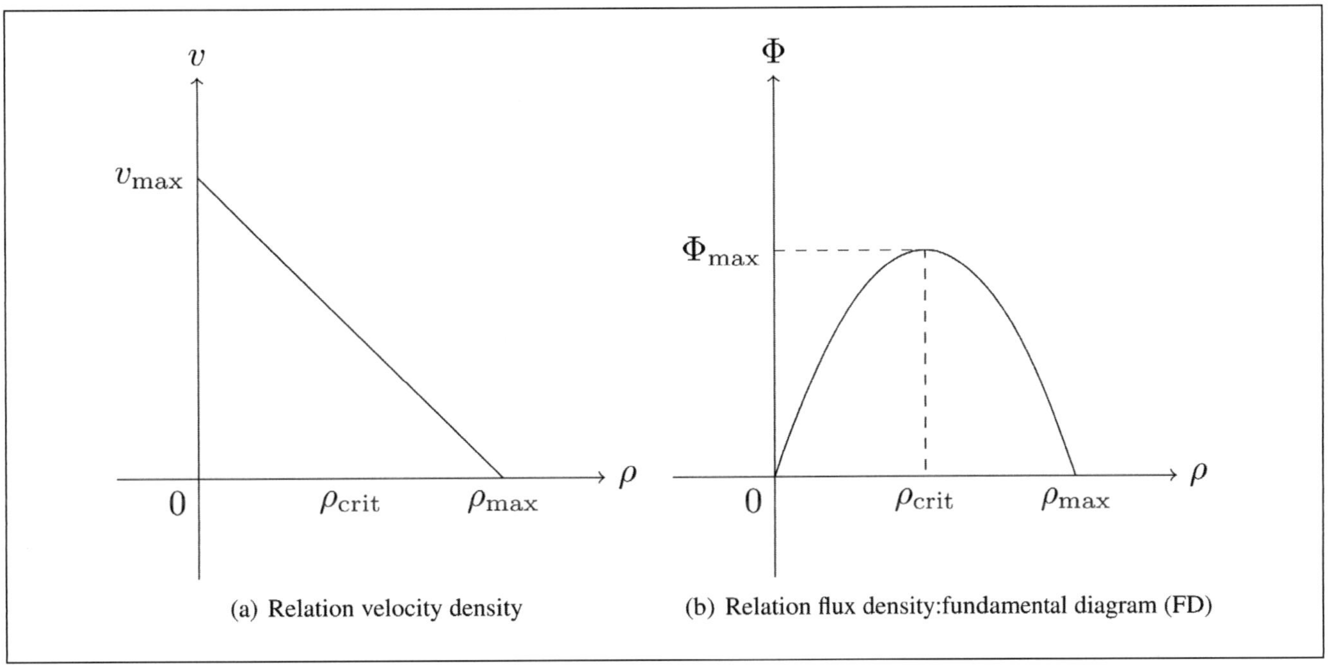

Figure 1. Speed and flow vs. density.

hypothesis that the velocity $v \in [0, v_{max}]$ depends only on the main traffic density $\rho \in [0, \rho_{max}]$. The presence of this relation ensures the existence of the FD (*2*). The FD gives the link between the main traffic density and the traffic flow (see Figure 1). In the LWR model the flux function is given by $\Phi(\rho) : (0, \rho_{max}) \rightarrow \mathbb{R} = v(\rho(t,x))\rho(t,x)$.

The evolution in time of the main traffic density is given by the following scalar conservation law:

$$\begin{cases} \frac{\partial \rho(t,x)}{\partial t} + \frac{\partial \Phi(\rho(t,x))}{\partial x} = 0 & (t,x) \in \mathbb{R}^+ \times \mathbb{R} \\ \rho(0,x) = \rho_0(x). & x \in \mathbb{R} \end{cases} \tag{3}$$

where the velocity and the flux are given by

$$v(\rho) = (1 - \frac{\rho}{\rho_{max}})v_{max} \tag{4}$$

$$\Phi(\rho) = \rho(1 - \frac{\rho}{\rho_{max}})v_{max} \tag{5}$$

The flux Φ is concave and attains one maximum, which we denote Φ^{max} at ρ_{crit}. Moreover, $\Phi(0) = \Phi(\rho_{max}) = 0$ (see Figure 1*b*).

Extension to Conservation Laws in Two Dimensions

In this section, we investigate how the LWR model can be extended to a 2D continuous model and what are the main differences.

It is possible to derive a 2D balance equation in a domain $\Omega \subset \mathbb{R}^2$ using the conservation of the mass. Let $A \subset \Omega$ be an arbitrary subdomain, $n_A(t)$ the number of vehicles within A at time t and $n_{in}^{\Delta t}$, $n_{out}^{\Delta t}$ the number of vehicles that respectively enter and exit A between t and $t + \Delta t$. Then:

$$n_A(t + \Delta t) - n_A(t) = n_{in}^{\Delta t} - n_{out}^{\Delta t}, \qquad \forall t \in \mathbb{R}^+ \tag{6}$$

dividing by Δt and passing at the limit when $\Delta t \rightarrow 0$, we obtain:

$$\frac{\partial(n_A(t))}{\partial t} = \Phi_{in}(t) - \Phi_{out}(t), \qquad \forall t \in \mathbb{R}^+ \tag{7}$$

where Φ_{in} denotes the inflow and Φ_{out} the outflow of subdomain A.

We define the density of vehicle $\rho(t,x,y) : \mathbb{R}_+ \times \Omega \rightarrow [0, \rho_{max}]$ and the flux of vehicle $\vec{\Phi}(t,x,y,\rho) : \mathbb{R}_+ \times \Omega \times [0, \rho_{max}] \rightarrow \mathbb{R}^2$. By definition of the density and the flux, we have:

$$\frac{\partial}{\partial t}(\iint_A \rho(t,x,y)\,dx\,dy) = -\oint_{\partial A} \vec{\Phi}(t,x,y,\rho) \cdot \vec{n}(x,y)\,dl, \qquad \forall t \in \mathbb{R}^+ \tag{8}$$

where $\vec{n}$ is the vector normal to the boundary and l the variable that follows the boundary of the subdomain.

Due to the divergence theorem, we have:

$$\iint_A \frac{\partial \rho(t,x,y)}{\partial t}\,dx\,dy = -\iint_A \nabla \cdot \vec{\Phi}(t,x,y,\rho)\,dx\,dy, \qquad \forall t \in \mathbb{R}^+. \tag{9}$$

Table 1. Comparison of Model Structure.

	1D model	2D model
Density	$[\rho] = veh/m$ (scalar)	$[\rho] = veh/m^2$ (scalar)
Velocity	$[v] = m/s$ (scalar)	$[\vec{v}] = m/s$ (vector)
Flux	$[\Phi = v\rho] = veh/s$ (scalar)	$[\vec{\Phi} = \vec{v}\rho] = veh/s/m$ (vector)
Equation	$\partial_t\rho + \partial_x\Phi(\rho) = 0$	$\partial_t\rho + \nabla\cdot\vec{\Phi}(t,x,y,\rho) = 0$

Note: *veh* = vehicle; *m* = space; *s* = time.

Equality (Equation 9) holds for all $A \subset \Omega$, therefore we get:

$$\frac{\partial\rho}{\partial t}(t,x,y) + \nabla\cdot\vec{\Phi}(t,x,y,\rho) = 0, \qquad \forall t \in \mathbb{R}^+, \forall(x,y) \in \Omega \tag{10}$$

This equation is completed with the following initial condition:

$$\rho(0,x,y) = \rho_0(x,y). \qquad \forall(x,y) \in \Omega \tag{11}$$

For readers interested in the existence and uniqueness of the solution for Equation 10, we refer them to the work of Kruzhkov (*30*).

Units and model structures between 1D and 2D are compared in Table 1.

The model described by Equation 10 is a direct extension in 2D of the LWR model. However, several differences remain in the expression of the flux that make the modeling problem more complex. To be valid, a first assumption is that 2D models have to describe a road network dense enough such that the flux, the velocity, and the density can be defined at any point of the plane as an extension of the road nearby. The flux in 2D models is a vector and, without loss of generality, we can split it into a flux magnitude m and a normalized vector of direction $\vec{d}_\theta$ such that $\vec{\Phi}(t,x,y,\rho) = m(t,x,y,\rho)$ $\vec{d}_\theta(t,x,y,\rho)$. We can note that the notion of direction does not exist in 1D, and specific attention should be paid to introducing the direction in the 2D model. The flux is defined as the product of the density and the velocity $\vec{\Phi}(t,x,y,\rho) = \vec{v}(t,x,y,\rho)\rho(t,x,y)$. Thus, in 2D, the velocity and the flux direction coincide. In the following we will refer to the velocity or flux direction when discussing either of them. The case where the flux (or velocity) function depends on time, space, and density corresponds to the most general case. In the following, we investigate whether each dependency is justified by traffic observations or whether some of them could be relaxed.

Derivation of the 2D Flux Magnitude

In this section, we discuss the relevance of the dependency of the flux magnitude with reference to time, space, and density.

First, we investigate the dependency of the flux magnitude with regard to the density. In 1D models, the velocity is defined as a decreasing function of the density, the higher the density on the road, the lower the velocity. This idea seems consistent with the representation of a 2D model. Thus, we consider Greenshields' FD as a starting point in the definition of the flux magnitude in 2D. The expression of velocity in Equation 4 depends on the density and also on two parameters: the maximum speed v_{max} and the maximum density ρ_{max}.

The maximal speed v_{max} could vary with respect to time, hence enforcing a time dependence of the flux. However, for simplicity we will not consider time dependence in this paper.

Finally, it is possible to have a space dependency if the parameters v_{max} or ρ_{max} change along the road considered. For example, these modifications occur at the point of the road where the speed limit or the number of lanes on the road change. However, variations in parameters require specific models and involve a more complex computation of the flux. In 2D models, vehicles run on an abstraction of the physical network in 2D. Thus, the notion of capacity and speed limit are different and depend on an interpolation of the underlying road network. Furthermore, the scale considered is larger in general. If the parameters v_{max} and ρ_{max} have even more reason to change in 2D models, the spatial variation of these parameters could be ubiquitous. Thus, it seems essential to develop a precise and accurate method to take into account these variations. As this problem could be very complex to handle, we have chosen to leave this aspect for future work and we do not consider space dependencies in relation to magnitude in the following. That is equivalent to saying that every road in the network considered is homogeneous with the same capacity and speed limit.

Variable Dependencies of Flux Direction

In this section, we discuss the relevance of the dependency of the flux direction with respect to time, space, and density. Contrary to the magnitude dependency, only a few comparisons can be done with the existing idea in 1D, as the notion of direction needs to be explicit only in 2D models. Thus, we suggest three ideas regarding what could affect the flux direction.

First, the flux direction should depend on the road of the underlying network. Indeed, we aim to design a model that allows the flux to go only in physically possible directions. That is to say, vehicles should not be able to move in a direction where there are no roads. Thus, we assumed that the velocity direction depended on the space variables.

A second idea is that the flux direction depends on the individual wishes of drivers. In reality, the flux direction is directly linked to the trajectory of each individual vehicle. However in practice, macroscopic models are not able to keep track of individual drivers, but we can still imagine that the direction of the flux could be linked to a global average trajectory. Thus, the flux direction might include dependency on the time and spatial coordinates. Indeed, the dependency of the flux direction with respect to space is necessary in order to describe the trajectory. The time dependency allows this trajectory to change during the simulation: for example, most of drivers may go into the city center in the morning and return to the suburbs in the evening. We chose not to introduce these dependencies in the model as we wanted a model that does not depend on external data feed in real time.

Another idea is that drivers may adapt their direction according to local traffic conditions. For example, drivers could change their planned itinerary in order to bypass congestion. This may lead to the flux direction being dependent on the density. We chose not to keep this dependency in the model.

The 2D Model for Traffic Considered

The model considered is based on a 2D conservation law, as is Equation 10. We recall that the flux function can be split into a magnitude part

$$m(\rho) : [0, \rho_{\max}] \to [0, \Phi_{\max}] = \rho \underbrace{(1 - \frac{\rho}{\rho_{\max}})v_{\max}}_{\text{velocity norm}}$$

and a direction vector

$$\vec{d}_\theta(x, y) : \Omega \to \mathbb{R}^2 = \begin{pmatrix} cos(\theta(x, y)) \\ sin(\theta(x, y)) \end{pmatrix}.$$

with θ the angle between the x-axis and the vector direction of the flux. Then, the flux can be defined: $\vec{\Phi}(\rho(t, x, y), \theta(x, y)) = m(\rho(t, x, y)\vec{d}_\theta(x, y)$. The equation considered for the model is:

$$\frac{\partial \rho(t, x, y)}{\partial t} + \frac{\partial\, m(\rho(t, x, y))\, cos(\theta(x, y))}{\partial x} + \frac{\partial\, m(\rho(t, x, y))\, sin(\theta(x, y))}{\partial y} = 0 \quad (12)$$

The expression of m is given by Greenshields' FD, (Figure 1*b*). An expression is also needed for the direction of the flux in order to be able to make predictions of the density propagation. Thus, we suggest in the next section a way to estimate the function d_θ using the information we have on the physical network as much as possible.

Velocity Direction Field Design from the Physical Network

Description of the Method

The aim of this section is to suggest one possible way to construct the function $\vec{d}_\theta$ using only the topology of the road network. We have seen that for any point $(x, y) \in \Omega$, we need to define the direction of the flux $\vec{d}_{\theta(x,y)}$. A network is a set of roads in which the link represents the roads, and the nodes the junctions.

Let $q \in \{1, \ldots, Q\}$ be the set of roads of the network with Q the number of roads. Each road q is described according to a parametric curve

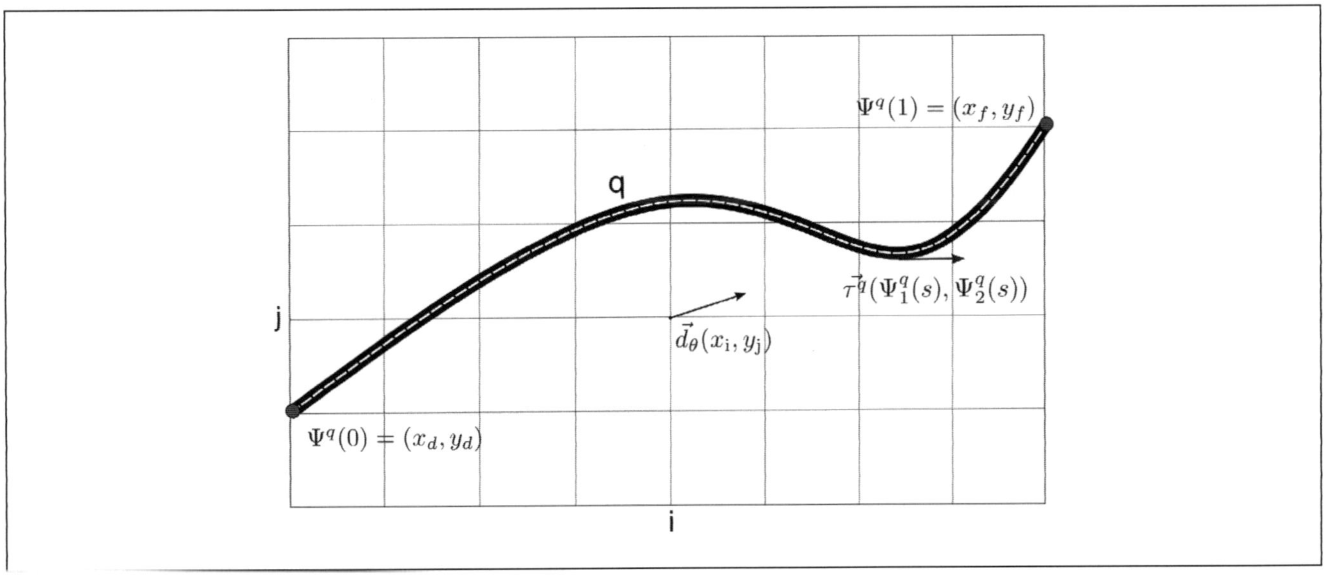

Figure 2. Representation of the road q and his vector tangent $\vec{\tau}^q$.

$\Psi^q : s \in [0,1] \rightarrow (\Psi_1^q(s), \Psi_2^q(s))$ that follows the road curvature from the initial point on the road, $\Psi^q(0) = (x_d, y_d)$, to a final point, $\Psi^q(1) = (x_f, y_f)$. For all points s of these parametric curves, we denote by $\vec{\tau}^q(\Psi_1^q(s), \Psi_2^q(s))$ the tangent vector of the curve that also represents the direction of the road. These notations are represented in Figure 2.

Consider a discretization of the space where I and J are respectively the number of discrete intervals for the x-axis and the y-axis. For a given point $(x_i, y_j)_{(i,j)\in\{1,\ldots,I\}\times\{1,\ldots,J\}}$; let $\vec{d}_\theta(x_i, y_j)$ be the directional vector for which the angle with the x-axis and the vector is $\theta(x_i, y_j)$ such that:

$$\vec{d}_\theta(x_i, y_j) = \begin{pmatrix} cos(\theta(x_i, y_j)) \\ sin(\theta(x_i, y_j)) \end{pmatrix}. \quad (13)$$

To estimate the vector $\vec{d}_\theta$, we use a spatial interpolation method called inverse distance weighting. The expression of the estimator is given in Equation 14:

$$\vec{d}_\theta(x_i, y_j) = \frac{\sum_{q=1}^{Q} \int_{s\in[0,1]} \rho_{\max}^q \; w(||(x_i, y_j) - (\Psi_1^q(s), \Psi_2^q(s))||) \; \vec{\tau}^q(\Psi_1^q(s), \Psi_2^q(s)) \; ds}{|| \sum_{q=1}^{Q} \int_{s\in[0,1]} \rho_{\max}^q \; w(||(x_i, y_j) - (\Psi_1^q(s), \Psi_2^q(s))||) \; \vec{\tau}^q(\Psi_1^q(s), \Psi_2^q(s)) \; ds||} \quad (14)$$

Thereby, the velocity direction $\vec{d}_\theta(x_i, y_j)$ is estimated as a linear combination of the direction $\vec{\tau}^q$ of every road $q \in \{1, \ldots, Q\}$ of the network. The weight coefficient of each road direction depends first on the distance between the road and the point (x_i, y_j). This dependency is evaluated with the weight function $w : \mathbb{R}^+ \rightarrow \mathbb{R}^+$ strictly decreasing. The weight function considered in the following is:

$$w : X \rightarrow e^{-\beta X} \quad \text{with} \quad \beta > 0.$$

We chose the exponential weights in order to give greater weight to close roads, with reference to far away ones. Furthermore, the weight coefficient can also depend of the capacity of the road $\rho_{\max}$ as roads with higher capacity can hold more drivers.

Example

In this section, we present the results of the estimation of a velocity field for an artificial network using the method described above.

Figure 3, a and b represent the result of the estimation of the velocity direction for $\beta = 10$ and $\beta = 50$, respectively. When the parameter $\beta = 10$, the direction field $\vec{d}_\theta$ gives only the global direction of the road network, whereas when $\beta = 50$, the direction field better describes the local direction of the road network.

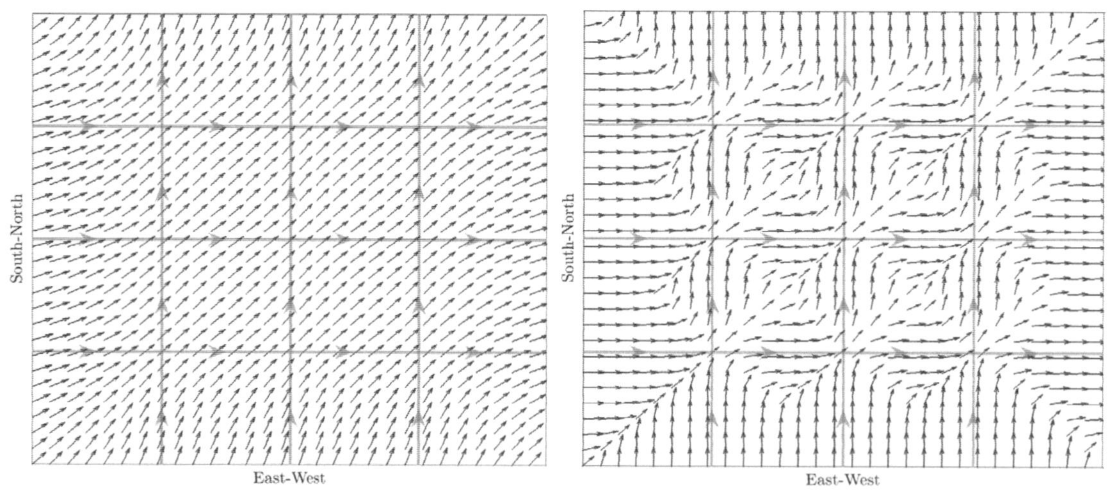

(a) $\beta = 10$ (b) $\beta = 50$

Figure 3. Estimation of the velocity direction for a 5 × 5 Manhattan grid.

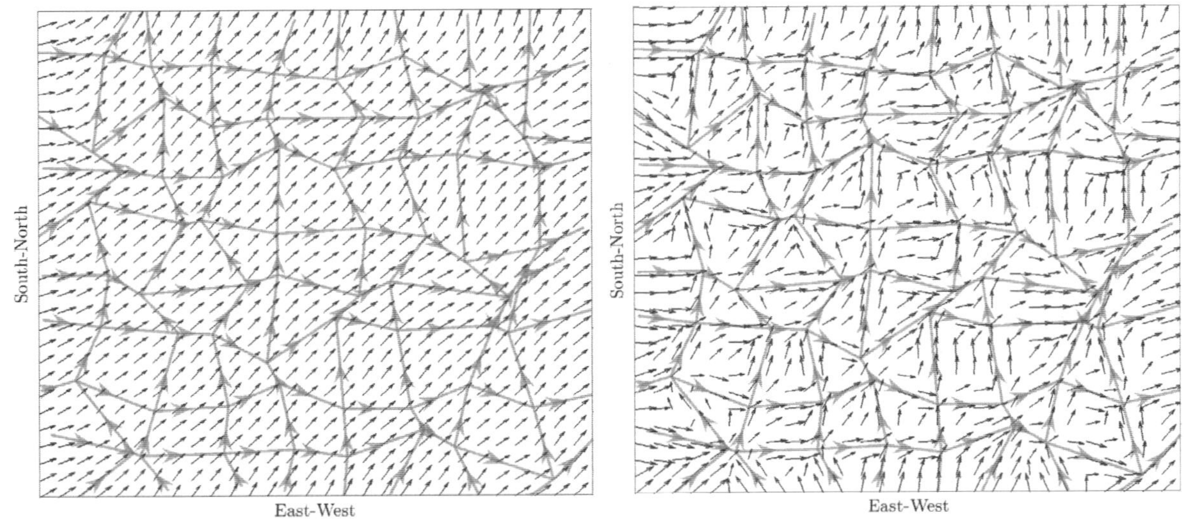

Figure 4. Estimation of the velocity direction for a less regular Manhattan grid for $\beta = 20$ (left) and $\beta = 100$ (right).

In order to test how adaptable the method suggested was, we also considered a Manhattan grid for which the position of each node has been slightly modified. An example for two different values of β is presented in Figure 4.

Numerical Method for Simulation

Choice of the Method

This section focuses on the description of the simulation method for Equation 12. The literature regarding numerical methods for the 2D conservation law is quite broad. However, only a few of these studies include a case of flux function with space dependency. In Lie, the author considers the case of quasilinear hyperbolic equation: $\frac{\partial u}{\partial t} + U(x,y)\frac{\partial f(u)}{\partial x} + V(x,y)\frac{\partial g(u)}{\partial y} = 0$ (*31*). It is possible to compute a stable approximation of a solution to this equation using dimensional splitting under the assumption that the functions U and V are Lipschitz continuous and bounded. We note that it is possible to expand the derivative of Equation 12 in order to reveal the quasilinear term and the source term: $\forall t \in \mathbb{R}^+, \forall(x,y) \in \Omega$,

$$\frac{\partial \rho}{\partial t} + \underbrace{cos(\theta(x,y))\frac{\partial\, m(\rho)}{\partial x} + sin(\theta(x,y))\frac{\partial\, m(\rho)}{\partial y}}_{Quasilinear} = \underbrace{-m(\rho)^*\left(\frac{\partial\, cos(\theta(x,y))}{\partial x} + \frac{\partial\, sin(\theta(x,y))}{\partial y}\right)}_{Source} \quad (15)$$

The principle of the splitting method is to separately compute the different terms of the equation. Thus, the formulation given by Equation 15 allows modification of the problem into the computation of an equation with a quasilinear term and an equation with a source term. The function $cos(\theta(.,.))$ and $sin(\theta(.,.))$ are bounded and are also Lipschitz continuous on condition that function θ is smooth. For the computation of the source term, we used the operator splitting method as presented in Toro, and Gosse (*32, 33*).

Description of Splitting Methods

Splitting methods were first considered by Godunov, but have since been formally introduced by Strang (*34, 35*). The concept of splitting is to divide the computation of the different terms of the equation into several numerical steps. For the general case, this method is called operator splitting. When the equation is divided with the derivative along different coordinates, it is called dimensional splitting.

Let $i,j \in [1,\ldots,I]^*[1,\ldots,J]$ be the discretization of the space and n the index for the time step. The iteration process for the computation of one time step is divided into three parts. First, we compute the propagation with respect to the first dimension. Then, we compute the propagation with respect to the second dimension using the updated density. Finally, we add the effect of the source term.

$$\rho^*_{i,j} = \rho^n_{i,j} - cos(\theta_{i,j})\frac{\Delta t}{\Delta x}(F^n_{i+\frac{1}{2},j} - F^n_{i-\frac{1}{2},j})$$

$$\rho^{**}_{i,j} = \rho^*_{i,j} - sin(\theta_{i,j})\frac{\Delta t}{\Delta y}(F^n_{i,j+\frac{1}{2}} - F^n_{i,j-\frac{1}{2}})\ \rho^{n+1}_{i,j} = \rho^{**}_{i,j} - m(\rho^{**})$$

$$\left(\frac{\Delta t}{\Delta x}(cos(\theta_{i+\frac{1}{2},j}) - cos(\theta_{i-\frac{1}{2},j})) + \frac{\Delta t}{\Delta y}(sin(\theta_{i,j+\frac{1}{2}}) - sin(\theta_{i,j-\frac{1}{2}}))\right)$$

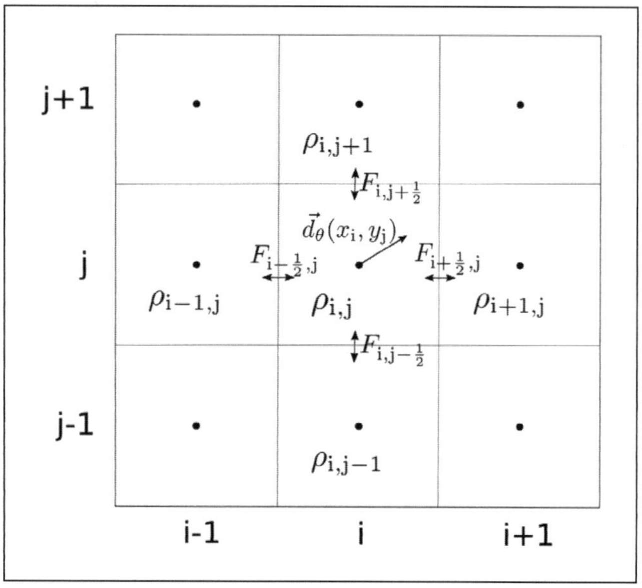

Figure 5. Spatial discretization for the numerical method.

For the notation, $F^n_{i+\frac{1}{2},j}$ corresponds to the Godunov flux between the cell, as we can see in Figure 5. Recall, the Godunov flux between a cell of density u to an adjacent cell of density v, for the function Φ of the FD, can be written as follows:

$$F(u,v) = \begin{cases} min(\Phi(u), \Phi(v)), & u < v \\ \Phi(u), & v \leq u \leq \rho_{crit} \\ \Phi(v), & \rho_{crit} \leq v \leq u \\ \Phi_{max}, & v \leq \rho_{crit} \leq u \end{cases} \tag{16}$$

The boundary condition is treated using ghost-cells.

Algorithm

In this part, we describe the numerical implementation of the previous methods. The algorithm is shown in Algorithm 1.

Simulation Results on an Artificial Manhattan Grid

In this section, we analyze the results of the simulations of the model presented for two scenarios. The scenarios represent the same network and thus have an identical estimation of the velocity direction $\vec{d}_\theta$. The network considered is built as a Manhattan grid on a square domain in which every road taken into account is oriented toward the north-east. The network and the velocity field can be seen in Figure 6.

Forward Propagation on an Artificial Manhattan Grid

In this section, we present the results of the simulations for an initial condition given by Equation 17:

$$\rho_0(x,y) = \begin{cases} \frac{1}{2}\rho_{max} & \text{if } \frac{1}{8}L_x \leq x \leq \frac{4}{8}L_x \text{ and } \frac{1}{8}L_y \leq y \leq \frac{4}{8}L_y \\ 0 & \text{otherwise} \end{cases} \tag{17}$$

where L_x and L_y are respectively the length of the domain along the x-axis and the y-axis. In Figure 7, we show the

Algorithm 1 2D model simulation

Define the parameters: v_{max} the maximum velocity, ρ_{max} the maximum density, T_{max} the time horizon, the constant coefficient CFL, dx and dy the spatial discretization and I and J the number of discrete intervals of the spatial discretization
Define the time step $dt = cfl^* min(dx, dy)/V_{max}$
Import or create the road network of the study case
Compute the velocity field d_θ ▷ Using the estimation method
Initialize the density ρ to a given value, the current time t to zero
while $t < T_{max}$ **do**
  Evalue the density at the next time step using the function Propagation_2D
  function PROPAGATION_2D $\rho^n, dt, dx, dy, I, J, v_{max}, \rho_{max}, d_\theta$
    Define the intermediate densities ρ^* and ρ^{**}
    for $i = 1$ to I and $j = 1$ to J **do**
      Update of ρ^* with flux along the x dimension computed with ρ^n
    end for
    for $i = 1$ to I and $j = 1$ to J **do**
      Update of ρ^{**} with flux along the y dimension computed with ρ^*
    end for
    for $i = 1$ to I and $j = 1$ to J **do**
      Update of the density value ρ^{n+1} due to the source term computed with ρ^{**}
    end for
  end function
  t = t + dt
end while

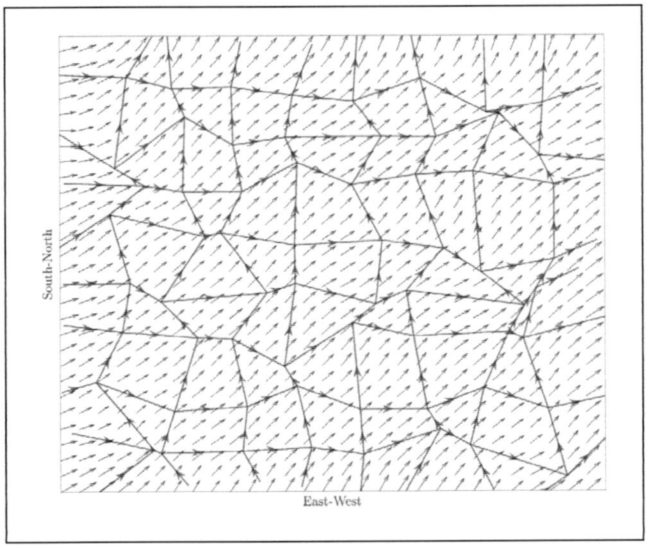

Figure 6. Network considered for the scenario (in black) and velocity direction estimated (in blue) for a parameter $\beta = 20$.

results for different times. First, we can remark that at the initial time, the initial conditions generate two main discontinuities. On the north-east, the discontinuity corresponds to a rarefaction wave, whereas in the south-west it is a shock wave. The shape of the shock wave is globally preserved.

Backward Propagation and Comparison with the CTM Model

In this section, we present the results of the simulations for an initial condition given by Equation 18:

$$\rho_0(x,y) = \begin{cases} \rho_{\max} & \text{if } \ \tfrac{2}{3}L_x \leq x \ \text{ and } \ \tfrac{2}{3}L_y \leq y \\ \tfrac{9}{16}\rho_{\max} & \text{otherwise} \end{cases} \tag{18}$$

The results at different times are given in Figure 8:

We can observe how the density over the whole network rises and the network is quickly filled.

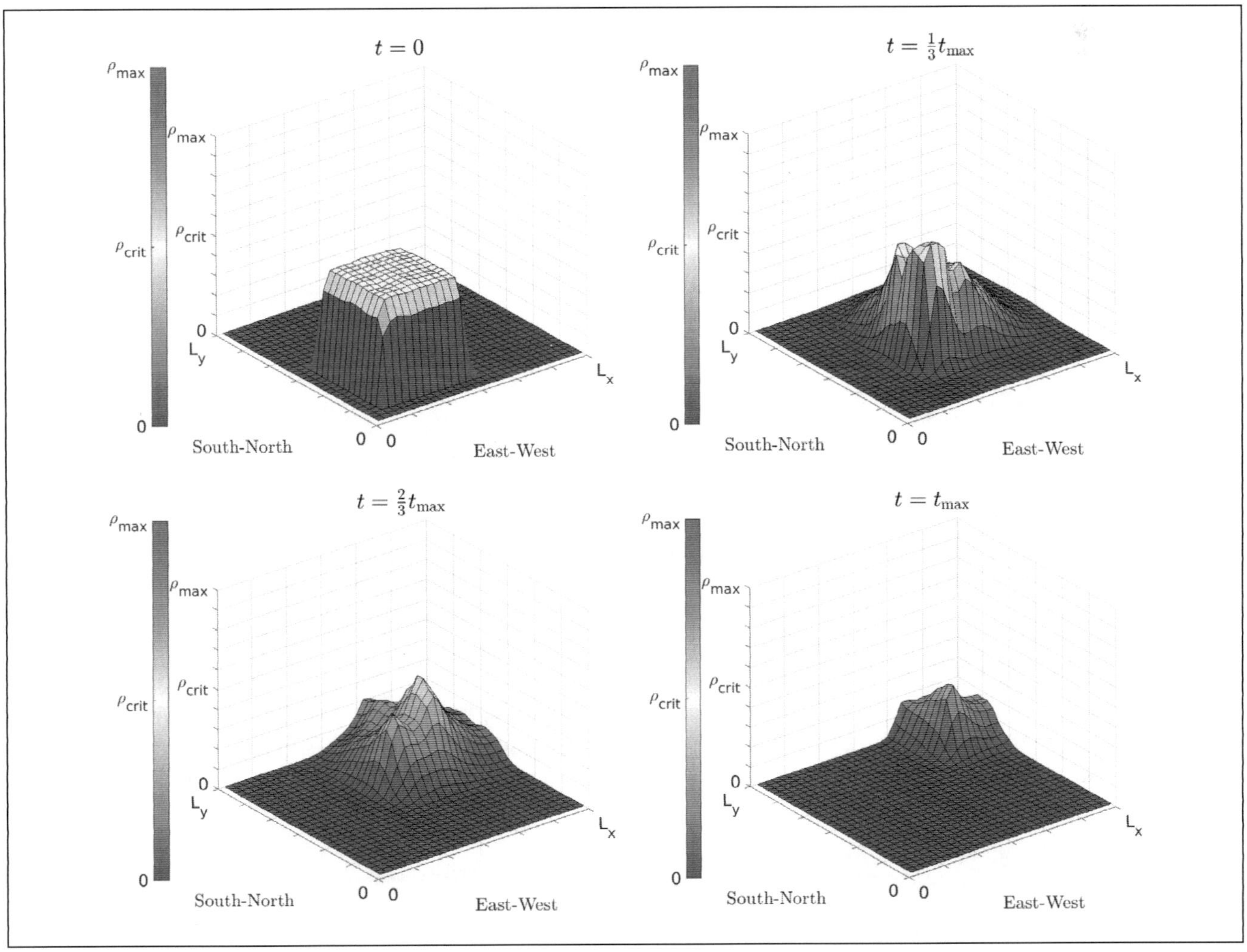

Figure 7. Simulation results at different times of a 2D model with initial traffic state given by Equation 17.

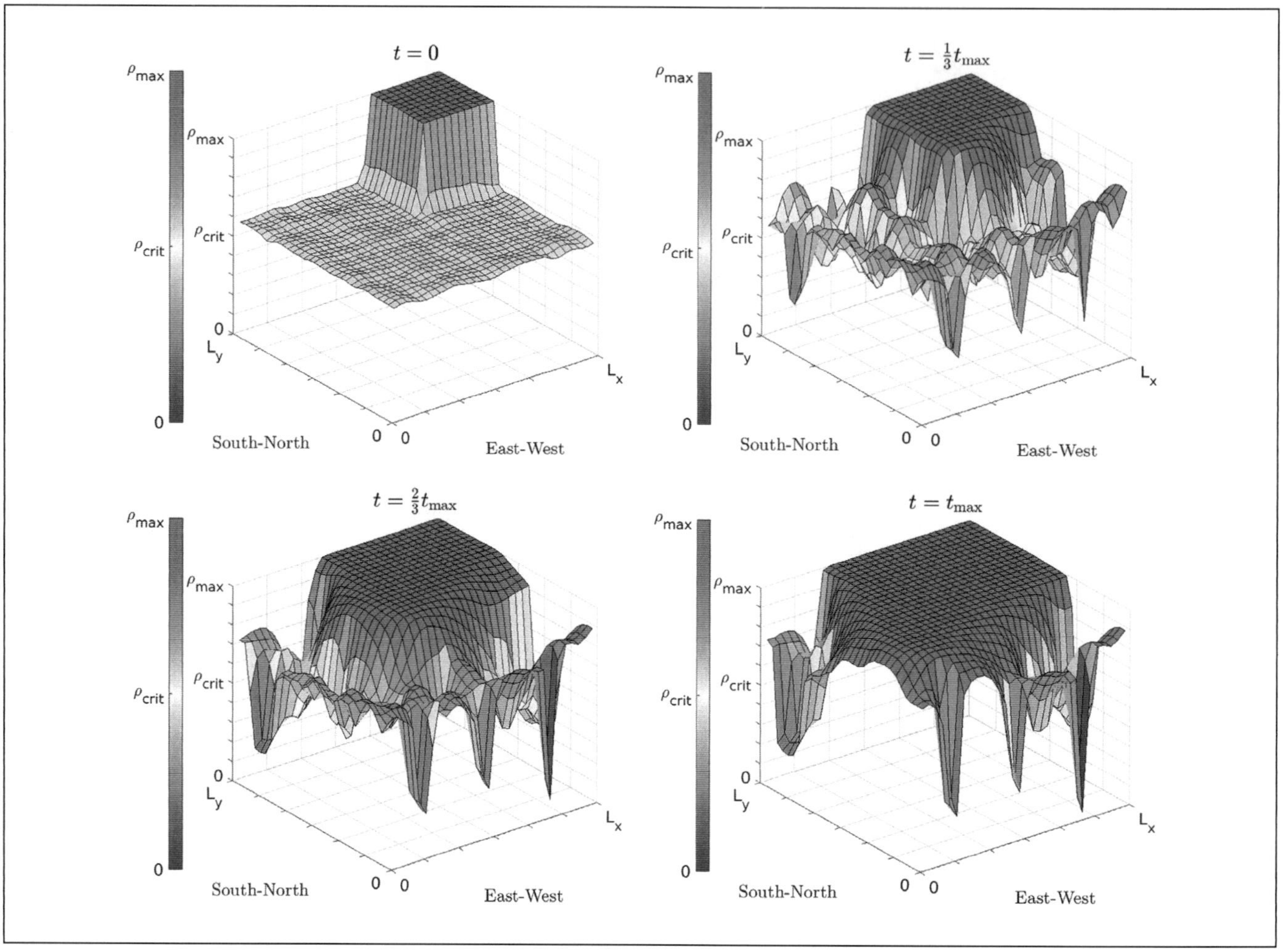

Figure 8. Simulation results at different times of the 2D model with initial traffic state given by Equation 18.

We have also simulated in parallel a CTM (*5*, *6*) on the same network and with the same initial condition, for comparison. The results of the simulation are presented in Figure 9:

We represent both simulations at the graph level. For the 2D model, the representation is only visual. Indeed, we have only completed the projection of the 2D grid coloring on the graph. Both in the 2D model and in the junction model, the wave propagates backward, as expected, with approximately the same speed of propagation.

Conclusion and Future Work

This article is an attempt to extend the LWR model in a 2D setting. Contrary to the 1D case, we underline the fact that we needed to complete the definition of the flux function with additional information. In particular, we focused on the flux direction and suggested a way to estimate this from the physical network. Finally, we presented the results of simulations which allowed comparison with the features observed on the LWR model in 1D. Future work will involve the extension of the model to heterogeneous networks, that is, capacity and speed limit variations with respect to space and time, and the possibility to describe flow in different directions simultaneously. A direct extension of this model employs the use of split ratios in the computation of the velocity field.

Acknowledgments

This project has received funding from the European Research Council under the European Union's Horizon 2020 research and innovation program (grant agreement no. 694209).

Author Contributions

The authors confirm contribution to the paper as follows: study conception and design: SM, MLDM, CC-d-W; data collection: SM, MLDM, CC-d-W; analysis and interpretation of results:

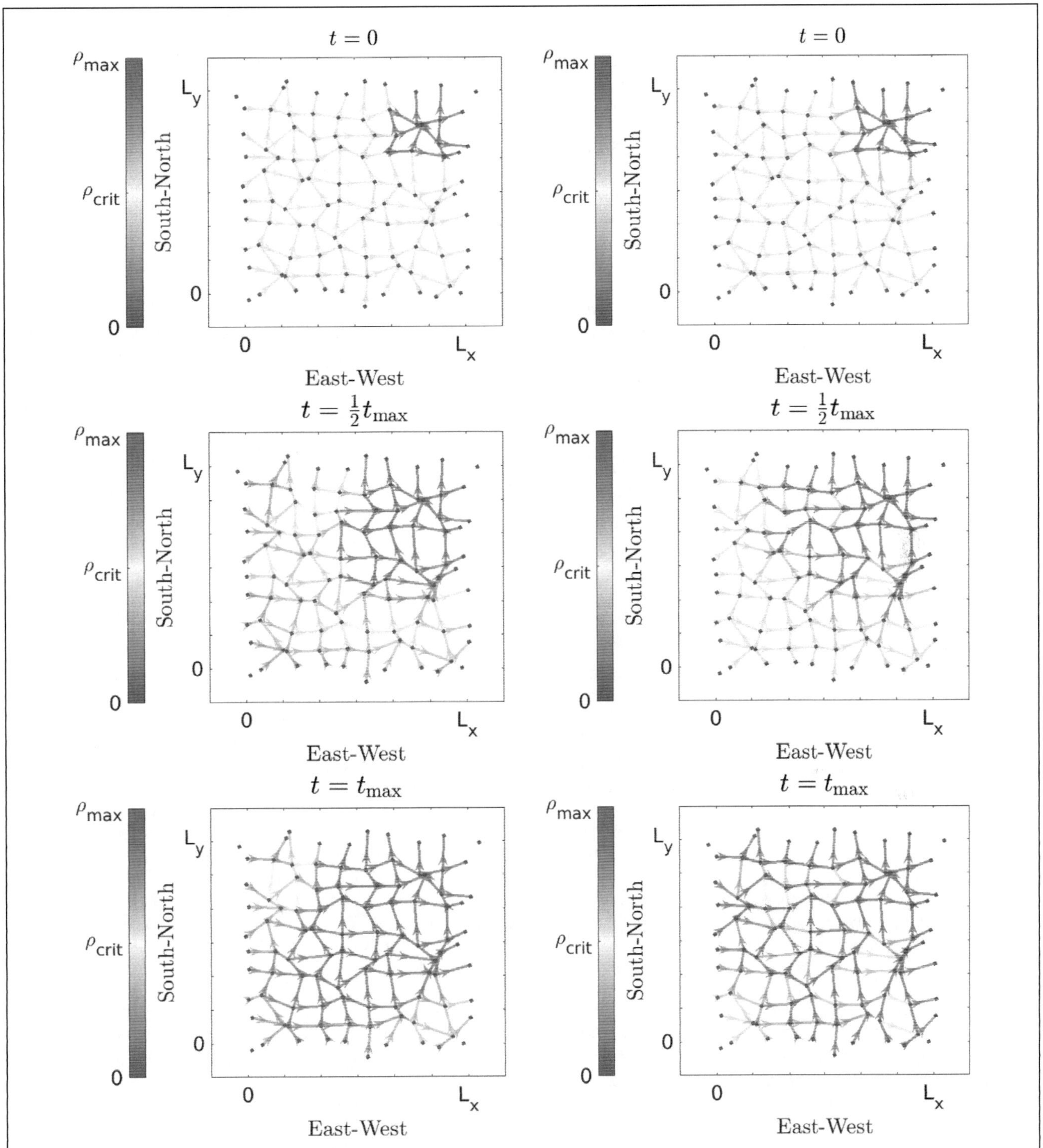

Figure 9. Comparison of simulation between 2D model (left) and CTM (right) for an initial traffic state given by Equation 18.

SM, MLDM, CC-d-W; draft manuscript preparation: SM, MLDM, CC-d-W. All authors reviewed the results and approved the final version of the manuscript.

References

1. van Wageningen-Kessels, F., H. Van Lint, K. Vuik, and S. Hoogendoorn. Genealogy of Traffic Flow Models. *EURO*

Journal on Transportation and Logistics, Vol. 4, No. 4, 2014, pp. 445–473.
2. Greenshields, B. D., J. Thompson, H. Dickinson, and R. Swinton. The Photographic Method of Studying Traffic Behavior. *Proc., Highway Research Board*, Vol. 13, 1934.
3. Lighthill, M. J., and G. B. Whitham. On Kinematic Waves. II. A Theory of Traffic Flow on Long Crowded Roads. *Proceedings of the Royal Society of London A: Mathematical, Physical and Engineering Sciences*, Vol. 229, 1955, pp. 317–345.
4. Richards, P. I. Shock Waves on the Highway. *Operations Research*, Vol. 4, No. 1, 1956, pp. 42–51.
5. Daganzo, C. F. The Cell Transmission Model: A Dynamic Representation of Highway Traffic Consistent with the Hydrodynamic Theory. *Transportation Research Part B: Methodological*, Vol. 28, No. 4, 1994, pp. 269–287.
6. Coclite, G. M., M. Garavello, and B. Piccoli. Traffic Flow on a Road Network. *SIAM Journal on Mathematical Analysis*, Vol. 36, No. 6, 2005, pp. 1862–1886.
7. Daganzo, C. F., and N. Geroliminis. An Analytical Approximation for the Macroscopic Fundamental Diagram of Urban Traffic. *Transportation Research Part B: Methodological*, Vol. 42, No. 9, 2008, pp. 771–781.
8. Geroliminis, N., and J. Sun. Properties of a Well-Defined Macroscopic Fundamental Diagram for Urban Traffic. *Transportation Research Part B: Methodological*, Vol. 45, No. 3, 2011, pp. 605–617.
9. Geroliminis, N., and C. F. Daganzo. Existence of Urban-Scale Macroscopic Fundamental Diagrams: Some Experimental Findings. *Transportation Research Part B: Methodological*, Vol. 42, No. 9, 2008, pp. 759–770.
10. Hajiahmadi, M., V. L. Knoop, B. De Schutter, and H. Hellendoorn. Optimal Dynamic Route Guidance: A Model Predictive Approach Using the Macroscopic Fundamental Diagram. In *2013 16th International IEEE Conference on Intelligent Transportation Systems-(ITSC)*, IEEE, 2013, pp. 1022–1028.
11. Leclercq, L., C. Parzani, V. L. Knoop, J. Amourette, and S. P. Hoogendoorn. Macroscopic Traffic Dynamics with Heterogeneous Route Patterns. *Transportation Research Procedia*, Vol. 7, 2015, pp. 631–650.
12. Beckmann, M. A Continuous Model of Transportation. *Econometrica: Journal of the Econometric Society*, Vol. 20, No. 4, 1952, pp. 643–660.
13. Ho, H., and S. Wong. Two-Dimensional Continuum Modeling Approach to Transportation Problems. *Journal of Transportation Systems Engineering and Information Technology*, Vol. 6, No. 6, 2006, pp. 53–68.
14. Helbing, D. A Fluid Dynamic Model for the Movement of Pedestrians. *arXiv preprint cond-mat/9805213*, 1998.
15. Hughes, R. L. A Continuum Theory for the Flow of Pedestrians. *Transportation Research Part B: Methodological*, Vol. 36, No. 6, 2002, pp. 507–535.
16. Jiang, Y. Q., P. Zhang, S. Wong, and R. X. Liu. A Higher-Order Macroscopic Model for Pedestrian Flows. *Physica A: Statistical Mechanics and Its Applications*, Vol. 389, No. 21, 2010, pp. 4623–4635.
17. Colombo, R. M., M. Garavello, and M. Lécureux-Mercier. A Class of Nonlocal Models for Pedestrian Traffic. *Mathematical Models and Methods in Applied Sciences*, Vol. 22, No. 4, 2012, p. 1150023.
18. Twarogowska, M., P. Goatin, and R. Duvigneau. Comparative Study of Macroscopic Pedestrian Models. *Transportation Research Procedia*, Vol. 2, 2014, pp. 477–485.
19. Kormanová, A. A Review on Macroscopic Pedestrian Flow Modelling. *Acta Informatica Pragensia*, Vol. 2, No. 2, 2014, pp. 39–50.
20. Della Rossa, F., C. D'Angelo, and A. Quarteroni. A Distributed Model of Traffic Flows on Extended Regions. *NHM*, Vol. 5, No. 3, 2010, pp. 525–544.
21. Jiang, Y., S. Wong, H. Ho, P. Zhang, R. Liu, and A. Sumalee. A Dynamic Traffic Assignment Model for a Continuum Transportation System. *Transportation Research Part B: Methodological*, Vol. 45, No. 2, 2011, pp. 343–363.
22. Du, J., S. Wong, C. W. Shu, T. Xiong, M. Zhang, and K. Choi. Revisiting JiangâĂŹs Dynamic Continuum Model for Urban Cities. *Transportation Research Part B: Methodological*, Vol. 56, 2013, pp. 96–119.
23. Jiang, Y.-Q., P.-J. Ma, and S.-G. Zhou. Macroscopic Modeling Approach to Estimate Traffic-Related Emissions in Urban Areas. *Transportation Research Part D: Transport and Environment*, Vol. 60, 2015, pp. 41–55.
24. Romero Perez, L., and F. G. Benitez. Outline of Diffusion Advection in Traffic Flow Modeling. Presented at 87th Annual Meeting of the Transportation Research Board, 2008, 08-1503.
25. Saumtally, T. *Modèles bidimensionnels de trafic*. Ph.D. thesis. Université Paris-Est, Paris, France, 2012.
26. Sossoe K.S., Lebacque JP. (2017) Reactive Dynamic Assignment for a Bi-dimensional Traffic Flow Model. In: Świątek J., Tomczak J. (eds) Advances in Systems Science. ICSS 2016. Advances in Intelligent Systems and Computing, Vol 539. Cham, Switzerland, 2016, pp. 179–188.
27. Chetverushkin, B., N. Churbanova, and M. Trapeznikova. Traffic Flow Simulation by 2D Macro- and Microscopic Models. In *Latest Trends on Urban Planning Transportation*, Wseas Press. Corfu Island, Greece, 2010.
28. Sukhinova, A. B., M. A. Trapeznikova, B. N. Chetverushkin, and N. G. Churbanova. Two-Dimensional Macroscopic Model of Traffic Flows. *Mathematical Models and Computer Simulations*, Vol. 1, No. 6, 2009, p. 669.
29. Lebacque, J., and M. Khoshyaran. First Order Macroscopic Traffic Flow Models for Networks in the Context of Dynamic Assignment. *Transportation Planning*, Vol. 64, 2004, pp. 119–140.
30. Kruzhkov, S. N. First Order Quasilinear Equations in Several Independent Variables. *Matematicheskii Sbornik*, Vol. 123, No. 2, 1970, pp. 228–255.
31. Lie, K.-A. A Dimensional Splitting Method for Quasilinear Hyperbolic Equations with Variable Coefficients. *BIT Numerical Mathematics*, Vol. 39, No. 4, 1999, pp. 683–700.
32. Toro, E. F. *Riemann Solvers and Numerical Methods for Fluid Dynamics: A Practical Introduction*. Springer Science & Business Media, Berlin, Germany, 2013.

33. Gosse, L. A Two-Dimensional Version of the Godunov Scheme for Scalar Balance Laws. *SIAM Journal on Numerical Analysis*, Vol. 52, No. 2, 2014, pp. 626–652.
34. Godunov, S. K. A Difference Method for Numerical Calculation of Discontinuous Solutions of the Equations of Hydrodynamics. *Matematicheskii Sbornik*, Vol. 89, No. 3, 1959, pp. 271–306.
35. Strang, G. On the Construction and Comparison of Difference Schemes. *SIAM Journal on Numerical Analysis*, Vol. 5, No. 3, 1968, pp. 506–517.

The Standing Committee on Traffic Flow Theory and Characteristics (AHB45) peer-reviewed this paper (18-01172).

Article

TRR
JOURNAL OF THE TRANSPORTATION RESEARCH BOARD

Transportation Research Record
2018, Vol. 2672(20) 262–276

DOI: 10.1177/0361198118784402
journals.sagepub.com/home/trr
SAGE

Improved Numerical Method for Aw-Rascle Type Continuum Traffic Flow Models

Saeed Mohammadian[1] and Femke van Wageningen-Kessels[2,3]

Abstract

Continuum traffic flow models are essentially nonlinear hyperbolic systems of partial differential equations. Except for limited specific cases, these systems must be solved numerically. Mathematical structure of continuum models can be different for each class of models. As a result, suitable numerical schemes for some classes may not be efficient for others. In this study, an improved numerical method is proposed for a class of second-order traffic flow models. The method is based on McCormack scheme, which is a widely-used method for non-homogeneous second-order traffic flow models, but fails to produce reasonable results for homogeneous models including Aw-Rascle type models which are the focus of this paper. It is shown that this is mainly due to spurious numerical oscillations. Smoothing methods to overcome these issues are studied and applied. Central dispersion and artificial viscosity (AV) methods are incorporated into the standard McCormack scheme and tested. To reduce numerical diffusion, a total variation diminishing Runge-Kutta time stepping scheme is applied. Results show the capability of the proposed methods, and especially the AV method, to eliminate the oscillations of the standard McCormack scheme as well as controlling numerical diffusion.

Traffic flow modelling has attracted growing research attention due to the need for optimizing the usage of either existing or future road infrastructure. The increasing and dynamic nature of demand on urban networks and the environmental, societal and economic costs associated with traffic jams are some reasons justifying robust studies of traffic flow, results of which could be advantageous in the decision-making process of traffic control strategies. Dynamic nonlinear phenomena such as stop-and-go waves and synchronization in congestion (*1*) require models that are able to reproduce them, while keeping the number of parameters limited (*2*). Furthermore, computer predictions should be acceptably accurate and computational costs must remain low when simulations are applied to real-world traffic scenarios.

Various techniques have been employed for traffic state estimation over the past decades including mathematical modelling (*3*). Continuum (macroscopic) approaches along with car-following and cellular automata models are the most successful methods for mathematical study of traffic flow. For detailed reviews of traffic flow models, we refer the reader to other studies (*4–8*). In this study, we focus on continuum models because of their low computational costs, simpler calibration compared to other approaches, and their high potential for incorporation in traffic control strategies (*9*). Continuum models reproduce the behavior of traffic flow by modelling it as a fluid, with aggregated variables such as average speed and average density. First-order continuum models assume that local speed is always in equilibrium with local density, while second-order continuum models define a dynamic equation for local speed (*10*). In general, second-order models can allow traffic states to fluctuate over equilibrium states introducing acceleration behavior and its associated phenomena such as reaction time and anticipation. Such fluctuations have been also associated with some complex phenomena such as traffic oscillations, hysteresis, and capacity drop

[1]School of Civil Engineering & Built Environment, Science & Engineering Faculty, Queensland University of Technology, Brisbane, Australia
[2]Department of Logistics, Tourism and Services Management, Faculty of Business and Economics, German University of Technology in Oman (GUtech), Oman
[3]Department of Transport & Planning, Delft University of Technology, Faculty of Civil Engineering and Geosciences, Delft, The Netherlands

Corresponding Author:
Address correspondence to Saeed Mohammadian:
saeed.mohammadian@hdr.qut.edu.au

which have been observed in empirical traffic studies (*11*). For a detailed review of such phenomena, we refer the reader to other papers (*12–17*). One of the benefits of second-order models is their ability to capture such non-linear traffic phenomena, which cannot be reproduced by single-class first-order models (*7*). However, the complex mathematical structure of such models, calls for different numerical schemes to first-order models (*6*).

Second-order continuum traffic flow models are classified as nonlinear system of hyperbolic equations (*18*). In the presence of a source term, they are also classified as non-homogeneous equations. The Payne-Witham (*19*, *20*), Kerner-Konhäuser (*21*), and the gas-kinetic-based (*22*) models are examples of non-homogeneous second-order traffic flow models because they include a relaxation acceleration source. However, in some other continuum models such as the Aw-Rascle model (*23*) and the Zhang model (*24*) there is no source term and they are classified as homogeneous equations. Some non-homogeneous models do not satisfy the anisotropy conditions (*25*, *26*), which means drivers do not react to their follower in traffic flow. However, homogeneous Aw-Rascle and Zhang models have favorable properties such as anisotropic behavior and it can be proven that solutions to important initial value problems exist (*27*).

One of the most common frameworks for solving non-homogeneous continuum models is to use the McCormack method because of its simplicity, limited computational costs and little associated numerical diffusion (*28*). Furthermore, the minimum supply demand method, which is commonly used for first-order models, has been adapted for second-order traffic flow models (*29*). In the context of hyperbolic systems, high-order methods have been proposed for computational fluid dynamics and heat transfer applications (*18*, *30*). By translating concepts from such studies, some complex methods have also been proposed for continuum traffic flow models (*31*, *32*). However, the McCormack remains one of the most commonly-used schemes in traffic flow modelling and simulation (*6*). As a second-order scheme, the McCormack scheme provides less numerical diffusion compared with first-order schemes (*3*). However, this scheme is associated with numerical oscillations. When applied to non-homogeneous models, numerical oscillations can be dispersed due to the effect of source terms. Therefore, results are usually only slightly affected by numerical oscillations (*28*). However, due to lack of source terms, this scheme fails to reproduce plausible results for the homogeneous models, as we will show in later section.

The contribution of this research is to adapt the McCormack scheme to be able to reproduce reasonable results when applied to homogeneous second-order traffic flow models, such as the Aw-Rascle model. Therefore, two smoothing methods are studied: central dispersion (CD) (*33*) and artificial viscosity (AV) (*34*). In order to increase order of temporal accuracy, total variation diminishing (TVD) Runge-Kutta time stepping (*35*) is also added to the method. Various numerical experiments are performed to evaluate the ability of improved scheme.

The rest of the paper is organized as follows: In the next section, the applied traffic flow models are introduced. The numerical methods, including the McCormack scheme and smoothing methods are then covered. Test cases are then introduced, and their results are discussed. Finally, some concluding remarks complete the study.

Continuum Traffic Flow Models

The general formulation of a second-order traffic flow model in conservative form (*10*) is

$$\partial_t \boldsymbol{U} + \partial_x \boldsymbol{F}(\boldsymbol{U}) = \boldsymbol{S}(\boldsymbol{U}) \tag{1}$$

where vector $\boldsymbol{U}$ indicates state variables, usually $\boldsymbol{\rho}$ the local density of vehicles (veh/m) and $\boldsymbol{v}$ the local macroscopic speed$\left(\frac{m}{s}\right)$.$\boldsymbol{F}(\boldsymbol{U})$ is the flux function and $\boldsymbol{S}(\boldsymbol{U})$ is the source term which can represent relaxation or anticipation. The system is non-homogeneous when the source term $\boldsymbol{S}(\boldsymbol{U})$ is nonzero. Payne-type and gas-kinetic-based traffic models (*20*, *22*) are classified as non-homogeneous models. The system in (1) can be written in quasi-linear form as

$$\partial_t \boldsymbol{U} + \boldsymbol{J}(\boldsymbol{U})\partial_x \boldsymbol{U} = \boldsymbol{S}(\boldsymbol{U}) \tag{2}$$

where $\boldsymbol{J}(\boldsymbol{U}) = \frac{\partial \boldsymbol{F}}{\partial \boldsymbol{U}}$ is the Jacobian matrix. Characteristic speeds equal the eigenvalues of the system and can be obtained by solving $\boldsymbol{det}[\boldsymbol{J}(\boldsymbol{U}) - \boldsymbol{\lambda I}] = 0$ for the eigenvalues $\boldsymbol{\lambda}$. In the presence of distinct and real characteristic speeds, the system is hyperbolic. Characteristic speeds can be studied to see if the model preserves the anisotropic nature of the traffic flow (*25*, *26*).

Aw-Rascle Type Traffic Flow Model

Aw-Rascle type of traffic flow models represent a generic family of homogeneous and anisotropic higher-order traffic flow models. For a section of road without on-ramps and off-ramps, Aw-Rascle type models can be written (*23*) in the formulation

$$\begin{cases} \partial_t \boldsymbol{\rho} + \partial_x(\boldsymbol{\rho v}) = 0 \\ \partial_t \boldsymbol{\rho}(\boldsymbol{v} + \boldsymbol{P}(\boldsymbol{\rho})) + \partial_x(\boldsymbol{\rho v}(\boldsymbol{v} + \boldsymbol{P}(\boldsymbol{\rho}))) = 0 \end{cases} \tag{3}$$

where v is vehicle speed (in m/s) and $P(\rho)$ is an increasing function of density to guarantee that the model is anisotropic and can be written as

$$P(\rho) = C_0^2\rho^{\gamma} - \psi \quad (4)$$

where C_0 is sonic velocity, ψ is a constant to regulate pressure, and γ is always greater than zero. The system has two real eigenvalues: $\lambda_1 = v - \rho P'(\rho)$ and $\lambda_2 = v$. Furthermore, except for at zero density ($\rho = 0$), the eigenvalues are distinct and thus the system is strictly hyperbolic.

Zhang Model

Zhang model is a special case of the Aw-Rascle type of models (*24*), in which the dynamic speed function is derived from a microscopic car-following model, leading to the pressure function

$$P(\rho) = -v_e(\rho) \quad (5)$$

where $v_e(\rho)$ is the equilibrium speed-density function, in this case Greenshields' model (*36*),

$$v_e(\rho) = v_{\max}\left(1 - \frac{\rho}{\rho_{\text{jam}}}\right) \quad (6)$$

with $v_{\max}$ representing the maximum vehicle speed and ρ_{jam} is the jam density.

Numerical Methods

In this section, the proposed numerical methods for solving Aw-Rascle type models are discussed. The McCormack method with smoothing is applied for spatial integration. Temporal integration is done using a TVD Runge-Kutta scheme.

McCormack Scheme

The McCormack scheme (see Pevret and Taylor (*37*) for example) is a second-order finite difference method that has been applied successfully to non-homogeneous higher order traffic flow models (*28*). It consists of a predictor and a corrector step:

Predictor

$$\tilde{U}_i^{n+1} = U_i^n - \frac{\text{dt}}{\text{dx}}\left(F_i^n - F_{i-1}^n\right) \quad (7)$$

Corrector

$$U_i^{n+1} = 0.5\left(U_i^n + \tilde{U}_i^{n+1}\right) - 0.5\frac{\text{dt}}{\text{dx}}\left(\tilde{F}_{i+1}^{n+1} - \tilde{F}_i^{n+1}\right) \quad (8)$$

As will become clear in the test case section, the standard McCormack method induces oscillations and is not suitable for homogeneous models such as Aw-Rascle type of models. Therefore, smoothing is applied. Two different smoothing methods are studied, as described in the following.

Central Dispersion Method

The first method for smoothing, the CD method, is based on the Lax Fredrichs scheme (*38*), and can be implemented on the updated results as

$$U_{i(\text{CD})}^{n+1} = (1 - S).U_i^{n+1} + S\left(\frac{U_{i+1}^{n+1} + U_{i-1}^{n+1}}{2}\right) \quad (9)$$

where S is a weight factor and $0 \le S \le 1$. With $S = 0$ there is no smoothing at all and the method reduces to the standard McCormack scheme. This method, in fact, acts as a hybrid scheme since it computes the traffic state as weighted average of results by the McCormack method and Lax-Friedrichs scheme. When choosing $S = 1$, the Lax-Friedrichs scheme substitutes the second step of the McCormack scheme as:

$$U_i^{n+1} = 0.5\left(U_{i-1}^n + U_{i+1}^n\right) + 0.5\frac{\text{dt}}{\text{dx}}\left(F_{i+1}^n - F_{i-1}^n\right) \quad (10)$$

However, since the Lax-Friedrichs scheme is diffusive (*38*), taking values for S close to 1 results in considerable numerical diffusion. In the test cases, a proper value for S was established by trial and error.

At the boundaries, U_{i+1}^{n+1} or U_{i-1}^{n+1} does not exist and smoothing is performed through

$$U_{i(\text{CD})}^{n+1} = (1 - S).U_i^{n+1} + S\left(\frac{U_{i+1}^{n+1} + U_i^{n+1}}{2}\right) \quad (11)$$

or

$$U_{i(\text{CD})}^{n+1} = (1 - S).U_i^{n+1} + S\left(\frac{U_i^{n+1} + U_{i-1}^{n+1}}{2}\right) \quad (12)$$

Artificial Viscosity Method

The second smoothing method that we study is the AV method (*34*). The method is based on introducing an artificial dissipative term to smooth out the numerical oscillations. With AV, the updated results in the McCormack scheme become

$$U_{i(\text{AV})}^{n+1} = U_i^{n+1} + \varphi_{i+\frac{1}{2}}\left(U_{i+1}^{n+1} - U_i^{n+1}\right) - \varphi_{i-\frac{1}{2}}\left(U_i^{n+1} - U_{i-1}^{n+1}\right) \quad (13)$$

where $\varphi_{i+\frac{1}{2}}$ is the AV at $(i+\frac{1}{2})$

$$\varphi_{i+\frac{1}{2}} = \kappa.\max\left(\varphi_{i,}, \varphi_{i+1}\right) \quad (14)$$

with κ representing the viscosity parameter and φ_i computed as

$$\varphi_i = \frac{|U_{i+1} - 2U_i + U_{i-1}|}{|U_{i+1}| + 2|U_i| + |U_{i-1}|} \quad (15)$$

The technique is similar to adopting flux-liming terms in high-order TVD Lax-Wendroff methods (*18*) and, as discussed in the following section, can provide robust results.

At the boundaries, U_{i+1}or U_{i-1} does not exist and the AV is computed using

$$\varphi_i = \frac{|U_{i+1} - U_i|}{|U_{i+1}| + |U_i|} \quad (16)$$

$$\varphi_i = \frac{|U_i - U_{i-1}|}{|U_i| + |U_{i-1}|} \quad (17)$$

Temporal Integration

To achieve higher order of temporal accuracy while avoiding numerical oscillations, a TVD Runge-Kutta scheme (*35*) is applied for temporal integration. At each time step, the new state variables are approximated as follows.

$$\begin{cases} U^{(p)}{}_i = U^{(n)}{}_i + \xi\left(U^{(n)}{}_i\right) \\ U^{(q)}{}_i = 3/4U^{(n)}{}_i + 1/4U^{(p)}{}_i + 1/4\mathrm{dt}\xi\left(U^{(p)}{}_i\right) \\ U^{(n+1)}{}_i = 1/3U^{(n)}{}_i + 2/3U^{(q)}{}_i + 2/3\mathrm{dt}\xi\left(U^{(q)}{}_i\right) \end{cases} \quad (18)$$

where $U^{(n)}{}_i$ the state in the i -th grid cell at time step n. p and q can be interpreted as intermediate time steps, used to compute the state $U^{(n+1)}{}_i$ in the i -th grid cell, at the next time step $n+1$. Furthermore, we use the change rate of effective flux

$$\xi\left(U^{(n)}\right) = -\frac{\mathbf{dt}}{\mathbf{dx}}\left(F_i^n - F_{i-1}^n\right) \quad (19)$$

$$\xi\left(U^{(n)}\right) = -\frac{\mathbf{dt}}{\mathbf{dx}}\left(F_i^n - F_{i-1}^n\right) \quad (20)$$

CFL Condition

To avoid numerical instabilities, the grid cell and time step size must satisfy the CFL condition

$$C_r = \max(\lambda_1, \lambda_2)\frac{\mathbf{dt}}{\mathbf{dx}} \quad (21)$$

Table 1. Physical and Numerical Parameters Used for Aw-Rascle and Zhang Models

Parameter	Value	Model
Length of road (**L**)	12000*m*	Both
Maximum speed ($\mathbf{v_{max}}$)	30*m*/*s*	Both
Maximum density ($\boldsymbol{\rho}_{\mathbf{jam}}$)	0.15*veh*/*m*	Both
$C_0{}^2$	$80m^2/s^2$	Aw-Rascle
ψ	$31.94veh.m/s^2$	Aw-Rascle
γ	0.5	Aw-Rascle
Critical density (ρ_{cr})	0.04*veh*/*m*	Aw-Rascle
Critical density (ρ_{cr})	0.075*veh*/*m*	Zhang
S	0.01	Both
κ	0.25	Both
dt	1	Both
dx	31.7*m*	Both
$\max(\lambda_1, \lambda_2)$	30*m*/*s*	Both
C_r	0.975	Both
dt(Reference solution)	0.0625s	Both
dx (Reference solution)	1.8*m*	Both

with C_r being the CFL number and $\max(\lambda_1, \lambda_2)$ representing the greater propagation speed. Numerical diffusion is lowest when the CFL number is close to 1, see Pevret and Taylor (*37*) for example. In the Aw-Rascle type models, the maximum of eigenvalues equals the maximum speed, is greater than backward propagation speed. In the numerical tests, the parameter values are chosen as in Table 1.

Numerical Tests and Discussions

In this section, numerical tests are performed for evaluation of proposed numerical methods for application on Aw-Rascle and Zhang models (*23*, *24*). The model parameters take values given in Table 1. Test cases differ in their initial conditions. They were chosen to reflect a wide range of realistic traffic scenarios, while also being demanding for the numerical methods. To confine effects of additional source terms, all test cases are performed on homogeneous road of length L. At the boundaries, the following conditions are applied:

$$U_1^n = U_2^n$$
$$U_I^n = U_{I-1}^n \quad (22)$$

where I denotes the last cell.

Case 1: Free-Flow Traffic with Transition from High to Low Densities

$$\rho(x,0) = \begin{cases} 0.46\rho_{\mathbf{jam}}\ for\ x < \frac{L}{2} \\ 0.1\rho_{\mathbf{jam}}\ for\ x > \frac{L}{2} \end{cases}$$

$$v(x,0) = v_e(\rho(x,0)) \tag{23}$$

Case 2: Congested Traffic with Transition from High to Low Densities

$$\rho(x,0) = \begin{cases} 0.9\rho_{\mathbf{jam}} \; for \; x < \frac{L}{2} \\ 0.55\rho_{\mathbf{jam}} \; for \; x > \frac{L}{2} \end{cases}$$

$$v(x,0) = v_e(\rho(x,0)) \tag{24}$$

Case 3: Queue Dissolution

$$\rho(x,0) = \begin{cases} 0.1\rho_{\mathbf{jam}} \\ \rho_{\mathbf{jam}} \\ 0.1\rho_{\mathbf{jam}} \end{cases}$$

$$v(x,0) = v_e(\rho(x,0))$$

$$for \; x \leq L/3$$

$$for \; L/3 < x \leq 2L/3$$

$$for \; L/3 > 2L/3 \tag{25}$$

Case 4: Free-Flow Traffic with Uniform Density and Non-Equilibrium Speeds

$$\rho(x,0) = 0.35\rho_{\mathbf{jam}}$$

$$v(x,0) = \begin{cases} v_e(\rho(x,0)) + 5 \\ v_e(\rho(x,0)) \\ v_e(\rho(x,0)) + 5 \end{cases}$$

$$for \; x < L/5$$

$$for \; L/5 < x < 2L/5$$

$$for \; x > 2L/5 \tag{26}$$

Case 5: Uniform Density with Non-Equilibrium Speeds

$$\rho(x,0) = 0.75\rho_{\mathbf{jam}}$$

$$v(x,0) = \begin{cases} v_e(\rho(x,0)) + 5 \\ v_e(\rho(x,0)) \\ v_e(\rho(x,0)) + 5 \end{cases}$$

$$for \; x < L/3$$

$$for \; L/3 < x < 2L/3$$

$$for \; x > 2L/3 \tag{27}$$

Case 1-3 include equilibrium traffic states that cannot be simulated through standard McCormack scheme. This is due to either existence of sharp transition from congestion to free-flow regime or formation of them due to oscillatory behavior of the method. In cases 4 and 5, initial non-equilibrium traffic states result in formation and propagation of contact discontinuity and expansion waves. Non-convergent formation of numerical oscillations in these cases result in failure of standard McCormack scheme which ultimately leads to physically unrealistic results.

Numerical Results for Standard McCormack Scheme and Discussion

In this section, behavior of standard McCormack scheme for simulation of Aw-Rascle type models will be investigated. We apply the method for some test cases, to illustrate why this scheme is not well-suited for these type of problems. Failure in these cases, implies that the method would not work well in any of the other cases either. Afterwards, we test both smoothing methods on all test cases. The standard McCormack scheme is applied to Zhang model, only for test case 1 and 2. Results are illustrated in Figure 1.

Results show oscillations around the transitions from high to low densities. These oscillations lead to discontinuous transitions from congestion to free-flow regime and ultimately result in failure of the method. This is because the Aw-Rascle type models are highly nonlinear and lack a dissipative term to smooth out numerical oscillations that are introduced by the McCormack scheme. Analogous behavior of the McCormack scheme has been reported when the scheme is applied to the shallow water equations with initial sharp transition from subcritical to supercritical flows, in the absence of a dissipative (friction-type) term or in the presence of a booster (topography-type) term for oscillations. We refer the reader to Tan (*39*) and Garcia-Navarro (*40*) for further discussion on shallow water equations. For this type of problem, smoothing methods are commonly applied to eliminate or reduce oscillations (*41*, *42*). It should be mentioned that the McCormack scheme can be well-suited for other traffic flow simulations, such as for non-homogeneous models and for specific cases with other initial conditions such as red-light modelling. However, it is not suitable to simulate a wide range of realistic traffic scenarios using a homogeneous second-order model.

Set Up of Numerical Tests on Smoothed McCormack with TVD Runge-Kutta Scheme

AV and CD smoothing methods are applied to decrease the oscillations inherent to the standard McCormack

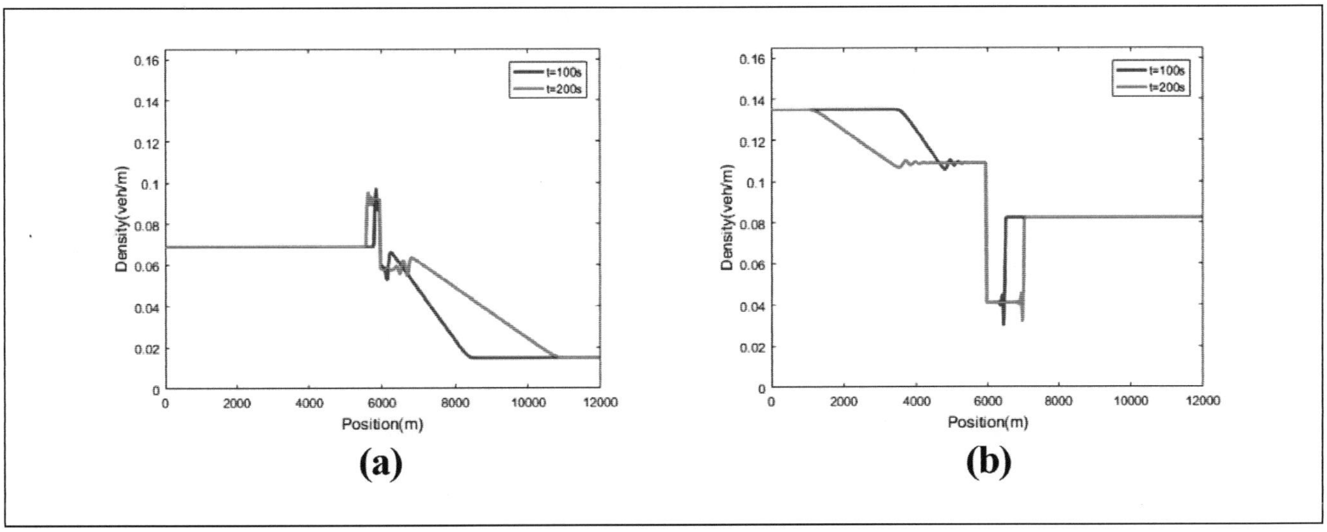

Figure 1. Results of the McCormack scheme for transition from higher to lower densities in uniform traffic condition for (*a*) Case 1, free-flow condition, and (*b*) Case 2, congestion regime.

scheme. The smoothing parameters are shown in Table 1. Several values were tested, and these turned out to give good results in terms of removing spurious oscillations, while keeping numerical diffusion low. Figure 2 illustrates an example of various smoothing parameters for both AV and CD methods. Choice of zero smoothing parameters in both methods reduces proposed method to the standard McCormack, which cannot provide reasonable solution. It is clear from Figure 2 that with the increase in smoothing parameters, numerical diffusion also increases which results in sooner propagation of traffic density. More elaborate testing would be beneficial for future applications. Results are compared with a reference solution that is obtained using a fine grid, with parameters as in Table 1. For the reference solution, AV with the same parameters in accordance with first-order explicit time stepping was used. The grid cell sizes are chosen such that the CFL number is equal to the CFL number of the methods on the normal grid. It is worth mentioning that exact solution for discussed models cannot be calculated using techniques such as method of characteristics (*18*), variational theory (*43*), or Hamilton-Jacobi representation (*44*). In fact, discussed models are both second-order and nonlinear and exact solution for such models cannot be obtained through analytical methods. However, we use fine-grid reference solutions through which our results are evaluated.

Solution of Riemann problem to the Case 1 is the formation and propagation of expansion waves to the right. This is because initial traffic state is of free-flow regime and the upstream density is greater than that of downstream. Density profile propagates in a slower rate for Aw-Rascle model than for Zhang model because of the different parameters taken for traffic pressure functions. As previously stated, the McCormack scheme can successfully simulate such cases. However, when the upstream density is very close to critical density, spurious oscillation leads to unreasonable results. Results from Figures 3 and 4 clearly show that smoothing methods can overcome the previously mentioned problem of the standard McCormack method and results. Furthermore, spatiotemporal evolution of density for both Zhang and Aw-Rascle model reveals that overall results provided by these smoothing methods are correct and only on a certain trajectory (the edge of the initial discontinuity) can some trivial oscillations be observed. These slight oscillations are also alleviated more effectively when AV is applied.

Being analogous to Case 1, the second case is also on transition from higher to lower densities. However, due to the dominance of congestion regime characteristic speeds, which determine the direction of traffic waves, are of negative signs and traffic density propagates leftward. Similarly, for the densities close enough to critical density, standard McCormack provides physically absurd results. According to Figures 5 and 6, overall results obtained from the two smoothing methods are highly acceptable compared with the reference solution. Although the general propagation of traffic density is correct for each of the smoothing methods, for Zhang model, obtained results from CD method are riddled with greater oscillations on the trajectory following by the edge of the discontinuity. This is because downstream density for Zhang model is closer to its critical density compared with Aw-Rascle model due to its different pressure function. AV smoothing provides less

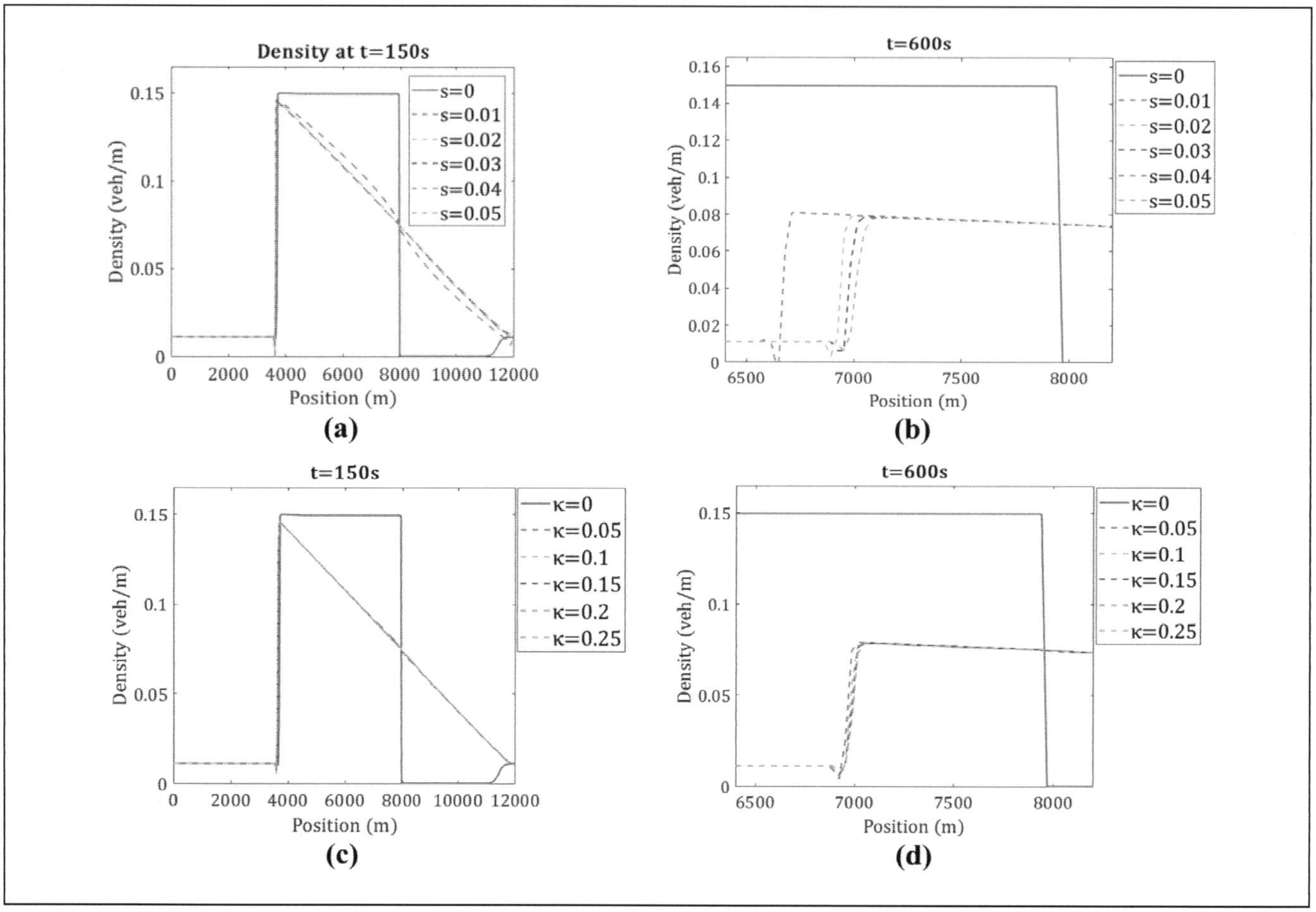

Figure 2. Effects of choice of smoothing parameter on numerical diffusion for case 3, initial queue problem: (*a*) and (*b*) CD method; (*c*) and (*d*) AV method.

oscillatory results in this case as well. Yet, CD method also provides acceptable results.

Case 3 is simply a condition in which free-flow condition is present in the upstream and downstream of congestion regime. Solution of this Riemann problem for the downstream discontinuity is expansion waves heading both to the left and right side because of discontinuous transition from congestion regime to the free-flow condition. Solution of the Riemann problem in the upstream is a shockwave which initially propagates leftwards, and after some points when free-flow condition is provided in downstream, this shockwave heads to the right side. This case cannot be simulated by standard McCormack scheme as discussed before. However, both smoothing methods can overcome oscillations to an acceptable point and produce reasonable results as shown in spatiotemporal evolutions of traffic densities shown in Figure 5 (*c, d, g, h*). Cross-sectional results (Figure 5 (*a, b, e, f*)), show that effects of oscillations are conspicuous when CD method is applied. Spatiotemporal results (Figure 5 (*c, d, g, h*)), however, reveal that there is slight difference in the overall results of CD method compared to AV, which also provides better results in these cases as well.

Simulation of non-equilibrium traffic state has been performed on Case 4 and Case 5. In these test cases, despite initial uniform density profile, local speeds are not in equilibrium with traffic density. Initial traffic density in Case 4 represents free-flow condition while in Case 5, congestion regime is dominant. Non-equilibrium local speeds result in formation of shock-waves as contact discontinuities in these cases and also expansion waves. Numerical oscillations in standard McCormack lead to the failure of scheme for simulation of such cases. However, both smoothing methods can successfully capture those waves and present reasonable results for spatiotemporal evolution of traffic flow as illustrated in Figures 6 and 7.

Results of Numerical Tests on Smoothed McCormack with TVD Runge-Kutta Scheme

Figures 3–7 show the test results, for both Aw-Rascle and Zhang model, comparing the smoothing methods

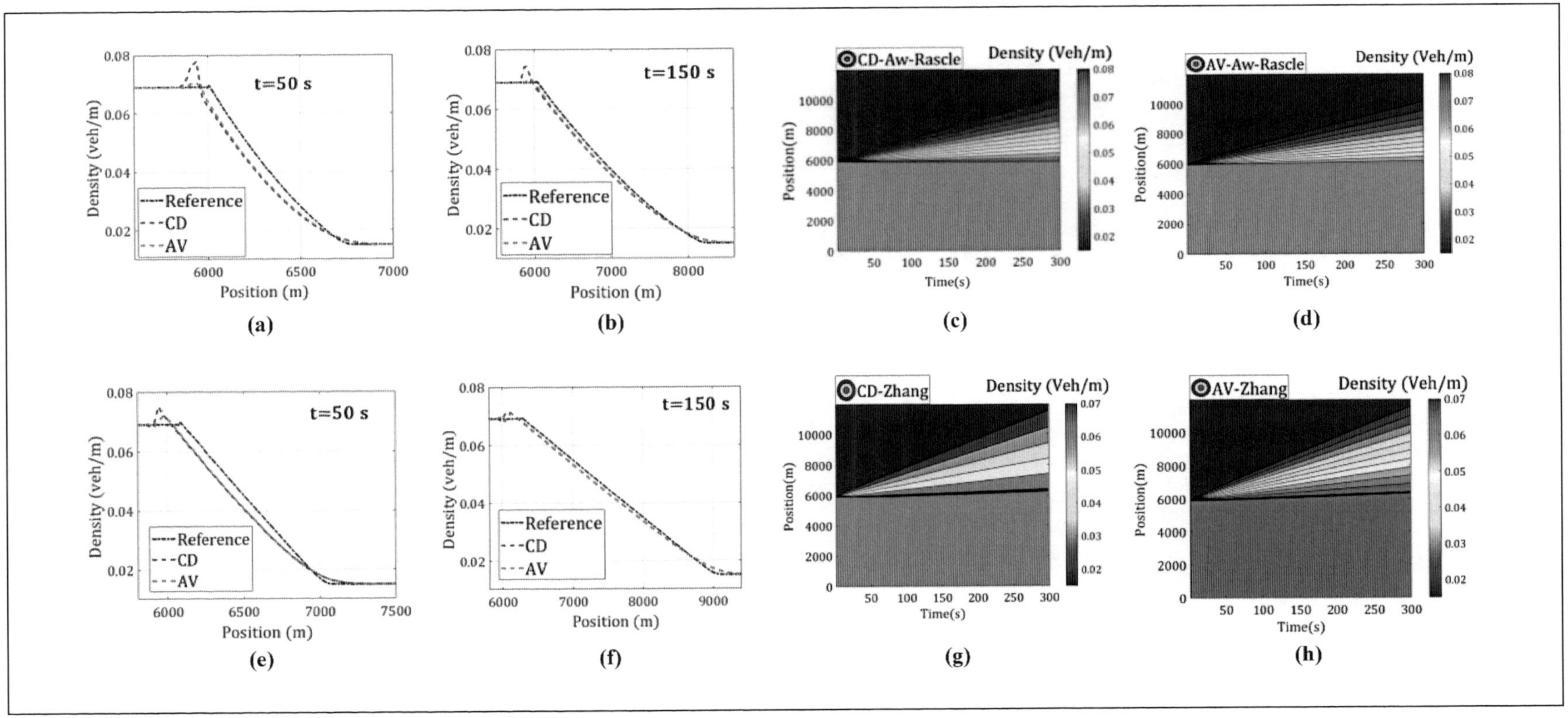

Figure 3. Case 1, free-flow traffic with transition from higher to lower densities: Aw-Rascle model results for (*a*) cross section at $t = 50$ s, (*b*) cross section at $t = 150$ s, (*c*) CD spatiotemporal evolution of local densities, and (*d*) AV spatiotemporal evolution of local densities; Zhang model results for (*e*) cross section at $t = 50$ s, (*f*) cross section at $t = 150$ s, (*g*) CD spatiotemporal evolution of local densities, and (*h*) AV spatiotemporal evolution of traffic densities.

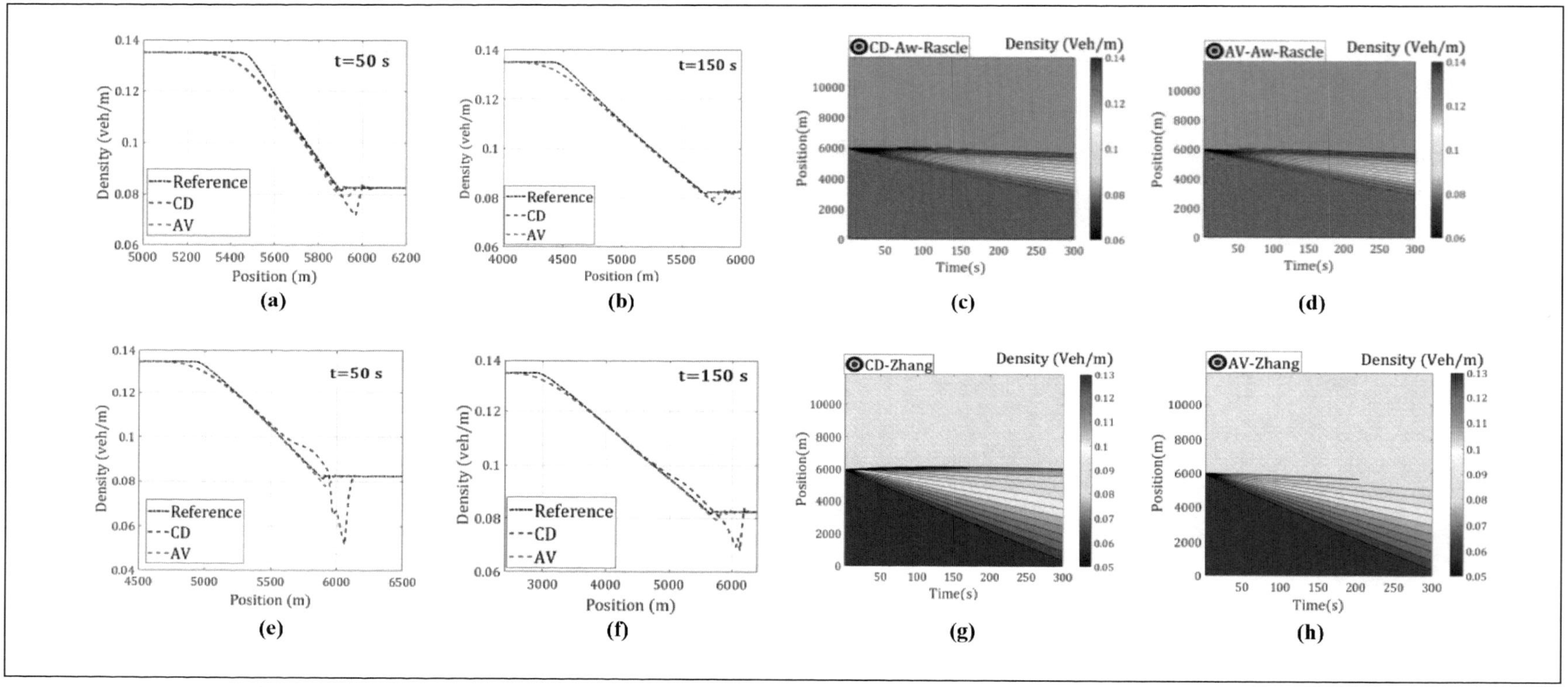

Figure 4. Case 2, congestion condition with transition from higher to lower densities: Aw-Rascle model results for (*a*) cross section at t = 50 s, (*b*) cross section at t = 150 s, (*c*) CD spatiotemporal evolution of local densities, and (*d*) AV spatiotemporal evolution of local densities; Zhang model results for (*e*) cross section at t = 50 s, (*f*) cross section at t = 150 s, (*g*) CD spatiotemporal evolution of local densities, and (*h*) AV spatiotemporal evolution of traffic densities.

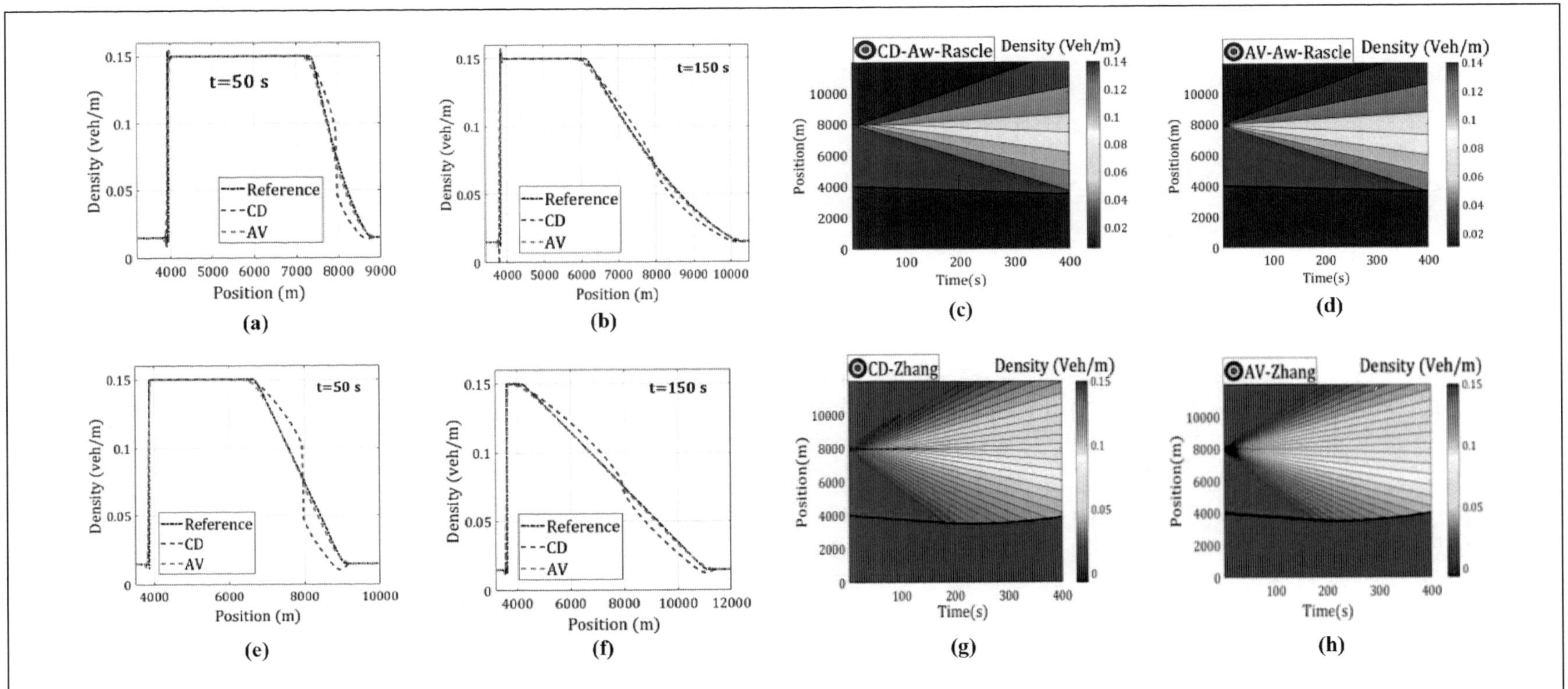

Figure 5. Case 3, queue dissolution: Aw-Rascle model results for (*a*) cross section at t = 50 s, (*b*) cross section at t = 150 s, (*c*) CD spatiotemporal evolution of local densities, and (*d*) AV spatiotemporal evolution of local densities; Zhang model results for (*e*) cross section at t = 50 s, (*f*) cross section at t = 150 s, (*g*) CD spatiotemporal evolution of local densities, and (*h*) AV spatiotemporal evolution of traffic densities.

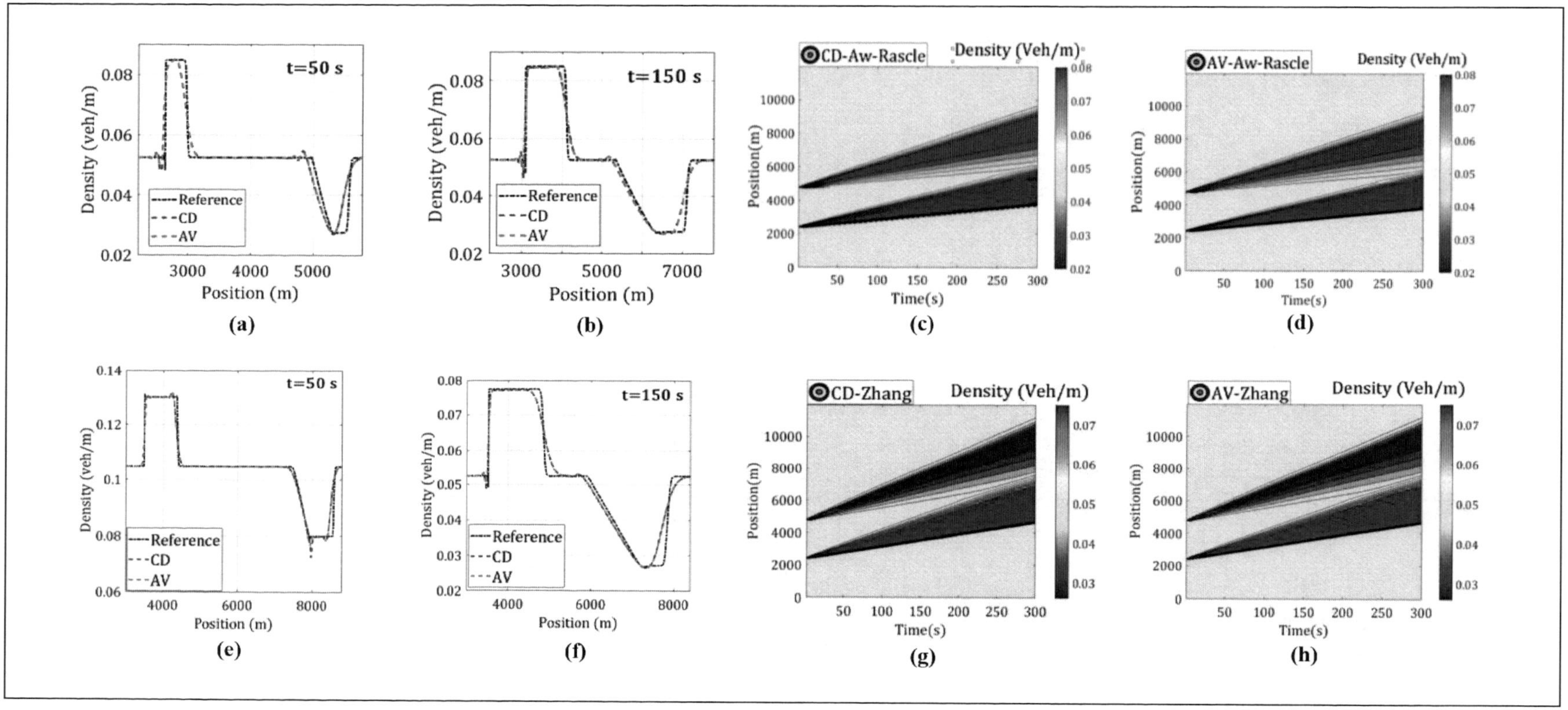

Figure 6. Case 4, Free-flow traffic with uniform density and non-equilibrium speeds: Aw-Rascle results for (*a*) cross section at t = 50 s, (*b*) cross section at t = 150 s, (*c*) CD spatiotemporal evolution of local densities, and (*d*) AV spatiotemporal evolution of local densities; Zhang model results for (*e*) cross section at t = 50 s, (*f*) cross section at t = 150 s, (*g*) CD spatiotemporal evolution of local densities, and (*h*) AV spatiotemporal evolution of traffic densities.

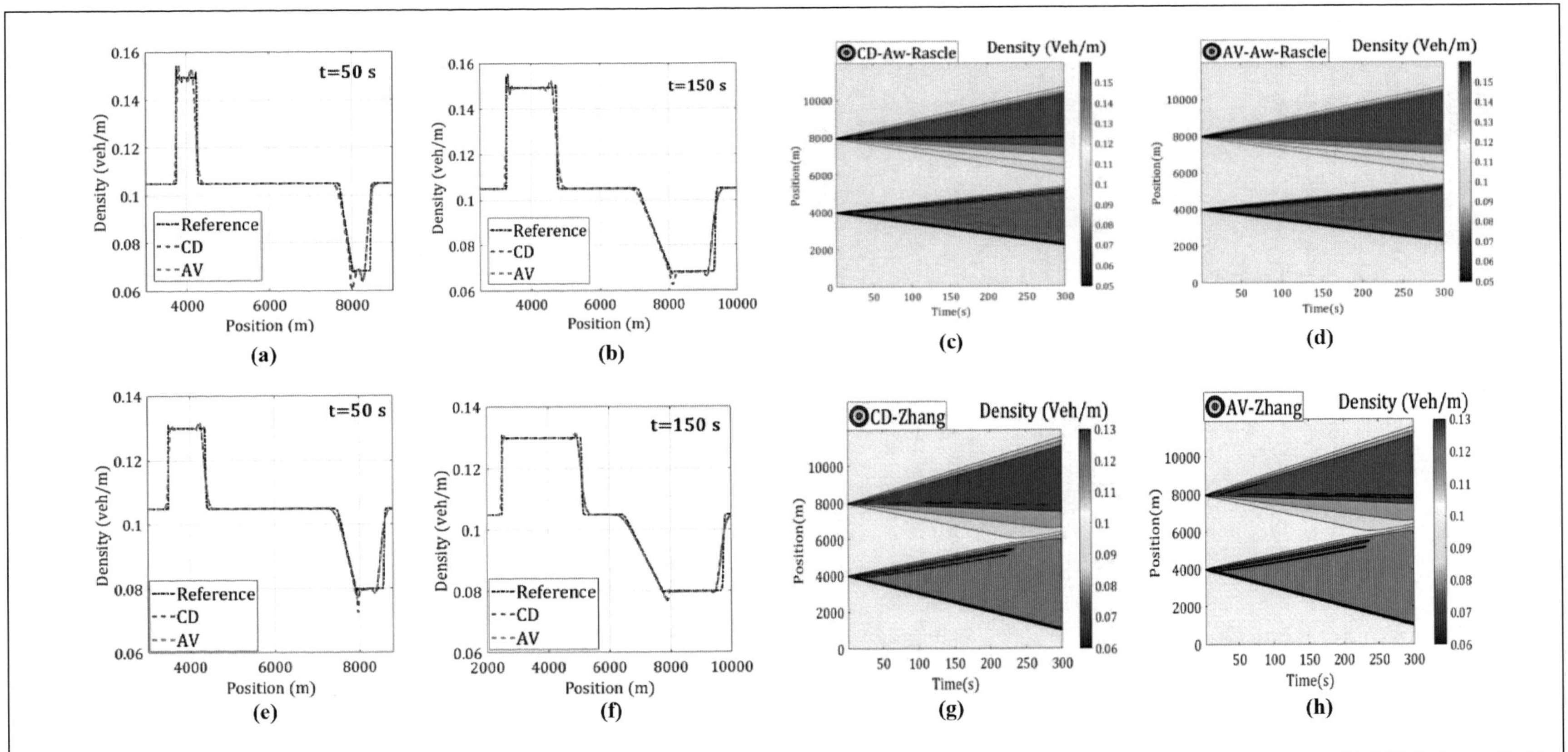

Figure 7. Case 5, congested traffic with uniform density and non-equilibrium speeds: Aw-Rascle results for (*a*) cross section at t = 50 s, (*b*) cross section at t = 150 s, (*c*) CD spatiotemporal evolution of local densities, and (*d*) AV spatiotemporal evolution of local densities; Zhang model results for (*e*) cross section at t = 50 s, (*f*) cross section at t = 150 s, (*g*) CD spatiotemporal evolution of local densities, and (*h*) AV spatiotemporal evolution of traffic densities.

with the reference solutions. Both numerical methods show little spurious oscillations, though there are more and stronger oscillations with the Zhang model, and with the CD. This is especially strong in Case 2: the Zhang model exhibits more oscillations than the Aw-Rascle model, and the AV performs better at reducing them than the CD method. Results also show that the Runge-Kutta time stepping provides a high-order temporal accuracy while avoiding numerical oscillations. Cases 2, 3, and 4, furthermore show differences in the continuous parts of the solution as well: i.e. the red (CD) line deviates from the reference solution further than the blue line (AV) does. In fact, Cases 1-3 show an almost exact overlap between the AV and the reference solution, except for at discontinuities. Case 4 and Case 5 show that AV is also susceptible to diffusion and the numerical solution is smoother than the reference solution. Results also show that the initial instabilities are dampened, and oscillations decrease over time; this is especially visible in the contour plots for cases 4 and 5 (Figures 6 and 7). Finally, the choice of the smoothing parameters influences the results as shown in Figure 2, which also indicates that the choice of $\boldsymbol{\kappa}$ (in AV) is much less sensitive than the choice of $\boldsymbol{S}$ (in CD) in terms of numerical diffusion.

Discussion and Conclusion

Aw-Rascle type traffic flow models are systems of homogeneous nonlinear hyperbolic equations. Because of their homogeneity, they require different numerical methods than the non-homogeneous higher-order traffic flow models. More specifically, the standard McCormack spatial integration scheme works fine for non-homogeneous models but does not for homogeneous ones. Therefore, two smoothing methods are proposed to reduce numerical oscillations inherent to the standard McCormack scheme. Furthermore, it is suggested to apply a TVD Runge-Kutta scheme to increase the order of temporal accuracy. Results show that both CD smoothing and AV method lead to more accurate results with less numerical oscillation. Diffusion is kept within reasonable limits in both methods compared with reference solution. Because of better performance at discontinuities (less numerical oscillations) and a lower sensitivity of the smoothing parameter, the AV method is preferred.

Future developments include application of the proposed numerical methods in simulation tools. Therefore, the methods will have to be extended to deal with nodes such as on-ramps and off-ramps. Furthermore, the methods can be studied further and applied to different models, such as other Aw-Rascle type models, the generic second-order model, traffic flow models in Lagrangian formulation and to both first-order and higher order models including multiple vehicle classes (e.g. cars and trucks). Non-homogeneous models with relative small source terms (e.g. due to large relaxation times) will also benefit from applying the smoothing methods in combination with the McCormack and TVD Runge-Kutta schemes. Other types of smoothing, including flux limiting, can be studied and compared with those presented here. Study on other flux limiting terms for smoothing is also suggested for comparison purposes. Accuracy of the methods, as well as their computational costs in terms of memory and computation time, can be compared in more detail.

Author Contributions

This is to confirm the contribution of the authors, Saeed Mohammadian, and Femke van Wageningen-kessels, in study conception and design, data collection, analysis and interpretation of results, and draft manuscript preparation. Both authors reviewed the results and approved the final version of the manuscript.

References

1. Helbing, D., A. Hennecke, V. Shvetsov, and M. Treiber. Micro-and Macro-simulation of Freeway Traffic. *Mathematical and Computer Modelling*, Vol. 35, No. 5–6, 2002 pp. 517–547.
2. Helbing, D. Traffic and Related Self-driven Many-particle Systems. *Reviews of Modern Physics*, Vol. 73, No. 4, 2001 pp. 1067–1141.
3. Seo, T., A. M. Bayen, T. Kusakabe, and Y. Asakura. Traffic State Estimation on Highway: A Comprehensive Survey. *Annual Reviews in Control*, Vol. 43, 2017, pp. 128–151.
4. Bellomo, N., and C. Dogbe. On the Modeling of Traffic and Crowds: A Survey of Models, Speculations, and Perspectives. *SIAM Review*, Vol. 53, No. 3, 2011, pp. 409–463.
5. Hoogendoorn, S. P., and P. H. Bovy. State-of-the-art of Vehicular Traffic Flow Modelling. *Proceedings of the Institution of Mechanical Engineers, Part I: Journal of Systems and Control Engineering*, Vol. 215, No. 4, 2001, pp. 283–303.
6. Treiber, M., and A. Kesting, Traffic flow dynamics. Traffic Flow Dynamics: Data, Models and Simulation, Springer-Verlag Berlin Heidelberg, 2013.
7. van Wageningen-Kessels, F., H. Van Lint, K. Vuik, and S. Hoogendoorn. Genealogy of Traffic Flow Models. *EURO Journal on Transportation and Logistics*, Vol. 4, No. 4, 2015, pp. 445–473.
8. Saifuzzaman, M., and Z. Zheng, Incorporating Human-factors in Car-following Models: A Review of Recent Developments and Research Needs. *Transportation Research Part C: Emerging Technologies*, Vol. 48, 2014, pp. 379–403.
9. Delis, A. I., I. K. Nikolos, and M. Papageorgiou. Macroscopic Traffic Flow Modeling with Adaptive Cruise Control: Development and Numerical Solution. *Computers &*

Mathematics with Applications, Vol. 70, No. 8, 2015, pp. 1921–1947.

10. Treiber, M., and A. Kesting. Macroscopic Models with Dynamic Velocity. In *Traffic Flow Dynamics: Data, Models and Simulation* (M. Treiber, and A. Kesting, eds.), Springer, Berlin, Heidelberg, 2013, pp. 127–155.
11. Treiterer, J., and J. Myers. The Hysteresis Phenomenon in Traffic Flow. *Transportation and Traffic Theory*, Vol. 6, 1974, pp. 13–38.
12. Laval, J. A. Hysteresis in Traffic Flow Revisited: An Improved Measurement Method. *Transportation Research Part B: Methodological*, Vol. 45, No. 2, 2011, 385–391.
13. Chen, D., S. Ahn, Z. Zheng, and J. Laval. Traffic Hysteresis and the Evolution of Stop-and-Go Oscillations. Presented at 92nd Annual Meeting of the Transportation Research Board, Washington, D.C., 2013.
14. Yuan, K., V. L. Knoop, L. Leclercq, and S. P. Hoogendoorn. Capacity Drop: A Comparison Between Stop-and-Go Wave and Standing Queue at Lane-drop Bottleneck. *Transportmetrica B: Transport Dynamics*, Vol. 5, No. 2, 2017, pp. 145–158.
15. Tampere, C., S. Hoogendoorn, and B. van Arem. A Behavioural Approach to Instability, Stop and Go Waves, Wide Jams and Capacity Drop. *Transportation and Traffic Theory*, Vol. 16, 2005, pp. 205–228.
16. Saifuzzaman, M., Z. Zheng, M. M. Haque, and S. Washington. Understanding the Mechanism of Traffic Hysteresis and Traffic Oscillations Through the Change in Task Difficulty Level. *Transportation Research Part B: Methodological*, Vol. 105, 2017, pp. 523–538.
17. Kontorinaki, M., A. Spiliopoulou, C. Roncoli, and M. Papageorgiou. First-order Traffic Flow Models Incorporating Capacity Drop: Overview and Real-data Validation. *Transportation Research Part B: Methodological*, Vol. 106, 2017, pp. 52–75.
18. LeVeque, R. J. *Finite Volume Methods for Hyperbolic Problems*, Vol. 31. Cambridge University Press, 2002.
19. Witham, G. *Linear and Non-linear Waves*. John Wiley, New York, 1974.
20. Payne, H. J. Models of Freeway Traffic and Control. In *Mathematical Models of Public Systems* (G. A. Bekey, ed.), Simulation Council, La Jolla, 1971, pp. 51–61.
21. Kerner, B. S., and P. Konhäuser. Structure and Parameters of Clusters in Traffic Flow. *Physical Review E*, Vol. 50, No. 1, 1994, p. 54.
22. Treiber, M., A. Hennecke, and D. Helbing. Derivation, Properties, and Simulation of a Gas-kinetic-based, Nonlocal Traffic Model. *Physical Review E*, Vol. 59, No. 1, 1999, pp. 239–253.
23. Aw, A., and M. Rascle. Resurrection of "Second Order" Models of Traffic Flow. *SIAM Journal on Applied Mathematics*, Vol. 60, No. 3, 2000, pp. 916–938.
24. Zhang, H. M. A Non-equilibrium Traffic Model Devoid of Gas-like Behavior. *Transportation Research Part B: Methodological*, Vol. 36, No. 3, 2002, pp. 275–290.
25. Daganzo, C. F. Requiem for Second-order Fluid Approximations of Traffic Flow. *Transportation Research Part B: Methodological*, Vol. 29, No. 4, 1995, pp. 277–286.
26. van Wageningen-Kessels, F., B. van't Hof, S. P. Hoogendoorn, H. Van Lint, and K. Vuik. Anisotropy in Generic Multi-class Traffic Flow Models. *Transportmetrica A: Transport Science*, Vol. 9, No. 5, 2013, pp. 451–472.
27. Lebacque, J. -P., S. Mammar, and H. Haj-Salem. The Aw–Rascle and Zhang's Model: Vacuum Problems, Existence and Regularity of the Solutions of the Riemann Problem. *Transportation Research Part B: Methodological*, Vol. 41, No. 7, 2007, pp. 710–721.
28. Helbing, D., and M. Treiber. Numerical Simulation of Macroscopic Traffic Equations. *Computing in Science & Engineering*, Vol. 1, No. 5, 1999, pp. 89–99.
29. Lebacque, J. -P., S. Mammar, and H. H. Salem. Generic Second Order Traffic Flow Modelling. *Papers Selected for Presentation at ISTTT17 in Transportation and Traffic Theory*, 2007.
30. Abgrall, R., and J. Qiu. Preface to the Special Issue "High Order Methods for CFD Problems". *Journal of Computational Physics*, Vol. 230, No. 11, 2011, pp. 4101–4102.
31. Delis, A. I., I. K. Nikolos, and M. Papageorgiou. High-resolution Numerical Relaxation Approximations to Second-order Macroscopic Traffic Flow Models. *Transportation Research Part C: Emerging Technologies*, Vol. 44, 2014, pp. 318–349.
32. Ngoduy, D., and R. Liu. Multiclass First-order Simulation Model to Explain Non-linear Traffic Phenomena. *Physica a-Statistical Mechanics and its Applications*, Vol. 385, No. 2, 2007, pp. 667–682.
33. Tannehill, J. C., D. A. Anderson, and R. H. Pletcher. *Computational Fluid Mechanics and Heat Transfer*, 2nd ed. Taylor & Francis, Oxfordshire, 1997.
34. Jameson, A., W. Schmidt, and E. Turkel. Numerical Solutions of the Euler Equations by Finite Volume Methods using Runge-Kutta Time-stepping Schemes. *AIAA Paper*, Vol. 1259, 1981, p. 1981.
35. Shu, C. -W., and S. Osher. Efficient Implementation of Essentially Non-oscillatory Shock-Capturing Schemes. *Journal of Computational Physics*, Vol. 77, No. 2, 1988, pp. 439–471.
36. Greenshields, B., W. Channing, and H. Miller. A Study of Traffic Capacity. In *Highway Research Board Proceedings*. Highway Research Board, National Research Council (USA), 1935.
37. Peyret, R., and T. D. Taylor. *Computational Methods for Fluid Flow*. Springer Science & Business Media, Berlin, 2012.
38. Thomas, J. W. *Numerical Partial Differential Equations: Finite Difference Methods*, Vol. 22. Springer Science & Business Media, New York, 2013.
39. Tan, W. -Y. *Shallow Water Hydrodynamics: Mathematical Theory and Numerical Solution for a Two-dimensional System of Shallow-water Equations*, Vol. 55. Elsevier Oceanography Series, Water & Power Press, Beijing, 1992.
40. Garcia-Navarro, P., F. Alcrudo, and J. Saviron. 1-D Open-channel Flow Simulation Using TVD-McCormack Scheme. *Journal of Hydraulic Engineering*, Vol. 118, No. 10, 1992, pp. 1359–1372.
41. Sun, M., and K. Takayama. Conservative Smoothing on An Adaptive Quadrilateral Grid. *Journal of Computational Physics*, Vol. 150, No. 1, 1999, pp. 143–180.

42. Swanson, R. C., and E. Turkel. On Central-difference and Upwind Schemes. *Journal of Computational Physics*, Vol. 101, No. 2, 1992, pp. 292–306.
43. Daganzo, C. F. A Variational Formulation of Kinematic Waves: Basic Theory and Complex Boundary Conditions. *Transportation Research Part B: Methodological*, Vol. 39, No. 2, 2005, pp. 187–196.
44. Laval, J. A., and L. Leclercq. The Hamilton–Jacobi Partial Differential Equation and the Three Representations of Traffic Flow. *Transportation Research Part B: Methodological*, Vol. 52, 2013, pp. 17–30.

The Standing Committee on Traffic Flow Theory and Characteristics (AHB45) peer-reviewed this paper (18-01677).

Sage Publications USA
sagepub.com

ISBN 978-1-5108-7825-9

CURRAN ASSOCIATES INC.
proceedings.com